Managed Ecosystems

The Mesoamerican Experience

Edited by

L. UPTON HATCH

MARILYN E. SWISHER

New York Oxford

Oxford University Press

1999

Oxford University Press

Oxford New York
Athens Auckland Bangkok Bogotá Buenos Aires Calcutta
Cape Town Chennai Dar es Salaam Delhi Florence Hong Kong Istanbul
Karachi Kuala Lumpur Madrid Melbourne Mexico City Mumbai
Nairobi Paris São Paulo Singapore Taipei Tokyo Toronto Warsaw

and associated companies in
Berlin Ibadan

Library of Congress Cataloging-in-Publication Data
Managed ecosystems : the Mesoamerican experience /
edited by L. Upton Hatch,
Marilyn E. Swisher.
p. cm. — (Topics in sustainable agronomy)
Includes bibliographical references and index.
ISBN 0-19-510260-6
1. Sustainable agriculture—Latin America. 2. Agricultural
ecology—Latin America. 3. Sustainable agriculture—Tropics.
4. Agricultural ecology—Tropics. I. Hatch, Upton. II. Swisher,
Marilyn E., 1948– . III. Series.
S473.9.T76 1998
333.7'098'0913—dc21 97-22125

1 3 5 7 9 8 6 4 2

Printed in the United States of America
on acid-free paper

Preface

New World tropics are a major frontier in the battle to preserve biodiversity and maintain relatively large areas of natural ecosystems in an undisturbed state. At the same time, population growth in Central and South America, combined with expectations of rising standards of living, causes increased human impacts on the natural ecosystems. Sustainable development is proposed as the solution to this dilemma by many. While the goals of sustainable development are readily identifiable, the institutional structures and social, biological, and agricultural science needed to make it a reality are elusive. Contrasting and often conflicting approaches are suggested. Some argue, for example, that indigenous knowledge and traditional agricultural systems represent the best alternative for meeting human needs and preserving natural resources. Others argue that biotechnology and employment of sophisticated agricultural technologies are the solution.

Scale is another important consideration. Some scientists are increasingly concerned about mesoscale phenomena and argue that we must gain an understanding of how to create sustainable landscape mosaics. Others see sustainability from a microscale, focusing on the individual household as the basic unit of study. Purely disciplinary research agendas fail to consider the crucial interactions among the biological, physical, and social sciences involved in the process of resource utilization and degradation. Designing sustainable resource-based human activities in agriculture, in forestry, and in the exploitation of aquatic systems requires enhanced interdisciplinary communication among agricultural, social, biological, and physical scientists. These scientists must rethink their research agendas and reconsider the institutional structures in which scientific inquiry occurs in order to break down the disciplinary boundaries that restrict the development of the scientific knowledge needed to make sustainable development a potential solution rather than wishful thinking.

Despite the perception held by many that large portions of New World tropics are relatively untouched by human activities, the reality is that even natural ecosystems in the neotropics are increasingly subject to human intervention, degradation, and management. The interactions between natural and human-dominated ecosystems grow as human activities increase in intensity and extent. Understanding the nature of the flow—and its potential impacts—between different ecosystems is therefore of increasing importance to the maintenance of the overall integrity of both biotic and nonbiotic natural resources at the regional level.

Managed Ecosystems: The Mesoamerican Experience provides the reader with an understanding of the interactions between the sociocultural and biophysical aspects of the management of the natural resource base of the New World tropics, focusing on interactions between agriculture and natural resources. It describes both the potential benefits and the liabilities of the major food, fuel, and fiber production systems in the region, stressing their long-term sustainability. Each system is examined through its potential impact on resource preservation and utilization at the regional scale.

This book is a departure from traditional treatments of either agricultural sustainability or natural resource management. While there are many books that address the sustainability of agricultural ecosystems, few specifically address the systems prevalent in New World tropics, and most restrict their discussion of sustainability to internal characteristics of the production system, ignoring the increasingly important role of flow between agricultural and natural ecosystems. Our discussion focuses on the interactions between human-dominated systems and natural systems at the regional scale. We contend that greater understanding of flows between systems and of the interactions between socioeconomic and biophysical factors is critical in developing agricultural ecosystems that are sustainable over the mid- to long term.

The book contains three parts. The first focuses on the relationships among economic institutional and sociocultural issues. The second part discusses some of the major kinds of natural resources and natural ecosystems that are affected by human-dominated ecosystems. The last part provides a description of major types of production systems in the region, presenting a balanced perspective on the long-term sustainability of each type of system, including socioeconomic, cultural, and biophysical aspects of sustainability.

The most important theme in the book is that understanding how agricultural ecosystems interact with the natural resource base depends on an enhanced appreciation for the interactions between biophysical and sociocultural-economic factors. We try to show the reader that simplistic approaches toward the evaluation of the sustainability of agroecosystems are inappropriate because they fail to take into account the complex web of positive and negative aspects of each system. We present a balanced view of these systems, stressing the need to enhance their positive aspects and reduce their negative aspects, rather than either advocating the innate superiority of one particular system or searching for idealized systems that do not, in fact, build on the systems that already exist over large expanses of the humid tropics.

There are two main audiences for our book. The first is biological, social, and agricultural scientists concerned with agricultural sustainability, particularly those with an interest in New World tropics. We expect the work to appeal especially to teachers and development specialists and, to a lesser degree, to researchers. The second audience is those individuals concerned with human impacts on natural resource management in New World tropics. We expect the work to be particularly useful to biological and social scientists who find themselves increasingly involved with and concerned about the role of agricultural production on natural resource management. Many of these scientists have relatively little or no training in agriculture, and our book gives them an overview of how those systems affect natural resources, particularly biotic, aquatic, and forest resources.

We expect an interdisciplinary audience. This book is relevant for many professionals in environmental and economic development areas. *Sustainable development* is a term increasingly used to imply that there is a scientific basis for development. Yet, there is little agreement over what the term means. More important, there is very little agreement about what kinds of scientific knowledge could contribute to sustainable development. The outcome of this ongoing discussion will have important implications for the research agenda of many scientists from a broad range of disciplines, including the social, biological, agricultural, and physical sciences. Our session will contribute to a greater understanding of the nature of research agendas in the coming decade.

We would like to acknowledge the assistance of several individuals. Lil Soto helped with much of the editing, especially with chapters that were originally written in Spanish. Carol Lee also assisted with the manuscript preparation. Her dedication to this book is greatly appreciated.

Contents

Contributors

William Ascher
Public Policy Studies and Political Science
Duke University
Durham, North Carolina

John R. Bort
Department of Anthropology
East Carolina University
Greenville, North Carolina

Mario A. Boza
Wildlife Conservation Society
San José, Costa Rica

C. Ronald Carroll
Institute of Ecology
University of Georgia
Athens, Georgia

Carlos L. de la Rosa
Director, Riverwoods Field Laboratory
Florida Center for Environmental Studies
Lorida, Florida

Richard F. Fisher
Forest Science Department
Texas A&M University
College Station, Texas

Don Werner Hagnauer
Retired Farm Manager, LaPacifica
Guanacaste, Costa Rica

Upton Hatch
Department of Agricultural Economics & Rural Sociology
Auburn University
Auburn, Alabama

Carlos A. Jiménez-Crespo
Animal Science Department
University of Costa Rica
San José, Costa Rica

Deborah Kane
Institute of Ecology
University of Georgia
Athens, Georgia

Jim Kettler
Graduate School of Environmental Studies
Bard College
Annandale-on-Hudson, New York

John C. Mayne
SARE Producer Grants Coordinator
University of Georgia
Griffin, Georgia

Raúl A. Moreno
Independent Consultant
San José, Costa Rica

Gabriela Soto Muñoz
Centro de Investigaciones Agronómicas
Universidad de Costa Rica
San José, Costa Rica

Luis L. Ovares
International Ocean Institute
Universidad Nacional Costa Rica
Heredia, Costa Rica

Hugh L. Popenoe
Director, Center for Tropical Agriculture
University of Florida
Gainesville, Florida

Catherine M. Pringle
Institute of Ecology
University of Georgia
Athens, Georgia

Dianne Rocheleau
Department of Geography
Clark University
Worcester, Maine

Martha E. Rosemeyer
Mulch-Based Agriculture Field Coordinator
for Central America
Miami, Florida

Vernon W. Ruttan
Department of Economics and
Department of Agricultural and Applied Economics
University of Minnesota
St. Paul, Minnesota

Frederick N. Scatena
International Institute of Tropical Forestry
U.S.D.A. Forest Service
Rio Piedras, Puerto Rico

Ken Schlather
Soil, Crop, and Atmospheric Sciences Department
Cornell University
Ithaca, New York

Denise Stanley
Department of Economics
University of Tennessee
Knoxville, Tennessee

Susan C. Stonich
Department of Anthropology
& Environmental Studies
University of California
Santa Barbara, California

Marilyn E. Swisher
Home Economics Department
University of Florida
Gainesville, Florida

Lawrence T. Szott
Program in Sustainable Tropical
Production Systems
University of Costa Rica
San José, Costa Rica

David Teichert-Coddington
Department of Fisheries and
Allied Aquaculture
Auburn University
Auburn, Alabama

Gilbert Vargas Ulate
Department of Geography
University of Costa Rica
San Jose, Costa Rica

Alexis Vásquez-Morera
Department of Soil Microbiology
University of Costa Rica
San José, Costa Rica

PART I

Globalization and Sustainable Resource Management: The Issues

1

Introduction

Upton Hatch
Marilyn Swisher

Economic development and resource conservation are often in conflict because of concurrent demands for an improved standard of living for a growing population and environmental quality. Globalization brings increasing biophysical and socioeconomic interactions. The potential irreversibilities associated with the loss of biodiversity and unique natural ecosystems to society are typically pitted against the opportunity of individuals to improve their economic well-being. Can resource management strategies be developed that balance these often contentious goals and perspectives?

The dual goals of development, particularly agricultural development, and resource protection and conservation in one region of the world, Mesoamerica, are the focus of this book. This region was selected because it exemplifies the processes and constraints that developing nations and regions throughout the world face as they attempt to reach these goals. At the same time, the region's small but highly diverse land area and culture permit an examination of how global changes and trends affect the development process at the local level.

Mesoamerica is a major frontier in the battle to preserve biodiversity and maintain large segments of natural ecosystems in an undisturbed state. As the land bridge between North and South America, the region claims a disproportionately large share of the world's unique ecosystems and biodiversity. As Boza (chapter 6) points out, the area represents only 0.5 percent of the world's land mass but has an estimated 7 percent of the earth's biodiversity (D. H. Janzen, 1986).

At the same time, population growth in the region is one of the highest in the world. This, combined with expectations of rising standards of living, cause increased human impacts on the natural ecosystems. Perhaps partly because of its juxtaposition to the affluent populations of Canada and the United States, these human impacts are subject to intense international scrutiny and intervention. Costa Rica is a prime example. While lauded by many as a model for "sustainable development," increasingly based on ecotourism, others point out that the nation is in many regards seriously ecologically compromised. Its cities, for example, lack sewage treatment systems and release untreated wastes into the same rivers that attract ecotourists. Similarly, Mayne (chapter 9) indicates that Costa Rica lost 17 percent of its forested area between 1977 and 1989 alone.

Unlike some other regions of the world, such as Asia, it is unlikely that policies and decisions made in Central America will have a significant impact on the global decision-making process. The small nations of Central America have limited ability to influence the direction of international decision making or to control how the decisions that are made affect their own economies, people, and resources. In this sense, Mesoamerica provides a nearly laboratory setting to examine how these global processes and trends influence and indeed constrain the strategies that the world's developing nations can implement to achieve the dual goals of development and conservation.

Many pose sustainable development as the strategy for achieving development with conservation both in Mesoamerica and elsewhere. However, while the goals of sustainable development are readily identifiable, the institutional structures and social, biological, and agricultural science needed to make sustainable development a reality are elusive. Contrasting and often conflicting approaches are suggested. Some argue, for example, that indigenous knowledge and traditional agricultural systems represent the best alternative for meeting human needs and preserving natural resources. Others argue that genetic engineering and sophisticated agricultural technologies are the solution (Hauptli et al., 1990). Scale also comes into play. Some scientists are increasingly concerned about mesoscale phenomena and argue that we must gain an understanding of how to create sustainable landscape mosaics. Others see sustainability from the microscale, focusing on the individual household as the basic unit of study.

The dual goals of sustainable resource management and sustainable agricultural development pose complex challenges for science. This book brings together authors from the biological, agricultural, and social sciences to examine these issues. Microscale research findings and development experiences are examined within the framework of macroscale trends and events. Concurrently, macroscale analyses are directed toward enhancing microscale research and development. The contributions in this book are not designed to provide a road map to sustainable development and resource use. Rather, the authors hope to foster discussion and thought about the importance of meso- and macroscale interactions to sustainability.

The Macroscale Environment

Often, scientists and other specialists tend to ignore the larger global trends that establish the framework within which development will occur. Macroscale, exogenous factors play a critical role in determining the use to which scientific knowledge is put, or even whether the knowledge is used. However, agricultural and biological research, in particular, tend to focus at the microscale; small plots or individual farm enterprises are the most common experimental units. Research questions, particularly in applied research, are often defined by some immediate problem: a "real world" problem of how to increase yield, for example. However, research at the microlevel can become irrelevant when placed in the larger context of macro political, economic, and environmental forces.

At the other end of the spectrum lies the large body of literature, normative in approach, that espouses one or another strategy for sustainable development based on philosophical considerations about the appropriate use of human and natural resources. While interesting and thought provoking, these approaches often appear to have little connection to the kind of immediate problems that face farmers or natural resource managers. With a focus on "what should be," rather than "what is," these approaches seem to many to have more basis in value systems than in science.

Three global trends will play a critical role in the coming decades in determining how agricultural development and resource management in Mesoamerica and the rest of the world occur: (1) globalization and free markets, (2) population growth and expectations of increased standards of living, and (3) concerns over environmental quality. Each of these three trends is addressed by the chapters in this book, where their impact at the micro- and mesoscale are examined repeatedly. To set the stage for these discussions, it is useful to examine the general nature and impact of these trends and how they serve as a backdrop for development in the twenty-first century.

Globalization and Free Markets

Macroeconomic and political relationships and rapid changes in communication and transportation technology are reducing the functional importance of political boundaries. The growth of free markets increases interdependence, resulting in an inability to isolate the effects of policy within national boundaries. As a consequence, policy decisions in one country will often have major consequences in other countries. For Mesoamerica, policies established elsewhere are apt to produce major changes at the national and regional level.

Globalization will affect the management of the natural resource base and choice of production systems throughout the world. There are many alternative scenarios for how the interaction between international and domestic markets and suppliers will alter the agricultural landscape and natural resource base in Mesoamerica. For example, export market opportunities may increase the production of nontraditional crops, resulting in greater degradation of the natural resource base. Alternatively, the Mesoamerican natural resource base may not be sufficient to support a production system that can compete in world markets, resulting in less demand for the region's products and a decline in foreign exchange earnings from agriculture (Moreno, chapter 17). Yet another potential outcome is that international suppliers will increasingly penetrate Mesoamerican markets for traditional crops, such as corn, decreasing the income potential from commercial agricultural production for local farmers.

Whatever the final result, globalization and the growth of free markets are major forces in shaping agricultural development and resource use in the region. Microscale research and national and regional development strategies will both play out against this backdrop. Aquatic, soil, biological, and human resources will all be mobilized to position the region as an actor in the international marketplace. Depending on how those resources are used, globalization may lead to greater or to lesser sustainability.

Population Growth and Standard of Living

Worldwide, human pressure on the resource base will increase in the twenty-first century because total demand for goods and services will grow as a result of both higher population and higher per capita resource use. Total demand for resources will tend to rise to accommodate population growth and enhance the standard of living in much of the less developed world (Ruttan, chapter 2).

The U. S. Bureau of the Census (1996), more conservative than some in its estimations, shows current worldwide annual population growth at 1.38 percent per year, down from peak growth rates of 2 percent per year in the late 1960s and early 1970s. It projects that growth rates will decrease to less than 1 percent per year by the year 2030. Nonetheless, the total human population will still increase dramatically because much of the world's current population is in the prereproductive or early reproductive phase (see Tietenberg, 1996, p. 101). Based on its projections, worldwide population will reach about 8.3 billion by the year 2030 and about 9.4 billion by 2050.

Gains in per capita caloric consumption since 1950 have been impressive. In 1969–1971, for example, worldwide daily caloric consumption averaged about 2,440. By 1990–1992, this had increased to 2,720 calories per day. This increase in caloric consumption was not, as some imply, simply a phenomenon of increased caloric intake in industrialized nations. Daily caloric consumption in this period increased from 2,140 to 2,520 in developing nations (FAO, 1996).

However, to maintain these levels of consumption, worldwide caloric production will need to nearly double by the year 2050. While caloric consumption, currently 3,350 in industrialized nations, could well stand to decrease somewhat (recommended caloric consumption is 2,100 to 2,500 calories per day), the developing world needs large, rapid increases in caloric availability. For Latin America and Asia, per capital caloric production must double to provide a basic caloric intake of about 2,400 per capita per day. For Africa, a fivefold increase is needed.

These increases in the demand for food will form the macroscale environment in which agricultural production occurs worldwide. The pressure to increase food production will not decline; rather, it will grow. By the year 2050, it is logical to expect that worldwide land, soil, and water resources will, in essence, be used to the maximum degree possible to meet the human population's basic food needs.

For Mesoamerica and other developing regions, however, local resources may or may not be devoted to meeting the basic food needs of the local population. Globalization may increase the access of people in industrialized nations both to more food resources and to resources from a greater portion of the globe. As a result, demand on the natural resource base in developing nations may well increase for both local and international consumption. Given the play of international free markets, a greater or lesser portion of the local natural resource base may be devoted to meeting basic, versus luxury, food consumption needs and to meeting local, versus international, consumption.

Environmental Quality

The concern about environmental quality on the part of the industrialized world has implications for resource use worldwide. The threat of global climate change and other apparent indications of declining global environmental quality have made it apparent to much of the developed world's population that their environmental quality is affected by environmental degradation in other parts of the world. As a response, decision makers in both public and private institutions have placed new restrictions on the kinds of activities that they are willing to fund in developing nations. For example, international development assistance programs now place a growing emphasis on protecting and conserving resources and have moved away from funding traditional agricultural development projects with a focus on production and yield. This rising demand for environmental quality in the industrialized world thus constrains the available policy options for development in less developed countries.

To some, the emphasis that many industrialized nations place on the preservation of environmental quality, particularly conservation of biodiversity and the ecosystem, is hypocritical. For them, the industrialized world destroyed most of its unique ecosystems during its own development process but now proposes to impose preservation on developing nations, greatly restricting the resource base for development. Others argue that the industrialized world only talks about environmental quality but is unwilling to pay its fair share of preserving natural resources in the developing world, either through direct assistance or by paying a fair market price for such products as tropical woods. They argue that protecting environmental quality is a real issue for citizens of the developing world, even a critical issue, but that demand for low-cost products from developing nations in the industrialized nations prevents poor countries from achieving adequate environmental protection.

Whatever the source of concern over environmental quality, the issue will probably remain a serious consideration for the foreseeable future. Certainly, public sector assistance from the industrialized nations will continue to be scrutinized in terms of the impact that development projects have on the natural environment. Private sector investment, although less amenable to public control, will also be the subject of scrutiny. Development will therefore occur against a background of concern about mid- to long-term environmental impacts.

Hypotheses

The authors offer seven hypotheses or questions for the reader's consideration. Each of the chapters that follows will provide the reader with insights about these hypotheses, allowing the reader to come to his or her own conclusion about their validity. Although these hypotheses are stated positively, the objective is not to serve as proponents of them but to foster discussion. The authors feel that they are important issues that must be addressed before realistic strategies for development can be formulated.

Hypothesis 1: Greater attention must be paid to meso- and macroscale phenomena in the development of sustainable agricultural systems

Most biological and agricultural research focuses on the microscale. However, phenomena at different scales interact, and microscale research cannot ensure the sustainability of agriculture or resource management. The impact of larger-scale phenomena on smaller scales is obvious. The sustainability of higher-level systems both imposes limitations on and provides support for the sustainability of all lower-level systems. If the planet as a whole is not sustainable, if its biophysical and socioeconomic systems fail, it is very clear that no single farm will be sustainable. A well functioning regional system of transportation, for example, allows farmers to get their products to market and thus helps ensure the economic sustainability of each farm in the region. Similarly, a poorly functioning regional system, such as a river that periodically floods because of deforestation and increased rates of discharge during storms, will obviously have negative effects on the sustainability of farms on its lower floodplain.

The converse is also true. The sustainability of the individual components of a regional system contribute to the sustainability of the higher level system. How well each individual farm-to-market road in a transportation system is maintained, for example, will help determine how well the overall regional system functions. In this sense, individual components in a larger-scale system do impose limitations and provide support for the sustainability of the higher-level system, just as the higher-level system does on the lower-level systems.

Nonetheless, the flow of information between the two scales is not equal. Lower-level systems, for example, can fail individually without causing failure of the higher-level system, although there will be some point at which a cascade effect is reached and the higher level system does fail. In contrast, failure in the higher-level system will mean failure for all of the lower-level systems.

The dilemma is that most people, including farmers, tend to be concerned with their immediate environment. Despite the fact that an individual's well-being is highly dependent on higher scale systems, most individuals devote their attention to lower-level systems more amenable to their immediate intervention. Yet, over the long term, the larger-scale phenomena may determine whether the individual's immediate environment is sustainable.

Research directed toward developing more sustainable agricultural ecosystems must therefore move beyond the farm and enterprise. Scientists must examine agroecosystems, certainly at the regional and perhaps even at the national and international scale. Over the long term, research about particular production practices will not be adequate to ensure that agricultural production as a whole is sustainable.

Hypothesis 2: Intensification of agricultural production on lands appropriate for agriculture is the best way to achieve the sustainability of agricultural systems in Central America

As discussed earlier, increases in world caloric population and consumption in the period 1969–1992 were impressive. Certainly, part of this improvement came about by the use of new agricultural technology. However, the increased area under production accounted for part of the change as well. While it is difficult, if not impossible, to sort out how much of the world increase in food production came from each of these factors, most authors (see Ruttan, chapter 2) would agree that an increase in the area used for agricultural production played an important role in increasing world food supplies.

Today, however, the best lands for agricultural production have for the most part already been brought into production. While some nations with lower population densities may have reserves of good agricultural land, Central America has already exploited its prime agricultural areas. In fact, many would argue that some of the land now in production, such as the Caribbean lowlands and very steep slopes (see Jiménez-Crespo, chapter 20, and Mayne, chapter 22), are inappropriate for intensive agricultural production, leading to problems of erosion, decreasing water quality, and loss of biodiversity. It is unlikely that the region can expect to increase food production significantly in the future through expansion of the agricultural frontier.

Agricultural production must be intensified on those lands that are appropriate for agriculture. This is the only approach that will permit us to simultaneously increase production, increase the standard of living of both farmers and consumers, and still preserve significant areas in a natural state. This suggests that the models for sustainability proposed by some, such as traditional slash-and-burn systems, which are low in productivity, do not in fact offer solutions to the problems of agricultural sustainability. In fact, the intensive production systems employed in industrialized nations may well offer more meaningful models.

Hypothesis 3: The Central American agricultural landscape is largely a result of the introduction of exotic species that produce less sustainable agroecosystems than indigenous species

The post-Columbian interchange of domesticated plants and animals between the Old and New Worlds had an enormous impact on both. New World crops that are staple food crops in Africa today include, for example, peanuts, maize, and manioc (Viola and Margolis, 1991). In many parts of Central Africa, these crops form the basis of the modern diet. Nonetheless, the impact of Old World crops and animals on the New World is even more striking. Both animal and plant production in Mesoamerica depend largely on species that are not native to the region. The region's landscape has undergone a transformation.

The impact of livestock, first brought to the New World on Columbus's second voyage in 1493 (Viola and Margolis, 1991, p. 94) is especially important. None of the domesticated ruminants now common in Central America are native to the region. Neither large herbivores nor the grasslands on which they feed were present at the time of European settlement. As a result, the introduction of livestock production in the region required the large-scale conversion of forested lands to savannas, a process that still proceeds today. Many argue that the attempt to produce livestock in an ecosystem not inherently suited to ruminants has led to loss of biodiversity, environmental degradation, and unwarranted investment of public funds in research and extension programs focused on animal production.

Exotic plant species have played nearly as large a role in the transformation of the region's ecology. Wheat, rice, yams, plaintains, bananas, and many commonly produced vegetable crops such as cabbage and potatoes are not native to Central America. Large-scale plantation production of exotic species played a critical role in opening areas for agricultural production, especially lowland, humid areas. African oil palm, bananas, and sugar cane are prime examples.

The recent emphasis on production for export of many international development organizations such as the World Bank and the U.S. Agency for International Development threaten to exacerbate this replacement of traditional crops and cropping systems by exotics. It would be more fruitful and less destructive to invest resources in improving the productivity of native species rather than encouraging more production of exotics. Research and development funds, therefore, should be directed toward enhancing the productivity and marketability of traditional crops, not toward replacing traditional crops and production systems.

Hypothesis 4: The Central American landscape is also a product of dependence on and domination by external cultures and economies, another factor reducing the sustainability of agriculture in the region

The transformation of the Central American landscape is also reflected culturally and economically. Production for export, for example, makes the region's agriculture dependent on foreign markets and tastes. Most inputs used in agriculture come from outside the region, making Central America highly dependent on continued foreign exchange earnings to maintain its agricultural base. Destructive production techniques and systems are employed, which result in environmental destruction and the mining of local natural resources to meet the demands of international consumers who have little concern over the fate of the region.

Even recent efforts at conservation may represent just one more imposition of external values and concerns on Central America's people. Is the demand for the preservation of natural ecosystems, for example, fed by the needs of local populations or by foreigners? Conservation of natural resources may in fact mean nothing more than an additional sacrifice in the standard of living for local populations, particularly if international donors are unwilling to support the cost of preservation.

A more appropriate model for development, therefore, would focus on increasing internal linkages both within nations and among nations in the region. An emphasis on food production to meet local demand would enhance the stability of food production and decrease dependence on both foreign markets and inputs from industrialized nations. Similarly, resource conservation efforts should focus on issues of immediate concern to local populations, such as water quality, rather than catering to the concerns of industrialized nations.

Hypothesis 5: From a local perspective, production for export does not mine natural resources any more than production for national, urban consumers. Both have negative impacts on the long-term sustainability of local resources and communities

Some argue, on the one hand, that it makes little difference to the rural population whether production is exported to national or to international markets. Unless products are being used at the local level, in the watershed, for example, local resources are "exported" to some degree. If fixed resources are used in a production process and the product is exported, the result is a net loss of resources to the local community. Further, the environmental degradation that occurs as a result of the production process will have a local impact, regardless of where the product is consumed. Even though transition to more use of renewable resources will decrease concerns about mining local fixed resources, production systems based on renewable resources may still be causing ecosystem degradation.

On the other hand, greater distance between production and consumption may make it more likely that the costs of ecological degradation caused by the production process will not be included in the prices. For the most part, consumers will not be sensitive to degradation that they do not experience in a direct way. When consumers in the temperate climates eat a banana produced in the tropics, are they concerned about the possibility that the water supply of a downstream community may be degraded as a result? Probably no more than consumers in the tropics who are eating bread produced from grain grown in the Midwest of the United States are concerned about groundwater contamination caused by grain production there.

One might hope that consumers, aware of negative effects of production in their own area, might realize that other similar effects are happening elsewhere. Similarly, education and income probably influence the degree to which consumers would be knowledgeable about, as well as willing to pay more for, less ecology-degrading production processes.

The diminishing economic importance of political boundaries through the globalization of the world economy—General Agreement on Tariffs and Trade (GATT), World Trade Organization (WTO), European Union (EU), and North American Fair Trade Agreement (NAFTA)—implies that it will be more and more likely that products will be sold outside the watershed and, indeed, outside the country. Lower living standards for people who are earning unskilled wages may imply that local producers will not have the education or income associated with the "luxury" of environmental quality. Thus, international trends would tend to suggest that concern for the environmental degradation associated with production will diminish as the distance to the final consumer and the poverty of the producer increase. All of these considerations point to the importance of multiscalar evaluation of alternatives for development, as Rocheleau (chapter 5) discusses. They suggest that both national and international economic policies are crucial components of long-term sustainability at the local level.

Hypothesis 6: Economic sustainability depends on prices that reflect scarcity. Therefore, for long-term sustainability, economic policy must address ways to ensure that prices do reflect resource scarcity

Two important commodities are scarce: fixed quantity, nonrenewable raw resources and natural ecosystems. If markets are functioning efficiently, the opportunity cost of using a fixed, nonrenewable resource in the present, as opposed to conserving it for the future, should be factored into the market price. However, if price excludes its opportunity cost, fixed resources will be relatively cheap and their use will exceed the efficient rate that would result if scarcity were fully reflected in price. Current consumer prices probably do not reflect the increasing scarcity that results from the production of consumer goods and services. For example, the most profitable mix of ingredients in fertilizer might be inefficient if the price of phosphorus, a fixed resource, does not adequately reflect its relative scarcity. Research programs therefore need to reflect the full cost of the use of inputs in production systems. Breeding programs, for example, should be based on the efficiency of scarce resource use, not on maximizing total production.

Similarly, current prices probably do not reflect the environmental degradation caused by production. The cost of using fixed resources, in the sense that the resources are unique ecosystems, should also be reflected in the price of the resulting commodities on the market. The argument that consumers are the ultimate cause of ecosystem degradation suggests that the price paid by consumers does not include an opportunity cost for the destruction of scarce ecosystems. Without intervention, this cost may not be included in the price until consumers think it is in their self-interest to protect scarce environments.

The price for a resource is based on the competition among various production processes for its use. This implies that the unregulated market will always result in the most degrading production processes being located in poor, especially scarcely populated, areas. Since the small number of local residents who will experience the degradation are poor, they will not be able to outbid the commercial process that causes it. Land use may therefore need to be regulated to protect poor, sparsely populated areas. Some would argue that this is especially true where these areas coincide with unique environments. "Green taxes" are frequently suggested as one way to impose the opportunity cost of scarcity, and they may be a fruitful avenue for research and analysis.

Hypothesis 7: We cannot know whether we are conserving what the future wants or needs. Efforts should therefore focus on nonreversible impacts

It is impossible to know the demands of the future. At the same time, there is a finite amount of public and private resources that can be invested in conservation and preservation. Setting priorities for conservation therefore poses a dilemma. Should we try to conserve everything? Should we focus on the conservation of scarce fixed resources and ecosystems?

One may argue that the best use of these scarce investment resources may imply that we concentrate on preserving large, contiguous, unique ecosystems rather than small, aesthetic ecosystems. The Peten, for example, is probably sufficiently unique and large to be considered one of the strategic environmental resources that should be preserved. However, what about some of the smaller Mesoamerican ecosystems? Similarly, what is the appropriate role of the private versus the public sector in preservation? If a dollar of public funds is available, should it be spent on preserving a large strategic resource? Or will it be more effective to invest in many small areas on private property? Scarce resources should be targeted at high-priority ecosystems, but what are the criteria for selecting these ecosystems for special attention?

Since it is impossible to know future technologies and consumer preferences, conservation efforts should focus on irreversibilities. This reasoning implies that we need to invest in research that teaches us about the physical and socioeconomic feasibility of reversing ecosystem degradation. The cost of restoration would then be the opportunity cost of using these resources in production systems.

Although the discussion in this volume is not confined to these seven hypotheses, they capture the essence of much of the debate related to the tradeoffs between environmental and resource conservation and economic growth and agricultural development. While the study area is largely confined to Mesoamerica, these analyses are relevant throughout the world. Scale, intensity, scarcity, and conservation are major issues regardless of geographic location. Furthermore, the use of nonindigenous species heavily dependent on inputs as the core of food and fiber production processes is a major contributor to environmental degradation throughout the developing world. The economic and cultural impacts of the industrialized world on these nations is an additional factor that must be considered in any serious discussion of long-term sustainability. This book provides a practical discussion of the complex interactions of these forces, which will be at the forefront of the debate over appropriate strategies to balance growth and conservation for many years to come.

2

Institutional Constraints on Agricultural Sustainability

Vernon W. Ruttan

Throughout most of human history, increases in agricultural output have been achieved almost entirely from increases in cultivated areas. We are in the closing decades of the twentieth century, rapidly approaching the time when all increases in agricultural output will have to come from increases in the intensity of cultivation on the lands already used for agricultural production. In the industrialized countries of the world, this transition began in the later years of the nineteenth century. For most developing countries, the transition did not begin until midcentury. This is an exceedingly short time in which to achieve the most remarkable transition in agriculture since it was first invented by neolithic women more than 10,000 years ago.

Concerns about Resources and the Environment

We are now in the midst of the third wave of social concern since World War II about the implications of natural resource availability and environmental change for the sustainability of improvements in human well-being.

The first wave of concern, in the late 1940s and 1950s, focused primarily on the quantitative relationship between resource availability and economic growth. The primary response to this first concern was technical change. The second wave of concern occurred in the late 1960s and early 1970s. The earlier concern about limits to growth imposed by scarcity was supplemented by concern about the capacity of the environment to assimilate the multiple forms of pollution generated by growth. The response to these fears was the development of institutions designed to force individual firms and other organizations to bear the costs that arose from the externalities generated by commodity production (Ruttan, 1971).

Since the mid-1980s these two earlier concerns have been supplemented by a third—the implications for environmental quality, food production, and human health of a series of environmental changes that are occurring on a transnational scale. Global warming, ozone depletion, acid rain, and other such issues pose problems for the sustainability of agricultural production and economic activity (Committee on Global Change, 1990; Committee on Science, Engineering and Public Policy, 1991). The institutional innovations needed to respond will be more difficult to design. They will, like the sources of change, need to be transnational or international.

It is of interest that with each new wave, the issues that dominated the earlier one were recycled. The result is that while the intensity of earlier concerns has receded, in part because of the induced technical and institutional changes, fears about the relationships between resource and environmental change and sustainable growth in agricultural production have broadened.

Recent historical trends in production and consumption of the major food grains could easily be taken as evidence that one should not be excessively concerned about the capacity of the world's farmers to meet future food demands. World wheat prices, corrected for inflation, have declined since the middle of the last century. Rice prices have declined since the middle of this century (Edwards, 1988; Pingali, 1988). These trends suggest that productivity growth has been able to more than compensate for the rapid growth in demand, particularly during the decades since World War II.

As we look toward the future, however, the sources of productivity growth are not as apparent as they were a quarter century ago. The demands that the developing economies will place on their agricultural producers from population growth and growth in per capita consumption, arising from higher income, will be exceedingly high. During the next several decades, growth in food and feed demand, the result of growth in population and income, will increase more than 4 percent per year in many countries. Many will experience more than a doubling of food demand before the end of the second decade of the next century.

Definitions of Sustainability

The concept of sustainability was apparently first advanced in 1980 by the International Union for the Conservation of Nature and National Resources (Lele, 1991; Pearce, Markandya, and Barbier, 1989, p. xii). Before the mid-1980s, it had achieved its widest currency among critics of "industrial" approaches to the process of agricultural development (Harwood, 1990, pp. 3–19). Proponents had traveled under a number of rhetorical vehicles, such as biodynamic agriculture, organic agriculture, farming systems, appropriate technology, and more recently, regenerative and low-input agriculture (Dahlberg, 1991).

Writing in the early 1980s, Gordon K. Douglass (1984, pp. 3–29) identified three alternative conceptual approaches to the definition of agricultural sustainability. One group defined sustainability primarily in technical and economic terms—that is, the capacity to supply the expanding demand for agricultural commodities in increasingly favorable ways. For this group, primarily mainstream agricultural and resource economists, the long-term decline in the real prices of agricultural commodities represented evidence that the growth of agricultural production has been following a sustainable path.

Douglass (1984) identified a second group that regards agricultural sustainability primarily as an ecological question: "For its advocates an agricultural system which needlessly depletes, pollutes, or disrupts the ecological balance of natural systems is unsustainable and should be replaced by one which honors the longer-term biophysical constraints of nature" (p. 2). Among those advancing this agenda, there is a pervasive view that present population levels are already too large to be sustained at present levels of per capita consumption (Goodland, 1991).[1]

A third group, traveling under the banner of "alternative agriculture," places its primary emphasis on sustaining not just the physical resource base but also a broad set of community values (Committee on the Role of Alternative Farming Methods, 1989). It often views conventional, science-based agriculture as an assault, not only on the environment, but also on rural people and rural communities. A major objective is to strengthen or revitalize rural culture and communities, guided by the values of stewardship and self-reliance and an integrated or holistic approach to the physical and cultural dimensions of production and consumption.

By the mid-1980s, the sustainability concept was diffusing rapidly from the confines of its agroecological origins to include the entire development process. The definition that has achieved the widest currency was that adopted by the Brundtland Commission: "Sustainable development is development that meets the needs of the present without compromising the ability of future generations to meet their own needs" (WCED, 1987, p. 43).

The Brundtland Commission definition represented a deliberate attempt to expand the concept of sustainability to take into account the need to respond to the growth in demand for agricultural and other natural resource-based products arising from population and income growth. However, the definition also raises the possibility that it may be necessary for those of us who are alive today, particularly those living in the more affluent societies, to curb our level of material consumption to accommodate growth in consumption in less affluent societies and avoid an even more drastic decline in the consumption levels of future generations. This is not a welcome message to societies that find it difficult to discover principled reasons for the contemporary transfer of resources across political boundaries in support of efforts to narrow the level of living between rich and poor nations or rich and poor people (Ruttan, 1989b).

Our historical experience, at least in the West, often causes us to be skeptical about our obligations to future generations. It was only a generation ago that Robert Solow (1974), one of our leading growth theorists, noted in his Richard T. Ely address to the American Economic Association: "We have actually done quite well at the hands of our ancestors. Given how poor they were and how rich we are, they might properly have saved less and consumed more" (p. 9). In most of the world the ancestors have not been so kind.

In spite of its challenge to current levels of consumption in the developed countries, it is hard to avoid a conclusion that the popularity of the Brundtland Commission definition is due, at least in part, to the fact that it is so broad that it is almost devoid of operational significance. The sustainability concept has undergone what has been referred to as "establishment appropriation." It is now experiencing the same "natural history" as earlier reform efforts. Initially a "progressive" rhetoric is advanced by critics as a challenge to the legitimacy of dominant institutions and practices. If the groups and symbols involved are sufficiently threatening, the dominant institutions will attempt to respond by "appropriating" or embracing the symbol themselves: "In so doing these dominant institutions—such as the World Bank and the agricultural universities—are typically able to demobilize the movement" (Buttel, 1991, p. 7). Buttel argues that sustainability has been embraced both by radical reformers and neoconservatives because it removes the focus from achieving greater participation of the poor in the dividends from economic growth to protecting an impersonal nature from the destructive forces of growth (p. 9). Runge (1992) presents a more positive perspective on the move by the traditional agricultural and development communities to embrace the sustainability concept. He visualizes sustainability as an integrative concept that can facilitate the synthesis of the research and policy agendas of the environmental, agricultural, and development communities.

Sustainable Agricultural Systems in History

It is not uncommon for a social movement to achieve the status of an ideology while still in search of a methodology or a technology. If the reform movement is successful in directing scientific and technical effort in a productive direction, it becomes incorporated into normal scientific or technical practice. If it leads to a dead end, it slips into the underworld of science, often to be resurrected when the conditions that generated the concern again emerge on the social agenda.

Research on new uses for agricultural commodities is one example. It was promoted in the 1930s, under the rubric of chemurgy, and in the 1950s, under the title of utilization research, as a solution to the problem of agricultural surpluses. It lost both scientific and political credibility because it promised more than it could deliver. Integrated pest management represents a more fortunate example. This term emerged in the 1960s as an alternative to chemical-intensive pest control strategies and was appropriated in the 1970s as a rhetorical device to paper over the differences between ecologically oriented and economically oriented entomologists. After two decades of scientific research and technology development, there are now sets of practices that come closer to meeting the definition of integrated pest management as visualized by those who had coined the term (Palladino, 1989).

We are able to draw on several historical examples of sustainable agricultural systems that proved capable of meeting the challenge of achieving sustainable increases in agricultural production. One example is the forest and bush fallow (or shifting cultivation) systems practiced in most areas of the world in premodern times and today in many tropical areas (Pingali, Bigot, and Binswanger, 1987). At low levels of population density, these systems were sustainable over long periods of time. As population density increased, short fallow systems emerged. Where the shift to short fallow systems occurred slowly, as in western Europe and East Asia, systems of farming that permitted sustained growth in agricultural production emerged. Where the transition to short fallow has been forced by rapid population growth the consequence has often been soil degradation and declining productivity.

A second example can be drawn from the agricultural history of East Asian wet rice cultivation (Hayami and Ruttan, 1985, pp. 280–298). Traditional wet rice cultivation resembled farming in an aquarium. The rice grew tall and rank and had a low grain-to-straw ratio. Most of what was produced, straw and grain, was recycled in the form of animal and human manures. Mineral nutrients and organic matter were carried into and deposited in the fields with the irrigation water. Rice yields rose continuously, though slowly, even under a monocultural system.

A third example of sustainable agriculture was the system of integrated crop-animal husbandry that emerged in western Europe in the late Middle Ages to replace the medieval two- and three-field systems (Boserup, 1965; van Bath, 1963). The "new husbandry" system emerged with the introduction and intensive use of new forage and green manure crops. These, in turn, permitted an increase in the availability and use of animal manures. And this, in turn, permitted the emergence of intensive crop-livestock systems of production through the recycling of plant nutrients in the form of animal manures to maintain and improve soil fertility.[2]

These three systems, along with other similar systems based on indigenous technology, have provided an inspiration for the emerging field of agroecology. But none of the traditional systems, while sustainable under conditions of slow growth in demand, has the capacity to respond to modern rates of growth in demand—in the 3 to 5 percent per year range—generated by some combination of rapid increase in population and in income (Hayami and Ruttan, 1985, pp. 41–42). In the presently developed countries, the capacity to sustain the necessary increases in agricultural production will depend largely on capacity for institutional innovation. If capacity to sustain growth in agricultural production is lost, it will be a result of political and economic failure. It is quite clear, however, that the scientific and technical knowledge is not yet available that will enable farmers in most tropical countries to meet the current demand their societies are placing on them, as well as to sustain the increases that are currently being achieved. In these countries, achievement of sustainable agricultural surpluses is dependent on advances in scientific knowledge and on technical and institutional innovations.

The Technological Challenge to Sustainability

One might ask why concern about the sustainability of modern agricultural systems has emerged with such force toward the end of the twentieth century? The first reason is the unprecedented demands that growth of population and income are imposing on agricultural systems. As stated earlier, before the beginning of this century almost all increases in food production were obtained by bringing new land into production. By the first decades of the next century, almost all increases in food production must come from higher yields—from increased output per hectare.

A second reason for concern about sustainability is that the sources of future productivity growth are not as apparent today as they were a quarter century ago. It seems apparent that the gains in agricultural production required over the next 25 years will be achieved with much greater difficulty than in the immediate past (Ruttan, 1989a; 1994). The incremental responses to the increases in fertilizer use has declined. Expansion of irrigated areas has become more costly. Maintenance research, the research required to prevent yields from declining, is rising as a share of research effort (Plucknett and Smith, 1986). The institutional capacity to respond to these concerns is limited, even in the countries with the most effective agricultural research and extension systems. In many developing countries there has been considerable difficulty in maintaining the agricultural research capacity that was established in the 1960s and 1970s (Eicher, 1993).

Within another decade, advances in basic knowledge may create new opportunities for advancing agricultural technology that will reverse the urgency of some of the preceding concerns. Institutionalization of private sector agricultural research capacity in some developing countries is beginning to complement public sector capacity (Pray, 1987). It is possible that advances in molecular biology and genetic engineering will soon begin to release the constraints on productivity growth in the major food and feed grains.[3] But advances in agricultural technology will not be able to eliminate what some critics tend to view as a "subsidy" from outside the agricultural sector. Transfers of energy in the form of mineral fuels, pathogen and pest control chemicals, and mineral nutrients will continue to be needed to sustain growth in agricultural production—and in much larger quantities—until well into the middle of the next century. Over the very long run, the scarcity of phosphate fertilizer and fossil fuels, reflected in rising real prices, is likely to become the primary resource constraint on sustainable growth in agricultural production (Chapman and Barker, 1991; Desai and Gandhi, 1990).

The third concern is about the environment spillover from agricultural and industrial intensification. The spillover effects from agricultural intensification include the loss of soil resources due to erosion; waterlogging and salinization; surface and groundwater contamination from plant nutrients and pesticides; resistance of insects, weeds, and pathogens to present methods of control, and the loss of natural habitats (Conway and Pretty, 1991). If agriculture is forced to continue to expand into more fragile environments, problems such as soil erosion and desertification can be expected to become more severe. Additional deforestation will intensify problems of soil erosion, species loss, and degradation of water quality and contribute to climate change. There can no longer be much doubt that the accumulation of carbon dioxide (CO_2) and other greenhouse gases—principally methane (CH_4), nitrous oxide (N_2O) and chloroflurocarbons (CFCs)—has set in motion a process that will result in a rise in the global average surface temperature and changes in rainfall patterns over the next 30 to 60 years. These changes can be expected to impose substantial adaptation demands on agricultural systems. The systems that will have the least capacity to adapt will be in countries with the weakest agricultural research and natural resource management capacity—principally in the humid and semiarid tropics (Lee, Norse, and Parry, 1994; Ruttan, 1992).

It should be apparent that a major issue over the next half century for most developing countries, including the formerly centrally planned economies, will be how to generate and sustain the advances in agricultural technology that will be needed to meet the demands that these societies will place on their agricultural sectors. This objective appears to be in direct conflict with the worldview of many of the leading advocates of sustainable development. Sustainable development is a concept that implies limits, both to the assimilative capacity of the environment and to the capability of technology to enhance human welfare: "To the sustainable development community the capacity of the environment to assimilate pollution from human production and consumption activity is the ultimate limit to economic growth" (Batie, 1989, p. 1085). But this is not a problem that has emerged only during the second half of the 20th century. Humankind has throughout history been continuously challenged by the twin problems of (1) how to provide oneself with adequate sustenance and (2) how to manage the disposal of what in recent literature has been referred to as "residuals." Failure to make balanced progress along both fronts has at times imposed serious constraints on society's growth and development (Ruttan, 1971, p. 707).

I differ in one fundamental respect from those who are advancing the sustainability agenda. It seems clear to me that the capacity of a society to solve either the problem of sustenance or the problems posed by the production of residuals is inversely related to population density and the rate of population growth and is positively related to its capacity for innovation in science and technology and in social institutions (Ruttan, 1971, p. 788). I am exceedingly concerned that the bilateral and multilateral assistance agencies, in their rush to allocate resources in support of a sustainability agenda derived more from developed than from developing countries' resource and environmental priorities, will fail to sustain the effort needed to build viable agricultural research in the tropics.

Three Unresolved Analytical Issues

In this section I will identify three unresolved analytical issues that must be confronted before a commitment to sustainability can be translated into an internally consistent reform agenda.

Substitutability

Our knowledge about the role of technology in widening the substitutability among natural resources and between natural resources and reproducible capital is clearly inadequate. Economists and technologists have traditionally viewed technical change as widening the possibility of substitution among resources—of fertilizer for land, for example (Goeller and Weinberg, 1976; Solow, 1974). The sustainability community rejects the "age of substitutability" argument. The loss of genetic plant resources is viewed as a permanent loss of capacity. The elasticity of substitution among natural factors and between natural and man-made factors is viewed as exceedingly low (Daly, 1991; James, Nijkamp, and Opschoor, 1989). When considering the production of a particular commodity—for example, the substitution of fertilizer for land in the production of wheat—this is an argument over the form of the production function. However, substitution also occurs through the production of a different product that performs the same function or fills the same need, for example, fiber-optic cable for conventional copper telephone wire or fuels with higher hydrogen-to-carbon ratios for coal.

The argument about substitutability, while inherently an empirical issue, is typically argued on theoretical or philosophical grounds. It is possible that historical experience or advances in futures modeling may lead toward some convergence of perspectives. But the scientific and technical knowledge needed to fully resolve disagreements about substitutability will always lie in the future (Constanza, 1989). The issue is exceedingly important, however. If, on the one hand, a combination of capital investment and technical change can continuously widen opportunities for substitution, imposing constraints on present resource use could leave future generations less well off. If, on the other hand, real output per unit of natural resource input is narrowly bounded—it cannot exceed some upper limit, which is not too far from where we are now—then catastrophe is unavoidable.

Obligations toward the Future

The second issue that has divided traditional resource economists and the sustainability community is how to deal analytically with the obligations of the present generation toward future generations. The issues of intergenerational equity is at the center of the sustainability debate (Pearce et al., 1989, pp. 23–56; Solow, 1991). Environmentalists have been particularly critical of the approach used by resource and other economists in valuing future benefits and costs. The conventional approach involves the calculation of the "present value" of a resource development or protection project by discounting the cost-and-benefit statement by some "real" rate of interest—an interest rate adjusted to reflect the cost of inflation. It is World Bank policy (but not always practice) to require a 10–15 percent rate of return on projects. These higher rates are set well above long-term real rates of interest (historically, less than 4 percent) to reflect the effect of unanticipated inflation and other risks associated with project development and implementation. An attempt is made in this way to avoid unproductive projects.

The critics insist that this approach results in a "dictatorship of the present" over the future. At conventional rates of interest, the present value of a dollar of benefits 50 years into the future approaches zero. "Discounting can make molehills out of even the biggest mountain" (Batie, 1989, p. 1092). Solow (1974, p. 3) has made the same point in more formal terms. He notes that if the marginal profit—marginal revenue less marginal cost—to resource owners rises slower than the rate of interest, resource production and consumption is pushed nearer in time and the resource will be quickly exhausted (Lipton, 1991).

A question that has not been adequately answered is whether, as a result of the adoption of a widely held sustainability "ethic," the market-determined discount rates would decline toward the rate preferred by those advancing the sustainability agenda.[4] Or will it be necessary to impose regulations—constraints on current consumption—in an effort to induce society to shift the income distribution more strongly toward future generations? It is clear, at least to me, that in most countries efforts to achieve sustainable growth must involve some combination of (1) higher contemporary rates of saving, that is, deferring present in favor of future consumption, and (2) more rapid technical change, particularly the technical changes that will enhance resource productivity and widen the range of substitutability among resources.[5] But will this be enough? I suspect not. Given the inability of economic theory to provide satisfactory tools to deal analytically with obligations toward the future, what should be done? My own answer is that we should take a strategic approach to the really large issues—how much we should invest to reduce the probability of excessive climate change, for example. We should continue to employ conventional cost-benefit analysis to answer the smaller questions, such as when to develop the drainage systems needed to avoid excessive buildup of waterlogging and salinity in an irrigation project.

Incentive-compatible Institutional Design

A third area where knowledge needs to be advanced is the design of institutions that are capable of internalizing—within individual households, private firms, and public organizations—the costs of actions that generate the negative spillover effects (the residuals) that are the source of environmental stress. Under present institutional arrangements, important elements of the physical and social environment continue to be undervalued for both market and nonmarket transactions. Traditional production theory implies that if the price to a user of an important resource is undervalued it will be overused. If the price of a factor, the capacity of groundwater to absorb pollutants, for example, is zero, it will be used until the value of its marginal product to the user approaches zero. This will be true even though it may be imposing large social costs on society.

The dynamic consequence of failure to internalize spillover costs are even more severe. In an environment characterized by rapid economic growth and changing relative factor prices, failure to internalize resource costs will slant the direction of technical change. The demand for a resource that is priced below its social costs will grow more rapidly than in a situation in which substitution possibilities are constrained by existing technology. As a result, "open access" resources will undergo stress or depletion more rapidly than in a world characterized by a static technology or even by neutral (unbiased) technical change.

The process is clearly apparent in U.S. agriculture. Federal farm programs encourage farmers to grow a small group of selected crops, to grow these crops on a continuous basis, and to use more chemical-intensive methods in production (General Accounting Office, 1990). The capacity of the environment to absorb the residuals from crop and livestock production has been treated as a free good. As a result, scientific and technical innovation has been overly biased toward the development of land substitutes—plant nutrients and plant-protection chemicals and management systems that reflect the overvaluation of land and the undervaluation of the social costs of the disposal of residuals from agricultural production processes (Runge et al., 1990).

The design of incentive-compatible institutions—institutions capable of achieving compatibility among individual, organizational, and social objectives—remains at this stage an art rather than a science. The incentive-compatibility problem has not been solved even at the most abstract theoretical level.[6] This deficiency in institutional design capacity is evident in our failure to design institutions capable of achieving contemporary distributional equity, either within countries or among rich and poor countries. It impinges with even greater force on our capacity to design institutions capable of achieving intergenerational equity.

An Uncertain Future

In closing I would like to emphasize how far we are from being able to design an adequate technological or institutional response to the issue of how to achieve sustainable growth either in agricultural production or in both the sustenance and the amenity components of consumption. In spite of the large literature in agronomy, agricultural economy, and related fields, there is no package of technology that can ensure the sustainability of growth in agricultural production at a rate that will enable agricultural producers, particularly in the developing countries, to meet the demands that are being placed on them (Board on Agriculture, National Research Council, 1991; Board on Agriculture, Science and Technology for Development, 1992; Rosenberg and Eisgruber, 1992; Vosti, Reardon, and von Urff, 1991). Sustainability is appropriately viewed as a guide to future agricultural research agendas rather than to practice (Ruttan, 1988). As a guide to research, it seems useful to adhere to a definition that would include (1) the development of technology and practices that maintain and/or advance the quality of land and water resources and (2) the improvement in the performance of plants and animals and advances in production practices that will facilitate the substitution of biological for chemical technology. The research agenda on sustainable agriculture needs to explore what is biologically feasible without being excessively limited by present economic constraints.

At present, the sustainability community has not been able to advance a program of institutional innovation or reform that can provide a credible guide to the organization of sustainable societies. We have yet to design the institutions that can ensure intergenerational equity. Few would challenge the assertion that future generations have the right to levels of sustenance and amenities that are at least equal to those enjoyed (or suffered) by the present generation. They also should expect to inherit improvements in institutional capital—including scientific and cultural knowledge—needed to design more productive and healthy environments.

My conclusion with respect to institutional design is similar to that which I have advanced in the case of technology. Economists and other social scientists have made a good deal of progress in contributing the analysis needed for course correction. But capacity to contribute to institutional design remains limited. The fact that the problem of designing incentive-compatible institutions—institutions capable of achieving compatibility among individual, organizational, and social objectives—has not been solved at even the most abstract theoretical level means that institutional design proceeds in an ad hoc trial-and-error basis and that the errors continue to be expensive. Institutional innovation and reform should represent a high-priority research agenda.

Notes

1. This view stems in part from a naive carrying-capacity interpretation of the potential productivity of natural systems (Raup, 1964).

2. In his study of sustainable agriculture in the Middle Ages, Jules N. Pretty (1990) notes, "Manorial estates survived many centuries of change and appear to have been highly sustainable agricultural systems. Yet this sustainability was not achieved because of high agricultural productivity—indeed it appears that farmers were trading off low productivity against the more highly valued goals of stability, sustainability and equitability" (p. 1).

3. For an argument that the results of genetic engineering can be expected to undermine sustainable methods of farming, see Hindmarsh (1991).

4. The question of the impact of a positive discount (or interest) rate on resource exploitation decisions is somewhat more complex than often implied in the sustainability literature. High rates of resource exploitation can be consistent with either high or low interest rates (Norgaard, 1991; Price, 1991). As an alternative to lower discount rates, Mikesell (1991) suggests taking resource depletion into account in cost-benefit analysis. For a useful commentary on the debate about the effects of high and low interest rates on sustainability, see Lipton (1991).

5. Norgaard and Howarth (1991) and Norgaard (1991) argue that decisions about the assignment of resource rights among generations should be based on equity rather than efficiency. When resource rights are reassigned between generations, interest rates will change to reflect the intergenerational distributions of resource rights and income. I interpret these arguments as saying that if present generations adopt an ethic that causes them to save more and consume less, the income distribution will be tilted in favor of future generations. This is, however, not the end of the story. A decline in marginal time preference lowers the rate of interest. Improvement in investment opportunities resulting, for example, from technical change, will increase the demand for investment and thus raise interest rates (Hirshleifer, 1970, pp. 31–45, 113–116).

6. The concept of incentive compatibility was introduced in a paper by Hurwicz (1972), where he showed that it was not possible to specify an informationally decentralized mechanism for resource allocation that simultaneously generates efficient resource allocation and incentives for consumers to honestly reveal their true preferences. For the current state of knowledge in this area see Groves, Radner, and Reiter (1987). For a detailed discussion of the difficulties of achieving incentive compatibility in the natural resources and in environmental policy and management, see M. D. Young (1992). For a set of case studies that illustrates successful efforts to reduce the transaction costs involved in institutional design and maintenance, see Ostrom (1990).

3

The Logic of Community Resource Management in Latin America

William Ascher

Community management of natural resources is an increasingly credible strategy for sustainable development. A strong belief in state control of natural resources prevailed for many years, followed by the more recent mania to privatize everything. The increased interest in community management of natural resources is a remarkable shift in thinking. This chapter begins by outlining the arguments for and against community resource management. It then discusses the requirements necessary for success. Finally, it describes several efforts to establish community management and some of the results of these strategies.

Any discussion of community management must begin by clarifying the term *community*. Much confusion over the scope and nature of community resource management exists because the term *community* has several different meanings. It is tempting to define a community as all persons living within a particular geographic area, delimited by history, geographic distinctions, or administrative boundaries. This definition leads to identifying villages or towns as communities in much of Latin America. On a larger scale, *municipios* (counties) or even districts would be defined as communities.

Perhaps the temptation to include all residents in the community comes from the optimistic view that everyone living in the same area will or should share in building community solidarity, common purpose, and mutual regard. For better or for worse, this is not an accurate view of the social and political realities in most Latin American areas, or elsewhere in the developing world. Moreover, a hidden implication of defining community by geographic boundaries is that "community governance" becomes the same as "local government." If the community is to manage natural resources and if the community is the same geographic unit as the jurisdictional boundary of local government, then the local government can take on the mantle of the community.

To define community according to geographic or administrative boundaries and community governance as local government discards two extremely important characteristics that most community management initiatives emphasize. First, community management is meant to be an aspect of civil society—activities by people outside the government and state apparatus. Community resource management is not just a modest shift from national government control to subnational government control. It is different from government control, at either the national or subnational level.

Second, community management usually means that some people, because of proximity, historical rights, or other special relationships to local resources, will be responsible for resource management rather than everyone residing in the area. As we will see, the logic of self-discipline in resource management hinges on reserving the rights to use resources to a limited number of people. It may seem unfair to reserve resource management to only certain elements of the population that live within a geographic area. After all, "community as everyone" seems egalitarian because it includes all people. In reality, including everybody means including the powerful and the wealthy, who typically dominate area affairs, either by controlling local government or by asserting their influence economically and socially. Paradoxically, community resource management is exclusionary but antielitist as long as the people who have the rights to manage resources are among the relatively low-income and low-power segments of the population.

To define community resource management as control by community groups that share common identification and interests is useful. The user groups often consist of the families that traditionally exploited the resources under customary rights. Many continued to exploit the resource illegally when state restrictions were imposed. These groups are often "indigenous people" who are economically, culturally, and politically out of the mainstream. Community resource management is often a restoration of traditional resource rights to a limited, often economically marginal group within a geographic area.

The strategy of community resource management takes on many additional dimensions and complications when the community is indigenous. Indeed, it makes sense to distinguish indigenous community resource management strategies from more ordinary strategies. This chapter covers both, but it is useful to clarify why the indigenous management strategy is so distinctive.

The term *indigenous* refers to peoples who have long histories of residence in particular areas and who are culturally distinct from the mainstream, Hispanicized culture. Of course, the history of population migration is typically very complicated. No one has been indigenous since the beginning of time. Many culturally distinctive Amerindian groups in the remaining relatively unexploited, "natural" areas of Latin America left the coastal areas and savannas at various times during the expansion of Europeans and Europeanized people into these areas. The Amerindians took refuge in more remote and economically less attractive regions. These include lowlands, such as the Amazonian regions of Brazil, Venezuela, Colombia, Ecuador, and Peru, and highlands, such as the altiplanos of Peru and Guatemala, where agricultural prospects were unattractive to the colonists. Nonetheless, it is useful to recognize that certain peoples have now resided in these areas for many generations. Their ways of life have been thoroughly adapted to where they live now. They view their uses of natural resources as traditional and customary practices going back indefinitely in time. They are often still separate from the mainstream Spanish- or Portuguese-speaking national cultures in legal and moral principles and other cultural aspects. Therefore, their beliefs about their rights and responsibilities in resource exploitation are distinct from those that the national government or its local agents may try to impose. In these respects, to consider them indigenous peoples, whose resource rights have a basis that is independent of the national legal system, is legitimate (Arango, Ochoa, 1992).

The coexistence of different systems and ideas of user rights and cultural separateness complicate the issues of community resource management by indigenous peoples. Their cultural separateness, often defined by language, is an important means of defining the boundaries between group members and outsiders. The resource exploitation practices of these people, often traceable to pre-Colombian practices, also complicate, and often enrich, their community resource management. Finally, indigenous groups usually have a much lower income than other people in the same area because they stay outside the national economy and rely on traditional modes of production. This is very important in thinking about the ethical implications of assigning or recognizing user rights.

Rationales for Community Resource Management

Two different trends are responsible for the rise in the popularity of community resource management. The first is the movement to empower and economically strengthen culturally remote, economically marginal groups. In many developing countries, these local groups, whether indigenous or not,[1] have organized themselves into grassroots organizations to press for political rights, economic improvement, and, commonly, cultural preservation. They have been helped in these efforts by a wide assortment of government agencies, nongovernment organizations, international organizations, and bilateral donors. These outside organizations typically justify their support of grassroots organizations in the name of the moral imperatives to alleviate poverty, preserve distinctive cultural heritages, and promote the political power of these groups. These three strands—economic, cultural, and political—are sometimes mutually reinforcing. Empowerment helps in economic efforts. Prosperity makes communities stronger and better able to defend their distinctive practices from both external and internal challenges.

The second trend is reliance on community groups as the linchpin of the conservation strategy. This trend is based on the growing awareness of both the limitations of government to police natural resources and the advantages that local people have. Governments rarely can marshall the manpower needed to guard forests, marshes, lakes, and other areas where natural resources may be overexploited if not adequately monitored. Government officials are often disliked and mistrusted by local people. The declaration of an area as a government-controlled reserve is often like a red flag. The people may be more willing to poach than if there were no government presence at all because the government is asserting its claims to police the area. In contrast, under community control, local people are more numerous in remote, "natural" areas than the government guards can be. The local people can frequently apply social pressure rather than more expensive policing devices, such as fines or imprisonment, that government control requires. Finally, the local community can organize itself to enforce rules that restrain resource use by its own members.

The choice of the community group to manage the resource is based on an image of local people, particularly indigenous people with traditional user rights, as willing and capable conservationists. This image is certainly not held by all. For example, many government officials still see uneducated rural people as unknowledgeable folk who despoil ecosystems out of ignorance, shortsightedness, or lack of civic consciousness. Yet, an increasingly popular view of traditional resource users is that they are more "in touch with nature" than are other people. According to this view, they recognize their dependence on the natural resource endowment, are wise in the lore of the wilderness, and have a deep respect and love for their natural surroundings.

The positive image of community members as willing and able to engage in serious conservation is sometimes, but not always, true. We therefore turn to the question of what a conservationist role for local, and particularly indigenous, people requires.

Premises of Conservationist Willingness and Capability

The willingness of traditional resource users to conserve the ecosystem requires them to have a strong emotional attachment to the existing ecosystem, a material interest in long-term exploitation of the ecosystem's resources, or both. They are presumed to value the particular elements of the resource endowment because these elements provide material rewards or spiritual value. Otherwise, the local people may be content to see the resource endowment converted into elements more suited to their own needs.

The emotional attachment to the land and particular features of the ecosystem is the easiest part of the equation to romanticize. The standard Western conception of nature is that it is under human dominion and that the conversion of the "natural" into the "artificial" serves human purposes. Non-Western cultures, including those of Latin America, often provide relatively high respect for nature and its components. However, economic necessities—feeding the family, buying medicine, paying unavoidable taxes—always loom large. Finally, even people with great respect for nature are not immune from temptation when immediate returns from resource exploitation are very high. Therefore, even when the local people could afford to use natural resources gradually and sustainably, they may exploit resources too rapidly to obtain high immediate returns. Rising aspirations for even higher incomes can undermine the discipline necessary for conserving resources for future use when some physical limits on how rapidly natural resources can be extracted are removed. In this setting, the availability and affordability of technologies that permit more rapid exploitation, such as chainsaws or more efficient fishing nets, become important.

Willingness also depends on whether the local people with traditional user rights can count on benefitting from resource use in the long run. They may liquidate the resources as quickly as possible, to ensure that they get as much of the benefits as possible, if they do not expect the resources to be available to them in the future. The likelihood that they will try to conserve the resources therefore depends on the security of their exclusive rights to use them. Local people can become alienated from the natural resources if their legal use is restricted. A common reaction to the government's assertion of control over natural resources is for local people to shift from being resource conservers to being resource despoilers in retaliation against the government, as well as to extract benefits before others can.

The decision to defer the benefits of exploiting natural resources also reflects the time horizons of the resource users. How valuable is income now as opposed to the possibility of greater income later? Patience is much more likely when people can afford to wait for greater incomes because their current incomes are adequate to cover necessities or because loans are available to cover emergencies. In short, willingness requires an economic structure that makes it rational for local people to conserve resources for their own future benefit.

The capability of local people to manage natural resources for long-term sustainability depends on three factors: (1) whether the carrying capacity of the ecosystem still presents the opportunity for traditional resource uses to be sustainable, (2) the skill levels of the community, and (3) the organizational strength of the community.

Carrying Capacity

Traditional resource uses, such as foraging livestock, fishing with traditional nets or traps, shifting (slash-and-burn) cultivation, and gathering nuts or bark, are sustainable at certain levels of intensity. The classic example is shifting cultivation. A small group of farmers may continue indefinitely to burn a small patch of forest, plant a few crops, and then go on to another forest patch. They continue this cycle until the group returns to the first patch, allowing time for regeneration. However, the group may exploit more land at some particular time, either because each farmer wants a greater yield or because the group is larger. When this happens, the cycle that returns them to a previously exploited patch will be shortened and the land will not have regenerated as fully. The farmers then have to exploit even more land to get sufficient yields. The group's desire either to sell or to consume more produce or simply its greater numbers can trigger this vicious cycle. Therefore, population growth and exposure to modern markets, which make increased production attractive, affect the sustainability of shifting cultivation.

Skill Levels

Knowing how to exploit natural resources sustainably cannot be taken for granted. Knowledge of techniques that are well adapted to local conditions can be essential, including how to tap rubber or resin without killing the trees, what crops will provide sufficient yield without using too much land or water, how to avoid killing female animals or birds if doing so leads to dramatic declines in their populations, and when to avoid the disruption of spawning cycles while fishing. Migration of people into an area can pose very serious risks to sustainability. The migrants add to the population pressure on systems of limited carrying capacity, and they often lack familiarity with local conditions and techniques. Moreover, migrants often underestimate the fragility of the ecosystem because they lack long-standing familiarity with its intricacies and have not experienced the results of its disruption.

Organizational Strength

Even if people are individually willing to use restraint in exploiting natural resources, they have to be well-enough organized to enforce discipline within the community and to exclude outsiders whose resource exploitation would exceed the ecosystem's carrying capacity. Within the community, individual responsibility can quickly dissolve when others seem to be taking advantage by overexploitation. The situation is even more demoralizing for conservationists when outsiders rush in to exploit resources that the community has managed carefully. Therefore, communities need to be organized so that they can perform the following 10 functions:

1. Gain recognition of user rights with sufficient autonomy and economic flexibility that resource managers have an incentive to conserve;
2. Establish clear boundaries of community membership;
3. Exclude outsiders from encroaching on the resources;
4. Establish clear rules of resource restraint within the community;
5. Provide adequate oversight and punishments for abuse of power by the leadership;
6. Mobilize sufficient expertise to develop and extract resources profitably and sustainably;
7. Punish use violations with sanctions that are appropriate for the severity of the violation;
8. Resolve conflicts within the community over resource use;
9. Arrange for appropriate rewards for the efforts of community members; and
10. Affiliate with broader associations of community groups to gain political strength and expertise.[2]

International connections are often invaluable in performing these functions. The international community can provide a crucial counterweight to the common reluctance of governments to allow nongovernment groups to pressure for recognition of their user rights. Governments may be deterred from suppressing grassroots or nongovernment organization (NGO) movements if such suppression would become known to the public and governments of the United States, Germany, the United Kingdom, Sweden, or other powerful or wealthy countries with a strong commitment to the freedom to organize and to conserve. Just as important, governments of developing countries are often very eager to allow local communities to undertake conservation efforts with funding provided by bilateral or multilateral donors.

Elements and Dilemmas of Community Control as a Conservation Strategy

Assuming that the 10 conditions favorable to community-controlled conservation do hold, what would the concrete elements of such a strategy be? What further dilemmas arise in choosing among strategic options? Five critical issues must be resolved for community resource management to be successful: (1) government recognition of user rights, (2) specification of the kinds of exploitation that will be prohibited or regulated, (3) the economic context in which community resource management will function, (4) government compensation for leaving some resources intact, and (5) organization of the user group.

The first step is for the government to recognize a community's right to use particular resources in a defined area. This rarely comes easily because government officials are often reluctant to relinquish the government's own power. In many circumstances, the community group has to lobby, agitate, or even occupy the land it claims to pressure the government into recognizing its rights. Organizing the community to fight for user rights is a distinctive challenge that often entails personal risk for community members.

If the community is indigenous, the government ought to invoke that community's historical (customary or traditional) use of these resources and acknowledge that this right preexists the government's recognition. This last element is very important for assuring the group (and signaling to others) that the government is not asserting its right to reassign or restrict the user rights at some future time. The government's role is more complicated when the community has no defensible claim to traditional or customary rights and is not indigenous or otherwise culturally distinctive. In this case, the government must specify the criteria for defining the boundaries of the group or accept the boundaries and self-definition offered by particular claimants.

One difficulty that often arises in the recognition of community rights is that complex histories do not always make it clear which specific people have the best claim to particular areas and resources. Historical migrations, evictions by the government or private groups, intermarriage, past reassignment of ownership by the government, and outright government land takeovers can result in multiple claims of virtually equal historical validity. It may be that no groups have plausible legal or tradition-based claims. This is commonly the case when the traditional resource users have left the area because the government excluded them from land put into government-controlled reserves or when the groups have become extinct. The government often cannot clearly identify the descendants of the traditional resource users, even though it wishes to recognize traditional user rights. Several groups may want to push their claim, or no one may be willing or able to do so. Even if some people are currently engaged in resource exploitation in forest reserves, its illegality may dissuade them from coming forth or disqualify them in the eyes of the government. Often the current resource users are recent migrants who have been unable to obtain agricultural land and therefore occupy hillsides, forests, marshes, or other marginal lands.

A pragmatic approach to determining who ought to be awarded user rights should begin by recognizing that relatively small groups provide the greatest potential for sustainable resource use. In small groups, each group member can gain more income from existing resources without putting undue strain on the system. Another consideration is that natural resource exploitation tends to be a modest income-generating activity. If high-income earners achieve control of resources, they are likely to want to exploit them more intensively to make their control worthwhile. High-income people are also more likely to have access to high-technology equipment that makes resource extraction more rapid. Moreover, the remaining "naturalness" of a particular area often reflects restraint by past resource users for the sake of their children and grandchildren. These people are therefore deserving of rights that will enable their descendants to enjoy the fruits of this earlier restraint. Finally, even if the claims of groups long in the area are conflicting, their claims are typically more politically viable than those of recent migrants, no matter how poor the latter may be. This implies that the government ought to select a small, low-income, well-defined group with a long history in the area when no clear-cut claimant to traditional user rights is present.

The second step is for the government to specify what kinds of exploitation will still be prohibited or regulated because of the potential for doing damage to others (so-called negative externalities). Consider the possibility that the government's watershed studies reveal that removing trees at high elevations may increase soil erosion and flooding at lower elevations, even though a community group has had unlimited rights to harvest trees in a particular area. Should the community's user right be restricted on the grounds that its actions are damaging to others? Perhaps surprisingly, the answer ought to be no. The appropriate approach is to recognize the group's right to extract the resource per se but prohibit or regulate any modes of extraction or processing that do damage to other people. If the government insists that the trees must remain, the user groups, with some justification, will believe that the government has reneged on its recognition of the group's rights. The user group will therefore feel less secure about whether it can count on benefitting from the land in the future. Given the practical difficulty of enforcing restrictions on timber harvesting, the result may be rapid deforestation. Potential beneficiaries, or the government on their behalf, ought to compensate the resource users sufficiently to achieve the desired outcome if society would be better off if the resource base is left intact.

This practical consideration is backed up by strong normative principles as well. The issue regarding the group's right to extract the resource is whether people who are holding user rights have an obligation to provide positive externalities—benefits that cannot be captured by those whose actions produce them. If serving society by providing positive externalities is a legal obligation placed on an individual or group (as opposed to the government), user rights and property rights have little meaning. In this case, the government can decide the highest societal use of any piece of land or resource and require that use, regardless of the return to the user or owner.

However, reducing the number of trees, marshes, or other resource endowments on one's property is qualitatively different from causing pollution or creating other damage through processing activities. These are negative externalities—costs that are not borne by those who produce them. Unlike the rights to develop or extract resources per se, harming others through such processing is not justified by user rights. The government has a legitimate and very important role to play in limiting the pollution from resource extraction or processing that impinges on other communities.

The third issue is the general economic context that the government establishes for the operations of community groups. The key challenge is to ensure that all members of the user group have economic incentives to conserve rather than to immediately extract natural resources. Governments often try to get heavily involved to continue their control over resource exploitation or because they believe that they can directly set prices, interest rates, or other market factors to ensure adequate economic returns for sustainable community resource management. However, experience shows that governments can create conditions that are conducive to the economic success of community resource exploitation but that they cannot engineer this success through direct interventions in the market. Providing cheap loans often attracts wealthier and more powerful groups, which capture the available credit, or encourages unsound economic activities by the user group. For example, the processing of pine resin or latex may seem profitable when the initial capital is cheap but may prove to be both ecologically and economically unsustainable. Similarly, governments should not take on the task of marketing the products extracted from natural areas, whether they are forest products, crops, fish, game, or livestock. Government purchasing agencies tend to be cumbersome and inefficient even when their personnel are well intentioned. In worse cases, they are often used to force resource producers to sell their output at low prices.

Certain government regulations that intend to control resource management less directly are also highly problematic. Some governments ban the export of raw products so that raw products are processed within the country. The logic is to have domestic workers capture more of the ultimate value of the final products. The usual effect is to reduce the sale price of the raw material because foreign demand is artificially eliminated. This eliminates the profit of the resource exploiter. Some governments also try to prohibit or limit intermediaries or moneylenders because they are believed to exploit naive local resource users. This often leaves no one to hedge against the risks of price changes in raw materials or to provide credit when resource users need it. The government should also avoid policies, such as price controls or subsidies, that make certain land uses artificially attractive.

The appropriate approach is to adopt policies that allow profits when resource development and extraction are truly productive and avoid policies that encourage unsustainable activities. There are many ways for the government to be helpful without being heavy-handed:

- Upholding appropriate boundaries
- Enforcing rules against spillover damage when communities cannot resolve these conflicts
- Helping the development of private credit institutions
- Providing credit at market rates
- Supporting research and technological development
- Providing technical assistance and training
- Providing market information
- Aiding diversification by easing the bureaucratic obstacles to entering into new activities or markets
- Entering into joint ventures with communities that lack sufficient internal resources
- Undertaking general poverty alleviation programs.

The fourth issue, implied in the concept of positive externalities, is whether the government should provide compensation for user groups to leave some resource intact. The sustainable uses of the resource base may be sufficiently attractive for the user group to engage in a high degree of conservation. However, the conservation achieved may fall short of the optimal level from the perspective of society overall, which, by definition, enjoys any positive externalities that do not accrue as benefits for the user groups. Such compensation can include (1) direct financial assistance for groups given jurisdiction over nature reserves, (2) subsidies, (3) outright payments for user groups to develop the resource endowment (e.g., turtle hatcheries, tree planting, and land terracing), and (4) payments for not extracting resources.

The fifth issue is how the user group ought to organize itself to accomplish the ten functions of community resource management previously listed. The challenges will vary, depending on whether the community has a long-standing, distinctive identity and how deeply the community has been involved in sustainable resource management. For nonindigenous communities, the biggest challenge is often to develop fair rules that simultaneously ensure that all community members will receive enough benefits from group membership to remain. If it is a well-defined indigenous community, the challenge of developing internal coherence may be minor, but adapting the community's internal decision-making rules to changing ratios of population and resources may be necessary. Sometimes the community has already been organized to fight for recognition of its user rights. This may result in a high degree of unity and community awareness, but it can also lead either to self-sacrifice or to self-indulgence, neither of which is conducive to long-term conservation. Sometimes, the victory in obtaining user rights will imbue the community with a positive sense of mission and self-sacrifice, especially if the victory was hard won. On the one hand, the problem is that everyone involved in resource management has to benefit materially for the arrangements to be sustainable. Enthusiasm and sacrifice cannot be sufficient motivations in the long run. On the other hand, a community's members may believe that they deserve immediate payoffs when the group has won the rights to exploit a resource. Therefore, both the means for everyone to benefit and the rules of restraint need to be present at the outset.

Cases

Many efforts to establish or restore community resource management in Latin America reveal the promise of community management but also show the challenges presented by each of the five issues analyzed above. This section briefly reviews four cases. The largely successful efforts of both indigenous and nonindigenous Mexican communities to expand community resource management during the 1980s illustrate obstacles and successful strategies for obtaining government recognition. The Colombian *resguardo* system highlights the dilemmas of the second issue, how to set restrictions on the community's resource uses and the relationship between restrictions and compensation. A comparison of resin-tapper cooperatives in Honduras illustrates the problems and importance of internal community organization. Efforts by the Honduran government to establish community resource management reveal the promise and problems of direct government intervention. Mayne's discussion of agroforestry in Guatemala (chapter 4) provides another example.

Colombia's *Resguardos*

Very large tracts of Colombia's Amazon have been made into protected indigenous reserves called *resguardos*, legitimized by changes in property rights under the 1991 constitution. Indigenous groups now have formal responsibility and jurisdiction to manage their own lands (Bunyard, 1989a,b). The Colombian government also provides financial assistance to the *resguardo* communities. This assistance is more generous than the very low income derived from the groups' current subsistence farming and hunting and gathering. The user rights of indigenous groups given authority over the *resguardos* have increased dramatically from the immediately previous situation. Their incomes are enhanced by the government transfers. However, the rights are restricted in several respects.

First, the *resguardos* are "inalienable." The user cannot lease, sell, or otherwise transfer the lands. The 1991 constitution explicitly recognizes the user rights and authority of indigenous peoples in "indigenous territories." It specifies that user rights are to be assigned by the councils of the indigenous groups within them.[3] However, the constitution limits the user rights of the indigenous peoples because the sale of *resguardo* is prohibited. *Resguardo* land is "collective and inalienable property" (República de Colombia, 1991, Article 329). The 1991 constitution ensures the claims of the indigenous groups but does not fully embrace the essential principle of putting resource management fully into their hands. Rather than relying on the wisdom of the indigenous people to pursue their own interests sustainably, the inalienability constraint limits the economic flexibility of the indigenous governance of the *resguardos*. The rights of indigenous peoples within their territories are not equivalent to those of a nation. Those living in "indigenous territories," presumably including *resguardos*, are still under Colombian law. Although Article 246 of the 1991 constitution states that "authorities of the indigenous peoples will be able to exercise jurisdictional functions within their territory, in conformity with their own norms and procedures," it qualifies this jurisdiction by adding, "only if they are not contrary to the Constitution and laws of the Republic."

The 1991 constitution also comes close to requiring indigenous leaders to preserve the existing economic modes of their communities. It states that "the exploitation of natural resources in the indigenous territories will be undertaken without impairment to the cultural, social and economic integrity of the indigenous communities. In the decisions that will be adopted with respect to such exploitation, the government will support the participation of the representatives of the respective communities" (República de Colombia, 1991, Article 33). This seems to give the government the right to oppose actions of community leaders and members if the actions are deemed damaging to some standard of social, cultural, and economic appropriateness. Economic practices, such as agricultural cultivation techniques or fishing practices, can easily be construed as sociocultural characteristics. This provision gives the government the right to interpret whether economic changes are threats to cultural integrity. Further, by dictating that *resguardos* cannot be sold or forfeited, the 1991 constitution treats indigenous groups as if they will not or should not advance economically beyond their current modes of production to activities that would call for selling or mortgaging parcels of land.

Second, the *resguardo* strategy has such a strong conservationist rationale that many Colombians, particularly within the government, view the rights of the indigenous groups as contingent on their pursuit of low-intensity resource use deemed acceptable by the government. There is a possibility that the government may decertify particular *resguardos* if the communities are held responsible for environmental degradation. An important but unresolved issue is whether this possibility will have positive or negative effects on the decisions made by the groups. One possibility is that the threat will keep the indigenous communities from resource exploitation that the government would find excessive. This would be good for the resource endowment. However, it could also keep the indigenous population from using the resources to their maximum potential for their own welfare. Another possibility is that the indigenous people will decide at some point that their economic prospects are so limited by the restrictions on resource use that compliance with the conservation rules is not in their interest. They may decide that the government has not really relinquished its control, unless the communities remain as economically backward "museum pieces." If so, the potential would exist for growing hostility to government conservation guidelines, whether explicit or implicit. Excessive, perhaps even vindictive, resource exploitation could be undertaken by indigenous people who have lost their confidence that a sustained resource base will benefit them in the long run.

These scenarios depend heavily on whether the indigenous people want to maintain their traditional ways of life and modes of production. The conception of indigenous people that underlies the government's approach presumes that they ought to remain as they are. It assumes that they value their culture, modes of living, and the subsistence economic activities that they currently undertake. Some indigenous groups are anxious to preserve their traditional lives. However, others are just as enthusiastic about entering the modern economy and taking advantage of increased productivity and income that may result if they transform the uses of *resguardo* land.

Has the Colombian government done a disservice to these groups by forcing them to choose between economic stagnation and the loss of their jurisdiction? One interpretation is that the government's unwillingness to provide unconditional recognition of the groups' user rights denies them the free option of deciding whether or not to embrace conservation because it is in their long-term interest to do so. The discussion is complicated by the fact that the *resguardo* groups and the government have agreed on what amounts to a financial compensation package. Therefore, another interpretation is that the government and the *resguardo* groups essentially have entered into contracts that bind the groups in exchange for the compensation. To complicate matters more, the contract conception is also questioned on two grounds. First, some argue that the government's bargaining power is greater than that of the indigenous peoples and that the latter essentially had to take what was offered. Second, some also argue that the government had no legitimate authority to "give" the indigenous groups conditioned rights to lands that were already theirs by tradition and custom.

Gaining Community Management Recognition in Mexico

The customary rights of Mexicans traditionally involved in resource exploitation, particularly rural lands and forests, received very uneven attention until the 1980s. In particular, over the years Mexico's 31 state governments had appropriated lands that "belonged" to local groups by virtue of traditional use, if not by official title. In other instances, governments made arrangements, often in a heavy-handed fashion, for commercial sectors to gain access to the resources on lands formally held by traditional users. These people often received only nominal income from resource extraction. The *ejidos*, communal property declared inalienable by the Mexican constitution, are theoretically the most secure environments for communal resource control. However, outsiders have gained access to resources through support from state governments and have taken advantage of the typically low economic capacities of the local groups. These capacities have been limited not only by the low savings and low technical expertise typical of these communities but also by the restrictions on using *ejido* land as collateral.

Community groups in many states, especially in the poorer southern states of Oaxaca, Chiapas, and Quintana Roo, began to protest these arrangements in the 1980s. Among the most visible protests were those directed against commercial logging operations. In challenging the concessions, these local groups were also challenging the wisdom and legality of the governments' arrangements. While there were myriad specific episodes, some cases reveal why these efforts became markedly more successful in the 1980s.

Mexico has long provided formal recognition of community resource management. Seventy percent of Mexico's forests are formally held by indigenous communities or other communal arrangements (Bray, 1991a, p. 16). Nonetheless, formal community jurisdiction had been severely compromised in practice by government actions. A movement toward effective community control has appeared only in recent years. The key problem was that state governments had frequently arranged for the assignment of user rights on *ejido* land. Especially in the 1950s, under an uncompromisingly progrowth policy of the central government, the state governments had arranged logging concessions for private timber companies on *ejido* lands. Often, the *ejido* leaders lacked the awareness, technical competence, or practical political clout to prevent or control these concessions, although the benefits for the *ejidos* were minimal. In the southern state of Oaxaca, for example, the foreign-owned logging firm Fábricas de Papel Tuxtepec (FAPATUX) secured a 25-year concession over a 261,000-hectare area in 1956:

> Its concession failed to give FAPATUX absolute access to community forests, requiring the company to negotiate yearly contracts with the communities. In these negotiations, however, FAPATUX clearly had the upper hand, frequently with the collaboration of the secretary of agrarian reform, using its legal standing as concessionaire to suppress the communities' attempts to assert their right. Communities were denied the right to sell their timber to other buyers, for example, and one community that wanted to set up a woodworking shop was told it would have to buy back its own pine from FAPATUX. (Bray, 1991b, pp. 14–15)

Similarly, in the southeastern state of Quintana Roo, the state government granted a concession to the state timber company, Maderas Industrializadas de Quintana Roo (MIQRO), in 1954. During its 30-year concession, MIQRO and the private logging companies associated with it made deals with *ejidos* to harvest mahogany and cedar selectively. However, the *ejidos* received little in return because MIQRO was bringing in its own workers and sharing little of the profits. MIQRO even excluded local people from access to their own forests (Bray et al., 1993).

The nationalization of FAPATUX in 1965 did little to alleviate the tensions between the enterprise and the local groups. In 1968, 15 communities boycotted the company by refusing to supply timber to the FAPATUX mills. The boycott lasted for 5 years but created only moderate problems for the company because of its ability to obtain timber from other communities. In 1981, the government of the state of Oaxaca tried to renew the FAPATUX concession, but this time a well-organized, multicommunity effort blocked the state and the company (Bray, 1991b, pp. 15–16). Outside help from Mexican nongovernment organizations enabled the communities to pursue legal claims against the infringement of *ejido* rights to decide the disposition of timber. Thus, FAPATUX failed to get the concession, although the community forestry operations, now organized on their own, still sell portions of their output to FAPATUX on better terms for the communities (p. 18).

In Quintana Roo, MIQRO also failed in its attempt to continue its concession. In 1983, with the backing of the state governor, ten *ejidos* secured the rejection of MIQRO's request to renew its concession. They then received permission from the federal Secretariat of Agriculture and Water Resources to manage the forests in their area (Richards, 1991). Once again, the *ejidos* could claim legality on their side, and MIQRO proved to be a paper tiger. Its connections in the earlier era were ineffective in the 1980s.

Why did these actions, and scores like them in other Mexican locales, succeed in the 1980s but not earlier? First, the local groups had gained capabilities and confidence from the support of local, national, and international nongovernment organizations. They also had developed the potential to extract and market the resources themselves through cooperatives. Second, the state and national governments had been much less receptive to these claims before, largely because they were much more politically secure and saw little reason to jeopardize their alliances with the commercial business interests involved. However, in the 1980s the dominance of the ruling Partido Revolucionario Institucional (PRI) party had become much more questionable. The potential of increasingly well-organized grassroots movements to challenge the legitimacy of the federal and state governments was rising in comparison to the capacity of the state-level business interests to mount a comparable challenge.

An extremely important aspect of these episodes is that they took place in many Mexican states, in many specific locales,[4] but without a high-profile national confrontation with the federal government. The pattern of challenging previously ensconced business interests, invoking previously unenforced laws, and forcing state governments to come to terms with the new mobilization of the communities was repeated many times. Yet there was no high-profile national movement, no need for a national decision, and no immediate threat to the legitimacy of the federal government. The result may have been a nationwide revolution in effective property rights, but it was conducted in a largely piecemeal fashion that did not challenge the central government. It was not lost on the federal government that these communities could have engaged in disruptive activities but typically eschewed them in favor of legal channels. The legitimacy of the state governments was often at stake. However, the overt challenge to national authority, like that mounted in 1994 by the Zapatista Liberation Army in the state of Chiapas, was largely absent. It is telling that Chiapas has been different from most other southern Mexican states. Community-based forestry initiatives were brought to a standstill in 1987 by the decision of the state government to reject new applications for logging permits (Bray, 1991a, p. 16).

Honduran Cooperatives: Internal Organization and the Capacity to Exclude

Comparing the fates of two Honduran resin-tapper cooperatives illustrates the logic of the 10 functions required of community organizing efforts. Pine resin has an international market for its industrial uses. Tapping pine resin has long been an important income-generating activity of economically marginal Honduran *campesinos* who live in or near the pine forests. A cooperative of resin tappers can serve several important purposes: laying claim to particular pine stands, enforcing these claims against possible encroachers, imposing restraint on the cooperative's members so that the trees are not overstressed by excessive tapping, and possibly organizing for the refining and shipping of the resin so that the community can gain more of the profit from the finished product. The first such cooperative, the Cooperativa San Juan, was founded in 1966 in the county (*municipio*) of Ojojona. It started impressively, growing rapidly from 60 members to more than 300 in the late 1970s. Roughly 100 other cooperatives followed.

In 1974, the Honduran government put all trees under government control. The threat to resin tappers was obvious. One resin-tapper cooperative, the Cooperativa Villa Santa–Los Trozos, convinced the state forestry agency, Corporación Hondureña de Desarrollo Forestal (COHDEFOR), to grant it a concession to use 22,000 hectares of national forest of highland pines in the mid-1970s. This included the rights to harvest and sell the pine trees after their useful resin-production years. This was accomplished with the technical assistance of university-educated advisors and lawyers and with the aid of the national federation of resin tappers, the Federación Hondureña de Cooperativas Agro-Forestal (FEHCAFOR), which had more than 2,500 individual members in the 1980s (Denise Stanley, 1991, p. 33).

This backing was important because of the frequent hostility that COHDEFOR had otherwise shown toward community forestry. Its hostility arose, in part, because the communities "get in the way" of the commercial logging that provides the fees that COHDEFOR needs for its operations. However, the Cooperativa Villa Santa–Los Trozos won the backing of the national federation and the NGOs that mobilized technical assistance. The cooperative convinced COHDEFOR, in effect, to recognize the exclusivity of the cooperative's claim on the area and even to intervene to prevent encroachments by outside loggers (Denise Stanley, 1991, p. 29).

By the early 1990s, most members of the cooperative had attained inheritable property rights and rights of transfer, which are overseen by the secretary of the cooperative. Members, now numbering around 200, can even engage sharecroppers, who divide the resin yield with the member who is claiming tapping rights on a particular land parcel. The tappers can also monitor encroachments by outsiders because they live near their tracts, which are largely fenced.

By securing its affiliation with broader associations of community groups, the Cooperativa Villa Santa–Los Trozos gained recognition of user rights with autonomy and economic flexibility. Both by its own efforts and with the support of the government agency, it established clear boundaries of community membership and was able to exclude outsiders from encroaching on its resources. The Cooperativa Villa Santa–Los Trozos avoided the misstep of overextending its operations. Through its own leaders and those who provided technical assistance, the cooperative mobilized sufficient expertise to extract resin profitably and sustainably.

The Cooperativa Villa Santa–Los Trozos was successful in building up its financial and administrative capacity because its members were willing to be taxed for each barrel of resin they produced. It had standing with COHDEFOR because the agency saw the cooperative as a well-organized movement that had enough backing from its members to be able to extract financial resources from them. With an economic base of this sort, a cooperative can send representatives to provincial or national capitals, hire experts, and provide modest payments for forest watchers. Although the Cooperativa Villa Santa–Los Trozos suffered through a major embezzlement by a leader, it survived this trauma because its members retained the expectation that they could still prosper from the cooperation. After the embezzlement, it strengthened the oversight by the cooperative's council to keep track of the financial actions of its officers. The Cooperativa Villa Santa–Los Trozos was also able to establish rules for its members to tap resin on a sustainable basis, with well understood, reasonable penalties for excessive resin extraction.

In contrast, the pioneering Cooperativa San Juan declined from its peak of 300 to less than 35 by 1991. Like the Cooperativa Villa Santa–Los Trozos, Cooperativa San Juan suffered from a major embezzlement—the president stole the equivalent of $50,000—but it never recovered. The volume of resin gathered by its members fell dramatically. This reflected not only reduced effort by the cooperative members but also, and more significantly, the theft of resin and resin-gathering equipment and the removal of pine trees by outsiders. This vulnerability reflects both legal and physical conditions. Fencing in the pine trees or creating other barriers to entry is not permitted because the land is formally owned by the *municipio*. It has not been willing to allow the cooperative members to exclude others by physical means. In addition, the cooperative members do not live close to their assigned parcels and therefore cannot monitor the theft of resin and of their equipment. Finally, as pine trees in the area became scarcer because of cutting by outsiders, the competition between resin tapping and other uses of the remaining trees increased. The inability to set legal boundaries for user rights and the physical inability to exclude outsiders have reduced the expectation of profit out of resin tapping for cooperative members.

Honduran Government Efforts

Finally, exploring the Honduran government's own efforts to establish community resource management is interesting. These efforts included several funded by foreign donors. The Honduran government, specifically COHDEFOR, was involved as a joint partner with bilateral agencies, NGOs, and community organizations. In contrast, one initiative, the government's own program of Integrated Management Areas (AMIs), received only modest start-up funding from the UN Food and Agriculture Organization. This program was a direct effort by the Honduran government to set up community resource management units that included all the residents of particular districts. The partnerships have worked. The government's largely solo effort has essentially failed.

While COHDEFOR has shown little intrinsic interest in community forestry management, funding by the Canadian International Development Agency (CIDA), beginning in 1987, propelled the only significant official project in the Honduran broadleaf forest involving community participation. The broadleaf (hardwood, nonconifer) forests are mixed species forests that are of little interest to commercial loggers. Given COHDEFOR's dependence on logging fees for its own budget and the fact that broadleaf forests offer few opportunities to collect sizable logging fees, COHDEFOR has usually shown little interest in these forests under any management arrangement. However, with CIDA providing more than $3 million of hard currency (U.S. dollars) for the project, it was attractive enough to COHDEFOR for the agency to introduce agroforestry farm systems and forest management conservation, with the community playing a significant role. It even moved to revise the stumpage fee system. In contrast to its overall highly negative judgment of Honduran forestry projects, an Abt Associates (1990b, p. 71) assessment for the U.S. Agency for International

Development judged the CIDA project as "promising." The even bigger Olancho Reserve Forest Development Project began in 1988 with funding from the U.S. Agency for International Development of more than $7 million; COHDEFOR was quite willing to build in community forestry components and to revamp inappropriate stumpage fee rules (p. 97).

The AMIs are forest areas of 1,000 to 10,000 hectares that the government has placed under community management for integrated forest uses, each with a resident technician provided by the government. This program started with great fanfare and modest financial support from the UN Food and Agriculture Organization in 1983, but essentially it has relied on government financing through COHDEFOR. By 1990, it was clear that the AMI initiative would not be significant. On the community side, local groups were not enthusiastic about undertaking the responsibility for resource management. Nor were they enthusiastic about working with a government technician whose role may have been intended to be simply technical assistance but was widely seen as the imposition of government control. The AMIs were not embraced by forestry communities (i.e., groups defined by their forest uses). They were based on the definition of community that encompassed all residents of that area. On the government side, COHDEFOR, probably viewing the AMIs as unwanted challenges to its authority over all forest land, devoted less than 1 percent of its budget to the AMI initiative. By 1990, the program had only 2,500 beneficiaries and overall performed poorly both in attracting participants and in enforcing forest management. An external assessment concluded that because of the limited resources devoted to the AMIs, "It is of little surprise . . . that AMIs have not functioned well" (Abt Associates, 1990a, p. 13).

The AMIs failed in part because they did not recognize community resource rights. Instead, they delegated government functions to local groups, with implicit oversight by the technicians and COHDEFOR officials, as official elements of the national government's forestry plans and projects. In other words, the Honduran government did not simply recognize forest users' rights to organize. Rather, it also tried, as the Mexican state governments tried, to involve local groups in new state institutions. Since each AMI covers a geographical area, it is responsible for representing the interests of everyone in that area rather than the possibly distinct interests of various types of forest users who may have wished to organize independently.

Conclusions

Conservation through community resource management has begun to fulfill its promise where governments have allowed it to do so and communities have organized themselves to motivate both participation and restraint. Putting resource management into the hands of low-income people with modest aspirations of wealth and modest technologies for resource extraction is a simple idea, but implementing it is complex. The paradox is that conservation requires resource extraction to be economically attractive to those with the secure rights to use the resources. Of course, it must be moderate, sustainable extraction. Another paradox is that reliance on low-income local people makes sense. However, people and cultures are not static. Initiatives to establish community resource management should not presume that resource users, even if indigenous, will be content with traditional economic practices if greater incomes can be obtained through other practices.

The dilemma for a government interested in launching a community resource management initiative is that the community's autonomy seems to be essential. The government can try to facilitate community management, by recognizing community user rights and providing various forms of assistance, but it cannot impose a community management system through government structures. Government restraint must go hand in hand with community restraint.

Notes

1. That is, they may be Europeanized and still be relatively poor people living in or near relatively unexploited areas.

2. This list is derived from Elinor Ostrom (1992), which specifies eight tools to ensure the effectiveness of self-governing resource management, and from additional organizational imperatives analyzed in Ascher (1994, chap. 3).

3. In contrast, the *reservas* are property of the state, assigned to indigenous peoples for the time being, but ultimately the government retains the constitutional authority to assign the land.

4. Bray (1991a) lists efforts in Chihuahua, Durango, Guerrero, Michoacan, Oaxaca, and Quintana Roo.

4

Agroforestry in Guatemala

John C. Mayne

The need for construction timber can deplete areas around towns or villages of trees, but the need for firewood can keep the same areas devoid of even woody shrubs. They are cut too frequently to produce sufficient biomass for sustainable growth. This problem is often exacerbated in ecological zones less hospitable to dense and rapid plant growth and/or where human population growth has exploded. Not surprisingly, food production is also a problem in these areas where much of the available land, even steeply sloping land, is in cultivation.

The Project

In the village of El Moral in the semiarid region of Guatemala, an agroforestry project was initiated in an attempt to provide the villagers with both firewood and maize. After three years the results were mixed, but the experience of El Moral points out the difficulty of making land use decisions in an area where land that should be left undeveloped is needed for conflicting uses by different groups.

The idea for the project came out of discussions between the village mayor and two Peace Corps volunteers. In discussing the need for arable land, it was mentioned that some steeply sloping uncultivated land owned by the *municipio* of Morazan lay just outside the village. El Moral was one of several villages inside the *municipio*. The villagers of El Moral were in the habit of climbing the hill and cutting firewood. The result was that there was no good firewood left to cut and the soil was eroded.

The mayor pointed out that people of the village wanted to plant maize on the land. He wanted to know if the volunteers, through their positions with the Department of Forestry (INAFOR, now called DIGEBOS), would petition the *municipio* for use of the land. The *municipio*, which controlled the land, had forbidden any use of it in the past. Nonetheless, people had been surreptitiously cutting wood and slowly denuding the land for years.

The land was so steep, up to 80 percent slopes, and the soil so thin that the land should not have been used at all. However, as often happens in developing countries, people live and raise crops on poor and marginal soils. Yields per hectare of land on these soils are lower than yields on more fertile soils. Farmers compensate by cultivating as much land as they can. Terracing the thin and rocky soil was not practical, so the volunteers suggested planting trees along contours to hold the soil and planting the maize in parallel rows between the rows of trees. With the mayor, they developed an agroforestry project for land outside El Moral.

The three of them met with the farmers in the community to see who was interested. Participating farmers were selected by the mayor and the community. The volunteers had no input in that decision. In subsequent meetings, the mayor, farmers, and one of the volunteers (the other had finished his service and returned home) surveyed the community to learn the firewood species preferred for different uses, what species were acceptable, and what species should not be used. These surveys showed that most of the people wanted fruit trees in addition to fuelwood species. An area at the bottom of the hill was designated for planting fruit trees (table 4.1).

The trees were direct-seeded to save labor and to keep the project as independent as possible. This saved the need for a nursery and the expense and administration that it would entail. Depending on the species, the seeds were pretreated by scarification, utilizing methods found to be successful from trials conducted the previous year in a nursery in the area. Unfortunately, it was a very dry year and only about 20 percent of the tree seeds germinated. Most of those were in only two species, *Leucaena diversifolia* and *Parkinsonia aculeata*.

The villagers did not try to direct-seed again, which would have guaranteed another poor result made worse by all the activity of maize planting, cultivating, and harvesting. Therefore, although they had originally attempted to set up the project without having to truck plants, with their heavy bags of soil, more than 30 kilometers from a cooperating regional Department of For-

Table 4.1 Tree Species Used in El Moral Agroforestry Project

Scientific Name	Common Name
Firewood Species	
Accacia spp.	Accacia
Cajanus cajan	Gandule, pigeon pea
Cassia siamea	Cassia
Casuarina equisetifolia	Casuarina, Australian pine
Eucalyptus camadulensis	Eucalypto, Eucalyptus
Gliricidia sepium	Madre de Cacao
Leucaena leucocephala K8	Leucaena
Leucaena diversifolia	Yaje
Melia azederach	Paraíso, Chinaberry
Parkinsonia aculeata	Palo Verde, Jerusalem Thorn
Prosopis juliflora	Prosopis
Sweetenia macrophyla	Caoba, Mahogany
Trema micrantha	Trema
Fruit Tree Species	
Anacardium occidentali	Marañón, Cashew
Mangifera indica	Mango
Psidium guajava	Guava, Guayaba
Tamerindus indica	Tamarindo, Tamerind

Source: From J. Mayne, 1997.

estry nursery, they were now facing that prospect. There would be a tremendous amount of labor involved. Each farmer would have to carry 5 to 10 trees in bags of soil up the long hill to be planted. Further labor would be involved in digging holes in which to plant the trees.

They did accomplish the labor-intensive planting. However, they also established a nursery on some property near the mayor's house, where an old well was cleaned out and made functional again. Once the trees were planted, the farmers planted their maize and *Cajanus cajan* seeds along the lines that divided each farmer's parcel from his neighbor's. They finished planting about halfway through the first peak of the area's bimodally distributed rainfall. During the dry interlude between the rains, the farmers replanted maize that did not germinate. They also planted trees from the newly established nursery near the mayor's house to replace those that had died or looked as though they would not survive.

The maize grew much faster than the trees, and it soon overtopped them. This slowed the growth of the trees, but they still enjoyed a nearly 70 percent survival rate at the end of the maize harvest. The next year, at the end of the dry season, another 15 percent of the trees had succumbed and were replaced. The trees were replaced not with trees trucked in from a regional nursery but from seedlings grown in the community nursery on what was now the mayor's land.

The second-year maize harvest was good, and the trees were well established. Some were 1.5 meters tall. The site, visible from most parts of the village, finally caught the attention of the *municipio* and the Department of Forestry. Both tried to claim the project because it was successful and appeared to have the potential to produce valuable firewood in the next few years.

The *municipio* claimed, correctly, that the project was on their land, and the Department of Forestry claimed, also correctly, that they had regulatory control over large tree plantings. Of course, the farmers wanted to keep planting maize, and in exchange for that, they were willing to tend the trees they had planted. The farmers appealed to the Peace Corps volunteer to intercede on their behalf. He was able to get the *municipio*'s permission to allow the farmers to plant maize for another year. By this time, however, he was due to complete his service and the volunteer selected to replace him was more interested in developing the nursery than in becoming involved in the local politics of the village.

During the next planting season, the agroforestry project to produce both food and firewood ended. The *municipio* did not renew the farmers' permission to plant maize and the Department of Forestry planted the whole site to *L. diversifolia* and *P. aculeata*, the two species that experienced the best survival in the original agroforestry plantings. The mayor was hired by the Department of Forestry to manage the firewood project and the nursery. He was able to hire several farmers to serve as helpers in the nursery and as night guards on the project. The farmers had no remaining role in the project. However, they had received two maize harvests and were promised firewood from the original trees they had planted and tended.

The hill outside El Moral was maintained as a firewood planting with the rights to the firewood owned by the *municipio* and the village of El Moral. The trees were guarded at night against cutting, and the village retained the use of the fruit trees, which were left undisturbed when the area was replanted with firewood trees. There was now firewood for the community.

Lessons Learned

The political and legal encouragements and constraints on communities and incorporated areas can have a great deal to do with the way that natural resources are used. Some members of El Moral wanted to grow maize on vacant land that had been denuded and were unconcerned about the soil erosion that could occur. They were interested in producing food for themselves for as long as they could. As the project grew and evolved with input from individuals from outside the community, their outlook changed because they also needed firewood. Growing maize was a shorter-term undertaking than the installation of a firewood planting. The farmers did not want, at first, to put effort into a multiyear project with no payoff for several years. Furthermore, they would have had to stand guard at night to protect young trees that would not yield firewood for several years.

In this case, the best land use strategy, given the extreme slope and erosion, was to plant the whole hillside in trees. The community agroforestry project was designed to protect the site against erosion and to allow the production of needed food and firewood for the village of El Moral. However, it did not have enough support from those with power and influence to overcome the alternate land use plans of the *municipio* and the Department of Forestry. Nevertheless, it could be argued that the project would be even more useful in keeping other areas from being cut if the entire hillside was planted in trees. The project served a valuable purpose because it allowed the few other areas not planted in maize to have a chance to grow without being cut frequently.

The project, as it evolved, led to the creation of a nursery in the village, operated, more or less, by the people of the village. The nursery was situated on the mayor's property and was in his charge. However, it was considered more a village nursery than one belonging to the Department of Forestry, which paid the mayor's and his helper's salaries and provided seeds and took many of the seedlings produced. The nursery also received seeds and bags from CARE (Cooperative for American Relief to Everywhere) through the Peace Corps volunteer assigned to El Moral. Also, because of the villagers' hands-on experience with seedlings and because the current Peace Corps volunteer was more interested in small-scale tree planting than large-scale projects, the nursery started producing fruit trees for home gardens and firewood trees for those villagers who had some space in which to plant them.

In the end, out of their attempt to use unused and degraded land, the villagers ended up with access to a Department of Forestry firewood planting near their village. They also gained the ability to produce firewood seedlings to plant on their own property. Although they had lost the opportunity to plant maize on the unused municipal land, they were producing fruit trees that would provide vitamin-rich food—but fewer total calories than the maize would have provided—and much-needed cash.

This case study of a Guatemalan village's efforts to use nearby degraded municipal land shows how many socioeconomic, political, and legal factors come into play when land use is an issue. In this case, a reduction in the pressure for firewood on adjoining lands was a fortunate consequence of a village's efforts to feed itself.

5

Sustaining What for Whom?

Differences of Interest within and between Households

Dianne Rocheleau

Structural adjustment is part of a broader economic restructuring throughout the world, and it is more than economic. Likewise, sustainable development is much more than an environmental "fix" for development (Blaikie and Brookfield, 1987; Peet and Watts, 1996). These parallel processes encompass a simultaneous restructuring of economies, ecologies, cultures, and political systems. Each of these elements constitutes a nexus of social relations between groups of people and between people and their physical and biological surroundings. We are witnessing the reintegration of each at different scales and under new relations of power (Rocheleau, 1995a).

Central America and the Caribbean are no exception. If anything, the region has set the trend for new international agrarian relations and new models of sustainable development and conservation (Deere, 1990b). Current regional changes in landscapes, livelihoods, and land use systems in Mesoamerica and the Caribbean are embedded in global processes of economic, environmental, and social change (De Janvry, Sadoulet, and Thorbecke, 1993; McAffee, 1987). These changes, in turn, are both anchored and expressed in everyday life within communities and households. Diverse actors are at work, often simultaneously, as producers, processors, resource users, resource managers, and consumers. All interact within complex social relations of power based on both conflict and affinity.

Repositioning the household within the global economy affects the position of the household in the community, of individuals within the household, and of both within local ecosystems. In the past, development planners treated the economy as a global and national entity, the environment as a global and regional concern, and class and ethnic differences as national categories. They often relegated gender to the microlevel, a detail of social life expressed solely within households and outside the economy. In fact, the process and the pattern of the current restructuring of livelihoods and landscapes are both global and local. Global economic change depends greatly on renegotiating the gender and class division of land, labor, cash, commodities, capital, markets, and social organization within local communities and households.

Only a few decades ago, many people in the region integrated all or most of their production and consumption activities at the local landscape level through household and community social units (Wilken, 1988). More recently, rural people have increasingly concentrated their production activities within private landholdings on small farms because of the commercialization of agriculture and the expansion of land markets. They have gained access to some commodities through local exchange and have traded some cash crops for consumer goods through national and international markets. The result has been to concentrate and localize production activities in household plots at a micro level and to expand trade and market linkages to national and international systems.

The latest wave of economic restructuring and the rise of nontraditional agricultural exports (NTAE) have pulled many more smallholders into more specialized, cash monocrop production, often displacing complex intercropping systems (Thrupp, 1995). This has, in turn, linked many farm households and entire communities more tightly into global markets as consumers of both imported staple foods and manufactured goods. As a consequence, farm households have decreased the flow of energy and the cycling of materials among their farms and between their farms and the surrounding landscape. They have dramatically increased the exchanges of energy and materials between their plots and larger national and international systems. Planetary-scale economic processes shape these flows.

Households vary greatly in their ability to participate in these restructured economies and ecologies on favorable, or at least viable, terms of control and exchange. Even those who cannot or choose not to participate in NTAE initiatives are affected. They may find themselves enmeshed in local landscapes and economies transformed by the spread of new commodities and land use systems to the farms of their neighbors. The processing and marketing channels for their established cash crops may disappear or switch to less favorable terms. The use and management of land, water, and forest resources may also change dramatically with the widespread adoption of new monocrops by surrounding farm households and large-scale commercial producers. This, in turn, may pressure nonparticipating smallholders to sell their land and migrate. Those who stay may enter into a land use system that puts them at a disadvantage with both their neighbors and their own prior land use and livelihood system.

These households and individuals are casualties of the privatization of land and related resources (Deere, 1990a), of increasing population densities and land shortages, and even of the "sustainable" intensification of land use. They have typically remained invisible. Researchers in another region, in Machakos District, Kenya, documented and debated this process recently. They found that privatization and land use intensification displaced smallholder farmers from use of the common woodlands and pastures for grazing and gathering. Many smallholders were unable to make ends meet without this crucial supplement to their subsistence base, and they sold or abandoned their holdings (Mbogoh, 1991). Their former lands may even have passed into the hands of wealthier farmers to be converted to orchards or other environmentally friendly and economically productive land uses (Rocheleau, Benjamin, and Duang'a, 1995; Rocheleau, Steinberg, and Benjamin, 1995).

Some have heralded this phenomenon in Machakos as economic and environmental recovery (Tiffen, Mortimore, and Gichuki, 1994). However, many displaced households went on to find and clear dry forest lands in "open" frontier areas, to seek work in plantations, or to find wage labor or refuge on the fringes of cities. Their displacement erased neither their poverty nor their environmental impact. Both problems were simply displaced to another jurisdiction, to be counted in another census or another environmental study. Meanwhile, among those who stayed, the hunger of some and the poverty of many remain, although shaded by the tree crops and woodlots of their wealthier neighbors and employers (Rocheleau, Benjamin, and Duang'a, 1995). This same socially divergent and sometimes polarized story has played out in many variations in Central America and the Caribbean (Arizpe, Paz, and Velásquez, 1993; Arizpe, Stone, and Major, 1994; Deere, 1990b; Rocheleau and Ross, 1995; Schmink and Wood, 1992; Stonich, 1993; Townsend et al., 1994).

A multiple-scale, "telephoto" analysis, coupled with a focus on the diversity of interests and experiences at each stop along the way, can make these invisible processes and people more visible. More important, once we map the diverse social terrain for sustainable resource management, we can involve the full range of user groups and interest groups in any one place in the planning and evaluation of land use changes. The major premise of this approach diverges from past resource management traditions. I argue that there is no single optimal land use or technology mix in a given place. Rather, there are many distinct visions of optimal environmental and economic futures and many possible combined outcomes. Both the visions and the composition of the various interest groups change over time, based on their own internal dynamics and in response to changes in the local, national, and international contexts (Rocheleau, Thomas-Slayter, and Wangari, 1996).

As a result, the objectives for development shift. Rather than searching for a single optimal solution, planners can facilitate a fair and effective process. It involves a constant negotiation of acceptable solutions among competing and converging interests in any one place and between local and larger systems. The history of households and user groups in resource management and sustainable development sets the stage for consideration of a new perspective focused on a diverse constellation of changing land user groups that operate under uneven relations of power (Rocheleau and Slocum, 1995; Slocum et al., 1995).

The History of Households in Environment and Development

The recent wave of economic and ecological change did not arise in a vacuum. As Ruttan notes (chapter 2), this can be seen as the third wave of social concern since World War II that is related to resource availability and environmental change. In each case, international and national agencies and nongovernment organizations have launched a set of initiatives, sometimes parallel and sometimes combined, to address related issues of agricultural production and environmental protection and conservation. Social differentiation has progressed from an invisible background to a focus of research and planning during this succession of initiatives. Today's research and planning attempt to address disparities between and within communities and households based on gender, class, ethnicity, and other dimensions of identity and difference. The multiple and overlapping interest groups among rural households and communities have emerged as a major concern of both social and environmental scientists and planners. This focus reflects both the internal evolution and the external critique of development and conservation efforts.

Immediately following World War II, the United States inaugurated the reconstruction of Europe and "international development" to modernize Africa, Asia, and Latin America (Escobar, 1995; Escobar and Alvarez, 1992; Esteva, 1992; Sachs, 1992). Modernization and the need to increase food production for growing populations merged with cold war political concerns to propel national and international institutions into the green revolution (Iasa and Jennings, 1981). Agricultural development programs in Mesoamerica, as elsewhere, relied heavily on technology transfer from the industrialized countries. Their goal was to establish the farm as a firm and the farmer as a business manager. The farm and the farmer (a man) were portrayed as ideal, universal types in initiatives to mechanize, specialize, and commercialize agricultural production. All were part of a broader agenda to modernize national economies.

Environmental conservation during this era proceeded as a distinct and parallel activity. It focused on preserving special species in special spaces (Rocheleau, Steinberg, and Benjamin, 1995). Parks and reserves were created in spaces apart from those used for agriculture, forestry and other production activities. The human interaction with environmental quality focused heavily on health and hygiene, particularly on latrines and the maintenance of clean, safe water supplies. These programs generally targeted undifferentiated households and municipalities through government agencies.

The next wave of global concern over environmental quality and food supply emerged in Central America and the Caribbean through parallel programs in food production and conservation. Several institutions in the region promoted Farming Systems Research and Extension programs directed at the farm family and the farm household (Lagemann, 1983). These programs sought to deal with farms as complex land use and livelihood systems and to bring the green revolution technologies to smallholders. The farming systems approach fitted commodity production to the complex production systems of small holdings in consultation with farm managers and their households (Chambers, Pacey, and Thrupp, 1989). By the mid-1980s, the farming systems approach incorporated emerging concerns for class, gender, and other differences between and within households (Feldstein and Poats, 1989; Moock, 1986; Poats, Feldstein, and Rocheleau, 1989; Poats, Schmink, and Spring, 1988; Rocheleau, 1987).

The new emphasis on gender reflected a major contribution from studies of the gender division of land, labor, and livelihoods in African agriculture (Berry, 1989; Carney, 1988; Davison, 1988; Moock, 1986; Schroeder, 1993; Stamp, 1989; Wilks, 1989). However, many national and international agencies in Central America and the Caribbean resisted the incorporation of gender analysis into agricultural and resource management work because of "regional cultural norms." They promoted an image of the household as a unit composed of a male farmer (household head), his wife, and their children, all sharing the products of the farm enterprise. These perceptions reflected the strong Spanish influence on the dominant national culture and North American "housewife" images on international agencies. It did not allow for differences either between or within the various cultural and class groups in countries in the region. In fact, women's and men's roles and interests in agriculture varied dramatically, depending on cultural and class differences (Arriagada, 1992; Guzmán, 1992). Much more direct involvement and recognition of women as farmers existed among indigenous and Afro-Caribbean peoples. Women's and men's roles and interests in agriculture and resource use and management also varied substantially according to class (Deere and Leon, 1987), occupation, location, life cycle (Momsen, 1993), and personal history (Rocheleau, Ross, and Morrobel, 1996; Rocheleau, Ross, Morrobel, Hernández, Amparo, Brito, and Zevallos, 1996).

During this period, conservation extended beyond species to encompass broader habitats and whole ecosystems. Conservation agencies increased consultation with and employment of neighboring communities. Conservationists treated park boundary communities as buffers and as potential park guardians against encroachment by other settlers or resource users (Anderson and Grove, 1987; Bonner, 1993). These communities became homogeneous targets of employment, environmental education, and recruitment to conservation enforcement, usually focused on the adult men in the community. Many programs employed young men or a few senior men. They hoped that these men could then guarantee the compliance of the rest of the community with conservation goals and environmental regulations. The concept of the unitary male-headed household was extended to the community as a whole, usually to little effect.

The third and current wave of environmental concern has spawned sustainable development (WCED, 1987). The unitary household has largely given way to homogeneous communities or groups, paying some attention to household dynamics. Sustainable development projects have usually added environmental criteria to agricultural and rural development efforts. They have also added development goals to environmental agendas. Overall these efforts have emphasized participation and have often sought to identify alternative technologies and production systems (Altieri, 1987). Many sustainable development initiatives have relied on the involvement of undifferentiated local people in conservation agendas through profit-making ventures (Scoones and Thompson, 1994). This has often resulted in community-based enterprises, largely in the hands of men from better-off households. This has been the strategy, for example, in Guanacaste, Costa Rica, where successful ranchers help restore the dry forest reserve and protect it from the encroachment of settlers (Primack, 1993).

Whatever their target or collaborator groups, sustainable development initiatives inevitably affect individuals, households, and communities. They also modify how people interact with their environment and with one another (Nygren, 1993). New technologies, land uses, and environmental regulations, however, have very different effects on different groups of households and on individuals within and across households. In some cases, what is positive for one group may spell social, environmental, or economic disaster for another.

The discussion that follows focuses on the elements of sustainable resource management and on the differences between households and individuals within local economies and ecologies. This requires a review of both the intrinsic and utilitarian values of sustainable resource management. After a brief review of the many elements of sustainability, the chapter delineates the structures of social stratification, based on multiple dimensions of identity, difference, and affinity between, within, and across households. This section concludes by identifying six domains where these social differences matter for rural resource management and vice versa.

The Varieties of Sustainability

To frame the discussion of differences, we must first define the nature of sustainability. We include both environmental and economic criteria, starting with the physical, chemical, and biological elements most likely to be subject to change by rural land use and resource management practices (table 5.1). The interests of any one group may be tied more to one or another of these aspects of sustainability within their local ecosystems. As resource managers evaluate the effects of a particular change, whether of a park or a new crop or an agroforestry system, it is possible to determine which of these elements will be affected and the relative importance of each to various groups of resource users and other interested parties.

The use values of ecosystem products and services (table 5.2) provide one reliable guide to the link between specific elements of the local environment and the interests of particular groups of people. The list of values includes the goods and services that support both subsistence and commercial production. The value attached to a given good or service on the list will vary by land user group. For example, at the community level, the members of a cooperative bakery might value energy supply more than food supply if they currently purchase food with their bakery earnings. Likewise, in decisions between cropping system options, households with a substantial cash income that purchase most of their food, on the one hand, might value a stable cash income from cash crops more than a favorable seasonal distribution of food. On the other hand, a poor family with no off-farm income that grows its own food might give high priority to seasonal food availability. Herbalists or religious leaders might choose to designate a particular patch of forest or range land as a local reserve for medicinal plants and/or religious observances rather than allow new settlers to convert the area to food plots.

Production and Resource Management Options

Throughout the region people in agrarian and forest landscapes face similar production and resource management problems and have a limited array of commonly employed options. These include several strategies (table 5.3) to maintain or improve production and/or resource management. Some of these options are compatible and can be combined; others are mutually exclusive. Each of these choices brings a distinct set of costs and benefits for different groups of resource users.

Six Domains of Change in Production and Resource Management

Each of the land use and technology options listed in table 5.3 affects the control of and access to six domains of resource use and management within communities (table 5.4): (1) land and related resources (use, tenure, and condition), (2) labor (availability and allocation), (3) cash and commodities, (4) other forms of capital, (5) markets, and (6) organizations. The effects can be tabulated based on differences between households and between individuals within households or in groups that cut across households within communities. For example, the adoption of NTAE crops in smallholder households where men engage in wage labor would probably both reallocate and produce an absolute increase in women's labor. In land-limited households, this would also lead to the partial conversion of land from food crop to NTAE production, resulting in a decrease in the amount and quality of food produced. Depending on the crop and the required or recommended management practices, the change could involve either a reduction or an increase in water pollution and soil erosion, with interests distributed unevenly between and within households. A change of cash crops might affect the gender structure of local markets or of local and regional organizations.

Diverse Actors in Shared Economies and Ecologies

As Ascher notes (chapter 3), a community is more than shared geographic space or joint occupation of an administrative unit. It is not necessarily equal to or even

Table 5.1 Elements of Sustainability in Rural Landscapes

Physical Integrity	Water supply—quantity, quality, timing	Runoff quantity and quality	Erosion and sedimentation	Slope stability
Chemical Integrity and Nutrient Cycles	Agrochemical pollution: Fertilizers, pesticides, herbicides 1. Pollution of air, water, soil 2. Exposure of rural people to spraying, tainted water, stored agrochemicals, tainted foods	Movement of nutrients in biogeochemical cycles		
Biological Health of Ecosystems at Global, Local, and Plot Scales	Biodiversity for agricultural production 1. Crop and livestock genetic diversity 2. Wild plants and animals as gene pools for domesticates 3. Wild plants and animals as potential future domesticates 4. Companion plants and animals (pollinators, seed treaters, seed dispersers of plants; pesticidal plants and pest predators 5. Fodder for livestock (from cattle to bees) 6. Soil microfauna and -flora that foster crop/fodder production	Biodiversity for local non-agricultural production 1. Commercial extractive products (honey and nuts, medicinal herbs, fiber, wood); regular fruit and vegetable supplements (wild leafy vegetables, roots, fruits); famine foods; specialty foods for sale (mushrooms, caterpillars, insects, rare fruits, nuts, flowers); game animals for food or for sale of meat, hides; game animals for sale of other products (horn, shell, bone etc.)	Biodiversity for its own sake	

Source: From D. Rocheleau

Table 5.2 Elements of Use Value Related to Sustainability

Cash Income (through environmental production and service)

1. Employment
2. Sale of products
3. Reduced spending on items otherwise purchased
4. Direct exchange of products for other goods

Food Supply (for people and livestock)

1. Amount
   Total annual supply
   Seasonal distribution
2. Quality
   Nutrition
   Taste, texture, dietary preference
   Ease of processing and preparation

Medicinal Products (human and veterinary remedies)

1. Amount, quantities available
2. Quality/species composition of wild plants
3. Accessability
   Distance, location, secure access

Energy Supply

1. Uses
   Cooking, home heat, light, amenity
2. Amount
3. Quality
4. Ease of access
5. Cost savings (cash saved)

Shelter and Infrastructure

1. Building materials
2. Shade
3. Protection from weather, animals, insect pests, and intruders
4. Definition of boundaries—fences and markers
5. Cultural and aesthetic expressions

Savings and Investment

1. Livestock
2. Land
3. Buildings
4. Business

Raw Materials for Crafts and Local Industries

1. Amount
2. Quality
3. Accessibility

Social Production

1. Funds, goods, services to meet social obligations
2. Species, places, products for religious and cultural observances

Source: From D. Rocheleau.

compatible with local government. Similarly, a household is more than a group of people that share home and hearth. Many distinct groups of households exist within communities, and different groups of people exist within and across households. The differences between households include some characteristics that are usually permanent, such as ethnicity, race, and religion (table 5.5). Other differences usually persist but are subject to change, including class, social status, political power, occupation, income sources, nuclear vs. extended family structure, and organizational affiliation. Yet other household attributes, such as life-cycle stage of the household and age and gender structure, change by necessity. The importance of any one of these household characteristics will vary from one community to the next, over time, and according to the land use changes at issue. For example, participation in labor-intensive vegetable cropping may depend on the number and ages of adult women, whereas timber farming may vary more by landholding size and the occupations of adult men in the household.

Individuals and households are differentiated into multiple and overlapping groups with respect to resource use, access, and control (see Mayne, chapter 4). Uneven relations of power between types of individuals or households result in differential access to land and related resources; labor, cash, and commodities; and other forms of capital, markets, and social organizations. Class, gender, and ethnicity are among the most significant differences between households and individuals within the communities in a region (Arizpe et al., 1994; Katz, 1992; Momsen, 1993; Townsend, 1993, 1995; Townsend et al., 1994). Each of these and several other dimensions of difference combine to in-

Table 5.3 Strategies to Maintain or Improve Production and Resource Management

1. Intensify labor to produce more on the same land with more labor.
2. Shift labor to nonfarm pursuits and wage labor. Reallocate household labor from domestic or self-employment to cash cropping; convert croplands to forest, pasture, gardens; or rent out croplands.
3. Intensify and improve management. Change specific soil, water, crop, and livestock management practices to get better economic returns and environmental quality with the same land and labor.
4. Commercialize and shift toward major cash crops or to NTAE crops to increase farm earnings and simplify labor allocation, production, and marketing.
5. Shift from annuals to perennials for more soil and canopy cover, to reduce tillage and to alter labor demands. Adopt agroforestry, social forestry, or tree cash-cropping systems.
6. Diversify (or maintain) multiple crops within fields and gardens to improve yields, pest control, and food self-sufficiency and to maintain plants of medicinal, cultural, and religious significance.
7. Rearrange land use on the same holdings according to ecological opportunism. Selectively intensify production on the least fragile, most fertile and well-watered areas. Reforest degraded slopes, ridges, stream banks; plant windbreaks. Combine species for positive mutual effects in pest control, soil fertility, weed suppression, and shading.
8. Rearrange plants in the same holdings into simpler, specialized (and less diverse) blocks. For example, plant salad vegetables in small gardens, staples in croplands, cash crops in separate monocrop plots, and commercial timber trees in monocrop woodlots.

Source: From D. Rocheleau.

fluence the distribution of control and access within all six domains of resource use and management shown in table 5.4.

How can this information serve resource management and social equity objectives? On a very practical level, careful attention to differences between households and individuals can help to define community, for community-based projects, where it is not yet understood. However, even within groups that are well defined and meet the criteria for well-organized communities as outlined by Ascher (chapter 3) and others, there may be significant differences between and within households. This stratification may have important social and ecological implications for sustainable development initiatives and their impacts on distinct groups (Thomas-Slayter and Rocheleau, 1995b).

Major Dimensions of Difference in the Region

Ethnicity

The major ethnic divides within the region include European, indigenous, and African and/or Asian ancestry, with various derivative cultural practices and linguistic differences. There are four main groups. The first includes Spanish, Portuguese, or mixed ancestry peoples strongly influenced by Spanish linguistic and cultural traditions. Afro-Latin peoples of mixed European and African descent combine cultural practices derived from Spain with those derived from different parts of Africa. Many of the predominantly indigenous people retain their own languages, religions, and related cultural practices. Sometimes, English-speaking peoples of African and/or Amerindian descent are present. The

Zambrana-Chacuey

An international nongovernment organization (NGO), Environment Development Alternatives (ENDA-Caribe), conducted a farm forestry program in the hilly region of Zambrana-Chacuey in the Dominican Republic. For the program, ENDA worked with a rural people's organization that met all the criteria for rural communities with potential for successful resource management. The Rural Federation of Zambrana-Chacuey consisted of roughly 800 members in 500 households who shared a 30-year history of successful land struggles. They or their parents had engaged in popular campaigns for land-to-the-landless, including land invasions and other forms of civil disobedience (Rocheleau and Ross, 1995).

While many well-organized communities focus on identity, the federation built successfully on affinities between very distinct groups across race, class, and gender lines. In 1993, the members were almost all smallholder by national standards. Nonetheless, they were substantially different in ways that matter for both adoption and outcomes of new forestry technologies and enterprises. Households presented distinct circumstances for the new timber crop based on the size, quality, and location of their land; livelihood strategies; and type and degree of federation affiliation. Individuals differed in their ability and interest in participating, depending on gender, personal income, occupation, and organizational affiliation (Rocheleau and Ross, 1995; Ross, 1995).

Table 5.4 Six Domains of Inter- and Intrahousehold Differences in Resource Use and Management

Resources	***Labor***	
Land	On-farm	Off-farm
Water	Individual	Individual
Crops	Family or household	Family or household (for wages or shared exchange)
Livestock—meat and milk	Exchange	
Wild plants	Hired	
Wildlife		
Cash and Commodities	***Other Capital***	
Cash	Infrastructure—houses, barns, corrals, fences, other buildings, water tanks, irrigation works	
Products of labor and natural resources	Equipment—carts, vehicles, plows, farm and building tools, looms, sewing machines, ovens, stoves	
Food, fuel, fodder, building materials, craft materials, and commodities for sale		
Markets	***Organizations That Mediate or Facilitate***	
Scale	Access to and control of public services	
Local, regional, national, or international	Access to and control of resources	
Types of commodities	Access to and control of labor	
Terms of exchange	Access to and control of markets	
	Access to and control of savings and investments	

Source: From D. Rocheleau.

last groups are a minority in some Spanish-speaking and bilingual countries (Dominican Republic, Belize, Nicaragua, and Honduras) and a majority in others (Jamaica, Bahamas, and Guyana). In many countries, the major division is between Indians and Ladinos, as in Mexico or Guatemala. In other cases, differences between peoples of primarily European or African descent and cultural orientation are more important, as in the Dominican Republic or Cuba.

Ethnicity has been widely recognized as differentiating regions and communities. Less well recognized, it also distinguishes between households and individuals within communities. This is especially true in frontier or settlement areas where people of different ethnic groups may farm in the same area or share forest, water, range, or coastal resources. Disaggregating community survey data by ethnic group has yielded strikingly different results, for example, in the population and poverty studies of smallholder farmers in the Lacondon Forest in Mexico (Arizpe et al., 1993; Arizpe and Valasquez, 1994).

Class

Class differences occur and matter throughout the region and are generally recognized as significant at the community and household level. The determinants and indicators of class differences may vary from one country to the next and one region to another. However, there are a few key elements of class differences that matter for resource management and sustainable production throughout Central America and the Caribbean. Landholdings (amount, quality, location, and tenure status), income (amount, source, stability, and diversity), other assets, and social and political status (table 5.5) are simultaneously determinants and indicators of class status. Each of these implies a certain power to manipulate and mobilize resources for production or to regulate resource management and use within a given community.

Gender

Gender differences have perhaps been least discussed and understood in the context of sustainable agriculture and resource management in the region (Silva, 1991). The gender division of all six domains of resource management (table 5.4) can be approached through an initial examination of the gender division of labor. Most people in the region adhere to one of three major forms.

First, where a strict gender division of labor exists, women work in the home, gardening, processing farm products, raising small livestock, and "helping" with specific farm tasks. Men do the main cultivation, for-

Table 5.5 Local Differentiation of Land User Groups: Differences That Matter for Resource Use and Management

Differences between Groups within Communities

Permanent Differences
- Ethnicity, race, gender

Persistent Attributes but May Change
- Class: Income, assets, social status, political power
- Landholdings: Size, quality, tenure status, location
- Occupation: Farmers, farm workers, herders and livestock keepers, gatherers, artisans, processors, vendors
- Employment: Unpaid labor, wage labor (by sector and location), casual labor, self-employed
- Culture
- Religion: Catholic, Protestant, African-derived, indigenous, traditional, evangelical, syncretic, politically engaged
- Organizational Affiliation: Affiliated or not, specific group, degree and strength of affiliation (NGO's, social movements, political groups)

Contingent and Changeable Attributes

Age or age set

Differences between Households

Permanent Differences
- Ethnicity, race, gender of members

Persistent Attributes but May Change
- Gender of household head
- Class and power relations: Income, assets, social status, political power
- Landholdings and homestead: Size, quality, tenure status, location
- Occupation(s): Farmers, farm workers, herders or livestock keepers, gatherers, artisans, processors, vendors, midwives, healers, and herbalists
- Employment: Unpaid labor, wage labor (by sector and location), casual labor, self-employed
- Culture
- Religion: Catholic, Protestant, Jewish, Muslim, syncretic, indigenous, African-derived
- Organizational affiliation: Affiliated or not, type of group, specific group, degree and strength of household affiliation, structure of household affiliation (by head of household)

Contingent and Changeable Attributes
- Life cycle or state of development of household: New, established, long-standing

Differences within Households—Attributes of Individuals

Permanent Differences
- Ethnicity, race, gender

Persistent Attributes but May Change
- Class and power relations: Personal income and access to household income, personal assets and access to household assets, social status and access to household contacts, political power and access to household contacts
- Personal landholdings: Size, quality, tenure status, location
- Individual occupation(s): Farmers, farm workers, herders or livestock keepers, gatherers, artisans, processors, vendors, midwives, healers, and herbalists
- Employment: Unpaid labor, wage labor (by sector and location), casual labor, self-employed individual cultural practices
- Religion: Catholic, Protestant, Jewish, Muslim, syncretic, indigenous, African-derived
- Organizational affiliation: Affiliated or not, type of group, specific group, degree and strength of personal affiliation, role in linking household (sole member, one of two members, etc.)
- Position in household: Relative age or seniority, authority, and production role (head, household or farm manager, provider, contributor, dependent)

Contingent and Changeable Attributes
- Migration status: Permanent, long term, seasonal, past, or never
- Relation to household as migrant (partial or full provider, regular or irregular contributor, exchange such as cash for food, separated with no contribution, or dependent)

Source: From D. Rocheleau.

estry, livestock production, and wage labor. However, it must be understood that this strict division can still be more ideological than real. For example, what is called "farming" and what is called "helping" may depend on who does it, men or women, as much as on the task itself (Flora and Santos, 1986). Second, flexible complementarity exists under uneven relations of power. Men have the advantage in choosing land use systems, farm enterprises and activities, and off-farm employment. Women complement those tasks as needed, fulfill their own domestic responsibilities; and may have the option of conducting other agricultural, commercial, or wage-labor activities of their own. Finally, sometimes labor in farm and forest activities is equitably shared by gender, with women having some fixed responsibilities for domestic labor, such as food preparation and child care.

Throughout most of the region the prevailing gender division of labor conforms to either the first or second version or a combination of the two. This depends in part on variations in culture and class and the degree and type of market influence, commercialization, wage labor, and migration that occur. The ability of households to maintain and adopt sustainable farming and resource management practices will depend in part on regional, community, household, and individual approaches to the gender division of labor (Fortmann and Rocheleau, 1985).

The structure and composition of the family also affects the ability of any given household to maintain or to adopt sustainable agricultural and forestry production practices and strategies (McCarry, 1990) (see table 5.5, part II). The practical capacity of the household as a labor force does not depend only on the total number of people or even on their ages. Producer-dependents ratios provide some indication of the balance between production capacities and consumption demand in different households. However, the labor contributions of children to domestic, subsistence, and commercial activities are often underestimated.

Though often ignored, the combined gender and age composition of the household is crucial to both adult and child labor allocation. In Honduras, Doxon (1988) found that women in large, young families with high dependency ratios were least able to allocate labor to gardening. The same households had the highest rate of malnutrition. Where labor is gendered, both gender and age composition and power relations within the household affect the ability of a household to adopt a particular technology. Where labor is gendered but flexible, the gender and age distribution of the current family affects the practical flexibility of women's labor allocation and their ability to participate in a new activity or task. This occurs especially where women may identify more strongly as either housewives or farmers or both. The gender and age distribution of a woman's children may also affect her choices to allocate more of her labor to farming or to domestic work. If a woman has three daughters aged 7 to 14 and a 4-year-old son, she may be more likely to allocate domestic work to the girls and to join her husband in agricultural tasks. Alternatively, the eldest daughter may become more of a farmer than she would if she had older brothers.

The gender division of labor is both complementary and flexible. However, it favors allocating male labor to commercial agriculture or wage labor to gain the highest returns. Women's labor (adults and female children), then flexes to meet the remaining demand. Once established as a farmer, a young woman may take that skill and identity into her own family and may lean more toward farming. Nonetheless, her decisions will still be contingent on the gender and age distribution of her children, the demands of domestic work and her husband's land use and labor allocation decisions (Rocheleau et al., 1996; Ross, 1995).

Overall, the new flexibility of international agribusiness has demanded an increasing flexibility by rural agricultural wage laborers, contract farmers, and other smallholder growers (Raynolds, 1994; Thrupp, 1995). Men's adaptation to the changing labor and commodity markets has placed increasing demands on women in the same households to (1) contribute labor to NTAE crops, (2) replace men's labor on the remaining traditional cash crops, or (3) replace men's land and labor on food crops by producing more household food on less cropland or in gardens. Women may contribute almost all of the labor required for NTAE crops in households where men already work as off-farm wage laborers. The flexible complementarity of gendered labor under uneven relations of power (Rocheleau and Edmunds, 1995) implies choice and flexibility for women. However, it can also mean that women are increasingly expected to be flexible in the interests of the household or the household head (Katz, 1992). Women's labor is often treated as intrinsically more elastic than that of men.

The net result of these gendered relations and processes is expressed in livelihood strategies at the household and community level, in the patterns of land use and cover in the local and regional landscape, and in the biotic composition of gardens and cropping systems. Changes in land use, technology, and cash crops reflect the prior gender relations of power within households and communities in the landscape. These new practices, in turn, may alter the gender division of land, labor, cash, commodities, markets, and social organizations.

Common Economic and Ecological Challenges to Diverse Actors

In what context do gender, class, ethnic, and other differences affect people's actions and interests in resource management in the region now? What are the changes that people face as individuals and as members of households and communities? Three major categories of change currently sweeping through the region are (1) globalization of markets and increasing links to local production systems, (2) population change resulting from changes in birthrates and major migrations into and away from rural communities, and (3) local environmental changes and local impacts of global environmentalism.

Globalization of Markets

Market globalization is expressed primarily in three forms in Central America and the Caribbean. The first is land purchases or seizure by international agribusiness. The second is the aggressive promotion of nontraditional agricultural exports (NTAE) and the resulting dramatic changes in traditional cash crop markets. The third is a rapid increase in the cost of food imports.

International agribusiness interests have long been part of the political and physical landscape in the region. Today they aggressively seek new lands to purchase, lease or put under contract to produce specific commodities, such as pineapples, citrus fruits, specialty fruits, root crops, and vegetables. For example, large international firms and national corporations have expanded their holdings in citrus to compensate for the loss of citrus groves in Florida. Multinational corporations continue to grow some crops, such as bananas, on plantations. However, they have also entered into more flexible kinds of lease arrangements coupled with small packing houses to produce and export crops like pineapple, as in the case of Dole in the Dominican Republic (Raynolds, 1994; Rocheleau and Ross, 1995).

Small farmers have three options in the face of expanded production by international agribusiness. They can sell out and move. They can resist direct buy-outs and the effects of agribusiness encroachment in their communities. They can affiliate with these same businesses through land leases, out-grower and contract farming arrangements, and wage labor. The ability of individuals and households to exercise these options differs dramatically, depending on the size and quality of their holding, occupation and income, and age and gender.

Beyond the spatial expansion of international agribusiness, the dramatic incursion of NTAE crops has replaced or displaced coffee, cocoa, bananas, and multipcropped food plots on smallholdings in many places (Raynolds, 1994; Thrupp, 1995). Terms of trade for traditional cash crops have declined dramatically in response to global market restructuring, such as the European Economic Community's decisions to purchase bananas from former colonies in Africa and Asia in preference to Latin America. The NTAE have filled the economic vacuum created by these changes in many cases.

The phrase *nontraditional agricultural exports* applies to a wide range of crops. In the Dominican Republic, for example, they include cassava and other traditional tubers destined for the growing urban market for "Latino" foods in the supermarkets of Puerto Rico and major U.S. cities. At the same time, small, flexible, mobile companies have promoted the export of Chinese vegetables for another segment of the mass market in the United States (Raynolds, 1994). The term simply implies any crop not traditionally grown as a cash crop for export in a given place. In fact, it refers more to a new flexible way of organizing export production and marketing fruits and vegetables than to any specific set of crops (Raynolds, 1994). Even when the crop is a traditional staple in the region, such as cassava in the Dominican Republic, the NTAE status will often change its position in the landscape and may lead to biodiversity loss. This occurs both because the genetic variation of the crop itself is narrowed and because species diversity decreases in local cropping systems as farmers shift to monocropping the NTAE crops.

Farmers have five basic choices with respect to NTAE crops. (1) They may replace current intercrops with NTAE monocrops. (2) Some elect to intercrop the NTAE with other food and cash crops. (3) Others split their fields between NTAE and traditional intercrops. (4) Some stay out of NTAE production altogether. (5) Another option is to adopt alternative versions of NTAE production based on independent linkages to national or international speciality markets through cooperatives or similar organizations.

Size and status of land holding, occupation and income sources, and organizational affiliation condition individual and household choices in large part. Organizational affiliation is especially important when farmers elect to develop independent market linkages, whether they try to exercise collective leverage on NTAE intermediary companies or to organize their own alternatives. Households and individuals connected to farmers' federations or producers' cooperatives are in a far stronger position to exercise any of these options. Women and poor men are likely to be excluded in practice, if not in principle, where such organizations are for men only and/or require individual (as opposed to family) dues or fixed production quotas (Campbell, 1990).

Even when those who are excluded do not actually want to participate, their exclusion from membership leaves them less able to influence collective decision making about local strategies for NTAE. Further, the structure of individual and household affiliations with rural organizations can affect not only the people directly concerned but also the landscapes and livelihood systems of the region for the future. The decisions taken now may determine the biotic composition of agroecosystems, the type and extent of land use and cover, and the nature and terms of local employment opportunities and linkages to global markets. These, in turn, will affect different groups of households and individuals in very distinct ways.

Finally, rural people find themselves increasingly enmeshed in international commerce to satisfy basic needs for food and clothing and a growing demand for consumer goods. As a result, they are increasingly vulnerable to decreases in the prices offered for traditional agricultural commodities and to increases in the cost of food. Often they must enter the NTAE ventures and/or wage labor to make ends meet. Price increases for food and other imports erode the ability of rural farmers to resist selling land to agribusiness or converting to NTAE monocrops.

The ability of individuals and households to cope with price increases for food imports depends largely on three factors: (1) the extent to which the household consumes imported foods, (2) the degree to which household and individual labor is already committed to NTAE, and (3) the availability of well-paid alternative employment for some members of the household. This varies substantially by gender. Married women are often pressured to provide more labor to NTAE if their husbands are already engaged in wage labor (Katz, 1992). For poor families, four responses to price increases are most likely. (1) They may increase acreage and/or time devoted to NTAE to make up the shortfall in cash income through increased production. (2) They can engage in wage labor on other farms to earn more cash. (3) They may produce more food on the farm in existing croplands or gardens. (4) Finally, they can produce more food on sharecropped or rented land.

Distinct environmental consequences follow from each of the production and related resource management options exercised by rural people as individuals and members of households, communities, and crosscutting groups. The distinct environmental outcomes are based on land and labor allocation; the type and degree of connection to local, national, and international labor and commodity markets; and the strength and structure of people's affiliations with social organizations of various types. The concrete expression of these social and economic choices include changes (positive or negative) in (1) soil and water pollution from agricultural chemicals and organic wastes; (2) quality, quantity, and timing of the water supply; (3) erosion rates in fields and stream banks; (4) sedimentation rates of rivers, lakes, and dams; (5) stability of hillslopes; and (6) biodiversity, including species and genetic diversity of crops, livestock, and the biota of the surrounding ecosystems. The different constraints, opportunities, decisions, and performance of the many land user groups contribute to the net effects of global markets on local, regional, and global ecosystems.

Population Increase, Poverty, and Environmental Change

Environmentalists have cited population as a driving force of environmental degradation throughout the region, although many researchers have contested this direct causal explanation (Arizpe and Velasquez, 1994; Palloni, 1994). Population increase is usually reported and analyzed as national averages of birth- and death rates and the percent annual increase in the total population. Yet, demographic changes in any given rural region reflect the net outcome of decisions taken from international to individual levels. The number of children in a household reflects constraints, choices, and the lack of choices by different people. Each set of actors operates on different incentives for production and conservation, all nested in social relations of power that give privileges to some decision makers and mute others. The apparently simple relationship among population, resource availability, and resource degradation in the aggregate becomes more complex and more amenable to practical change as we focus on the community, the household, and the many individuals and distinct interest groups within and across households (Arizpe et al., 1993; Sen 1994).

The relations among poverty, education, employment, and family size are contingent on ethnicity, class, gender, and the histories of particular peoples and places. Likewise, the relationship between population growth rates and environmental change is contingent on and changes radically from aggregate to individual scales (Arizpe and Velasquez, 1994; Palloni, 1994). The relative importance of different elements that connect population and environmental change may also shift unpredictably as context changes.

Any demographic shift at the aggregate level will have consequences at the scale of individual smallholder households and of the individuals—women, men, and children—within them. Changes in family composition must be addressed jointly with changes in land use and livelihood strategies if they are not to work at cross purposes. Fewer children may mean more land per household member. However, it also means less labor per household unit, whether for subsistence or commercial agriculture or for wage labor to supplement household income. For example, a reduction in the number of children in rural smallholder families could militate against maintaining or adopting labor-intensive technologies for soil conservation, organic agriculture, or tree horticulture. Large-scale migration and population decrease may lead to stagnation or even abandonment of agricultural production (García-Barrios and García-Barrios, 1990; García-Barrios and Taylor, 1995). Under such circumstances, smallholders may sell their land to agribusinesses, land speculators, or local largeholder ranchers. None of these alternatives necessarily implies improved resource management and environmental quality. Alternatively, households short on labor may revert to less labor-intensive and more damaging practices, such as short fallow shifting or farm forestry that requires little labor.

There is real impetus for demographic or land use change where rural people have exceeded the capacity of their land to absorb labor under a given land use system. Yet, just as aggregate and individual ends may differ, so may short- and long-term incentives for small vs. large families. The long-term incentive for large families is strong when adult children can still subdivide the land or purchase land, settle nearby, and specialize in complementary types of subsistence and commercial production, even when returns to children's labor as children are not high. As one successful woman farmer in Puriscal, Costa Rica, noted: "We owe it all to our 14 children; we couldn't have done it without them" (Rosa Rodríguez,[1] personal communication, March 1983). Two sons specialized in tobacco production. Two others grew maize and beans. One daughter-in-law raised chickens and sold eggs. Another son and daughter-in-law bred cattle, and yet another couple specialized in dairy cattle and milk products. The state, communities, and local social networks or markets must provide similar opportunities if smaller families are to replace such successful deployment of children as economic and land management partners. They will need to develop other mechanisms that can offer farm households the chance to realize the same simultaneous economies of scale, diversification and complementarity as Rosa's enterprising family.

Where the returns to agricultural labor still favor large families, even over the short term, any move to decrease family size must incorporate equal or greater returns to land, with less labor (e.g., timber as a cash crop) or greater freedom, and returns to women's labor (farm or nonfarm wage labor or small enterprises at the individual or group level). External markets combine with gender, class, and ethnicity differences within and between households to forge the microdemographic scenarios that drive the broader trends of population change at the national level and that determine the net results of population change in environmental terms. Birthrates often fall rapidly where women enter the paid labor force in large numbers and on viable terms or where they gain access to higher levels of education. This reflects both the opportunity cost of the children relative to paid employment, a rather limited economic assessment of personal and family decisions, and women's increased financial autonomy. These may improve their bargaining position in the household (Jones, 1986) and give them an increased ability to make their own reproductive choices (Hartmann, 1987).

In contrast, both organic farming and many of the fruits and vegetables currently promoted as NTAE have begun to show unanticipated environmental effects distributed differently according to social relations between and within households. These production systems tend to draw more heavily than previous cropping systems on the elastic but still finite resources of unpaid household labor, especially that of women and children. Either of these options that link smallholder farms to international markets could have unforseen social and environmental consequences. The implications for population dynamics are not promising where new labor demands keep children in general and girls in particular out of school. A system based on the continuing availability of unpaid family labor will keep girls from attaining higher levels of education and will prevent women from entering into paid employment and small businesses. Katz (1992) has noted that Guatemalan women working on their husband's NTAE cash crops as unpaid family labor were less able to produce food on their own plots and were unable to participate in their own income-generation enterprises. These conditions are unjust, probably untenable over the long term, and favor larger as opposed to smaller families.

Sustainable Development, User Groups, and Environmental Change

Green revolution technologies and highly mechanized approaches were both differentially adopted by and had distinct effects on people of different classes and on women and men. National and international planners measured the success of the green revolution agricultural technologies in terms of national per capita food production. However, many critics traced its failures to the selective exclusion of poor smallholder and women-headed households and to the displacement of local foods, resources, and employment opportunities important to the poor and to women in general. Success at the national level was experienced in very divergent ways within rural communities. The net results for rural people ranged from profit and growth to hunger and privation, depending on differences between and within rural households (Harriss, 1982).

In spite of this experience, many researchers and policy analysts seem to assume that sustainable agriculture and locally based resource management are exempt from this effect. They apparently do not expect these technologies and resource management strategies to differentiate between and within households, either in their adoptability or in their effects. However, differences between communities, between households, and within households have affected not only their ability to adopt the technological packages of the green revolution but also their capability to participate in organic agricultural production, forestry, agroforestry, soil conservation, and a host of similar sustainable technology and land use changes (McCarry, 1990; Sachs, 1993; Thomas-Slayter and Rocheleau, 1995a).

Some communities will be less able than others to adopt particular cropping systems or production strategies. For example, in the Dominican Sierra, farmers on the dry hillslopes of the Pananao valley commented: "You can't believe what they want us to do now . . . organic peanuts . . . after so many years of tobacco, peanut, and cassava this soil is exhausted. How can they expect us to do this? Nobody could grow anything in these soils without fertilizers." In this case an entire community in one small basin is unable to consider organic production because of its previous production history, sandy soils, steep slopes, and dry climate (field notes, 1993).

A household's ability to enter into new sustainable land use systems and the broader changes adopted by its neighbors depends on endowments of land and related resources; available labor; access to cash, commodities, and markets; and organizational affiliation. For example, to plant new tree crops, households with little land might need to replace their food croplands, patio gardens (or solars), or remnant forest and pasture plots with the new crops. They might also replace their food crops with the new cash crop on their best plots and displace their food crops to more fragile lands or to rented lands formerly in forest and pasture, with a net negative effect on the surrounding watershed. Largeholders might have to make less difficult decisions concerning replacement, but they might also expand both cash crop and food crop production for their hired labor onto steeper lands currently in pasture or forest.

One very graphic example of this phenomenon has been documented in the Central Mountains of the Dominican Republic. In the Sierra region, a sustainable development program based on the replacement of shifting cultivation with improved varieties of tree crops (coffee) led to exactly this response. The pattern of planting by both smallholders and largeholders was influenced by the structure of their social and economic relations to each other, as well as by intrahousehold demands for land, labor, food, and cash. Largeholders expanded their already substantial coffee acreage with the new caturra variety, a Brazilian dwarf variety grown without shade, on their best lands. They also expanded and displaced food crops to forest and pasturelands, often on steep slopes, to produce food for their hired labor (both local smallholders and seasonal migrant labor from the lowlands). Smallholders planted the unshaded coffee on their best lands, and they, too, displaced food crops to forest, pasture, and steeper slopes or planned to produce less food and to buy more from the small store owned by the largeholder family.

Ecological models of land use change projected that an increase in coffee acreage under these social circumstances would result in a net increase in erosion at the watershed level. It would also increase smallholders' dependency on largeholders for coffee processing, market connections, wage labor, and food. Both results ran counter to the program's goals, which were to reduce erosion and increase smallholders' income. The structure of interhousehold relations between smallholder and largeholder farmers in the community and the within-farm structure of food production for the work force in both types of households affected the outcome (Rocheleau, 1984). Eventually, the falling price of coffee reversed the trend by discouraging the expansion of coffee in the area.

Local Impacts of Global Environmental Initiatives

Several distinct approaches to environmental and social problems find a place under the broad umbrella of sustainable development. It encompasses a range of land use strategies from sustainable agriculture and community-based resource management to conservation collaboratives, reserves, and parks. Sustainable agriculture has tended to focus on the household, women's groups, and/or other community groups. The other approaches have usually concentrated on the undifferentiated community or male heads of household.

Agricultural alternatives to the green revolution include such divergent approaches as organic food production on smallholdings (see Soto, chapter 18) and industrial use of trees to provide nitrogen in cash crop plantations. The first addresses health and environmental concerns, whereas the second represents a kind of biologically correct and economically efficient factory farming. Other sustainable agriculture approaches include intensification based on intercropping and perennials, integrated pest management (IPM), agroforestry and social forestry, and soil conservation techniques, all of which can apply to either smallholder or large-scale commercial farms. There are also broad movements to maintain and restore the crops' genetic and species diversity and to build on local knowledge, skills, species, and priorities. Sustainable agriculture programs have targeted smallholder households, and they have often pursued the same strategies as green revolution and farming systems programs.

Community-based resource management has emerged as a major innovation in the management of forests, water, and other resources to maintain rural communities and the integrity of their surrounding ecosystems. In the Amazon basin, the Rubber Tappers' Union developed extractive reserves to maintain their forest homes and livelihoods based on collecting latex and Brazil nuts and other ecologically sustainable economic pursuits (Schmink, 1994; Schmink and Wood, 1992). Other well-known examples of community-based resource management include (1) the Kuna Camarca of Panama, an effort to preserve both the rainforest and the Kuna culture; (2) the community-centered landscape design and forest construction of the Kayapo in the

Brazilian Amazon (Posey, 1985); (3) the local biosphere reserves of the Mexican countryside (Gómez-Pompa, Fores, and Sosa, 1987; Gómez-Pompa and Kaus, 1992); (4) the management of common resources in marine fisheries and aquaculture (M. García, 1992); and (5) forestry, agroforestry, and garden systems throughout the Amazon (Anderson, 1990). Mestizo, indigenous, and Afro-Caribbean forest, farming, and coastal peoples throughout Central America and the Caribbean have developed common property regimes for resource management based on community use and control (Oldfield and Alcorn, 1991). In some parts of the Caribbean, the community's access to former forest flora persists largely in private lands and in the dooryard and patio gardens of women (Kimber, 1988; Pulsipher, 1993; Rocheleau, Ross, Morrobel, and Hernández, 1996). In each of these cases, differences between user groups based on class and gender were mediated through differential division of land and related resources, labor, cash, other forms of capital, markets, and organizational affiliation.

Larger-scale conservation collaboratives have sprung up in many locations, from Costa Rica to Bolivia to Ecuador. They also affect people very differently, depending on gender, class, and other differences. The arrangement between Merck Pharmaceuticals and the Costa Rican government is one example of a state and corporate collaborative. Merck pays the government a flat fee annually to maintain a large expanse of forest and for the rights to all patents on pharmaceuticals and other products developed through genetic prospecting by the company. The state will receive a percentage of the patent royalties. The major benefit of the agreement to local communities, according to promotional materials, is employment of local men as forest guides and botanical experts. Neither the community nor the household is an acknowledged player, although entire communities, households, and specific groups within them will be affected by changing terms of access to forest resources.

Some development critics decry the ethical and political problems with such approaches, whether on farms in forests or at a global scale (Peet and Watts, 1996; Schroeder, 1993). However, Arturo Escobar (1995) cites the potential for the convergence of specific cultural groups' interests in forest preservation and territorial integrity with those of environmentalists and biotechnology firms in the preservation of forest biodiverstiy. In the Pacific rainforests of Colombia, the Afro-Colombian forest communities are considering the feasibility of such an option to define and maintain their homelands and lifeways (Escobar and Pedrosa, in press). The choice of boundaries and terms of territorial integrity reflects the overall emphasis on ethnicity, race, and cultural politics. However, differences of gender, class, age, and occupation all matter to the many residents and users of the forest and, ultimately, to the forest itself, based on daily practices of use, harvest, seed collection, and planting.

Many environmental organizations and national agencies still resort primarily to demarcation and expansion of national and international parks and reserves to conserve endangered wildlife and ecosystems. Parks claim nonhuman species, future generations of people, distant taxpayers, and tourists as their constituents (McNeely, 1994; McNeely et al., 1995; Western and Pearl, 1989). They do not claim to serve local people, although they do, in fact, selectively serve some and disserve others (Bonner, 1993; Rocheleau, Schofield, and Mbuthi, 1992). Local people's everyday use of renewable resources, such as water, deadwood, herbs, wild foods, small game, and fiber from park areas, becomes illegal. At the same time, commercially or politically connected local residents may secure contracts to divert water and forest resources from neighboring areas to provision the park's tourist facilities. The well-educated and bilingual residents may secure employment (see also Sodikoff, 1996). Park employment may favor men or women, depending on the nature of the work. Often, however, men are hired as "family breadwinners" when park managers and concessionaires want to establish relations with local communities.

Sustainability with Unanticipated and Differential Consequences

Two cases, one in the Dominican Republic and the other in Mexico, illustrate the ways in which both sustainable development and conservation initiatives can play out very differently within and across households, with consequences at the community or regional level. Each also shows that the perceptions of outsiders about the identity, intentions, and environmental impact of local producers and resource managers may also influence both the content and consequences of national and international policy on sustainable development.

The Forest Micro-Enterprise Project

The Forest Micro-Enterprise Project was founded as a collaboration between ENDA, an international NGO, and the Rural Federation of Zambrana-Chacuey. Zambrana-Chaucey is a hilly region at the edge of the Cibao Valley in the Dominican Republic. The farmers of the federation have planted small blocks and rows of timber trees on their farms and have constructed a community sawmill to process and market their timber. In spite of strong community and institutional organization, gender and class have influenced the selection or self-selection of participants, the choice of products included in the program, the structure of public participation and decision making, and the distribution of costs and benefits of these forest-based enterprises (Rocheleau and Ross, 1995).

Gender ideology played a major role wherever women were unwillingly integrated into or excluded from the new forest enterprise, based on *Acacia mangium*. Women and men accustomed to varying types and degrees of partnership in tree production suddenly found themselves cast as timber foresters and salad gardeners, respectively. Their role was based on uneven relations of power and the gender biases or misunderstandings of local practices by a dominant national culture and international forestry and agricultural professionals. The unanticipated social and environmental impacts included the replacement of highly diverse tree and crop assemblages in women's gardens and shared croplands with monocrop timber plots. Sometimes, biodiversity decreased, although forest cover increased. Similarly, as men's income increased, women's income or food supplies decreased. Women and households with very small holdings were usually largely excluded or disadvantaged in the first phase, not by design but through the failure to address gender and class differences in project policy and in everyday practice. The federation and ENDA took action in the next phase to remedy the inequities of both project participation and project impacts and to promote greater species diversity and timber species amenable to intercropping (Rocheleau, Ross, Morrobel, and Hernandez, 1996).

The Ban on Cutting Trees in the Lacondon Forest

Regulating forest clearing now affects both indigenous and colonial communities in Lacondon Maya areas. The effects may vary substantially between and within households in these two types of communities. Even communities that appear to be very homogeneous with respect to their households may find themselves differentiated rapidly and dramatically by the abilities and inclinations of different households to deal with new realities imposed by market or regulatory changes. For example, forest dwellers whose families had the same rights to shared land and related resources and who lived in very similar homes might suddenly be in very different positions as wage laborers who reside in deforested settlement communities. They could be stratified according to their children's literacy, their ability to speak Spanish, or their established contacts with petty traders. These factors, of little consequence for resource management in the past, suddenly become crucial for present and future access to land, wage labor, and markets (Arizpe et al., 1993, 1994; Townsend et al., 1994).

In areas where both men and women formerly farmed, women migrants may find themselves in new forest cooperatives designed by professionals from the dominant culture. They may encounter very difficult circumstances. For example, they might be expected to provide food for their families without any allotment of personal land to grow food crops. Women may be newly constructed as unpaid employees (or helpmates) of their husbands, a role more unfamiliar and more undesirable than their previous ones as both household contributors and independent producers (Townsend, 1995; Townsend et al., 1994).

Implications for Research and Development

The differences among people, both within and between households, have crucial implications for research and project design and for the development of the institutional frameworks required to define and promote sustainable development (Collins, 1991). The inclusion of diverse land user groups at the local level will require that both planners and researchers embrace complexity (Moser, 1989). Many of us may opt to forego simplistic, although convenient and accepted, research and design frameworks for the sake of clarity, effectiveness, and equity. Variability within and between households will be a useful indicator by itself of the range of environmental options and outcomes. Differentiation and the distribution of power within and between nuclear and extended families may also be useful for framing technologies and policies that are environmentally and socially effective.

Both our questions and our methods may need to change radically to understand these complex social relations within communities and their articulation with larger systems. We may find ourselves combining approaches, such as participatory rural appraisal, with household surveys, life histories, and landscape land use histories. Anthropologists and sociologists, on the one hand, may use remote sensing imagery to find and map the resources used by particular groups. Ecologists, geographers, and land use planners, on the other hand, may need to refer increasingly to detailed sketch maps and local taxonomies and to combine these with local and individual environmental and social histories (Rocheleau, 1995b,c; Ross, 1995; Slocum et al., 1995). We may begin our research planning or evaluation tasks with a very different set of questions. How are individuals and households—or multiple and overlapping groups that cut across households and communities—different in resource use, access, and control? How are uneven relations of power expressed in differential access to land, labor, commodities, cash and markets, and social organizations?

Conclusion

Throughout the region people have experienced a profound shift in the division of land, labor, markets, and organizational affiliation, as members of communities, households, and groups that cut across boundaries. Based on gender, class, age, ethnicity, and other dimensions of difference, rural people find themselves repositioned relative to one another; to the larger economy; and to the ecosystems that support them, from their homesteads to the global environment. This phenomenon requires a telephoto lens to observe the interactions within and between ecological, economic, and social systems from the individual to the planetary scale. The repositioning phenomenon also demands a research framework that illuminates the complex patterns embedded in multiple and overlapping domains of resource use, access, and control—a kind of kaleidoscopic view that both complicates and clarifies. Development of sustainable and just economies and ecologies will require a telephoto lens of observation across scales, with a kaleidoscopic analysis focused on the critical differences between the actors at each scale along the way.

Note

1. A pseudonym is used to maintain the anonymity of survey participants.

PART II

Challenges to Managing the Natural Resource Base in a Shrinking World

6

Biodiversity Conservation in Mesoamerica

Mario A. Boza

Mesoamerica is the region that extends from southern Mexico, including the states of Chiapas, Tabasco, Campeche, Yucatan, and Quintana Roo, to Panama. All of the Central American nations—Guatemala, Honduras, El Salvador, Nicaragua, Costa Rica, and Panama—are included in the 768,989 square kilometers comprised in this bioregion. This area corresponds to only one-half of 1 percent of the total land area of the world. However, the biological diversity of this small region is extraordinarily high. D. H. Janzen (1994) estimates that this small area contains approximately 7 percent of the entire planet's biodiversity.

Many factors contribute to Mesoamerica's biological wealth. First, this region serves as a bridge between North and South America. The North American pines reach their furthest southern extension in Bluefields, Nicaragua, while the forests of Mesoamerica are inhabited by both North American fauna, such as deer and beaver, and by South American species like sloths, anteaters, and monkeys. High mountain chains further enhance biological wealth. Tajumulco Volcano in Guatemala reaches a height of 4,210 meters. The Andean highland vegetation complex reaches its most equatorial extension in the Talamanca range of Costa Rica and Panama. North and South American species meet in Mesoamerica, enhancing plant biodiversity and leading to the evolution of many locally endemic species. Flores and Gerez (1989) offer an interesting comparison between the United States and Central America. The United States, on the one hand, has a land area of 9.4 million square kilometers, with 22,000 plant species. Central America, on the other hand, with a land area of only 0.5 million square kilometers, has 18,000 to 20,000 plant species.

The region's biological wealth includes many economically important species. Many primitive species of corn, beans, cacao, tomato, cotton, pepper, and forage legumes are indigenous to Mesoamerica. These plants play a vital role worldwide as sources of food, feed, and fiber. Preservation of the primitive species in Mesoamerica provides an important pool of genetic information that can contribute to the improvement of cultivated varieties of the species, for disease resistance, for example. *Pimenta dioica*, or black pepper, is from Chiapas in Mexico. The various *Dioscorea* species, the plants from which cortisone was originally extracted, also originate in this region. Preservation of Mesoamerica's biological wealth, therefore, is not simply a question of preservation of biodiversity for biodiversity's sake. It may also play a critical role in humankind's ability to maintain the genetic viability of the many foods, fibers, and fuels on which we depend.

The Areas of Major Importance

The areas that should be protected in Mesoamerica can be classified into three groups. The first group includes 17 large, contiguous areas considered to be of primary importance for the preservation of biodiversity in Mesoamerica. These areas vary in size from a minimum of about 100,000 hectares to some that include several million hectares, either currently or potentially protected. It is estimated that these 17 areas include 75 percent of the biological diversity in the region today. The second group includes areas that are usually smaller, sometimes measuring only a few hundred hectares (table 6.1). They are significant because they are sites that protect habitats or species of special regional or even worldwide importance. Beaches that serve as breeding grounds for sea turtles are one example. These areas may protect another 15 percent of the biological diversity of the region. Finally, there are other protected areas in each country whose conservation is primarily of local or national interest. These areas, with national and commercial forests and other agroforestry systems, could protect the other 10 percent of the region's biodiversity. The following are the

Table 6.1 Areas of Secondary Biodiversity in Mesoamerica

Site	Size (in hectares)	Special Features
Laguna de Términos, Mexico	700,000	Wetlands and mangroves
Central Chiapas, Mexico		Pine and oak forest
Flamingo National Park, Mexico		Dry and thorn forest
Río Lagartos, Mexico	55,350	Coastal dunes, mangroves, natural salinas, wetlands and lowland dry forest
Chuchumantanes and Bisis Caba, Guatemala	200,000	Pine and dry forest
Manchon-Guamuchal Multiple Use Area, Guatemala	13,000	Dry forest, mangroves, and wetlands Migratory birds
Yojoa Lake, Honduras	40,000	Aquatic fauna. Surrounding wetlands excellent bird habitat
La Muralla National Park, Honduras	15,000	Cloud forest containing various coniferous species
Cerro Verde National Park, El Salvador	6,500	Dry forest
Delta del Estero Real Natural Reserve, Nicaragua	55,000	Mangroves
Masaya Volcano National Park, Nicaragua	5,100	Dry forest (disturbed by volcanic activity)
Lake Nicaragua, Nicaragua	826,400	Tenth-largest freshwater lake in the world. Includes 300 islands
Tempisque Conservation Area, Costa Rica	19,083	Wetlands critical to migratory birds. Also dry forest, mangrove, and mixed forest
Las Baulas de Guanacaste National Park, Costa Rica	500	Beach is the third most important site in the world for the marine turtle *Dermochelys coriacea*
Ostional National Wildlife Refuge, Costa Rica	162	With Nancite Beach in Santa Rosa National Park, constitute the two most important areas in the world for the marine turtle *Lepidochelys olivacea*
Sierpe Mangroves Forest Reserve, Costa Rica	22,000	Most extensive Pacific Coast mangroves in Costa Rica
Coco Island National Park, Costa Rica	2,400	500 km from the continental land mass. Extraordinary marine richness
Chiriquí Golf, Panama		Most extensive Pacific litoral in Mesoamerica
Eastern Azuero Peninsula, Panama		Most representative area of dry forest in Panama
Coiba Island National Park, Panama	270,000	Tropical humid forest and very humid premontane forest
Taboga Island, Panama	257	Largest colony of *Pelecanus occidentalis* in the world

Source: Boza, 1994. Biodiversidad y desarrollo en Mesoamérica. San José, Costa Rica. Paseo Pantera and COSEFORMA/GTZ Projects.

17 areas in the first group considered of primary importance for biodiversity conservation.

Sian Ka'an Biosphere Reserve, Mexico

Sian Ka'an Biosphere Reserve, on the east coast of the Yucatan Peninsula, includes 528,147 hectares, of which 120,000 are marine. About one-third of the area is in tropical forest, another third in wetlands, and the final third in coastal and marine ecosystems. Most of the reserve is subject to frequent inundation, providing an excellent habitat for birds. Some 336 bird species are found there. Many aquatic organisms are also found in Sian Ka'an, and the reserve serves as a breeding ground for 4 species of sea turtles (*Chelonia mydas, Eretmochelys imbricata, Caretta caretta* and *Dermochelys coriacea*).

Some 800 fishermen live in the transition zone inside the reserve, which also contains some livestock.

La Selva Maya, Mexico, Guatemala, and Belize

The 2,200,000 hectares in La Selva Maya make up the largest contiguous block of lowland primary forest under legal protection in Mesoamerica. Although one block, this protected area in Mexico, Guatemala, and Belize includes many subunits that are subject to different levels of protection and are legally the responsibility of three nations. It includes two biosphere reserves, five national parks, three biotopes, and several reserves and conservation areas. More than 60 percent of the area is covered by evergreen forests, representing an immense biological diversity of tree species. The area is also important culturally because it includes one region where the Maya civilization evolved and contains many archeological sites of great importance. The Calakmul Reserve, for example, includes 524 archeological sites. Accelerated colonization and destruction of the surrounding forests and concessions to timber interests are the principal threats to La Selva Maya. While timber extraction is selective, involving only two to three trees per hectare, the practices used cause significant damage to the rest of the forest and understory community. The logging roads become access ways for settlers.

Montes Azules Biosphere Reserve, Mexico

Montes Azules Biosphere Reserve includes 331,200 hectares, protecting the heartland of the Lacandon Forest, which measures a total 1,500,000 hectares. The Lacandon Forest is the largest single block of tropical rainforest in North America and includes 20 percent of the total biodiversity of Mexico, although accounting for only 0.16 percent of Mexico's national territory. Thirteen vegetation types are present in the reserve, including seasonally inundated forests, mixed tropical rainforests, pine forests, and cloud forests. It is estimated that Montes Azules has 3,000 plant species, 106 mammal species, 306 bird species, 109 amphibian and reptilian species, and some 800 butterfly species.

Sierra de las Minas Biosphere Reserve, Guatemala

The nucleus of Sierra de las Minas Biosphere Reserve includes 103,000 hectares. However, the total area increases to 230,000 hectares if all of the forested areas in the buffer zone around the reserve are included as protected areas. Sierra de las Minas includes at least 60 percent of the cloud pine forests of Guatemala and accounts for a high percentage of Guatemala's total biodiversity. Mahler (1993) says that, because of its geographic isolation and range of elevations, the reserve contains some 885 species of birds, mammals, amphibians, and reptiles, including the quetzal, Guatemala's national bird. He also points to the reserve's important role in supplying clean and abundant water for Guatemala's population. More than 63 permanent streams originate there. A large part of this reserve remains under private ownership, at least 40,000 hectares within the nucleus, a roadblock to preventing exploitation of forest resources. Although extraction theoretically occurs under a management plan, it has a severe impact on biological diversity. An additional problem is the presence of subsistence farmers on the southern slopes of the range.

The Coral Reefs and Atolls of Belize, Mexico, and Honduras

Coral reefs and atolls extend from the eastern point of the Yucatan Peninsula in Mexico southward 240 kilometers into the Gulf of Honduras; this reef system is the most extensive on the continent. The reef, which parallels the coast at a distance of some 12 to 30 kilometers, includes three elements. The coastal zone, where the water depth is 3 to 5 meters, extends from the beach to the breaker line and is underlain by sand. The marine pastures (*Thalassia testudinum* and *Syringodium filiforme*), important to turtles and manatees, grow in this zone. The second zone includes the reef crest, which is shallow and forms the breakwater. The frontal zone, from 5 to 50 meters deep, is characterized by abrupt slopes and is the principal growth zone for the corals, some of which form colonies up to 10 meters high. This spectacular reef system rivals the famous Great Barrier Reef of Australia in complexity and has been proposed as a World Heritage Site (Carter et al., 1991). Overfishing threatens some species in this region. The use of prohibited fishing techniques, water contamination from sewage outflows, agricultural chemicals, ship discharge, hurricanes, and excessive development of tourism facilities are additional threats to aquatic species. The area under protection is inadequate, particularly for marine ecosystems.

Maya Mountains, Belize

The Maya Mountains make up a complex of protected areas and areas under forestry management in southern Belize. It includes 450,000 hectares, the central block of which has 10 forest reserves, 4 natural reserves, and Chiquibul National Park. Elevations in the complex range from 300 to more than 1,000 meters. It is characterized by high rainfall and has both pine and broadleaf forests. The forests in the reserve include several broadleaf forest types and constitute the most extensive block of broadleaf forest still intact in Belize. The zone also has cultural importance. It includes the Caracol Archaeological Reserve, which is the most important Mayan site in Belize and one of the four most important Mayan sites in Central America.

Pico Bonito National Park, Honduras

Located in North Central Honduras near the Caribbean coast, Pico Bonito National Park covers a total area of 107,300 hectares, 49,800 hectares in the nucleus and 57,500 hectares in the transition zone. The park has five life zones, ranging in altitude from nearly sea level to 2,435 meters at the top of Bonito Peak. Relief varies from coastal plains to hilly landscapes to medium to high mountain chains. Vegetation ranges from cloud forest complexes abounding in palms and ferns to pine-oak associations on the drier southern slope. The five life zones provide habitats for an abundant, varied fauna, including, for example, three species of monkeys (*Alouatta palliata, Cebus capucinus*, and *Ateles geoffroyi*), anteaters, peccaries, tapirs, and many bird species.

La Mosquita, Honduras and Nicaragua

La Mosquita includes approximately 1,000,000 hectares that make up seven separate protected areas in Honduras and Nicaragua. These seven areas are contiguous or very close to one another and stretch like an arch from northeastern Honduras to northern Nicaragua. The entire region is commonly called the Mosquito Coast. The region includes low mountain chains up to 1,500 meters in height, as well as coastal zones with mangroves, freshwater coastal lagoons, littoral vegetation, gallery forests, and wetland forest complexes that are periodically or permanently inundated in fresh water. Fauna is abundant, and the zone maintains populations of species threatened or in danger of extinction that require large extensions to feed and reproduce, such as the puma (*Felis concolor*), jaguar (*Panthera onca*), tapir (*Tapirus bairdii*), peccary (*Tayassu pecari* and *T. tajacu*), and agouti (*Agouti paca*). The region's cultural importance springs from both its archeological interest (more than 80 archeological sites) and its occupation today by indigenous groups, such as the Miskito, Tawahka-Sumu, and Pech, and by the Garifuna community. The area under protection is inadequate and should include the marine portion of the Cayos Miskitos Reserve, 250,000 hectares in northern Nicaragua. Although the Mosquito Indians are the principal human group occupying the region, the wealth of marine resources attracts fishermen from other parts of Nicaragua, neighboring countries, and the Caribbean who use very destructive fishing methods. Fishing is a serious threat to commercial species of fish, mollusks, crustaceans, and even dolphins.

International System of Protected Areas for Peace (SI-A-PAZ), Nicaragua and Costa Rica

A region of 700,000 hectares in Nicaragua and Costa Rica, made up of forests, wetlands, rivers, streams, estuaries, lakes, and marine waters, is part of the International System of Protected Areas for Peace (SI-A-PAZ). The region has extremely high precipitation—between 5,000 and 6,000 millimeters per year—and is primarily characterized by extensive wetlands. No fewer than 11 habitat types have been identified, including littoral vegetation complexes dominated by *Cocos nucifera* and *Coccoloba uvifera*; high, very humid forests; ridge forest complexes; wetland forest complexes; rafia stands (*Raphia taedigera*); and herbaceous wetland and lake communities. The fauna is rich and diverse. Monkeys, frogs, birds, and freshwater fish are particularly abundant. Unplanned colonization in the northwestern portion of the Indio-Maiz Reserve on the Nicaraguan side of this area and enormous deforestation and contamination from extensive banana plantations on the Costa Rican side threaten this region.

Guanacaste Conservation Area, Costa Rica

The Guanacaste Conservation Area includes three national parks, one wildlife refuge, and a biological corridor. The area connects Guanacaste National Park on the northwestern mainland of Costa Rica with Rincon de la Vieja National Park in the mountain chain that extends through north central Costa Rica. If proposed biological corridors are aggregated to the existing protected areas, Guanacaste would include 120,000 hectares of terrestrial habitat and an additional 70,000 hectares of marine ecosystems. This conservation area protects the largest extant territory of dry forest in Mesoamerica. It contains numerous other diverse vegetation complexes, such as mangroves and the cloud forests that lie on the skirts of the three volcanoes in the area. The 15 habitats provide a rich and diverse fauna. Some 140 mammal species, over half of which are bats; 300 bird species; 100 species of amphibians and reptiles; and more than 10,000 insect species, including some 5,000 species of diurnal and nocturnal butterflies and moths, are found here.

Arenal Conservation Area, Costa Rica

Elevations in the Arenal Conservation Area in Costa Rica range from 600 to 2,000 meters. The area has two national parks, one protected zone, two forest reserves, and two private reserves. These units comprise 80,000 hectares. If proposed corridors are put into place, the total area under protection would increase to 125,000 hectares. The area has precipitation that ranges from 2,500 millimeters per year at lower elevations toward the west to 5,000 millimeters per year at higher, more eastern sites. According to the Holdridge (1967) classification, the Arenal Conservation Area includes five life zones: premontane wet forest, tropical wet forest, premontane belt transition forest, premontane rainforest, and lower montane rainforest. Cloud forests in the conservation area include some 300 orchid species and 200 fern species.

The Central Volcanic Cordillera Conservation Area, Costa Rica

Like the other conservation areas in Costa Rica, the Central Volcanic Cordillera Conservation Area includes several units under different forms of protected status. Today, four national parks, the Guayabo National Monument, the Grecia Forest Reserve, and the privately held La Selva Biological Station (belonging to the Organization for Tropical Studies) have put 70,000 hectares under protection. Projected additions and corridors would increase this to 180,000 hectares. The evergreen forests, which monopolize the area, are characterized by extreme density and floristic complexity. They vary with topography, drainage, temperature, cloud cover, and the impact of eruptions from the active volcanoes in the region.

La Amistad Biosphere Reserve and World Heritage Site, Costa Rica and Panama

More than 1 million hectares are included in La Amistad Biosphere Reserve and World Heritage Site, including national parks, forest reserves, biological reserves, wildlife reserves, protected zones, protected forests, and indigenous reserves. The reserve includes the most extensive mountain system of Costa Rica and Panama and is the region of greatest biodiversity in Costa Rica. More than 10,000 species of vascular plants, 4,000 species of nonvascular plants, 900 lichen species, 1,000 fern species, and 1,000 orchid species are found in La Amistad. Many habitats that exist in few or no other parts of Central America occur here. These include, for example, the paramo, a chaparral vegetation complex occurring in zones above 2,900 meters above sea level, highland oak (*Quercus* spp.) communities characterized by enormous individuals, and dense fern associations (*Lomaria* sp.). La Amistad also has great cultural significance, containing more than 200 archeological sites and trails that humans have used for more than 3,000 years (de Mendoza, 1994).

Osa Conservation Area, Costa Rica

A total of 80,000 hectares are included in the Osa Conservation Area, including Corcovado National Park, the Isla del Cano Biological Reserve, the Golfito National Wildlife Refuge, the Golfo Dulce Forest Reserve, and a group of privately owned natural reserves. Proposed additions and corridors would increase the area to 150,000 hectares, along with 50,000 hectares of marine habitat. The Osa vegetation is characterized by a very high degree of endemism and a high affinity to South American vegetation. The principal habitats are montane forests, which contain the greatest variety of both flora and fauna in the conservation area. They include cloud forests dominated by oaks, tree fern complexes, coastal forests, wetland forests dominated by rafia palm (*Raphia taedigera*), herbaceous wetlands, and mangrove swamps. Fauna is equally rich and diverse. Corcovado National Park, for example, protects the largest population of scarlet macaws (*Ara macao*) in Costa Rica. Illegal gold mining in the beds and stream banks of the rivers, illegal hunting, and deforestation by settlers are the principal problems on the Osa Peninsula.

Bocas del Toro and El Cope National Park, Panama

Bocas del Toro and El Cope National Park, on the Caribbean Coast of Panama, includes the Almirante and Chiriqui lagoons, the Valiente Peninsula, five islands, and two cays. The biodiversity of the islands in this system is of intense scientific interest because the degree of evolution that they illustrate is rare in islands so young and so near a major land mass. Each island has a specific flora and fauna and each is home to various endemic species or subspecies, such as the giant arboreal rat (*Tylomys*) and the dwarf sloth (*Bradypus*). Four species of sea turtles (*Chelonia mydas, Eretmochelys imbricata, Dermochelys coriacea*, and *Caretta caretta*) use the beaches in the area, and other endangered species such as the manatee and jaguar are present.

The Proposed Interoceanic Park of the Americas, Panama

The original proposal for the Interoceanic Park of the Americas included the creation of a national park of some 62,000 hectares that would occupy the border of the Panama Canal from the Caribbean to the Pacific. Although approved by the Panamanian government, implementation has not occurred because of administrative complications. Several protected areas already exist in the proposed park, including five national parks. The proposed park would include these and other lands, totaling 260,000 hectares. Flora and fauna in the region exhibit high biological diversity. On Barro Colorado Island, which measures only 1,564 hectares, 1,369 species of vascular plants, 93 mammal species, 366 bird species, and 90 species of amphibians and reptiles have been identified. The forests in this region have been preserved largely because of the presence of the Panama

Canal Zone. Since these lands reverted to Panamanian jurisdiction, uncontrolled settlement and rapid deforestation have become increasing threats to the forests and to biodiversity in the region. Invasive grass species are another threat.

Daríen National Park, Panama

Daríen National Park, with 579,000 hectares, is the largest national park in Panama. On the southern border, the park connects to the Colombian Los Katios National Park, which includes an additional 72,000 hectares. The forests of the Daríen are dominated by immense, emergent trees, such as *Cavanillesia platanifolia, Anacardium excelsum*, and *Ceiba pentandra*. Lianas are very common, and the palms, such as *Sabal allenii*, are very abundant forest floor species. Animal diversity is high, with a total of 499 bird species, of which 203 are migratory, and more than 130 mammal species, including 5 cat species. Settlement and subsequent deforestation are major threats. These problems may be exacerbated by the construction of the Pan-American Highway through the Daríen.

The Loss of Biodiversity

There are 17.4 million hectares, or 36 percent of the total land area of Central America, in forest. The estimated deforestation rate is 354,000 hectares per year. Pasos (1994) estimates that more forests have been destroyed in the region between 1950 and 1990 than in the previous 500 years. All of the countries exhibit the same pattern, and what has happened in Costa Rica is a good example of the general situation. Costa Rica lost 23 percent of its forested area between 1973 and 1988 (Pasos, 1994). Today, it has about 650,000 hectares in national parks and equivalent reserves, some 400,000 hectares in secondary forests capable of generating forestry products, about 200,000 hectares in productive primary forests, and about 75,000 hectares in planted forests of different ages. These areas total about 1,325,000 hectares, or 26 percent of the country's area. Yet, no less than 50 percent of its area (2,500,000 hectares), because of topography and rainfall, should be under permanent forest cover according to land use capability classification. The enormous effort that would be required to reforest, through plantings or natural reforestation, to meet land use capability classification recommendations is clear. Along with deforestation, forest fires, land and water contamination, illegal hunting and the selective extraction of plant and animal products are the principal causes for biodiversity loss in Mesoamerica.

The environmental situation and trends in Mesoamerica are, in large part, a reflection of traditional, nonsustainable policies applied in the region. These policies have created a model of development that is not viable. Solorzano et al. (1991) provide an excellent example in their recent study of the lack of valorization of natural resources in Costa Rica. Vargas (chapter 7) shows how ineffective government policy has been in Costa Rica, for example. Substantial changes in these policies, above all in policies that direct investment and disinvestment, are needed to eliminate unsustainable practices in agriculture, natural resource use, population growth, waste management, urban growth, energy production, and other areas. The primary precondition for true socioeconomic progress in the region is that natural resources, including biodiversity, be considered, managed, and protected as tools for sustainable development. The direct and indirect consequences of a failure to do so have resulted in a generalized social crisis in the region. This crisis is manifested in poverty, malnutrition, delinquency, social inequalities, destruction of natural resources, environmental pollution, poor health, and overall a poor quality of life.

The Inadequacies of the Protected Areas

Settlement and subsequent deforestation gravely threaten most of the 17 areas of major biodiversity in Mesoamerica. The greatest tragedy is that natural resources of great value are destroyed to achieve very short-term gains for a small percentage of the population (see Vásquez-Morera, chapter 11). In the majority of the existing protected areas, high rainfall impedes the development of sustainable agroecosystems. The soils are of low native fertility and are quickly and easily degraded and eroded, leading to the rapid abandonment of many settled areas and the destruction of yet more forest resources on virgin lands.

The problem of settlement in protected areas is especially severe. Lack of legal registration of protected lands as national property, protection by the government, and demarcation of protected land boundaries all contributed to this phenomenon. A large part of the protected areas created by law exist only on paper, having no administrator, personnel, or installations. MacFarland and Morales (1994) estimate that 30 to 40 percent of all the protected areas in the region are protected only on paper.

Settlement is one major cause of deforestation generally and of invasion of protected lands in particular. In all of the Mesoamerican countries, rapid population growth, unjust systems of land tenure, and misdirected development policies promote agriculture at the expense of forest ecosystems. Accelerated rural population growth, combined with lack of land, concentration of land ownership in the hands of a few, and the degradation of farmlands provoke movement and settlement in forested areas, especially by young couples.

Pasos (1994) also cites speculation as a cause of deforestation in protected areas. Speculators who know where national parks or other reserves will be sited purchase and deforest with the intent of seeking indemnization later. The problem is especially severe in areas where protection is provided on paper but where there is no demarcation, regulation, or institutional presence. Ugalde and Godoy (1992) also mention drug traffic as a grave threat to protected areas, in addition to the illegal extraction of shrimp and fish and the threat of the importation of toxic wastes. Illegal extraction of mineral resources is also a problem in some protected areas.

In some countries, such as El Salvador and Nicaragua, some functionaries of the institutions charged with conservation believe that virtually the entire system of national parks and equivalent reserves are threatened (Ugalde and Godoy, 1992). Honduras provides a good example. According to the Environmental Profile for Honduras (SECPLAN, 1989),

> Various protected areas economically important for water production, the development of ecotourism, fisheries production and conservation of biodiversity, are suffering visible deterioration in face of the lack of energetic action by the government. Spontaneous settlement, timber exploitation, the expansion of coffee and cardamon production, mariculture with assistance from the government and international agencies, the construction of roads, pesticide abuse, the uncontrolled exploitation of wildlife, and the lack of effective land titulation programs continue to be problems not only around the protected areas but within many of them.

What is happening in the southern part of the Maya Biosphere Reserve in the Peten provides an illustrative example. This reserve is being penetrated by roads, along which farmers remove forest cover to plant crops and ranchers clear forest to create pastureland. Although the problem is apparently just now beginning, it will threaten the integrity of the reserve if it continues. Timber extraction, even when it is selective and does not result in total loss of forest cover, encourages deforestation by farmers. They take advantage of the road system to come into the area, principally from southern and eastern Guatemala. Immigration into the region is estimated at 250 persons per day (Heinzman and Reining, 1990, Whitacre et al., 1993).

Ugalde and Godoy (1992) also mention several other factors that threaten the integrity of the protected areas, including development projects, environmental disasters, and inadequate networking. With respect to development projects, tourism complexes or construction to facilitate mineral extraction near or around protected areas pose a threat. Where government agencies such as the park service have jurisdictional authority and the legal and economic means needed to take action in transitional zones around protected areas, they may be able to negotiate steps to reduce or limit negative impacts. However, where there is no institutional control over the transition zones, the only recourse is for national and international nongovernment organizations (NGOs) and local communities to pressure for appropriate government intervention. Environmental disasters may be natural, man-made, or both. Whatever the cause, ways to estimate damage to habitat and to put mechanisms for recuperation into place are needed. Lack of networking between local communities and the institution responsible for the protected area is a common problem. Lack of clarity regarding jurisdictional rights among local institutions causes duplication in management and diminishes authority.

MacFarland and Morales (1994) point to inadequate planning and design of protected areas as major constraints. Particularly important are the failure to include the principles of conservation biology and land use capability classification in establishing the boundaries of protected areas and the failure to establish corridors between protected zones. These authors also discuss the growing tendency for conservation NGOs to displace and/or replace government institutions in the administration of protected areas. In some cases, these organizations grow increasingly bureaucratic in nature, consuming ever more funds that should be devoted to environmental education and protection in maintaining the organization itself. In other cases, NGOs distance themselves from government agencies, ignoring official government policies. To these problems could be added the problem of stability. In Costa Rica, for example, there are more than 300 environmental NGOs, many of which survive only a few years. Often, organizations are created because of a particular opportunity for funding.

MacFarland and Morales (1994) also cite personnel problems in both government and nongovernment organizations. Public officials are often new to the job, have little experience and receive little training. The economic crisis of the last 10 to 15 years has led to low salaries, lack of incentives and promotion potential, poor working conditions, and lack of training opportunities for government employees. One result is that many of the best trained professionals now work for NGOs. Add the continual change in personnel where there is no civil service protection or the reverse, systems in which it is virtually impossible to remove professionals who do not perform adequately, and the difficulties of government management of protected areas become even clearer.

The development of protected areas lags behind habitat destruction throughout the region. More than 300 protected areas, including more than 10 million hectares (about 20 percent of the region's total area), have been proposed for Central America. However, in 1991 only 173 protected areas were either established or under management (Ugalde and Godoy, 1992). In reality, Costa Rica and Panama have the highest percentage of national territory under protection, 12 and 13 percent, respectively.

The protected areas are also insufficient in number, are small, are not representative of the total biodiversity of Mesoamerica, and are isolated and fragmented. The size of the protected areas is a serious limitation. Most are not large enough to ensure the viability of the populations of species that they contain, especially the large cats and raptor birds. The problem is demonstrated by SI-A-PAZ. It would be necessary to put more than half a million hectares under absolute protection to ensure the survival of the jaguar (*Panthera onca*), puma (*Felis concolor*), eagle (*Harpia harpyja*), tapir (*Tapirus bairdii*), and macaw (*Ara macao*) populations in the zone. The small size of protected areas is particularly evident in Central America's small nations. Seventy percent of the protected areas in Central America are 10,000 hectares or less. Of the 173 protected areas that existed in 1991, 122 accounted for a total of only 350,000 hectares (Ugalde and Godoy, 1992). National parks of a half million hectares or more are rare; most have 30,000 to 70,000 hectares.

Many zones of endemism and unique ecosystems are not well represented in the current system of protected areas (Ugalde and Godoy, 1992). To cite only a few examples, dry and semiarid ecosystems are poorly represented, as are humid highland montane forests, Nearctic vegetation, and coastal wetlands. Furthermore, the protected areas are fragmented and isolated. In all nations in the region, the protected areas are spread throughout the national territory, limiting the free movement of species that migrate daily or seasonally. Fragmentation threatens many species, especially large species and species with large territorial needs (see Carroll and Kane, chapter 8). Reduction of genetic variability in the population, increasing dominance by species tolerant of humans, and alteration of basic ecological processes all contribute to the problem.

Threats to Biodiversity Outside the System of Protected Areas

Besides the threat that deforestation in protected areas poses, biodiversity is threatened by the general process of deforestation in Mesoamerica. Settlement is a key culprit in the process. Sometimes, the government directly or indirectly promotes settlement in forested lands by the landless. In other cases, settlement is promoted simply by building roads into forested regions, sometimes by the government, sometimes by lumbermen. However it occurs, settlement usually results in occupation of lands that are not suitable for agriculture, which are planted primarily in subsistence crops like manioc, corn, beans, or plantains. In yet other cases, settlement and deforestation are promoted by land speculators who live in the capital cities. Once lands are deforested, they get them at low cost, often converting land use to extensive grazing systems for cattle production. The situation is aggravated where national policies permit the landless to occupy and acquire legal title to "unimproved" forested lands. Ranchers and farmers deforest their own lands to achieve a more intensive land use than forestry, such as the production of export crops.

Overexploitation of forest resources, principally for the harvest of fuelwood, is another important cause of deforestation. Firewood serves as the principal household energy source in 72 percent of homes in Central America (Camino, 1993). Mendez (1988) indicates, for example, that the principal cause of coniferous forest destruction in the Altiplano of Guatemala is firewood harvest.

Forest fires are another source of forest destruction. This is an especially serious problem on the Pacific side of Mesoamerica because of the prolonged dry season. Sometimes, fires are purposefully started to clear land for agriculture. In many other cases, fires set on pasturelands for weed management escape from control. Because forests have little perceived value, most farmers are unlikely to expend much effort in trying to put out these "escaped" fires.

Lack of clear title to the land is another force that provokes forest destruction. Small farmers and indigenous people often do not have clear title to the lands that they use. Frequently, these lands are the legal property of absentee landowners or of the government. Commonly, land must be under cultivation or in pasture to secure legal title. Often, forested lands cannot serve as collateral for loans, and very often forested lands have lower value than cleared lands. In summary, forested lands are often viewed as "belonging to no one" or "underutilized," creating a situation in which there are no incentives for protecting and managing forested lands for mid- to long-term gain.

Even when landowners have clear title to their land, low timber prices and the lack of incentives in the forestry sector lead to the elimination of forests. Traditionally, neither credit nor extension programs have promoted reforestation and management of primary and secondary forests. Also, low prices for standing wood impede landowners from developing strategies to obtain benefits over the long term, turning forests into a "mine" for short-term gain. The system of selling wood by stumpage is a poor one that encourages waste and corruption. Additionally, in many cases the timber companies contribute to forest deterioration because they do not develop sustainable management plans. They contribute to settlement by not safeguarding the integrity of the areas that are put under their supervision through concessions (Camino, 1993).

Sometimes, forest services also play a role in forest deterioration and destruction. Although every country in Mesoamerica has a forestry law and a forestry service, lack of adequate control is the norm and forest resources are lost. There are multiple reasons for this state of affairs: low salaries, few professional positions, lack of vehicles and operating expenses for travel, legal and bureaucratic roadblocks and excessive paperwork, fear in the face of the pressure that the interested parties can bring to bear in seeking permission to utilize forest resources, complacency in the face of deforestation and forest loss, a lack of a global vision of the services' mission, strong interest groups, corruption, and similar problems.

Mangrove destruction is another serious problem. Accelerated loss of mangroves is caused by conversion to agriculture and animal production; the establishment of population centers; tourism development on the coasts; and more recently, aquaculture. Mangrove loss is very disturbing because these ecosystems are some of the most productive in the world, capable of supporting very complex and rich trophic chains because of the continual flow of nutrients into the system.

Additionally, there are new causes for forest loss: war and structural adjustment. According to Camino (1993), while war sometimes reduces the rate of deforestation in some sites, the movement of refugees into other areas accelerates deforestation there. Camino also argues that structural adjustment and the need to increase exports has accelerated forest loss in recent years because they displace farmers who produce basic grains to hillside and agricultural frontiers.

In many areas, wildlife is disappearing rapidly because of illegal hunting. Specimens are hunted for food, commerce, hides, skins and feathers, medicine or rituals, zoos, biomedical research or pets, and sport (Redford and Robinson, 1991). Some severely impacted species include the guatusa (*Dasyprocta punctata*), peccaries (*Tayassu pecari* and *T. tajacu*), the tapir (*Tapirus bairdii*), the crested guan (*Penelope purpurascens*), the green iguana (*Iguana iguana*), the agouti (*Agouti paca*), and Virginia deer (*Odocoileus virginianus*). Habitat destruction and direct and indirect impacts on groups and individuals have reduced some populations severely, even driven some to extinction throughout the region. In Honduras, for example, 10 reptile species, 10 bird species, and 16 mammal species are in danger of extinction (SECPLAN, 1989). Elizondo (1994) says that 18 mammal species and 76 bird species are threatened with extinction in Costa Rica.

Specific examples abound. Aranda (1991) discusses the rapid reduction of populations of wildlife in Chiapas, Mexico, from illegal commerce in skins and hunting to obtain meat. Fragoso (1991) shows the impact of hunting on the tapir in Belize. A comparison of regions where hunting is allowed with regions where hunting is not permitted revealed two tapirs per kilometer of river in the former and eight in the latter. A similar study (Glanz, 1991) in Panama compared the number of diurnal mammals in unprotected areas with the number in Barro Colorado, which is protected. In Barro, 7.04 mammals per square kilometer were found, whereas only 0.49 mammals per square kilometer were discovered in the unprotected sites. Lageaux (1991) says that 100 percent of the eggs of the turtle *Lepidochelys olivacea* deposited on the Honduran Gulf of Fonseca on the Pacific Coast were collected and sold during the last 40 years.

Both migratory and resident birds are severely affected. Some of the most important vegetation types for migratory birds are humid and dry broadleaf forests, pine forests, and mangroves. Birds that reproduce in North America during the summer and overwinter in Mesoamerica are highly susceptible to effects from deforestation, forest fragmentation, and forest alteration. Nearly 250 bird species migrate to Mexico, Central America, and the Caribbean. Of these, at least 23 depend on primary forest for habitat, making them particularly vulnerable to alterations in tropical forests. Another 37 are also vulnerable, although they can inhabit moderately disturbed sites.

The neotropical psitacids are highly threatened by the combination of habitat destruction and alteration, pesticides, killing to protect crops, and illegal hunting and commerce. The trade in wild parrots has reached alarming proportions, for example. At least 30 percent of the 140 parrot species that exist in the hemisphere are threatened with extinction. The international pet trade delivers parrots from developing to industrialized nations in massive numbers. In Honduras, from 1982 to 1985, a minimum of 62,170 psitacids was legally hunted and sold, a legal quota with no basis in scientific information (Thomsen and Brautigam, 1991). Illegal trade is also a problem, such as that between Mexico and the United States.

Overexploitation of aquatic species is also rampant. The overexploitation of marine species for local consumption and/or exportation has been well documented. Shrimp (*Penaeus* spp.), octopus (*Octopus* spp.), dolphins (*Tursiops* spp. and *Delphinus* spp.), marine turtles, and some fish species are rapidly disappearing from the seas that border on Mesoamerica.

In summary, despite the development of a large number of protected areas in Mesoamerica, the loss of biodiversity remains a critical problem. The protected areas are often small, isolated, and fragmented. Those that do exist are subject to deforestation, forest alteration, and other direct impacts that reduce the quality of the biological and physical resources in the reserves. Problems with reserve design, management, and legal status abound. Phenomena outside the reserves, such as settlement, forest resource extraction, natural disasters, and illegal hunting, threaten the integrity of the reserves directly and indirectly.

Working toward Preserving the Biodiversity of Mesoamerica

If current trends in the destruction of forests and other ecosystems continue until 2005 or 2010, only fragments of most of the large wild areas that exist today will remain. Even these could disappear by the year 2015.

Nonetheless, despite the enormous environmental destruction that has occurred in the region, as much as 75 percent of the original biological wealth of Mesoamerica may remain. Although you will encounter ample disturbed areas, it is still possible to fly over the Selva Maya and look down on an immense green expanse that extends uninterrupted for more than 2 million hectares, the largest block of pristine forest in Mesoamerica. This experience can still be enjoyed in other areas as well, over the Río Plátano-BOSAWAS, SI-A-PAZ,[1] La Amistad, and the Daríen, for example. Most of Belize's natural ecosystems remain intact, and forest still covers 40 percent of Panama's national territory.

The loss of this extraordinary natural wealth in Mesoamerica would be an ecological disaster, one that would compromise the social and economic well-being of current and future generations of Mesoamericans. We should and must take urgent, forceful action to avoid this disaster. The decade 1995–2005 will be our last opportunity to take decisive action—to consolidate the protected areas that have been declared; to maintain other, unprotected ecosystems in a natural state; to develop biological corridors between nearby protected areas; and to consolidate and develop the buffer zones around protected areas on a sustainable basis. To these ends, we conservationists who work in Central America are using all of the technical, scientific, legal, and administrative tools at hand. We are seeking help from the appropriate authorities, from the population as a whole, and from the private sector. International conservation organizations that work in the region are giving us meaningful technical and financial support.

We are also taking—and should take—all actions that can have an effect, either direct or indirect, on conservation and restoration of biological diversity. Some of these are (1) to strengthen conservation NGOs, (2) to consolidate protected areas, (3) to design plans for new protected areas and for biological corridors, (4) to train both government and nongovernment personnel, (5) to encourage wildlife management, (6) to create ex situ conservation strategies, (7) to develop the forestry sector, (8) to develop sustainable activities in buffer zones, (9) to create private nature reserves, (10) to encourage ecotourism, (11) to conserve and manage watersheds, (12) to improve environmental legislation, (13) to increase research in existing ecosystems, (14) to conduct rapid ecological assessments, (15) to find ways to add economic value to conservation efforts, (16) to promote environmental education, (17) to involve more sectors of society in conservation, and (18) to encourage family planning. We are optimistic. We are convinced that the people of the region will have the will and the power to avoid a final tragedy for the precious natural inheritance of Mesoamerica.

Notes

This chapter refers to the present problems of the protected areas that preserve a high percentage of the biodiversity in Mesoamerica. A book about possible solutions to many of these problems has been published in Spanish with a summary in English: *Biodiversidad y Desarrollo en Mesoamérica* (San José, Costa Rica: Proyecto Paseo Pantera, M. A. Boza, 1994, 240 pp.) This publication can be acquired by writing to Dr. Archie Carr, Regional Coordinator, Wildlife Conservation Society, 4424 NW 13th Street, Suite A-2, Gainesville, FL 32609.

1. BOSAWAS is formed mixing the names of three places in the area: the Bocay River, the Saslaya mountain, and the Kininuwas mountain. SI-A-PAZ = Sistema de Areas Protegidas para la PAZ (System of Protected Areas for the Peace).

7

Protecting Natural Resources in a Developing Nation

The Case of Costa Rica

Gilbert Vargas Ulate

Any nation's development should be based on the rational use of natural resources, the objective being to use them optimally. The definition of areas that will be used for the development, protection, and exploitation of natural resources is necessary. Furthermore, it should be done in a way that will provide both immediate and long-term benefits to the population. To make this a reality, both the ecological dimension (the natural environment) and the cultural dimension (the historical, social, economic, and technological environment) should be considered.

Nonetheless, the kind of planning that permits the renovation and sustainable use of natural resources has been lacking in developing nations. To understand why natural resource use is poorly planned and executed, our starting point should be the Latin American reality, especially an understanding of Latin America's socioeconomic and environmental problems. In a developing nation like Costa Rica, any attempt to analyze natural resource protection policies must take as its starting point the characteristics of underdevelopment and their historical evolution, which include land tenure, marginalization, social impoverishment, farmers without land or economic assistance, low productivity of traditional agriculture, high population, and illiteracy. An understanding of the context in which resource use planning occurs is the key to developing sound management in the future, not the simple imposition of development models that do not fit Latin American conditions.

In this chapter I use Costa Rica as an example of the historical development of resource use planning in Central America. The first part of the chapter identifies and describes six historical periods in which different policies toward resource use were put into place. The second part of the chapter illustrates the demographic, social, and cultural factors that influenced policy formation during these periods. Finally, I conclude with some perspectives on the future direction of resource use policies.

Historical Periods in the Protection of Natural Resources

Legal protection of natural resources in Costa Rica began in the first years of independence when various laws and decrees were passed to protect the natural environment. Beginning in 1821, the protection of forest resources can be divided into six periods.

Preliminary Conservation Legislation, 1821–1850

From 1821 to 1850, Costa Rica's population was concentrated in the Central Valley, where most county and municipal seats were found (Sanders, 1962). Outside the Central Valley, villages, which were called villas, were formed, for example, Canas, Quesada, Liberia, Esparza, Puntarenas, and Limón. These early urban aggregations led the federal government to pass a series of laws and decrees aimed at maintaining and protecting the natural resources of the communities. The municipalities were responsible for enforcement.

The first protectionist measures appeared in 1928, Decree No. 161 of the Constitutional Assembly of the Free State of Costa Rica. This decree called for the municipalities to "conserve and replant the natural vegetation and the commons" (Porras and Villarreal, 1986). The "commons" referred to communal lands of the municipality used for grazing or farmland. Decrees No. 21 in 1833 and No. 32 in 1941, which amplified this legislation, urged the municipalities to replant trees and care for the quality of water in rivers and springs. They encouraged the owners of pastures to plant trees that would provide lumber or firewood in the fence lines. Decree No. 35 was also passed in 1841, in a chapter of federal law dealing with agriculture. This decree mandated that 10 manzanas (1 manzana equals 0.7 hectares) of all common lands should be reserved for trees that would provide wood for construction. It also stated that anyone who wanted to cut a tree should secure permission from the local political authorities, should replant a tree, and should cut only full-grown trees.

The Expansion of Coffee and Abandonment of Conservation, 1850–1900

In the second half of the nineteenth century, very few protectionist measures were undertaken. The expansion of coffee farms occurred in this period, well described by Cardoso (1973), Facio (1972), Hall (1976), and Vega (1975). Conservation policies and protection were not the objectives of the upper class, who sought to concentrate their landholdings. Capital was centralized and concentrated through three monopolies: credit, product processing, and marketing of the product. President Bernardo Soto, demonstrating very advanced thinking for his time, declared a zone of 2 kilometers of the slopes of the crater of Barva Volcano as inalienable property of the state. In 1889, he made privatization of land in this zone invalid. However, these were the only measures taken to protect the natural environment during the second half of this century.

Proliferation of Ineffective Legislation, 1900–1940

Concern about the environment revived after 1903 when the legislature passed Decree No. 36. This decree asked the executive power to appoint a commission, consisting of two lawyers and two agricultural scientists, to develop as soon as possible a forestry code for submission to Congress in the following legislative session. This project, however, did not come to fruition (Porras, 1980). Later presidents pursued similar goals. In 1909 President Cleto Gonzalez declared the Fire Control Law, and in 1913 Ricardo Jiménez declared the zone for 2 kilometers around the crater and lake of Poás Volcano to be inalienable state property. Forest guards were established in 1906 to manage and control the exploitation of forested lands belonging to the state, and rules concerning the use of state forests were developed in 1930. The forest guards were named by the secretary of housing and worked under the provincial governors. They guarded the springs and rivers that provided water for potable water systems, industry, and electrical plants, and they watched over tree cutting on private and municipal lands and near springs. The state security's functions, according to Law No. 4 of 1923, were to prosecute fraudulent exploitation of national forests and control forest exploitation to comply with the stipulations of the state.

The word *reforestation* appears for the first time in Costa Rican legislation in 1923 as well. The justification for the Reforestation Law of that year argues that measures and means to ensure compliance with the forestry laws were urgently needed to maintain and protect springs and forests. This law prohibited individuals and especially municipalities from transferring property rights; using the property as mortgage security; or exploiting a border 200 meters wide near rivers, springs, waterholes, or areas with steep slopes. Municipalities were also instructed to consult with the Department of Agriculture about the adequate use of land and to request permission both to use and to carry out reforestation on their lands.

The first outcries against environmental degradation and cries for conservation and the protection of natural resources also occurred at the beginning of the twentieth century. Great naturalists such as Anastasio Alfaro, Carlos Werckle, Philip and Amelia Calvert, Henrie Pittier, Vicente Lachner, and José Fidel Tristán, among others, voiced their concern over the status of Costa Rica's environment and the need for its protection. The words of naturalist Alberto Brenes (1953) are an example:

> The forests have been robbed of their best trees, because here [referring to Cañas, Guanacaste] as in all parts of the country trees are cut without foresight, without need, without any criteria, and without control. It is useless to repeat what has been said so many times about the vandalistic and needless exploitation of the forests in general, that it will bring the ruin of our agriculture, the only source of national wealth. (p. 134)

In 1935, President Ricardo Jiménez promulgated the National Forest Use Law. Under this law, it was the state's function to specify which sectors of the state forests could be exploited. The right to use common lands could be obtained only through a public auction conducted by the general inspector of the Secretariat of Housing. Several requirements were put into place: (1) the exploited area could not exceed 500 hectares, (2) use fees would be based on the type and quantity of timber extracted, (3) only fully grown trees could be cut, (4) the user should plant two trees of the same species for each tree that was cut, and (5) any on-site timber sale was prohibited. The same law called into existence a commission made up of the chief of forestry, the director of the Department of Agriculture, and a technical advisor. This commission had the responsibility for identifying areas within the national forests that should be conserved in a pristine state because of their flora, fauna, and natural beauty. Using this mandate, the commission declared the island of Capo a conservation area in 1935, based on the presence there of the largest indigenous cementary in Costa Rica.

President León Cortés Castro passed further legislation, the General Law for Public Land Use, in 1939 (Articles 6, 7, 8, and 9). This law dictated that the forest could not be exploited: (1) in a band 500 meters wide along navigable rivers, (2) in a 2-kilometer-wide zone along each side of the Pan-American Highway, and (3) in an area 500 meters in diameter around the headwaters of rivers. This law also prohibited private ownership of certain lands in the watershed of the Banano River to protect the city of Limón's water supply; it also set aside zones 2 kilometers wide around Poás Volcano and its lake, Barva Volcano, Irazú Volcano, and the Zurquí highlands. An additional 2-kilometer-wide zone along the frontiers of Nicaragua and Panama were put under protection.

During this period, legislation was specific and repetitive. Execution of the laws was ineffective and inefficient because a collection of institutions was responsible for enforcement. Despite the executive power's attempts at conservation, the principal national objective continued to be unplanned agricultural colonization of forested state lands. Colonization occurred in the San Carlos, Sarapiquí, Guatuzo, and Parrita lowlands; the Turrubares highlands; and the Nicoya Peninsula (Sanders, 1962). Such laws as the Reforestation Law and the Public Lands Use Law were often the mechanisms used to expand the agricultural frontier legally.

Influence of International Legislation and the Creation of Protected Areas, 1940–1969

This period is characterized by the marked influence of international legislation and the beginning of a consciousness of the need to create protected areas by public officials. Before 1940, the idea of a natural protected area as a management unit did not exist in Costa Rican legislation. With Costa Rica's participation in the Washington Accord for the Protection of Flora, Fauna and Scenic Panoramas of the Americas (1942), the concept of a natural protected area appeared in the country. The Legislative Assembly ratified the accord 16 years later with Law No. 3763.

This accord made two important contributions to the protection of natural resources in Costa Rica. First, it defined four categories of management that provided the basis for establishing the first protected areas: national parks, national reserves, natural monuments, and conservation areas. Second, for the first time, elected officials created management categories and tried, although rudimentarily, to control and protect them. The protected areas established were the Poás Volcano National Park and Irazú National Park, created by José Figueres in 1955. In 1961, in Law No. 2825, the government of Mario Echandi declared a zone 2 kilometers wide on each side of the Pan-American Highway, in the Talamanca Mountains, to be a national park to protect the oak forests there. The first biological reserve in the country, Cabo Blanco, was also established in 1961, and the first forest reserve, Río Macho, was established two years later.

Natural protected areas were created in this period, but the institutions that could effectively administer them were not. In 1961, the Land and Colonization Institute (Instituto de Tierras y Colonización, ITCO) was given the dual functions of protecting forest resources and carrying out agricultural colonization. However, ITCO never complied with its role as a protector of forest resources. In fact, it used forested lands to develop new settlements for farmers. Typically these settlements became unproductive after only three or four years because of poor soils, topography, and climate. These policies accelerated the process of transforming much of the country into pastureland.

Uncoordinated and Heterogeneous Legislation, 1969–1986

The new Forestry Law, passed in 1969 under the government of José Joaquín Trejos, began the legal process of administrative organization of forest resource management. This law established that the forest heritage owned by the state is the property of the executive power, under the Ministry of Agriculture. These lands include national parks, biological reserves, forest reserves, protected areas, and forest nurseries. In 1970, José Figueres Ferrer established the Wildlife Conservation Law, created the Wildlife Department as an administrative unit, and gave the wildlife refuges status as protected areas. In 1977, Daniel Oduber declared the new Reforestation Law, in which it was explicitly recognized that the reduction in forested areas was caused by the expansion of cattle ranching. Ranching was cited as a factor in provoking erosion, destroying watersheds, and creating acute problems of environmental pollution.

From 1969 to 1986, protection of forest resources fell under three laws, the Forestry Law and its amendments, the Wildlife Conservation Law, and the Reforestation Law. The Ministry of Agriculture had responsibility for administering the protected areas through the National Forestry Service, the Wildlife Department, and the National Park Service. Virtually all of the protected areas that exist in Costa Rica today came into being or were consolidated into their current form during this period.

From a quantitative perspective, Costa Rica's creation of protected areas can be considered a success. Nonetheless, there are serious constraints to protecting these natural areas, and the quality of protection is inadequate. There are not enough park guards or trained personnel to carry out the functions of conservation and protection. Basic infrastructure is also inadequate. The size of the protected areas is another serious problem.

Ecological and geographic variables were ignored when they were established. Straight lines cut across the headwaters of waterways and produce problems in managing their boundaries. Furthermore, many protected areas do not properly belong in the protection category to which they have been assigned.

Two important events in this period produced criticism, reflection, and an increased conscience regarding the use of natural resources in Costa Rica: the First National Congress for the Conservation of Natural Resources in 1974 and the Costa Rica 2000 Symposium in 1977. The symposium recommended the creation of a National Institution for Natural Resources (INDERENA) (OFIPLAN, 1977). The recommendation was forwarded to the Legislative Assembly in 1978. However, it was not discussed, nor was any plan of action developed under the administrations of Rodrigo Carazo (OFIPLAN, 1979) or Luis Alberto Monge (MIDEPLAN, 1982).

Forces for Integration, 1986–Present

This period began with the creation of the Ministry of Natural Resources, Energy, and Mines (MIRENEM) under President Oscar Arias Sanchez. The functions of protecting, using, conserving, and developing forest resources and protecting wildlife that the Ministry of Agriculture previously held passed to the new ministry. Under MIRENEM, the executive power became responsible for the forests and the fauna of the nation. The National Park Service and National Forestry Service administered the forests. The first has responsibility for the national parks and biological reserves, and the latter the protected areas and forest reserves. The Department of Wildlife has sole responsibility for fauna, administers the Wildlife Reserves, and controls hunting and fishing.

In 1992, protected areas were further consolidated into a National System of Conservation Areas (SNAC). The nine regional areas are (1) Central Volcanic Cordillera Mountain Range, (2) La Amistad, (3) Arenal, (4) Guanacaste, (5) Osa, (6) Bajo Tempisque, (7) Central Pacific, (8) Tortuguero Plains, and (9) Coco Island. The areas fall under the jurisdiction of the National Park Service. The objective in their formation was to ensure integrated development of the protected areas, taking into account both the surroundings and the area's inhabitants.

A strategy for conservation and sustainable development in Costa Rica was expanded in 1989 (Quesada, 1990). This has been a long and continuous process of defining objectives and determining the policies necessary to reach a level of economic and social development that is compatible with the natural heritage and will be a lasting development that will not destroy the ecological potential of the nation's resources. However, to date, this exercise has remained at a theoretical level and has not been applied.

Environmental legislation in Costa Rica is represented by many varied and disperse laws and decrees that refer to natural resources, environmental protection, and the struggle to prevent pollution. Similarly, many offices, services, and institutions have been created that are concerned with environmental protection and resource use. Porras (1981) and Ballar González (1985) agree that most of these laws, decrees, and institutions have encountered problems that still face the state. They add that the only thing gained from this legal and administrative proliferation has been greater duplication of functions and activities.

Current Management Categories

Currently, five management categories exist under Costa Rican law: national parks, forest reserves, biological reserves, natural wildlife refuges, and protected areas. Together they make up 19 percent of the national territory (figure 7.1). Of these areas, timber extraction and agricultural use, with previous permission and development of an environmental impact statement, are permitted only in the protected areas and forest reserves.

The national parks, biological reserves, and wildlife refuges are devoted to protection, recreation, and research, the last only when they do not conflict with conservation objectives. The national parks comprise regions or areas of historical significance, natural scenic beauty, or flora or fauna of national or international importance. Devoted to recreation, public education, tourism, and scientific research, they account for 413,906 hectares, or 8 percent of the national territory (figure 7.2).

Forest reserves include forested areas that have the potential for timber production. They are protected but subject to rational use, based on technical evaluations of how much timber can be extracted annually. Approximately 7 percent of the country's area is in forest reserves, a total of 337,347 hectares.

Biological reserves may include freshwater and saltwater areas. They include regions where biological communities include one or more species of animals or plants, terrestrial or marine, that are important to preserve in their natural state because of their characteristics and/or their risk of extinction. These areas include less than 1 percent of the national territory, or 31,357 hectares.

The national wildlife refuges are devoted to the protection of wildlife and to research about the flora and fauna, especially those in danger of extinction. There are 111,168 hectares, or about 2 percent, of the national territory in wildlife refuges.

Protected areas include forests or forested lands set aside to protect the soil, maintain and regulate hydrologic flows, or serve as climatic or environmental regulators. These protected areas border springs, rivers, canyons, lakes, or lagoons for a radius of 5 to 60 meters.

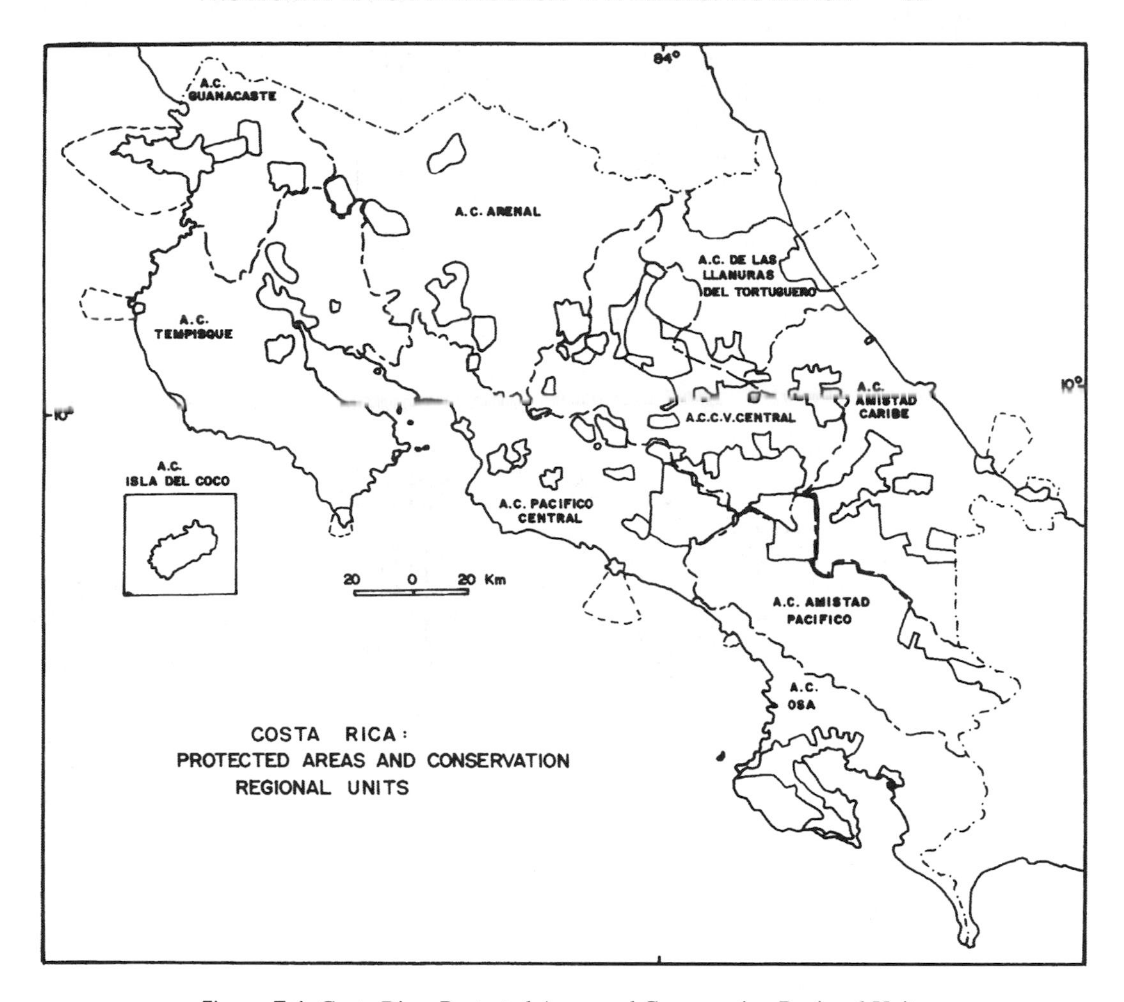

Figure 7.1 Costa Rica: Protected Areas and Conservation Regional Units

There are 19 protected zones in the country, including about 1.5 percent of the national territory, a total of 79,030 hectares.

An Overview of the Problems Costa Rica Faces Today

The forested area in Costa Rica in 1800 was estimated at 46,501 square kilometers, or about 91 percent of the national territory (figure 7.3). The enormous transformation of this area can be divided into two historical periods: the first from 1800 to 1950, and the second from 1950 to the present.

Unplanned Agricultural Expansion on the Periphery of the Central Valley, 1800–1950

More than 27 percent of Costa Rica's forests disappeared in the 150 years between 1800 and 1950, an average of 91.9 square kilometers (9,198 hectares) per year. This transformation was a direct result of government policies. The state promoted agricultural settlement in areas outside the Central Valley, in regions such as San Carlos, Valle del General, Turrubares, Nicoya, Sarapiquí, Tilarán, and Limón (Sanders, 1962).

Between 1850 and 1880, coffee cultivation increased greatly in importance on both small and large farms (Cardoso, 1973; Hall, 1976). Nonetheless, coffee production was insufficient to absorb the growing labor pool of small farmers, or their offspring, who entered the labor force during this period. Youths, in particular, responded to economic pressure by migrating to areas open to settlement outside the Central Valley. Climax forests were destroyed to clear land for settlement.

Land ownership in the Central Valley also became more concentrated in this period (G. Vargas, 1986), which produced pressure on the government to find new land for farmers. This pressure was alleviated slightly by

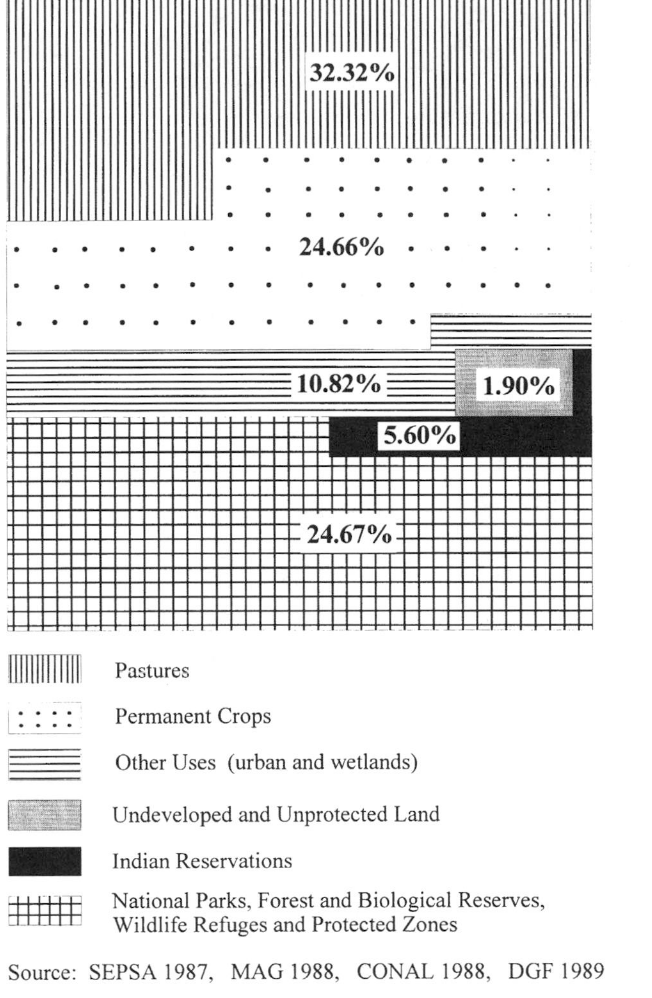

Figure 7.2 Costa Rica: Relation between Land Use and Protected Areas

converting the landless into settlers and by expanding the agricultural frontier in outlying regions of the nation. Nonetheless, economic problems were persistent and the pressure for land remained, resulting in a series of laws: the Head of Family Law (1903), the General Public Lands Law (1930), the Public Information Law (1941), the Abandoned Lands Act (1942), and many others. These laws and decrees held in common the strategy of giving the landless lands outside the Central Valley—to use the national public lands as a means to reduce economic and agrarian problems in the country. Agricultural settlement outside the Central Valley in this period was not "spontaneous." Quite the contrary. The landless were pushed out of the Central Valley by social and economic pressure due to capitalist agrarian development and state policy (G. Vargas, 1986).

Forest clearing also occurred under the auspices of large landowners. Very large cattle ranches were established in Guanacaste, developed by clearing tropical dry forests (Sequeira, 1985). In 1882, one of the most massive land use transformations occurred. The government of Tomas Guardia gave a contract to Soto-Keith, permitting the North American Minor C. Keith to construct a railway from San José to the Atlantic. Under this contract, Keith received property rights to 6.3 percent of the national territory (324,000 hectares). Large-scale banana production began on these lowland Caribbean lands, destroying many hectares of very wet tropical rainforest (G. Vargas, 1979). Banana production persisted in the region until 1925, when production declined because of bacterial and fungal diseases (see Soto, chapter 18). Later, in 1942, under the Cortes-

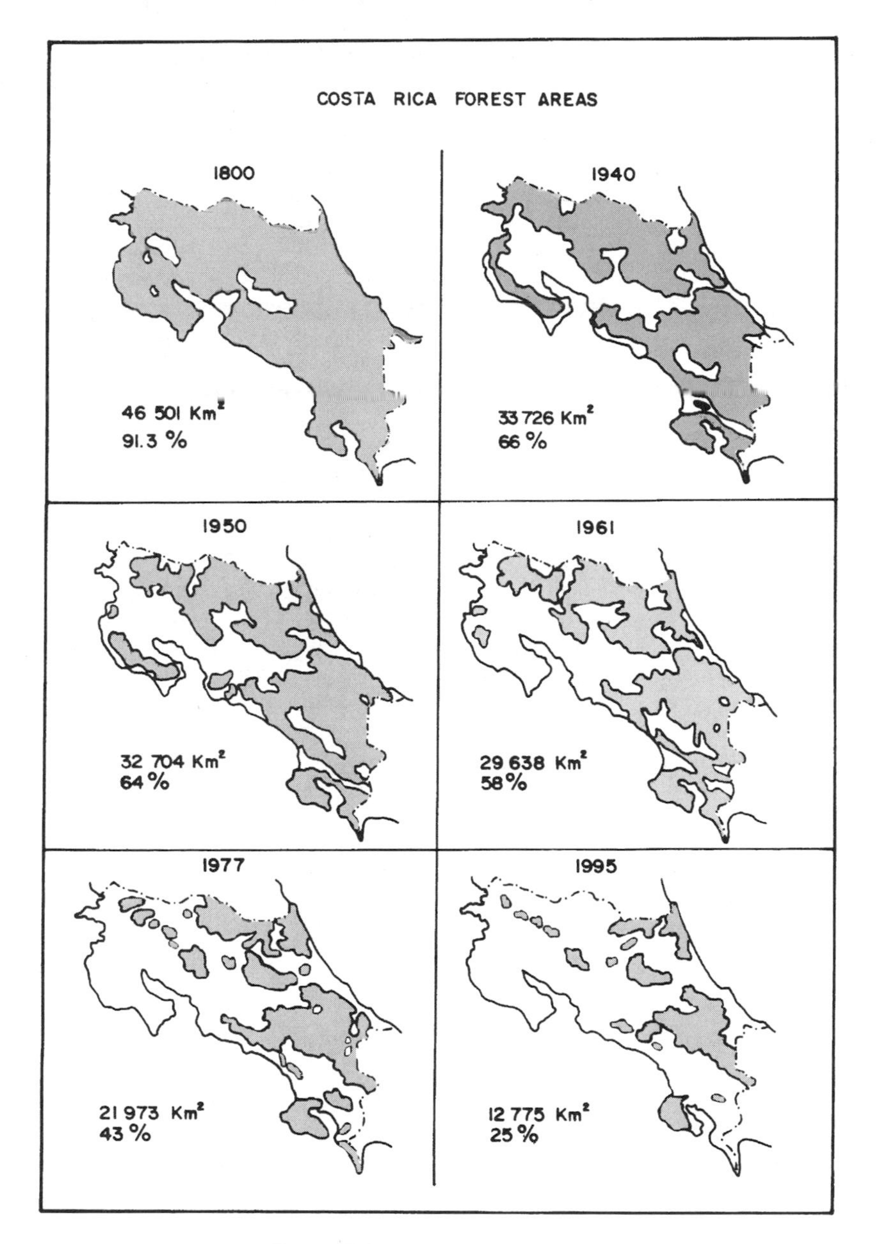

Figure 7.3 Costa Rica Forest Areas

Chittenden contract, the United Fruit Company established large operations in the Parrita Valley and in the region around the Golfo Dulce in the southern Pacific. These plantations resulted in the loss of another 3.2 percent of the national territory, or 165,000 hectares (G. Vargas, 1979).

Mining during the nineteenth and beginning of the twentieth century also had serious environmental impacts in Costa Rica. Two periods were especially important, the first from 1821 to 1840 in the highlands of Aguacate and the second from 1890 to 1920 in Abangares (Granados, 1983). State policies, such as low taxes, were openly protective of the mining interests. Some companies, such as Abangares Mining Syndicated Limited, received concessions of 6,000 hectares. Controls over gold extraction were virtually nonexistent (Araya, 1973; Granados, 1983).

Large quantities of hardwoods were needed to shore up the tunnels used in gold mining and to produce energy, especially in the dry season. Further environmental damage resulted from releasing the waters used in extraction and hydroelectric production. Granados

(1983), using municipal records, shows that a common set of problems occurred because of mining operations. Sedimentation and contamination by cyanide in rivers were major problems. For example, the municipality of Abangares declared the local rivers unusable in 1922 because of excessive chemical contamination. The effects of this activity have never been completely documented, but the damages from this period are still observable today.

Closing the Agricultural Frontier and Privatization of Public Lands, 1950–Present

In the period 1950–1990, only 37 years, 39 percent of the forested area in Costa Rica was cleared (19,929 square kilometers), an average of 539 square kilometers (53,862 hectares) per year. The causes for this accelerated deforestation are not to be found solely in farmers' activities or the growing population. Deforestation is a critical problem that has its roots in multiple historical, social, demographic, economic, and political factors.

Starting in 1950, social groups with political influence and high social status promoted the expansion of ranching in Costa Rica. Government expenditures to assist ranchers increased 36 percent during this period. During the same time, government expenditures for traditional agricultural production declined 34 percent (G. Vargas, 1981). The average farm size increased between 1950 and 1970 in older areas of agricultural settlement. At the same time, the number of small farms declined, displacing small farmers, renters, and laborers to marginal production regions like the steep slopes on the central mountain range (G. Vargas, 1981). Between 1950 and 1984 the area in pasture increased by 64 percent, from 937,644 hectares in 1950 to 1,651,561 hectares in 1984. Pasture expansion was closely related to deforestation. Forest areas declined 61 percent, from 3,270,400 hectares to only 1,277,500 hectares in the same period.

Another factor that favored deforestation was the deficient agrarian policies carried out by ITCO, reorganized and renamed the Institute for Agrarian Development (IDA) in 1982. This state institution did not prove capable of meeting its principal objectives of distributing rural lands, partitioning wealth equally, and developing Costa Rica's agricultural sector. Thus, ITCO and IDA became an escape valve, parceling out forested lands to the landless without providing the technical, economic, and marketing support needed by the settlers. In many cases, settlers abandoned the lands handed out by IDA after three or four years because they become unproductive and severely eroded. Both ITCO and IDA have avoided true redistribution of land and the accompanying pressure from large landowners, preferring instead to carve settlements out of such protected areas as Palo Verde National Park, Corcovado National Park, and Carara Biological Reserve.

Annual deforestation rates in Costa Rica were estimated by OFIPLAN (1979) at 60,000 hectares per year from settlement and commercial logging. Commercial logging is regulated under the Forestry Law, administered by the Forestry Service. This institution permits commercial exploitation and is responsible for supervision and for conducting environmental impact studies. Nonetheless, the service lacks both operating funds and personnel to meet its regulatory obligations. Despite the national and international outcry over deforestation, the Forestry Service increased the number of logging permits each year from 1985 to 1989 (table 7.1). Legal deforestation amounted to 24,381 hectares per year during this period. The service estimated that illegal logging took an even greater toll during the same period, perhaps 30,000 hectares per year. Taken together, legal and illegal logging accounts for OFIPLAN's estimate of 60,000 hectares of deforestation annually.

Perspectives for the Future

Of the 25 percent of forested land area that remains in Costa Rica (figure 7.3), 24.65 percent lies in protected areas. The agricultural frontier has been closed since 1960. The only way to reduce the pressure to invade and deforest protected areas is to put into practice a policy, based on democratization, that ensures true property redistribution to marginal farmers combined

Table 7.1 Area and Amount of Permits Released by Dirección General Forestal, 1985–1989, in Costa Rica

	1985	1986	1987	1988	1989
Number of permits	751	802	999	1,023	1,101
Area (ha)	27,222.7	28,620	22,375	33,345	34,720

Source: Dirección General Forestal, 1985–1989; Departamento de Permisos, San José.

with needed technical and economic assistance. Although this appears to be a serious, even dangerous, policy for the nation, failure to put such a policy in place will eventually lead to the destruction of the protected areas.

The concept of sustainable development adopted by MIRENEM in recent years (Quesada, 1990) proposes development based on the optimum use of the natural environment. This idea will be workable only if the rural population can satisfy its needs for food, land, credit, and housing. Involving the local community in conservation efforts is imperative. Only then can development integrate the rural population through a productive process that generates a higher quality of life for those who must protect the resources.

8

Landscape Ecology of Transformed Neotropical Environments

C. Ronald Carroll
Deborah Kane

The American tropics contain slightly more than half of the world's tropical moist forests (Somner, 1976), but these forests are being rapidly converted to other land uses. Between 1980 and 1990, 32 countries in Latin America lost approximately 830,000 square kilometers of forest area for an annual deforestation rate of 0.9 percent (FAO, 1990b). Central America is losing approximately 3,400 square kilometers of lowland and montane forest each year (Nations and Komer, 1983). Much of the early deforestation that followed European contact occurred during the 19th century with the expansion of rubber plantations in Brazil and the spread of cattle ranching throughout the American tropics. In the latter part of the century, additional forested lands were converted to cocoa, coffee, and banana plantations.

Deforestation rates throughout the American tropics have accelerated in the last four to five decades through increased emphasis on large-scale export agriculture and beef production. Nearly two-thirds of all forest clearing in Central America has occurred since 1950 (Parson, 1976). Approximately one-third of the cleared land is used for beef production on permanent pasture (Leonard, 1987). By the 1960s, virtually all tropical dry forest land had been converted, largely for cattle and cotton production. Only a few protected areas of tropical dry forest currently exist, almost all of them in northwestern Costa Rica and southwestern Mexico. On unprotected land, small patches of degraded forest are all that remain of the tropical dry forest that once extended along much of the Pacific coastal lowlands from southern Mexico through northern Costa Rica and from central Ecuador southward. Thus, a dramatic and ecologically catastrophic change in the neotropics was largely brought about in the past 50 years or so through the conversion of land, from supporting natural forest ecosystems to supporting extensive and intensive economic extraction.

In this chapter, we interpret the major ecological consequences of this massive transformation of the neotropical environment from the perspective of landscape ecology. In particular, we discuss the ecological consequences that result from the historic conversion of extensive and continuous forest ecosystems where habitat boundaries are blurred to the contemporary neotropical mosaic landscape in which discrete, small forest patches are scattered across a matrix of agricultural lands. We briefly examine two important contemporary uses of land in the neotropics to illustrate how the principles of landscape ecology can assist this interpretation. We first examine lands used primarily for economic gain: the effects of beef production, riparian development, and shrimp mariculture in mangrove ecosystems. Next, we examine buffer management zones around protected areas as an example in which ecological principles and socioeconomic needs are jointly considered.

Principles and Application of Landscape Ecology in Mosaic Environments

Harrison and Fahrig (1995) summarize the main components of landscape spatial patterns in mosaic environments as the

- Amount of habitat in the landscape
- Mean size of habitat patches
- Mean interpatch distance
- Variance in patch sizes
- Variance in interpatch distances
- Landscape connectivity

These authors go on to offer the following generalizations about landscape pattern and ecological processes:

- The survival of populations is related to the amount of habitat.
- For the same total amount of habitat, increasing average patch size increases the probability of population survival, and this positive effect of increasing patch size outweighs the negative effect of increasing interpatch distance that results from having fewer but larger patches.
- Increasing interpatch variance in patch size increases the probability of regional survival.
- Landscape spatial pattern is relatively unimportant when habitat patches are short-lived or when disturbance rates are high.
- When habitat is ephemeral, regional population survival increases with increasing patch size.

We would make two additions to this list. First, interpatch movement is inhibited when the matrix environment is very different from the habitat patch; for example, a matrix of pastureland surrounding a patch of forest inhibits movement between forest patches more than if the matrix consisted of tree plantations or second growth. Thus, the nature of the matrix environment and the interpatch distance jointly influence the connectivity of habitat patches. Second, the quality of the patch in terms of its influence on survival and reproduction of the residents influences both the size of the aggregate population that is found in all patches and, probably, the likelihood that individuals will either settle in suitable patches or attempt to leave patches of low habitat quality. For our purposes, habitat is the spatial area that contains the resources that a population of a particular species requires for survival and reproduction. High-quality habitats contain sufficient resources so that reproduction generally exceeds mortality. In this sense, habitat patches of high quality are referred to as "source" habitat patches, which generate excess births over deaths and therefore contribute to the pool of potential migrants, whereas low-quality patches are "sinks" in which deaths exceed births and populations are maintained through migrants from source habitats. In mosaic environments, such as characterizes most of the neotropics, population dynamics is strongly influenced by the spatial relationship between source and sink habitat patches (Pulliam and Dunning, 1994).

A habitat patch refers to the circumscribed habitat in order to distinguish habitat area from nonhabitat area; the latter is often referred to as the matrix within which habitat patches are embedded. Just as habitat may vary in quality for particular species, the matrix will vary in terms of ecological costs (such as probability of mortality, energetic demands of a hostile environment, etc.) that are imposed on individuals that are attempting to move between isolated habitat patches. Generally, when the matrix environment is very different than the environment of habitat patches, such as with forest patches surrounded by a matrix of pasture, the matrix is a formidable barrier to migrants between habitat patches. As the matrix becomes more similar to the habitat patches (such as old forest patches in a matrix of second-growth or mixed agroforestry land), the matrix may not inhibit migration for at least some forest-dwelling species. As the rural landscape is converted to agricultural uses, natural patches of habitat are reduced and isolated and their internal quality degraded. Agricultural lands become the matrix in which these habitat patches are embedded. In contrast to the complex structure of natural forest, agricultural lands are open and relatively simple in structure. Thus, agricultural land, as matrix, is a migration barrier to forest-dwelling animals, thereby effectively increasing the isolation of the remaining forest habitats.

The boundary zone between a habitat patch and the surrounding matrix environment is termed an *ecotone*. The ecotone between matrix and habitat patches may become sites of increased mortality on forest-dwelling species from generalist predators (Andrén, 1995; Burkey, 1993), which frequently hunt in the ecotones between forest patches and agricultural land. These heavily hunted ecotones become death zones for vulnerable species such as ground-nesting birds. As the average size of forest patches decrease, the killing zone represented by the ecotone becomes an increasing fraction of the forest habitat (Gibbs, 1991). Narrow, sharply demarcated ecotones between habitat patches and a very dissimilar matrix environment may have lower generalist predator pressure than in broad ecotones that separate habitat patches from a more similar matrix environment. For example, Gibbs (1991) found more predator pressure from generalists in forest patches that were surrounded by tree plantations than in those that were surrounded by the very dissimilar pasture matrix.

Some animals migrate seasonally, and these populations have requirements for more than one habitat type (Carroll and Meffe, 1994). The most common reason is that resources seasonally decline in one region, and animals must migrate to other regions where resources are at least adequate for survival. Hummingbirds move from lowlands to mountain slopes, following seasonal flowering patterns (Stiles, 1980). Some frugivorous birds have similar migration patterns, following the phenology of fruit production. Some lepidopteran adults avoid the dry season in tropical dry forests by migrating to more moist forests. For species that have seasonal migration, the spatial pattern of habitat patches is highly structured. That is, population survival is dependent on sufficient habitat quality and abundance in more than one region. An important consequence for these seasonally migratory species is that habitat destruction in one region can have negative effects on the biodiversity of more distant regions, even when habitat there is relatively undisturbed.

Climate change, resulting from deforestation or through longer-term effects of accumulated "greenhouse" gases, can affect landscape ecology in complex ways. Deforestation may have contributed to increased temperatures in parts of Central America. Some evidence supports claims of recent temperature increases there. In lowland Honduras, the average temperature between 1972 and 1981 was 21.7°C. Through 1982–1990, the average had increased to 28.7°C ($p < .05$) (calculated from data given in Almendares et al., 1993). However, the long cyclical effects of the El Niño–Southern Oscillation (ENSO) on weather, especially in the Pacific regions of northern South America and Central America, make interpretation of recent changes in the weather pattern uncertain. Whether or not regional increases in temperature have resulted from deforestation, there is ample evidence that local air temperatures over deforested regions are considerably higher than those over forest canopy. For example, in Guanacaste National Park, air temperatures above the forest canopy are lower than those above anthropogenic grasslands, the latter averaging 55°C (Kramer, 1996), which exceeds the upper critical thermal maxima for many invertebrates.

One link between increased air temperature and landscape-scale processes is through the effects on soil moisture of increased rates of evaporation at higher temperatures. Spatial variation in water stress affects flowering phenology; that is, greater spatial variation in water stress leads to greater spatial variation in flowering phenology (Augspurger, 1983; Rathcke and Lacey, 1985; Reich and Borchert, 1982). When temperatures increase with no concomitant increase in rainfall, the spatial variance in soil moisture that is needed to support flowering is expected to decrease because increased evaporation will eliminate flowering in those sites that were already in the lower moisture range. The expected results of increased temperature and lowered soil moisture are fewer flowers for pollinators and truncated flowering periods. As increasing numbers of individuals of some flowering species miss flowering periods, the average effective distance between floral resources for pollinators will increase. Thus, an additional effect of increased regional temperatures is to further isolate and reduce floral resources. This effect will be especially egregious for those species of plants that require outcrossing, but for any pollination system, increasing the effective isolation of plants and lowering the population size of individuals in flower will most likely lead to decreased within-population genetic variation.

Other likely consequences of increased isolation and smaller floral resources include negative impacts on specialist pollinators—central place foraging bees; hummingbird specialists on large floral patches; and such traplining species as some Euglossine bees (D. H. Janzen, 1971), Heliconine butterflies (L. E. Gilbert, 1975), and hummingbirds (see Bronstein, 1995, for an insightful discussion of the landscape ecology of pollinators). Figs, which are important sources of fruit during seasons of generally low fruit production, may be especially affected by average decreases in population size and increased intertree distance. Bronstein, Gouyon, and Gliddon (1990) have demonstrated through simulation studies of an African fig that 95 trees in asyncronous fruit production represent the minimal population size that is needed to maintain populations of specialist fig wasp pollinators. Decreased spatial variance in flowering as a result of decreased average levels of soil moisture could generate critically lethal gaps in fruit phenology, leading to local extinction of fig wasp populations and subsequent failure in fig fruit production.

In summary, agricultural development in the tropics has resulted in the following landscape changes to forest habitat patches, all of which act to reduce population size and biodiversity:

1. The number of habitat patches has decreased.
2. The average size of habitat patches has decreased.
3. The interpatch variance in patch size has probably decreased (because fewer larger patches exist).
4. The average habitat quality of patches has decreased, thereby reducing the ratio of source to sink habitat patches.
5. The average interpatch distance has increased.
6. Because the agricultural land matrix is very different from the forest habitat patches (from the perspective of potential migrants), the effective interpatch distance is even greater.

Landscape Ecology Perspective in Three Different Patterns of Land Use

Ecologically Dysfunctional but Economically Attractive Landscapes: Beef Production, Riparian Croplands, and Shrimp Mariculture in Central America

Cattle Production. The rural landscape of the neotropics has been transformed by cattle ranching more than by any other production system. With the minor exception of llamas, vicunas, and large capybara rodents, the American tropics lack the large herds of native ungulates and other vertebrate grazers and browsers that are so characteristic of the African savannas. Instead, cattle, grazing largely on grasses of African origin in pastures maintained through periodic burning, represents the American tropics' analogue of the rich African savanna ecosystem. The complex relationships between ungulates and the landscape have been intensively studied in African savanna lands. For example, Harris and Fowler (1975) have argued that the establishment of national parks may disrupt cycles of interactions between elephants and hunters that act to maintain the mosaic of forest, brushland, and grasslands. The elements of their argument are that elephants' browsing maintains open grassland and when indigenous ivory hunters periodically depress populations of elephants, the forest and brush encroach into the grasslands. Then the forest and brushland create more habitat for the tse tse fly (*Glosina* spp.) vector of sleeping sickness, and people and their cattle leave the fly–infested region. As ivory hunting decreases, elephant populations recover and begin to destroy woodlands and brushlands; the grasslands expand, thereby continuing the cycle.

The edaphic physical environment is also important in producing the savanna landscape. Belsky (1995) has shown that trees and bush are largely excluded from large areas of sodic soil. Thus, native browsers and grazers, in combination with other biophysical factors, maintain the mosaic of grasslands, brushlands, and woodlands that characterizes the African savanna landscape.

In contrast, the introduction and great expansion of cattle ranching in the neotropics has led not to a mosaic of forest and grasslands but to a great expansion of nonnative grasslands and concomitant losses of forested lands. When forest patches are left on cattle ranches, they are usually invaded by cattle and the habitat quality of small patches is degraded. Grasslands are usually maintained through frequent burning, and because productivity for cattle is generally low, cattle ranches tend to be extensive operations. As discussed earlier, the large expanse of grass creates an environmental matrix that is very different from any remaining small forest patches; the forest patches are further degraded by cattle and fire and isolated, thereby leading to population declines of forest-dwelling species. Declining soil moisture will probably reduce the spatial variance in flowering and seed set, leading to further decline in the biodiversity of plants and of the animals that use them.

Riparian Croplands. Rivers, especially in the seasonally dry tropics, have many important ecological features that influence landscape-scale processes and patterns. Beyond their value per se as wetland ecosystems, riparian, or gallery, forests may provide a habitat corridor that connects coastal lowlands to mountain uplands and riparian bottomlands and provides wet refugia in seasonally dry tropical forest. However, several pressures from economic development degrade the ecological value of riverine environments. Most large tropical rivers have one or more sizable cities along their banks, usually located either in the upland stretches of the river (e.g., the Tarcoles River, which flows through part of San José in Costa Rica) or at the mouths of the river (e.g., the Queveda River, which flows through the major coastal city of Quayaquil in Ecuador). Some rivers may flow through several cities and towns; for example, the Guayabamba River in Ecuador flows through the highland capital of Quito and the coastal city of Esmeraldas. Unregulated urban discharges into rivers have major polluting effects. The Guayabamba River from Quito to its mouth is often referred to as the sewer of Quito, and farmers avoid using water from this river even for irrigation.

The riparian forest corridor is frequently broken by human settlements and by agricultural lands for crop exports. Along the Montagua River in the seasonally dry forest region of Guatemala, virtually all of the riparian forest has been replaced by irrigated lands for melon production or for cattle ranches. High-value crop production not only directly results in deforestation of riparian forests but also contributes large volumes of pesticides to the river. This is especially true for export crops that require high cosmetic standards for marketing, such as melons. In some cases, ecologically important estuarine ecosystems are the ultimate recipients of these toxins.

Shrimp Mariculture in Honduras. Before the 1980s, nearly all of the shrimping performed in Honduras was done by artesanal fishermen who harvested shrimp on a very small scale, primarily for local and domestic consumption. These small-scale extractive shrimping operations probably had little impact on the estuarine and mangrove environments. However, in the early to mid-1980s pond-raised Honduran shrimp exploded onto the international market. Shrimp now compete with bananas and coffee as Honduras's most important export (see Stonich, Bort, and Ovares, chapter 24; Stonich, 1992).

Shrimp farms developed quickly in coastal areas along Honduras's Gulf of Fonseca to meet the large and rising international demand. By 1992, approximately 20,000 acres in the Gulf of Fonseca had been converted into shrimp ponds (see Stanley, chapter 26). In the same year, Honduras exported approximately 15 million pounds of farm-raised shrimp tails to the United States, earning about $30 million (Wille, 1995). The National Fisheries Institute predicts that shrimp sales will increase faster than any other seafood, reaching an estimated 1 billion pounds by the year 2000, up from 567 million pounds in 1989 (Egan, 1990). Most of the growth predicted by the Institute will probably be supplied by farm-bred imports of the shellfish.

The most threatening ecological cost of shrimp aquaculture in Honduras may be the clearing of mangrove forests and other coastal ecosystems to make room for mariculture operations. Mangrove forests serve as nurseries and spawning grounds for many fish species and crustaceans. Mangroves help to build soil along shorelines by holding sediments in place. As silt traps, mangroves play an important role in water purification. In addition, mangroves provide essential biotic matter, through leaf shedding and decomposition, to sustain a complex food chain of organisms fundamental to artesenal and industrial fishing. Although several factors foster coastal destruction, shrimp farming is seen by many as the prominent threat.

In a recent Agency for International Development (AID) report, Tropical Research and Development (TRD) found that of all the plausible causes for declines in high-quality mangrove cover in the Gulf of Fonseca (Vergne, Hardin, and DeWalt, 1993), the loss of 2,132 hectares, or 44 percent, of the 4,839 hectares lost since 1982 can be directly attributable to the occupation of former forest lands by shrimp farms. An additional 2,174 hectares of dwarf or stressed stands of mangrove were also occupied by shrimp farms. Furthermore, from 1982 to 1992, the aggregate land area occupied by constructed shrimp farms increased from 1,064 hectares to 11,515 hectares.

While the actual construction of shrimp farms directly contributes to mangrove destruction, it has indirect effects as well because shrimp farm construction can alter local hydrology, on which mangrove health is extremely dependent. In fact, tidal inundation, sedimentation patterns, and the range of ambient salinity are the most important environmental factors in determining mangrove placement and structure. Therefore, to alter local hydrology patterns as a result of building shrimp farms poses a threat almost as great as actually clearing a mangrove stand for the farm itself.

Shrimp farms typically use ponds, about the size of football fields, as holding pens in which shrimp larvae grow to maturity. Ironically, 60 percent of the shrimp larvae used to stock the ponds typically come from mangrove forests. Wild shrimp collectors, or *laveros*, use a variety of capture methods, ranging from handheld pushnets to large bag seines, to catch shrimp larvae, which they sell to the shrimping industry. It is estimated that for every shrimp postlarva caught, collectors also catch up to five young fish and shellfish, which are discarded on beaches or otherwise destroyed (Vergne et al., 1993). The destruction of habitat and incidental catch are of great concern because they both result in decreased biodiversity.

In addition, water quality in the Gulf of Fonseca is declining rapidly. One of the largest concerns is organic loading from shrimp pond effluent. Ponds constantly accumulate organic load in the form of shrimp excretion, uneaten food, and dead plankton, all of which are drained on a regular basis. The negative effects on natural waters of an excessive introduction of organic waste can be significant. While this is a problem with shrimp farms worldwide, the morphology of the Gulf of Fonseca exacerbates the effects in Honduras. For example, many farms are located on the upper reaches of tidal creeks, which contain relatively small amounts of water even in the rainy season and almost no freshwater inflow or tidal exchange during the December–April dry season. As a result, there is not sufficient water to adequately dilute the organic matter pumped out by the shrimp farms nearby.

In addition, water in the Gulf of Fonseca is being contaminated by pesticides used in upland farms. Numerous upland farms are producing melons, another nontraditional crop, for the lucrative U.S. and European winter markets. To do so, large areas of land have been deforested and severely degraded. Watershed deterioration has accelerated the siltation of rivers and mangrove areas, and pesticide-contaminated runoff is now thought to affect shrimp production.

One of the most serious social problems in Honduras is the perception that wealthy, powerful shrimp farmers have been able to appropriate and destroy common resources for private gain. At issue are seasonal lagoons and estuaries typically used by local fishermen. For example, seasonal lagoons, which are fed mainly by rainwater, develop each year on barren mud flats behind the mangrove fringe. Artesanal fishers move into the lagoons just as shrimp and fish congregate in the pools of water, and they are able to catch a significant amount, often in the tens of thousands of pounds. Conflict over the use of these ponds is now rampant.

Estuary and lagoon fishermen claim that seasonal lagoons provide fewer fish and shrimp than they once did. They blame this reduction on the shrimp industry for destroying and contaminating the areas they once fished. In addition, lagoons possess many favorable characteristics for the construction of shrimp ponds. Concessions were originally given out to large shrimp farms with little thought to public needs, and a number of farms were built on or at the edge of what used to be seasonal lagoons. As a result, the conversion of formally public lands into private farms means local fishers are no longer able to fish in many of the areas they once used. Shrimp farms now employ armed guards to keep local people away from a resource they once claimed as their own. The most violent confrontations in the Gulf of Fonseca, and even one death, have occurred between shrimp farms and local fishers who are trying to exploit estuaries and lagoons.

Ecologically Functional but Economically Problematic Landscapes: Buffer Zone Management and the Case of Podocarpus, Ecuador

Buffer zone management, or the practice of working with communities who live in or near protected areas, has recently been extolled as both a sustainable development strategy and a sustainable conservation strategy. A significant problem in creating parks or protected areas has often been the need to either evict local people or curtail their use of park resources in the process. However, because the preservation of protected areas ultimately depends on the support of local people, buffer zone management techniques embrace people-centered development alternatives. The concept is based on the premise that poverty alleviation is critical in reducing pressure on protected areas and seeks to balance the economic and cultural needs of local people, the environment, and future generations. While encouraging limited exploitation of natural resources, buffer zones make it possible for people to remain in the vicinity, living and working next to protected areas. Health care, clean water, schools, and other services are often provided to compensate for lack of access to park resources. In addition, activities that involve ecological principles in the management of natural resources are supported, as well as economic development efforts. When properly implemented, buffer zone management techniques can prove to be extremely effective in maintaining biological diversity and increasing the standard of living in communities near protected areas.

In the southeast corner of Ecuador, the project Proyecto Agroecológico Podocarpus (PAP) was initiated in 1990 by the Foundación Maquipucuna, a small Ecuadorian nonprofit, nongovernment organization. The goal of PAP is to create, in collaboration with local people, a well-managed buffer zone adjacent to Podocarpus National Park. To accomplish this goal, PAP works with eight different communities, all situated along the Jamboe River along the northeast portion of the park.

Podocarpus National Park was established in 1982 and covers an area of 146,000 hectares. Establishment of the park had considerable international support in order to conserve important natural resources and heritage for future generations. However, because of its modest size, irregular form, and broken topography, it is particularly vulnerable to agricultural and wood extraction activities. In addition, the area had been colonized by settlers some 20 to 30 years before the park was officially declared a protected area. While the majority of communities found in this part of Ecuador are located outside the park or border the outer perimeter, some communities actually exist inside the park itself. With technical assistance from PAP, communities are now working to establish ecologically sound agroforestry systems that can provide necessary protection to the park while at the same time raising the standard of living in the communities and establishing viable alternatives for income generation.

The project area is presently inhabited by approximately 185 families, spread out over the eight different communities. Each community has a public school, although the majority of schools have only one teacher, who is in charge of all the grades. Individual farms are typically 50 hectares in size but could range anywhere from 30 to 100 hectares. Each farm is a complex structure, including agroforestry and second growth mixed with small, intensive food plots. Some farms are beginning to produce seedlings of native tree species to be sold for use in reforestation and agroforestry projects. The majority of farmers have legal title to their land. However, to gain legal control of land, the Ecuadorian government requires farmers to put that land into "productive" use. Therefore, to show "productive" use, numerous farmers cleared their plots for cattle raising and/or subsistence agriculture activities.

Inadequate housing, limited electricity, insufficient diets, lack of potable water, minimum educational opportunities, few medical services, and poor roads are all representative of life in the eight communities, as in much of rural Ecuador. The two main sources of income in the region are wood extraction and cattle production. Trees are felled with chainsaws and planks cut on the spot, and then they are transported by mules for sale in nearby markets. Cattle serve a dual purpose as they provide both milk and meat, which can be sold in the local market. Before PAP began to work in the area, agricultural production was limited to very small subsistence plots.

Initial funding for PAP was extremely limited. Nonetheless, the project has been able to accomplish a great deal, in large part because of the level of community support it has built over the years. Communities are involved in every stage of the project, from planning and implementation to monitoring and evaluation. In this context, the emphasis is on a true collaboration and not on artificial or cosmetic participation. Because communities feel a sense of ownership over project activities, both positive and negative, PAP initiatives are more likely to continue once official funding has stopped. By being completely involved in the process, community members also learn to carry on without the project's assistance and ultimately become self-sufficient.

The PAP staff found that an excellent way to ensure the greatest amount of participation was to hire one member from each community to act as a liaison between the community and staff. Referred to as promoters, these people are chosen for their leadership skills and ability to organize and motivate others. All of the promoters chosen have also been long-standing community members with strong records of service. Promoters attend monthly meetings with the PAP staff and are responsible for keeping the community informed of project activities and goals. Promoters also serve as reference points in the community, answering questions and dispelling misperceptions as needed. In addition to informing the community about PAP activities, promoters also inform PAP about community needs and various issues. Originally, all of the promoters chosen were male, but the PAP staff soon found it necessary to address the gender bias.

Buffer zone management is most effective and successful when all stakeholders, both male and female, are included in project decision making. To accommodate the needs of both men and women, every effort was made by PAP to hold community meetings at times convenient for both, for example, in the evening when chores had been finished. Attendance records at community meetings indicated that about 45 percent of the participants were female and 55 percent were male. On the surface, these statistics would indicate that women are fairly well integrated into the participatory process. However, further examination showed that while the statistics were representative of attendance levels, they were not in fact representative of participation levels. Often, the women attending PAP meetings were more casual observers than active participants. Based on this realization, PAP decided to hire a woman promoter to work specifically with women to address issues particularly relevant to women. Female-only meetings were held on occasions because the women had commented that they felt more comfortable speaking in public at smaller, women-only gatherings. In addition, female extension agents and project staff were hired to further incorporate women into all phases of the project.

With the assistance of the promoters and the enthusiasm of community members, PAP has successfully introduced alternative income-generating activities to each of the eight communities. For example, PAP has provided technical assistance in the establishment of tree nurseries. All of the species grown in the nurseries were native species and chosen by community consensus. The project provides technical assistance and training in the actual construction and management of the nurseries, as well as seeds for the first year only. In return, communities agree to replant 40 percent of all surviving trees in the buffer zone around the Podocarpus National Park. Trees were selected by the communities, which should help minimize erosion; maintain soil fertility; and produce fuel, fruits, woods, and forage. Trees replanted in the project area will thus protect the integrity of the park, as well as provide a sustainable resource for the community. In the long run, planting trees for production of high-quality wood will also increase income possibilities.

The other 60 percent of surviving seedlings are used as the community wishes. Currently, because of new national reforestation regulations, there is a high degree of demand for tree seedlings throughout Ecuador. As a result, community members have been able to find buyers for the remaining 60 percent of their stock, resulting in increased income for the community that is not the result of ecologically destructive practices. In addition to providing technical assistance in constructing and maintaining the actual nurseries, the PAP staff has trained community members in seed collection and preparation so that the nurseries will eventually be self-sufficient.

Organic gardens have also been established on an individual, community, and school-run basis. Technical assistance to community members has been provided in all aspects of organic gardening, ranging from organic compost and companion planting to the construction of living fences. Currently, most of the produce from the gardens is for home consumption, although produce from school-run farms serves to increase the nutritional level of schoolchildren. In the near future, PAP hopes to establish larger communal gardens, which might function under the same 40-60 split as the nurseries. Forty percent of the produce would be used for home consumption, while the remaining 60 percent could be sold in the local market for increased income. Income generated from the gardens could be reinvested in seeds for future crops and used to purchase home necessities.

Agroforestry has received considerable attention as a promising form of land use that is potentially sustainable and highly adaptable to the needs of small-scale producers. The project hopes to promote agroforestry systems in on-farm demonstrations. Ideally, model farms would be located close to the one road that passes through all eight communities so that they will be highly visible to community members from each town.

Although PAP is in the early stages of development, it serves as a good example of the ecological and economic costs and benefits of buffer zone management. From the perspective of landscape ecology, buffer zone management has the following potential traits that benefit biodiversity:

- Minimal use of pesticides
- Incorporation of native species
- Maintenance of heterogenous and structurally complex landscape
- Creation of an environment adjacent to a protected area that is more similar to the natural forest habitat than many other types of economic land use.

Conclusions

The lowland neotropical landscape has been radically transformed from nearly continuous forest, with occasional small agricultural clearings, to a matrix of agricultural lands in which are scattered a few patches of forest and, rarely, large protected forest blocks. The tropical dry forest has been much more throughly transformed than the wetter forests. In the tropical dry forest regions of Pacific drainage over much of Central America, the nonforested lands are mostly extensive, low-quality cattle ranches. The principles of landscape ecology, as applied to explanations for biodiversity in spatially heterogeneous environments, may be more appropriate in the dry forest region than the conventional approaches of population and community ecology. In addition to these general principles, we argue that the large thermal differences between the grassland matrix and forest patches suggests that grasslands may impose a kind of physiological barrier to some potential migrants between forest patches. These thermal differences are largest between evergreen forest and grasslands.

As agricultural transformed the dry forest region, the once continuous riparian forests have commonly been replaced by high-value export agriculture, such as melon production. Thus, migrational corridors between lowlands and uplands have also been disrupted. Along shallow bays and estuaries, especially around the Bay of Fonseca, shrimp mariculture has fragmented the mangrove forests. In some regions of Central America, especially along the southern Pacific Coast of Honduras, cattle ranches have replaced dry forests, export agriculture has replaced riparian forest, and shrimp mariculture has replaced much of the mangrove forest.

In contrast to these economically viable but ecologically disastrous forms of development, buffer zone development around protected areas may be described as ecologically more benign but economically problematic. The Proyecto Agroecológico Podocarpus in southern Ecuador emphasizes small-scale vegetable production, slopeland protection, and the commercial production of tree seedlings for sale to restoration projects.

9

Neotropical Forests

Status and Prediction

John C. Mayne

The status of forests in Central America today is a result of millennia of exploitation. (Wiseman, 1978). Even the commonly used term, *forest resources* implies use and exploitation. Perhaps more important, it was not only the forest resources that were exploited but also the land on which they grew.

In pre-Hispanic Central America, forests of all types covered nearly all of the area and indigenous peoples of the region used the forests and cleared the land. While knowing the extent of their usage is difficult, it is safe to assume that in areas of high population density the forest was cleared for agriculture, housing, and other human activities. The same demands for the land and forests exist today. However, higher population and more intense human activities and the consequent deforestation have increased, so that what remains of Central American forests may be gone by the early part of the next century (Nations and Komer, 1987) (table 9.1).

Forest Type and Location

The Food and Agricultural Organization of the United Nations (FAO) uses several criteria to classify the forests of Central America: height of the forest formations, the extent to which they drop (or retain) their leaves during the dry season, and temperature and rainfall of the areas in which they grow. The FAO-UNESCO (1975) lists the major forest formations as tall forest, medium forest, low forest, conifer forest, and oak and deciduous forest. Within these groupings, forest types can be broken down further into evergreen, semievergreen, and deciduous.

The tall forest is a diverse formation comprising trees taller than 30 meters. These trees are found in hot climates with average annual temperatures greater than 20°C and average annual rainfall greater than 1,200 millimeters. They are found on the slopes of the Gulf of Mexico and the Pacific. The two major forest types in this formation are tall evergreen forest and the tall semievergreen forest.

The former is very dense, and epiphytes and lianas are abundant. Although some species may shed their leaves during flowering, the vast majority of the trees in this formation remain green. This formation is found in areas with more than 1,500 millimeters of rainfall and a very short dry season or none at all. The major areas in which this formation occurs are parts of Nicaragua, Costa Rica, and Panama. Trees typical of the tall semievergreen forest lose 25 to 50 percent of their leaves during the severest time of the dry season. Leaf loss occurs in isolated pockets, such as canyons, where rainfall is lower than in the areas occupied by tall evergreen forests, although still greater than 1,200 millimeters annually. The mean temperature is at least 20°C.

The medium forest consists of the medium semievergreen forest and medium deciduous forest. The former ranges in height from 15 to 30 meters and, like the tall semievergreen forest, is found in areas with average annual temperatures over 20°C and an average annual rainfall greater than 1,200 millimeters. The major areas in which this formation is found are parts of the Peten in Guatemala, Honduras, Nicaragua, Costa Rica, and Panama.

Seventy-five percent or more of the trees in the medium deciduous forest lose their leaves at the apex of the dry season. This forest grows in a climate with a mean annual temperature of more than 20°C, annual average rainfall of around 1,200 millimeters, and a clearly defined dry season. This forest type commonly occurs in association with low deciduous forest or savanna. The medium deciduous forest type is now found primarily only in remnants, as gallery forest, because it grows on moderately deep soils that are amenable to agriculture.

Table 9.1 Population of Central America, in 1000s

Country	1935	1950	1970	1980	1990
Belize	53	67	—	—	182
Costa Rica	551	801	1,798	2,111	3,015
El Salvador	1,531	1,868	3,454	4,524	5,252
Guatemala	1,975	2,805	5,111	6,839	9,197
Honduras	1,042	1,428	2,704	3,439	5,138
Nicaragua	728	1,060	2,021	2,559	3,871
Panama	546	797	1,468	1,789	2,418
Total	6,426	8,826	16,556	21,261	29,073

Sources: Forest resources assessment 1990, FAO Forestry Paper 112 (Rome: FAO, 1993c); *Statistical Abstract of Latin America for 1957* (Los Angeles: Committee of Latin American Studies, Regents of UCLA, 1959); *Statistical Abstract of Latin America*, vol. 30, UCLA Latin American Center Publications (Los Angeles: University of California, 1993).

Note: Totals for 1970 and 1980 do not include Belize.

Twenty-five to 50 percent of the trees in the low semievergreen forest lose their leaves in the dry season, depending on its severity. This is in contrast to the low deciduous forest, in which all or most of the trees lose their leaves in the dry season. The latter grows in areas with an average annual temperature of over 18°C and a long dry season.

Of the conifer forests, the pine forest is the major formation that occurs in Central America and it is found in a wide variety of climates and at altitudes from 300 to 4000 meters. The greatest extent of this forest type is found in Honduras and Guatemala.

Oak and deciduous forests occur in localized areas in Central America. They are too small for the FAO to delineate accurately on their 1:5 million scale map.

The location of Central American forests can be seen in figure 9.1. This map is useful in illustrating general forested areas because it has been drawn so that the region is divided into three broad ecological zones. The first, lowland areas with an extended dry season, occur primarily in the northern Peten of Guatemala, the Pacific coastal regions of Guatemala, El Salvador, Nicaragua, the Puntarenas region of Costa Rica, and parts of the Pacific coast of Panama. The second zone, lowland areas with a short or no dry season, includes the lower Peten extending down to the Altiplano of Guatemala and a band along the Pacific coast inland of the first zone, the eastern half of Honduras, and the eastern two-thirds of Nicaragua, as well as most of Costa Rica, Panama, and Belize. In fact, a large percentage of the remaining forests in Costa Rica and more than half of those in Belize, Guatemala, Nicaragua, and Panama are found in this zone (table 9.2). The third zone, comprising the highland areas, occupies the lower third of Guatemala and western half of Honduras, excluding the Pacific coastal regions of both countries. A section running north to south in the center of Costa Rica also falls in this zone. The largest percentages of forest in Costa Rica, El Salvador, and Honduras are found here (table 9.2). These three ecological zones correspond to the regions where forest has been cut in the past (zones 1 and 3) and the zone where it can still be found today—and where it is therefore now most threatened (zone 2).

Because of their rich soils, lengthy rainy seasons, and relatively abundant precipitation, the lowland areas with extended dry seasons have long since been converted to agriculture. In Guatemala, this area supports large landholdings planted in sugar cane, bananas, and coffee. Most of El Salvador lies in this ecological zone, and as a result, most of the forest in this country has been cut. The land is now used for export crops and subsistence agriculture (FAO-UNESCO, 1975; Leonard, 1987).

Deforestation

The major areas in Central America now undergoing deforestation are the eastern and Caribbean lowland portions of Honduras and Nicaragua, although deforestation continues throughout all Central American countries. It is the forests in the wettest and most inaccessible areas of Central America that have been left undisturbed the longest. Nevertheless, they, too, are now being cut as demand for land and timber, coupled with the scarcity of these commodities in less remote areas, make the lowland forests more desirable.

The challenge in determining deforestation rates lies in identifying where and what types of forest exist throughout Central America. Field surveys of forested areas are slow and time consuming to complete and are conducted only rarely. Estimates are frequently made.

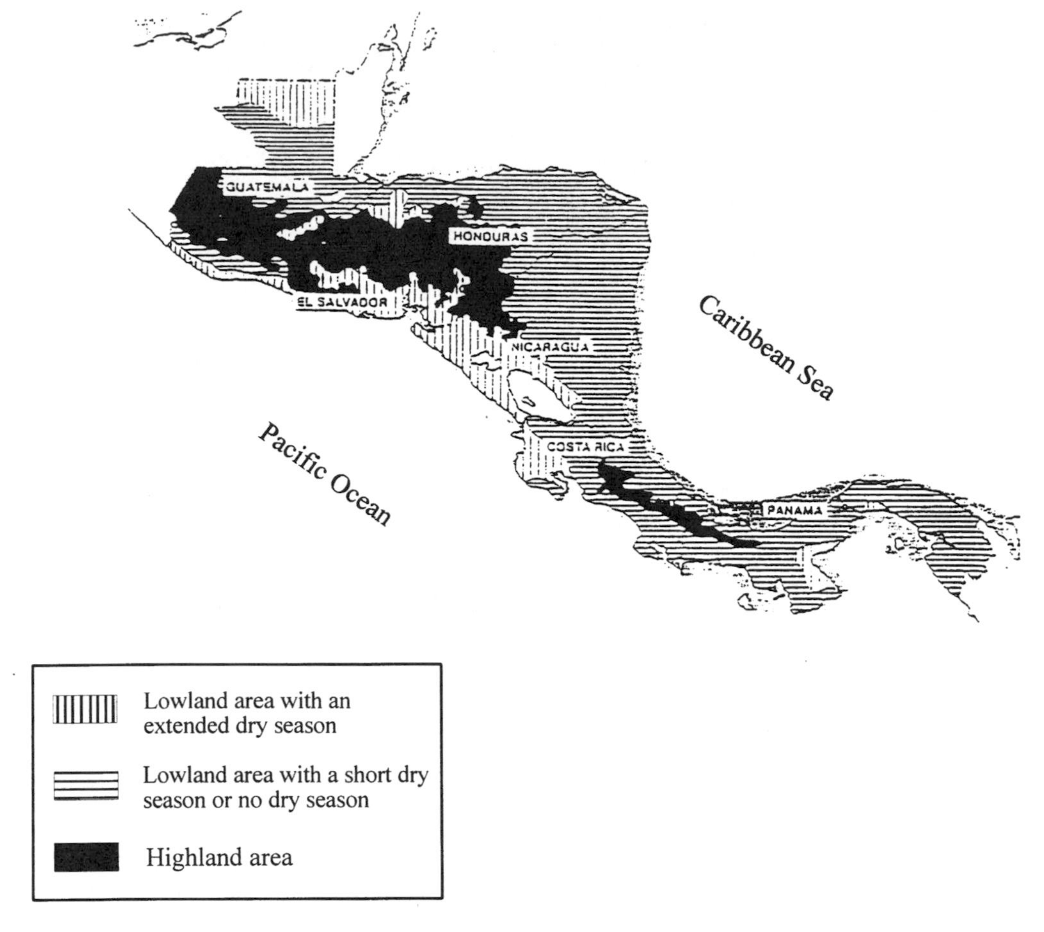

Figure 9.1 Ecological Zones

However, the errors inherent in estimates may differ widely among countries (Lanly, Singe, and Janz, 1991).

Because in-depth forest surveys are not repeated frequently, existing maps and data may often depict forest that no longer exists. However, they are still useful. When compared with older maps, they allow us to see trends and directions in deforestation. Figure 9.2, published in 1987, illustrates where and how much forest was lost between 1950 and 1985. More forest has disappeared (table 9.3) in the following decade (WRI, 1992b). The map in figure 9.2 shows the progression of deforestation during a 35-year period and illustrates the current trend toward deforestation of lowland areas. In the future, satellite imagery will be an increasingly helpful tool to detect changes in land use over time, such as deforestation, without the need for on-the-ground surveys.

The different definitions of the term *forest*, used by organizations that collect land use data, also contributed to the difficulty of determining deforestation rates and is one reason that differing estimates are found in the literature. Forest is sometimes broken down into closed or open (FAO, 1981), or forest or woodland (WRI, 1992b), or it is simply reported as forest/woodland (World Bank, 1993). Because of this problem, deforestation rates have also been determined by measuring the increase of pasture and agricultural land areas. More accurate and timely data can be found for these land uses (Houghton, Lefkowitz, and Skole, 1991).

The World Resources Institute (WRI) reported the following percentage changes in forest and woodland from 1977 to 1989: Belize, 0; Costa Rica, 17.0; El Salvador, 31.6; Guatemala, 17.0; Honduras, 18.8; Nicaragua, 23.5; Panama, 19.4. The FAO (1993b) reports even higher deforestation rates in localized areas and ecological zones. It does warn, however, that the scale it used (1:5 million) to arrive at these numbers must be kept in mind when using this information on a country level (table 9.4). An additional source of the variation found in these reports is the very definition of the term *deforestation*. According to the FAO, an area that is not in agriculture and that retains as little as 10 percent cov-

Table 9.2 Zones of Forest Formations, 1990

Country	Tropical Rainforest		Moist Deciduous Forest		Hill and Montane Forest		Total Forest
	100s of ha	Percent	100s of ha	Percent	100s of ha	Percent	
Belize	1,957	98	39	2	16	1	1,996
Costa Rica	625	44	0	0	802	56	1,428
El Salvador	33	26	12	10	79	64	123
Guatemala	2,542	60	1,615	38	69	2	4,225
Honduras	1,286	28	437	9	2,882	63	4,605
Nicaragua	3,712	62	348	6	1,953	32	6,013
Panama	1,802	58	67	2	1,249	40	3,117
Total	11,957		2,518		7,050		21,507

Source: Forestry Resources Assessment, 1990, FAO Forestry Paper 112 (Rome: FAO, 1993c).

Note: Numbers may not tally because of rounding.

erage by tree crowns is considered a forest. By this definition, these areas would have to have less than 10 percent crown cover or a change in land use would have to occur for them to be considered deforested. However, many areas that have been seriously damaged by logging retain more than 10 percent crown cover. Such forests are called degraded and are sometimes reported in deforestation figures and sometimes not (FAO, 1993b; Mather, 1987). In spite of the differences in deforestation rates reported by different sources, deforestation is obviously occurring at a rapid rate. Countries like Haiti, which has lost nearly all of its natural forest, and El Salvador (table 9.4) exemplify unchecked deforestation.

Causes of Deforestation

There are many causes of deforestation in Central America, each seemingly dependent on and causal to the other. In attempts to explain this interdependence, authors have classified the causes as direct and indirect (Leonard, 1987) or proximate and ultimate (Shaw, 1989). In the end, however, one truth is evident: the loss of forest is due more to the need for land than the need for timber or other forest products.

The major causes of deforestation in Central America are the unequal distribution of population and land ownership and tenure patterns. All other stresses on the land and forest arise from these root causes. This is not to trivialize deforestation by commercial timber operations. Even if Mesoamerica's population were small and evenly distributed, commercial concerns would still log significant forested areas, as has happened in the past throughout Central America (Tucker, 1992) and is still happening today in sparsely populated Belize. Nonetheless, if the driving desire for land were not present, the problem of deforestation in Central America would not be of the crisis proportions that it is today.

The triumvirate of road construction, cattle ranching, and subsistence agriculture has caused the major loss of forest in the last 40 years in Central America. The relative importance of each has varied in different regions and instances. As often happens, shifted cultivators (subsistence farmers who are forced to move to a new area, as opposed to shifting cultivators, who practice a particular style of migratory agriculture) move into an area and clear patches for cultivation. After several years, because of the loss of soil fertility and the buildup of weeds, insect populations, and pathogens, the shifted cultivator moves on, often leaving the cleared area to be used by ranchers. This scenario can work in several ways. A farmer can own the land and then sell it to a rancher. In other cases, the renter or squatter does not own the land. Nonetheless, they can often "improve" the land by seeding it for pasture and receive a little money for the effort before moving on, leaving it for the ranchers (Leonard, 1987; Nations and Komer, 1987; Utting, 1993; Weinberg, 1991).

However, it is not always the shifted cultivator who comes into an area first. Larger areas of forest are also cut and the land converted to pasture. Roads are built to facilitate timber removal and cattle operations. The new roads provide access to uncleared, formerly unavailable land, and subsistence farmers move in. In yet another version of the triumvirate, roads are constructed to help develop a region, and this brings the subsistence farmers and cattle ranchers. It does not really matter which comes first. The outcome is that forest is cut for what is nearly always a short-term activity. Ironically, this makes the soils unsuitable for the very activities that justified forest clearing.

In Central America, most soils that are suitable for long-term agriculture have long since been converted from natural forest to export crop production (FAO, 1993b) (figure 9.1). Much of the remaining forest is on soils that are unsuitable for long-term pastures or agri-

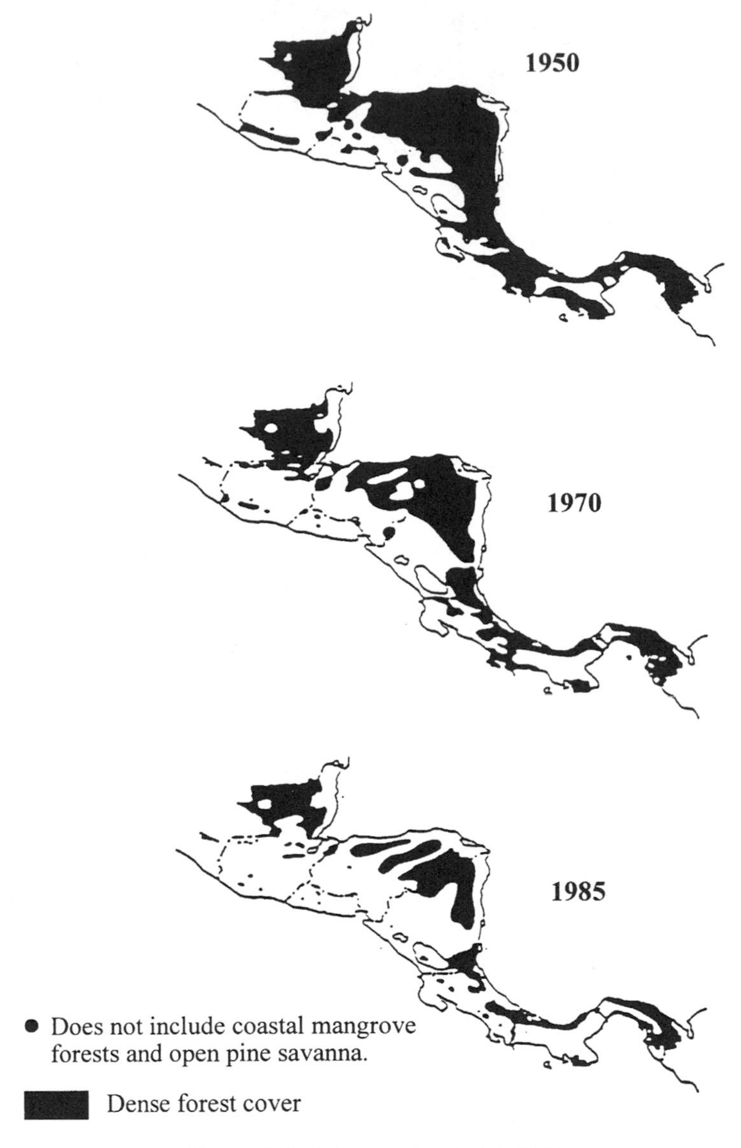

Figure 9.2 Deforestation 1950–1985

culture (FAO-UNESCO, 1975). Many tropical soils, which support complex forest ecosystems, can maintain those diverse ecosystems because the soil plays only a limited role in their fertility. Most of the nutrients are held in the plant biomass. When the natural forest vegetation is removed, so, too, are the nutrients. Furthermore, the barren soils are then susceptible to nutrient leaching and/or erosion. Still other soils are composed of sesquioxides of iron and aluminum and irreversibly harden when they dry after exposure to the sun.

When forests are cut, water resources are affected as well. When a watershed is left bare, it can become eroded and lose its ability to store water. A forested area acts as a sponge that holds water and releases it slowly over time. When the forest is cut, much less

Table 9.3 Area in Natural Forest and Plantations (in 1,000s of hectares)

	Natural Forest		Plantations	
Country	1953	1990	1965	1990
Belize	2,050	1,996	1.1	3
Costa Rica	3,925	1,428	2.4	40
El Salvador	721	123	1.7	6
Guatemala	2,455[a]	4,225	1.0	40
Honduras	4,874[b]	4,605	—	4
Nicaragua	6,450[a]	6,013	0.1	20
Panama	5,270[c]	3,117	1.0	9
Total	28,740	21,507	7.3	122

Sources: World Forest Resources, Results of the Inventory Undertaken in 1953 by the Forestry Division of FAO (Rome, 1955); Evans, J. 1992. *Plantation Forestry in the Tropics*, 2nd ed. Clarendon Press, Oxford; Forestry Resources Assessment, 1990, FAO Forestry Paper 112 (Rome: FAO, 1993c).

Note: Terminology on forests differs between 1953 report and 1990 report, which may affect areas reported; 1965 source data reported to the nearest one-hundredth hectare.
[a]1952.
[b]1954.
[c]1947.

water moves into the atmosphere through evapotranspiration. Furthermore, because of the loss of trees and other plants and their network of roots, changes in soil texture and structure may also occur. The changes increase the runoff into ephemeral and higher-order streams. Downstream, the rivers become filled with eroded soil and they also flood with greater frequency and severity.

The cutting of the forest for subsistence agriculture did not cause these problems when population densities were low. Areas burned, cleared, and cultivated were small enough and far enough apart that the forest could regenerate without major soil degradation. Only after population densities increased to the point where this system could not support the population did soil degradation take place on a large scale.

Table 9.4 Deforestation by Zone, 1981–1990

	Tropical Rainforest		Moist Deciduous Forest		Hill and Montane Forest		
Country	100s of ha	Percent	100s of ha	Percent	100s of ha	Percent	Total Forest
Belize	4.9	0.0	0.1	0.0	0.0	0.0	5.0
Costa Rica	21.7	43.8	0.0	0.0	27.9	56.2	59.6
El Salvador	0.8	26.4	0.3	9.9	2.0	63.7	3.1
Guatemala	48.9	60.2	31.1	1.9[a]	1.3	1.6	81.3
Honduras	31.2	27.9	10.6	9.5	69.8	62.6	111.6
Nicaragua	76.6	61.7	7.2	5.8	40.3	32.5	124.0
Panama	33.5	52.0	0.1	0.2	30.8	47.8	64.4
Total	217.6		49.4		172.1		439.0

Source: Forestry Resources Assessment, 1990, FAO Forestry Paper 112 (Rome: FAO, 1993c).

Note: Numbers may not tally because of rounding.
[a]Determined from table 9.3.

Fuelwood and Timber

Although pressure for land is a major reason that Central America's forests are falling, demand for timber and, to a lesser extent, fuelwood has played a role. Fuelwood cut in Central America is used almost exclusively for internal consumption (FAO, 1994a). Nearly 75 percent of the households use fuelwood for cooking (Martínez, personal communication, 1989). In rural areas firewood—the predominant form of fuelwood—is often cut from trees that are not killed. If the trees are allowed to regrow, which depends on pressures, this use of fuelwood does not contribute to deforestation. In drier areas and where population density is high, the demand for firewood is so great that the trees are killed, resulting in deforestation (see Vargas, chapter 10). In areas of Guatemala, for example, people spend half a day every few days in just collecting firewood (Zanotti, personal communication, 1985). Although the deforestation due to firewood collection is a problem particular to specific areas of Central America, it is not the major cause of deforestation across the region (Leonard, 1987) (table 9.5).

Besides firewood, charcoal is often used in the cities. It is much lighter than firewood so is therefore much easier to transport. However, much wood energy is lost in the charcoal production process. Consequently, the fuelwood used to manufacture charcoal is extensive. Charcoal production contributes significantly to deforestation around major metropolitan areas.

Because of the volume and the methods of extraction, the timber industry has caused more deforestation than has the fuelwood industry. Generally, pine forests have fared better than broadleaf forests, although there are exceptions. Pine is not as desirable as nonconifer species for firewood, although a great deal more is used in Guatemala than in the rest of Mesoamerica (table 9.6). In some areas, particularly from Nicaragua north, the soils on which natural stands of pine grow are mostly sandy and acidic (not good agricultural soils), so there is less intrusion by colonists. Furthermore, fires favor the regeneration of pines over nonconifers so that pines sometimes move into disturbed broadleaf forest areas (Leonard, 1987).

Economics and politics have kept Central American pine forests from being decimated in modern history, although they were exploited in the past. In colonial times, the forests were exploited by Peru by the shipbuilding industry and for construction (Parsons, 1955). By the nineteenth century, many stands were degraded. In fact, pine from the southern United States was exported to Central America during the nineteenth century (Fickle, 1980). In the early part of the twentieth century, when southern pine forests in the United States had been overexploited, U.S. interests looked toward Central America as a source of pine. However, in large part because of the unstable politics of the region, the resistance to large U.S. timber interests, and the resurgence of the southern U.S. pine industry in the 1920s, Central American pine forests never became a significant source of pine timber and navel stores for export out of the region (Tucker, 1992).

Nevertheless, in most areas natural pine forests are declining. Even when selective cutting is practiced, the demand for wood products can damage the future of the forest by damaging the genetic base. This occurs because loggers traditionally take the best trees. In the forestry industry, this is called hygrading and is a common practice in selective pine harvesting. Hygrading means that the most desirable trees of a species are removed and the less desirable trees are left. Seed for regeneration then comes from those trees that are genetically less desirable.

Unlike pine, which grows in stands, there are often only a few individuals of any one broadleaf tree species per hectare, and they are generally of higher eco-

Table 9.5 Firewood Production (in 1,000s of metric tons)

	1981			1992		
Country	Conifer	Nonconifer	Charcoal	Conifer	Nonconifer	Charcoal
Belize	—	79	—	—	126[a]	—
Costa Rica	—	2,228[a]	14[a]	—	3,022[a]	19[a]
El Salvador	15[a]	3,712[a]	21[a]	17[a]	4,365[a]	24[a]
Guatemala	7,338	1,594	—	9,139[a]	2,003[a]	—
Honduras	1,301[a]	2,642[a]	—	1,871[a]	3,800[a]	—
Nicaragua	262[a]	2,120[a]	—	359[a]	2,906[a]	—
Panama	—	723[a]	—	—	910[a]	—
Total	8196	13,098	35	11,386	17,132	43

Source: FAO Yearbook, Forest Products, Forestry Series 27, Statistics Series 116 (Rome: FAO, 1992).

[a] FAO estimate.

Table 9.6 Roundwood Production (in 1,000s of cubic meters)

	1981		1995		1992	
Country	Conifer	Nonconifer	Conifer	Nonconifer	Conifer	Nonconifer
Belize	4	113	11	156	18[a]	170[a]
Costa Rica	—	3,384	—	3,412[a]	—	4,192
El Salvador	59[a]	3,795[a]	45[a]	3,896[a]	72[a]	4,456[a]
Guatemala	7,464	1,670	8,225	1,817	9,242	2,014[a]
Honduras	2,337	2,672[a]	2,334	3,070	2,393	3,826
Nicaragua	607[a]	2,655[a]	393	2,572	459	3,106
Panama	—	922[a]	—	952[a]	—	1,028
Total	10,417	15,211	11,008	15,875	12,184	18,802

Source: FAO Yearbook, Forest Products, Forestry Series 27, Statistics Series 116 (Rome: FAO, 1992).

[a] FAO estimate.

nomic value than pines. Therefore, all identifiable members of a species are often logged in an area (Kemp, Namkoong, and Wadsworth, 1993). In addition, consequent damage to surrounding smaller trees occurs when the larger trees are removed. Trees along the removal path are often damaged or destroyed as well (see Boza, chapter 6). These degraded forests, as mentioned earlier, are not always reported in deforestation statistics.

In wet, primarily broadleaf forest areas, logging can cause severe damage to the ecosystem by human, animal, and machine activity in wet soils, in addition to collateral damage to other trees. This causes further damage by disrupting the soil's physical properties and by eliminating many different plants and habitats. This activity and the resulting habitat loss cause severe perturbations in animal populations.

The loss of plant biological diversity that results when a forest ecosystem is destroyed is incalculable. Plant species are destroyed before they are even discovered by researchers. Beyond the value of the functions of these species in the ecosystem, medicinal and other uses from plant products represent real economic potential for those countries in which these forest resources lie. For example, the National Biodiversity Institute (INBio) of Costa Rica and the chemical company Merck & Company have formed a partnership to exploit the forest for plant species that have the potential for commercial development. Merck pays INBio to search and collect plant, insect, and other organisms to be screened for potentially useful drugs. Should Merck produce a profitable drug from the material it receives, it will pay INBio royalties, a proportion of which go to the government for conservation efforts (WRI, 1993).

This is an example of one country's recognition of the economic value of its plant resources. It is one of many schemes to keep tropical forest ecosystems from being destroyed by finding higher-value and less destructive uses for the forest (D. H. Janzen, 1991). The goal is to achieve high utilization and low exploitation of forests rather than have forests that are overexploited and underutilized (Myers, 1995).

Preservation Strategies

Many different ideas for preserving natural forests have been proposed. They range from management techniques for sustainable forest use to substituting plantations for natural forests to finding substitutes for forest products to revamping the laws and economic incentives for forest clearing in Central American countries. Sustained yield management is a way to extract products, primarily timber, from a natural forest and leave the rest of the forest undisturbed. This form of management has been proposed to reconcile the demand for forest products with the need for maintaining a relatively intact forest and concomitant biological diversity. However, the actual use of sustainable forest management does not bode well for this approach. Less than one-eighth of 1 percent of tropical forest worldwide is logged sustainably where timber is extracted commercially (Poore, 1989). Others argue that it is possible to create management systems based on deferred profits and fiscal restraint to provide benefits that accrue beyond the lifetime of the managers. This solution, too, is problematic when the people who manage forests are themselves faced with severe economic hardships in the present (see Ascher, chapter 3).

Plantations of pine (*Pinus* spp.), teak (*Tectona grandis*), mahogany (*Swietenia macrophylla*), and other tropical tree species, native and exotic, have been established in Central America. These plantations are viewed as a way to produce forest products in a managed system without continually destroying tropical forest ecosystems (see Fisher, chapter 23).

The arguments in favor of plantations generally stress two factors. The first is land availability. Plantations can often be established on land that has already been cleared. They can also be established on land that is more accessible to roads, thereby avoiding the new road construction that often encourages settlement by colonists (Evans, 1992). The second factor is high productivity. In addition to the rapid growth of tree species used in plantations, the even spacing and full stocking of the site make management easier and the plantation more productive than natural forests.

Arguments against plantations focus not so much on the benefits of plantations as on the fact that they make up a very small percentage of forested land (table 9.3). Furthermore, deforestation results only in part from cutting forests for timber production (Colchester, 1990, 1993; Nations and Komer, 1987). Forests in Central America are being destroyed every day. Even if plantations were replacing what is being lost on a tree-for-tree basis—which is *not* happening—the remaining natural forest ecosystems would continue to fall because of commercial timber production and population pressures.

Although plantations are not a solution to deforestation in Central America, they do have the potential to take some pressure off natural forests for wood products. Central American countries have a demand for construction materials, furniture, and other tropical wood products. The United States is also a market for tropical hardwood products, although it also imports tropical hardwoods from Asia (FAO, 1994a). Although the area given to plantations is still a small percentage of tropical forests, it has increased an order of magnitude in most Central American countries in the last 25 years (table 9.3).

Sometimes tropical forests are given a reprieve from overexploitation when a substitute for a forest product is found or economies in other countries lessen the demand. For example, at the end of the sixteenth century, logwood, a hardwood used for dye, was being harvested. Logwood grew in the Yucatan and down through present-day Belize and the Caribbean coasts of Honduras and Nicaragua. The timber was exported to Europe until the early years of the twentieth century, when chemical dyes became more desirable. Extraction of logwood virtually ceased (Tucker, 1992). Technologically derived replacements for wood products such as plastics also have the potential to take the pressure off natural forests.

Additional forests could be preserved if the timber being cut, no matter the reason, could be more fully utilized. Only a small portion of the timber cut in Central America is used for commercial purposes (Leonard, 1987). Antiquated milling practices and a disregard for the reduction of postharvest losses contribute to the inefficient use of the timber that is harvested.

Furthermore, every Central American country needs to discourage the cutting of forests for cattle ranching or for other short-term purposes. According to Strasma and Celis (1992), inappropriate land use policies, often promulgated when populations were smaller, can be found throughout Central America. In many countries there are legal inducements to ranchers to clear land or trade land close to population centers for larger, more remote and undisturbed parcels that they can clear and convert to pasture (Leonard, 1987; Strasma and Celis, 1992; Utting, 1993).

Protected Areas and National Parks

Protected areas and national parks are promoted to protect all types of Central American forests, along with everything that lives in them (see Boza, chapter 6). In Costa Rica, more than one-fourth of the country is in some type of protected status and 10 percent is in national parks (Barborak, 1992). In the approximately 1.4 million hectares of the Maya Biosphere Reserve, Guatemala has the largest single protected area in Central America (CONAP, 1990). More than half of this area is in multiple-use and buffer zones. Certain land use activities are permitted in these zones, and some unpermitted land uses are also sure to be practiced. Nevertheless, there are some good examples of the utilization of nontimber resources by the people who live in the reserve. Allspice, harvested from *Pimienta dioica*, and chicle (latex), extracted from *Manilkara zapota*, are products that have been harvested in the area for 30 and more than 90 years, respectively (Reining and Heinzman, 1992). The rationale is to allow these activities in some areas and to place other areas of forest off-limits.

The goal is commendable but the reality is not always encouraging. Many protected areas exist only on paper (Utting, 1993). This can cause problems as residents, knowing the land on which they live is to become protected, cut the forest before enforcement becomes a reality. Then there is the problem of what people who find themselves living within a newly protected area will do for employment if their current activities are not permitted (see Ascher, chapter 3). Several schemes have been suggested, such as hiring these people as park guards. Such plans are a start, but not everyone can work as a guard. Still other people are moved to other areas of the reserve where agricultural activities are permitted. Whatever the problems inherent in setting aside and policing large areas of forested land, the goal of protecting dwindling ecosystems is worth the effort.

The improved productivity of subsistence farmers will allow more food to be grown on the land area already in agricultural production. It will also keep farmers from becoming shifted cultivators when soil fertility decreases. Efforts to improve the productivity of subsistence farmers must be supported and redoubled by regional governments and by regional and international conservation organizations. Many farmers are displaced or descended from displaced farmers. As a result, often neither the individual farmer nor the community has the knowledge needed to farm the land efficiently and sustainably. Subsistence farmers are also recipients of technical information provided by chemical and fertilizer companies, either through government extension agents or company representatives. Each country must encourage efficient, low-cost methodologies that are less dependent on costly commercial fertilizers and pesticides. Further, they must promote efficient and self-sufficient techniques such as agroforestry, mixed cropping, terraces, small stands of firewood species, and nitrogen-fixing trees and crops. These technologies and practices will not only increase productivity but also reduce erosion and improve fertility (see Szott, chapter 21).

For farmers to make the financial effort to improve the land by building terraces or planting trees, they must have some guarantee that they will benefit from their investment. It takes several years to begin to see improved yields from terraced hillsides or to be able to cut firewood from new woodlots. Realistically, if we expect farmers to practice methodologies that will reduce erosion, protect the watershed, and reduce the cutting of forest, Central American governments must devise legal and tax schemes to enable continuity of ownership or, at the very least, encourage both the renters and the owners of land to practice sustainable methodologies.

10

The Geography of Dryland Plant Formations in Central America

Gilbert Vargas Ulate

Regional studies of vegetation distribution in Central America are rare, particularly studies of the region's dryland herbaceous and shrub complexes. However, while regional studies are lacking, important research, touching either directly or indirectly on these vegetation types, has been conducted at the national level. This chapter calls on previous regional studies, data sets, research, and other technical information available for each of the Central American nations. This information is pooled to provide a Central American regional analysis with a geographic perspective and an emphasis on semiarid formations.

The areas in which these formations are found coincide with the most densely populated parts of Central America. They account for 81 percent of the cities with populations of more than 35,000. Important agricultural areas are also included, often characterized by subsistence farming. In the driest regions, socioeconomic problems and energy deficiencies are growing because of continual deforestation. El Progreso, Zacapa, and Solola in Guatemala; Chalatenango in El Salvador; Choluteca and Nacaome in Honduras; Matagalpa, Jinotepe, and Masaya in Nicaragua; La Cruz and Sardinal in Costa Rica; and Chitre, Las Tablas, and Norte de Penonome in Panama demand large quantities of fuelwood. Part or all of this demand could be met by native species. Research programs to select native species that regrow rapidly after pruning, many of which occur in herbaceous and shrub formations, are needed. These species could be cultivated as living fences or in small rotational parcels. Further studies of the potential of these species for fuelwood would be an excellent area of research that could contribute to socioeconomic well-being, ecological balance, and maintenance of biodiversity in Central America.

Many authors have contributed to our knowledge of these plant formations, including ICATA (1984) and Stanley and Steyermark (1945) for Guatemala and Lundell (1945) and Hartshorn (1984), among others, for Belize. Piñeda's (1984) geographic study in Honduras provides an excellent complement to the work of Silliman and Hazelwood (1981) and of Hilty (1982) for El Salvador. For Nicaragua, the studies by Parsons (1955) and Terán and Incer (1964) are good resources. Tejeira (1975) provides an excellent study for Panama. Several previous studies in Costa Rica are important, including Gomez (1986); Gordon, Baker, and Opler (1974); Holdridge et al. (1971); and G. Vargas (1987a,c). Geologic, climatic, edaphic, vegetation, and land use analyses also exist for all of the Central American nations, providing a complement to other environmental information.

Some authors have also devoted attention to Central America in larger regional studies of Middle America, which includes Mexico and the Antilles, as well as Central America. Among those are West and Augelli (1976), who focus on cultural factors. Unfortunately, they treat the relationship between human activities and vegetative formations only summarily. Sorre's 1928 work is a good comparative analysis and an important source of information about the dry thorn, herbaceous, and shrub formations in the isthmus. Lassere's (1977) work is a synthesis, emphasizing socioeconomic aspects, but does not have the depth of the material cited previously.

Environmental Characteristics

The intertropical zone of Central America is an isthmus whose maximum width of 540 kilometers is found between Cabo Gracias a Dios and the Gulf of Fonseca and whose minimum width of 64 kilometers occurs at the Panama Canal. The most notable topographic feature of the isthmus is its extensive, complex system of mountain ranges, especially the volcanic axis that runs along the isthmus from northwest to southeast. This system of mountain ranges, reaching altitudes of 2,500 meters, separates two very distinct coastal belts on the Caribbean and Pacific sides of the isthmus. The Caribbean coast is a wide lowland, where rainfall is high (2,000 to 5,000 millimeters per year) and evenly distributed throughout the year, interrupted only occasionally by one or two dry months. The Pacific coast, in contrast, is a narrow lowland with low rainfall (less than 2,000 millimeters per year), unevenly distributed during the year. A five- to six-month dry season is the norm for this coast.

The vegetation formations in the isthmus are a response to these climatological characteristics. Very humid evergreen formations occupy the Caribbean lowland, whereas dry to very dry semideciduous formations are characteristic of the Pacific coast. Between the base and the peaks of the mountain ranges lies a progression of vegetation types, corresponding to altitudes that range from lowland to high montane vegetation complexes.

Dry herbaceous and shrub formations cover a total of 139,732.6 square kilometers, or 27 percent of the total land area in Central America. These vegetation types occur predominantly in the coastal lowlands on the Pacific side of the isthmus (11 percent), the interior volcanic uplands (8 percent), and the Caribbean coastal and Peten lowlands (7 percent). Protected intermontane valleys account for another 3 percent (figure 10.1). These areas include both lowland and premontane areas and five life zones: (1) dry tropical forest, (2) very dry tropical forest, (3) tropical wet forest, (4) premontane wet forest, and (5) premontane dry forest.

These vegetation complexes are a bioclimatic phenomenon, and their plants suffer from a periodic water deficit. Lack of available water in the root zone prevents the plants from carrying out their physiological functions normally during the dry season. Plants adapt to these unfavorable conditions by both physiological and structural mechanisms, such as longitudinal root development either at the surface or underground, water-retaining leaves and spines, and thick stems resistant to fire. Either climatological or edaphic conditions or both can cause a water deficit.

Climatological aridity may be caused either by low precipitation or by the fact that the annual precipitation, though sufficient for normal plant growth, is not distributed evenly during the year. The six-month dry season, characteristic of much of Central America, is an example. The dry vegetation of the Caatinga and the Cerrado in Paraiba State in Brazil are examples of vegetation types that develop where annual precipitation is low, between 300 and 700 millimeters per year. The most representative examples of plant adaption to uneven rainfall distribution are the Pacific lowlands of Central America and the Venezuelan *llanos* (Tamayo, 1956).

The structural and textural characteristics of soil are closely tied to the soil's ability to retain water, and some will produce edaphic aridity. A soil with high sand and gravel content, for example, has a high percolation rate and low moisture-retention capacity, resulting in edaphic aridity. The Mosquito Coast of Honduras and Nicaragua and the Peten lowlands are examples of the vegetation types that can result under these conditions. Although rainfall in these areas is greater than 2,700 millimeters per year, distributed evenly throughout the year, plants in these areas suffer from a water deficit. The savannas in Zaire (Koechlin, 1961) and Surinam (Heyligers, 1963) are other examples.

In Central America, climatological and edaphic aridity may act together, sometimes accentuated by topographic factors that further reduce rainfall by protecting intermontane valleys from moisture-laden winds. This situation is very common in the central highlands and protected intermontane valleys of the isthmus, which are the region's driest zones.

The ecological factors that are most important in the development and distribution of dry herbaceous and shrub formations in Central America are climatic factors, geologic and edaphic factors, topography, and human activities, all of which may interact. The role of each of these factors is described below.

The principal mountain chains and highlands of Central America reach altitudes of more than 2,500 meters. They are oriented from west to east and from northwest to southeast, forming a barrier to the dominant northeast trade winds. Localized shadow effects from rain make the deep valleys very arid, even when they lie on the Caribbean side of the isthmus. Precipitation in areas sheltered from the trade winds—the protected valleys, central highlands, and Pacific lowlands—ranges from 470 to 1,900 millimeters, whereas the Peten, Belize, the Mosquito Coast, and southeastern Costa Rica receive average annual precipitation of 2,010 to 2,900 millimeters. In both cases the yearly rainfall is bimodal, with two maximums and two minimums, a pattern typical of an equatorial regime modified by trade winds. On the one hand, the feature that is most important in creating arid conditions in the protected valleys, central highlands, and Pacific lowlands is the uneven distribution of rainfall. In the Peten and Belize lowlands and on the Mosquito Coast, on the other hand, very limited moisture-retention capacity of the soil, rather than total precipitation or distribution of precipitation, produces aridity.

The mountain chains in Central America, because of their orientation and elevations of more than 2,500 meters, exercise strong control over the movement of the two major circulation systems that affect the region, the northeast trade winds and southwest equatorial winds. Data from 23 climatological stations in Central American were analyzed. Of these, 19 show a prolonged dry period between November and April. These 19 stations also show the impact of mountainous barriers on northeast trade winds. Aridity on the Pacific side occurs because the dry winds sink earth-

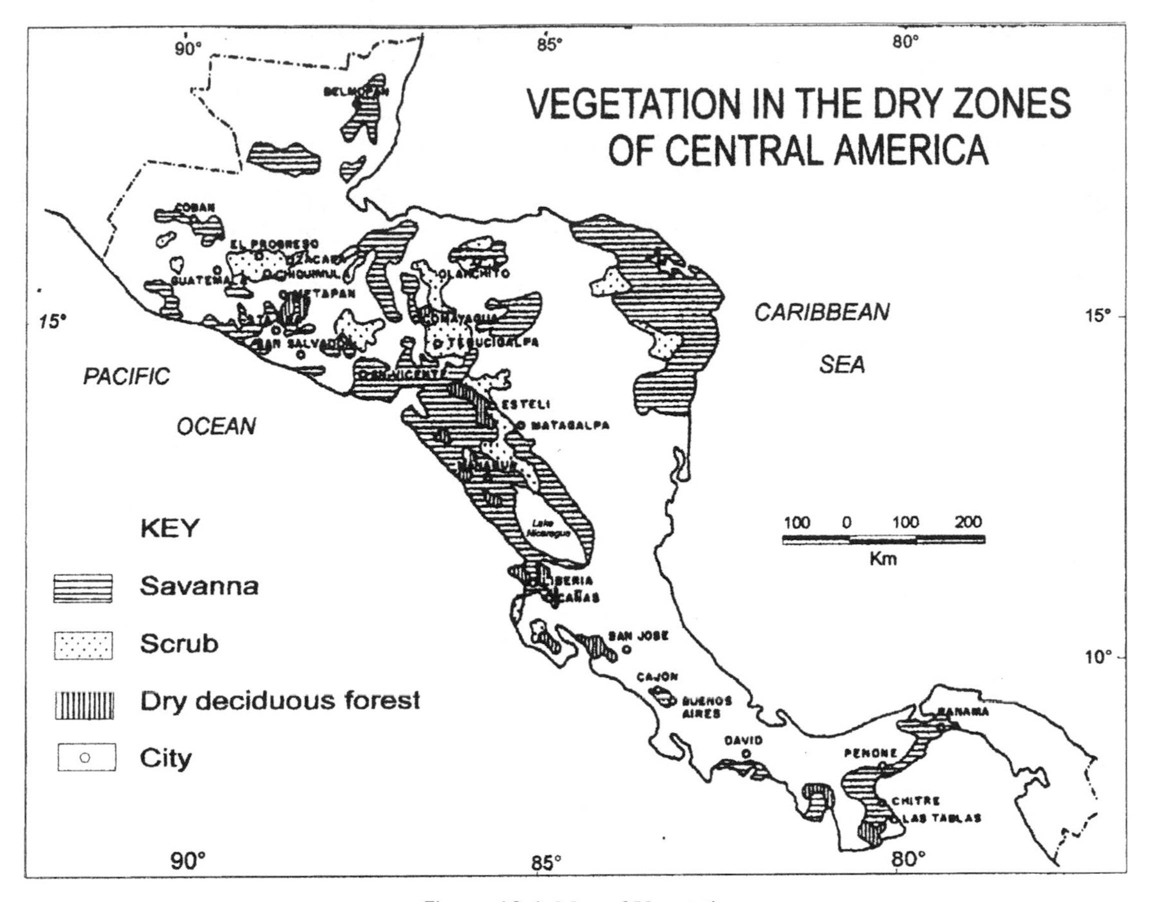

Figure 10.1 Map of Vegetation

ward after having released their moisture on the Caribbean side. At these sites, the first annual rainfall maximum occurs in May and June, the result of humid equatorial winds from the southwest that are displacing the northeast trade winds. Throughout the Pacific lowlands of Central America, the second rainfall minimum, commonly called the *canicula* or *veranillo*, occurs during July and the first days of August. During this period, subtropical high-pressure cells gain strength, permitting the northeast trade winds to circulate over Central America. Equatorial winds predominate again during September and October, producing the second rainfall peak.

The duration of the dry season in Central America is variable. Nonetheless, regions with a five- to six-month dry season predominate (Darlin, 1982). Five parts of the isthmus are the most arid. (1) Zacapa, El Progreso, and Chiquimula in Guatemala receive only 400 to 600 millimeters of rainfall annually. The water deficit is 1,000 to 1,400 millimeters per year, and on the average, there is rainfall only on 65 days each year. The aridity in these areas is due to the blockage of the northeast trade winds by the Sierra de las Minas. Aridity is accentuated by the complicated relief of Guatemala, which produces local circulation patterns that often run counter to the dominant patterns. An example is the middle section of the Motagua River basin, where winds blow parallel to the river and do not produce rain (ICATA, 1984). (2) In Olanchito, Honduras, in the Aguan River valley, rainfall is 912 millimeters annually, with a water deficit of 850 to 900 millimeters. The valley's profundity, in combination with the barrier of the Sierra Madre de Dios, produces aridity. (3) In Honduras, Tegucipalpa and the surrounding area and some valleys to the north of the metropolitan area have a desertlike appearance (Silliman and Hazelwood, 1981). These areas are part of an interior mesa protected by a system of mountains, the Sierra de Comayagua, Sierra de Opalaca, Sierra de Sulaco, and Cordillera de Montecilla. Precipitation is only 900 millimeters, and the hydraulic deficit is 800 millimeters. (4) The area around La Guija Lake in Metapan, El Salvador, is a continuation of the very arid region of Motagua Valley and El Progreso, described earlier. (5) In Nicaragua, the

Matagalpa Meseta and Esteli lie in the rain shadow of the Cordillera Central. The first two of these areas lie on the Caribbean side of the isthmus, and the last three are Pacific sites.

Data from the remaining four meteorological stations, of the 23 data sets analyzed, show precipitation of more than 2,000 millimeters per year and an insignificant dry season. In three cases, the stations are found on the Caribbean side of the isthmus, and the high rainfall is due to the unobstructed northeast trade winds. The rainfall peaks in November, December, and July are characteristic of the Caribbean precipitation regime. The fourth station is found on the Pacific Coast in southwestern Costa Rica. The high precipitation at this site is caused by the incursion of southwest equatorial winds. However, the vegetation at all four sites consists of dry herbaceous and shrub formations because soil edaphic conditions produce aridity, although rainfall is abundant (figure 10.2).

South of the Panama Canal, rainfall increases to more than 3,000 millimeters annually. This high rainfall regime extends to the Pacific Coast of Colombia. In contrast, a dry zone extends from north of the Azuero Peninsula in Panama to the Costa Rican border. This arid region lies in a large rain shadow, isolated from both the northeast trade winds and the southeast equatorial winds (Darlin, 1982).

The areas occupied by dry herbaceous and shrub formations coincide directly with regions underlain by igneous rocks, limestone formations, and marine and continental sediments with high sand, gravel, or salt contents (figure 10.3). The igneous rocks were originally deposited in fissures. They underlie small mesas with very abrupt escarpments, such as the small mesas of Progreso in Guatemala; Tegucigalpa and its environs in Honduras; Esteli and Matagalpa in Nicaragua; and Santa Rosa, Liberia, and Cañas in Costa Rica. However, much larger regions are also underlain by igneous rocks. About 70 percent of the land area of El Salvador and the highland and piedmonts of the Pacific volcanic cordillera rest on granite formations. These deposits are very acidic. Most were deposited as lava and are well consolidated, but some are porous airborne tuffs (Weyl, 1960). The granites in Guatemala, Honduras, El Salvador, and Nicaragua date from the Tertiary (mid-Pliocene), whereas those in Costa Rica are more recent

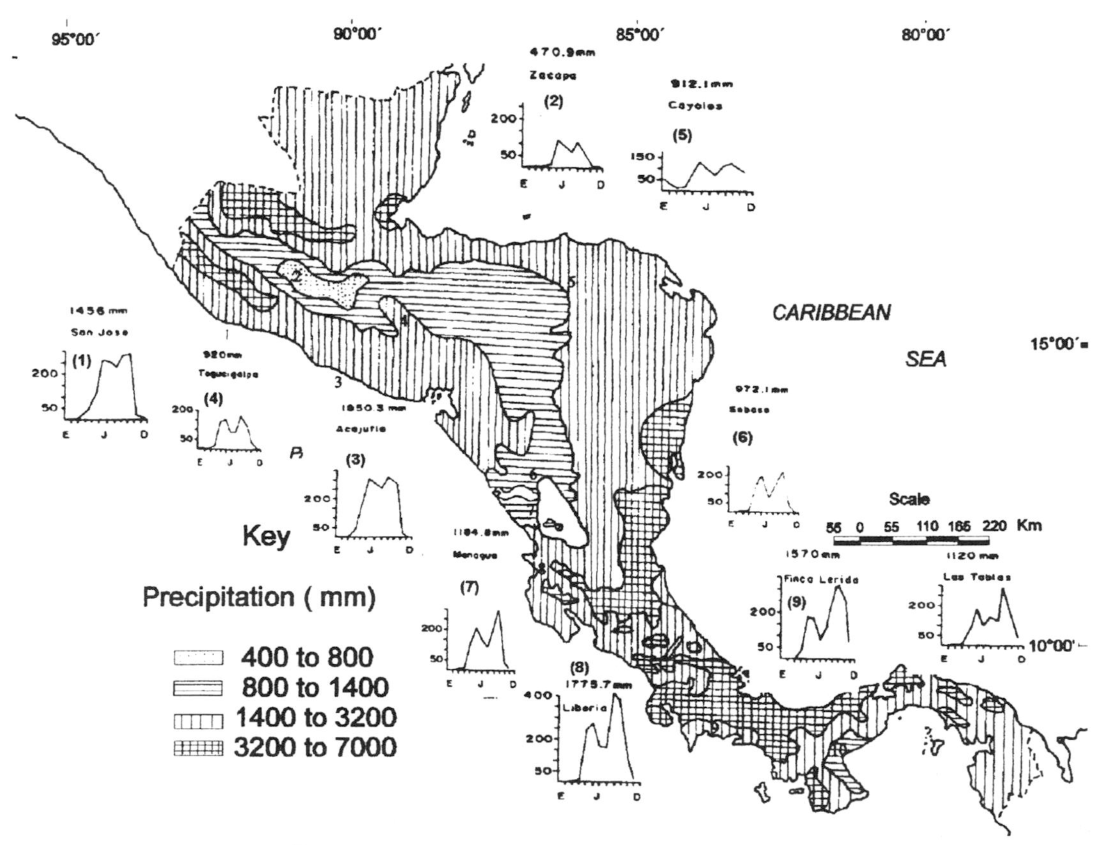

Figure 10.2 Precipitation Patterns in Mesoamerica

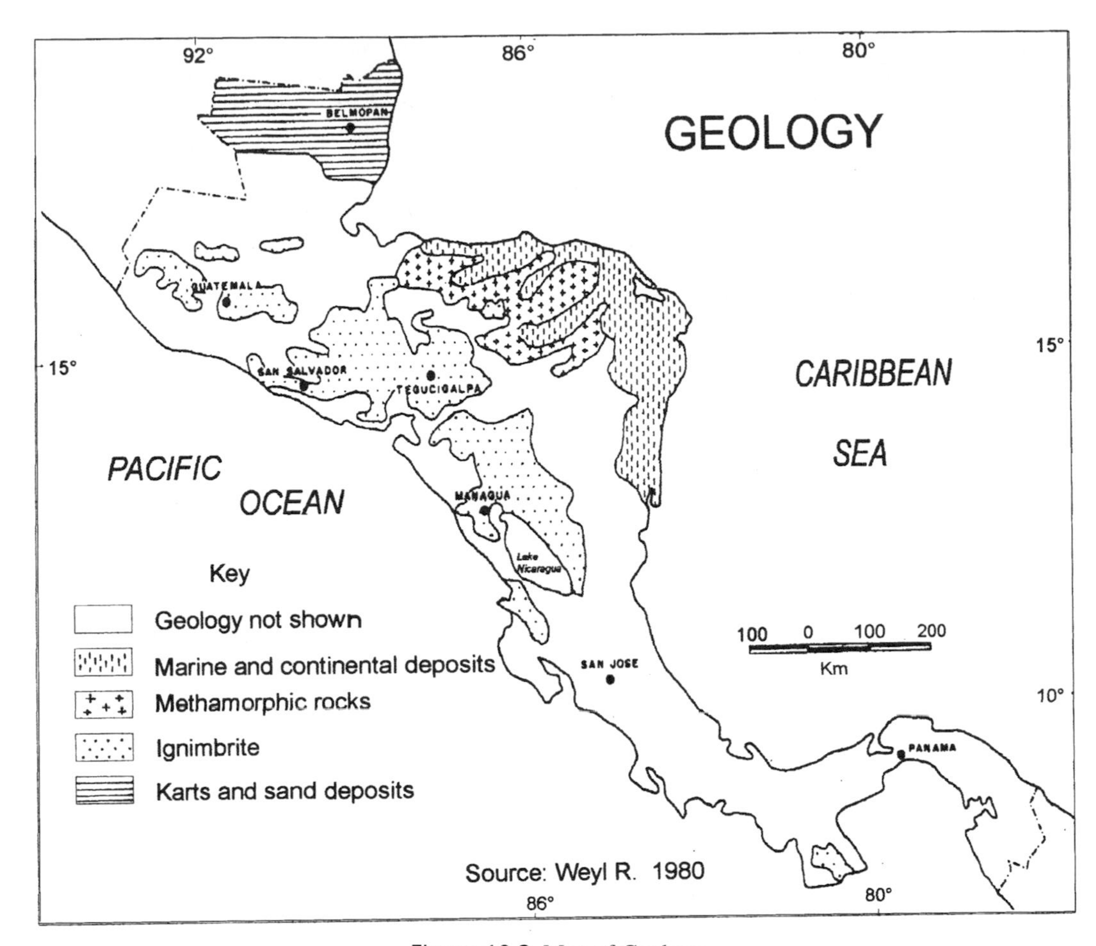

Figure 10.3 Map of Geology

Quaternary deposits (Weyl, 1980), deposited in fissures extending from the volcanic cordillera of Guanacaste (Weyl, 1960).

The Azuero Peninsula in Panama is a unique case. Eocene and Oligocene sediments made up of sandstones, limestone, and continental tuffs form the basement rock on the peninsula. However, surface deposits are the product of fluvial and volcanic action. Rivers have deposited tuffs, granites, and conglomerates on the plain (Cedeño, 1982).

The most extensive limestone formations in Central America occur in the regions of Alta Verapaz, around Lake Chichén Itzá, and in Belize. Calcitic limestone domes form a typical karst topography, dating from the Mesozoic (Weyl, 1980). Highly eroded Cretaceous limestones occur in the Chamalecon and Ulua River valleys.

Marine and continental sediments are found on the Mosquito Coast in Honduras and Nicaragua and in the lower watersheds of the Aguan, Patuca, and Chamalecon Rivers. The Mosquito Coast deposits are made up of sands and gravels with moderate salt concentrations (Weyl, 1980). These materials may have been transported during the Wisconsinan glaciation (11,000 years ago), and the terraces in the region may be correlated with changes in sea level during the glaciation. The Coco and Patuca Rivers have deposited large quantities of gravel of igneous origin as well (Parsons, 1955).

Soil types in Central America are a result of parent material, rainfall patterns, and topography. Typic ustropepts, lithic ustropepts (inceptisols), and lithic ustorthents (entisols) dominate in the mesas, piedmonts, and mountain escarpments underlain by igneous formations. These are young, shallow soils with weakly developed profiles. Their pH is acid to very acid. The soils contain a high percentage of sand and exhibit low moisture-retention capacity, either because of excessive drainage caused by high porosity or excessive surface runoff when they are compact and rocky. Vertisols (vertic ustropepts and typic pellusterts) occupy small depressions and flat-bottomed valleys in the areas underlain by granite. These have very high clay content, as well as a high montmorillinite content. They are heavy, textured soils that drain poorly, exhibiting large cracks when dry and excessive stickiness when wet. These soil types are also found in the littoral lowlands in association with ustic dystropepts and ultic haplustalfs (alfisols), which are deep, rocky, clay soils that are dry for more than 90 days per year and strongly susceptible to wind erosion.

In the Peten of Guatemala and Belize, tipic calcioquolls, tropic fluvaquents, and tipic trapaquepts cover a calcareous base. In some cases, these are sandy soils prone to droughts because of excessive drainage. Impermeable vertisols sometimes occupy depressions.

Ustoxic palehumults and plintic palehumults occupy the area from Buenos Aires in southern Costa Rica to David in Panama. These ultisols are residual, reddish soils on undulating relief. They are deep, acidic soils with a high clay content covered by a thin cap of laterite. These soils are poorly drained and have low native fertility. They are severely degraded soils that have been highly altered by human activities.

Soils of fluvial-marine origin occur on the Mosquito Coast and in the lower watersheds of the Congo, Patuca, and Ulua rivers. The topography is flat, and these sandy, excessively drained soils contain an abundance of lithic fragments, such as gravel and pebbles, deposited by the rivers. In some sites moderate salinity is observed. They are classified as typic tropopsamments, fluvenitc tropopsamments, and lithic tropopsamments (entisols).

Humankind both transforms and maintains its environment. In Central America, in both the pre-Hispanic epoch and since Spanish colonization, the highest population densities in the region have been on the Pacific side and in the interior highlands of the isthmus. Humans have thoroughly transformed these regions, and their actions have played a major role in creating and maintaining the grass and shrub vegetation characteristic of these areas.

Dryland Vegetation Formations

Grasses and shrubs constitute most dryland vegetation formations in Central America. These formations cover 25 percent of the isthmus's area, a total of 128,428 square kilometers. Grasslands or savannas are a herbaceous formation of unbroken grass cover, reaching a maximum height of 2 meters, maintained by annual burning, and containing dispersed woody shrubs. They are called either grass savannas or shrub savannas, depending on the preponderance of shrubs in the landscape. These two types of savanna are not distinguished here. Thickets consist of dense shrubs that branch from the base of the trunk. Spiny species form an important component of the vegetation, depending on aridity. Grass savannas may evolve into shrub savanna, thickets, or dry secondary forest under ecological conditions favorable to these vegetation types. Dry deciduous forests are included in this analysis because they are the vegetation formation from which the majority of herbaceous and shrub formations in Central America evolved under human influence.

Grasslands

Central America's grasslands can be classified into three types, depending on topographic, edaphic, and climatological factors: (1) the Pacific grasslands, (2) the grasslands of protected valleys and highlands, and (3) the lowland grasslands of the Peten, Belize, and Mosquito Coast. The most important herbaceous components in all three types of savanna are the grasses. A great variety of native species occur in each of the Central American nations. The genera *Andropogon, Sporobolus, Paspalum, Cynodon, Aristida,* and *Trachipogon* are especially important, the last on the most arid and poor sites. Despite the abundance of native species, the uniformity of the savanna is due to the almost continuous cover of the African pasture grass *Hyparrhenia rufa* (jaragua). Other families also form significant components of these grasslands, including Leguminosae such as *Crotalaria, Zornia*, and *Mimosa*; Rubiaceae such as *Diodias* and *Borreria*; Compositae such as *Wedelia* and *Melampodium*; and many Ciperaceae genera.

The Pacific grasslands extend from the littoral and piedmonts of the volcanic cordillera to the interior highlands. Rainfall ranges from 1,100 to 1,954 millimeters annually, with a dry season of five to six months. The soils are very poor, of granitic origin. Two floristic types are distinguished, depending on topography: (1) the ridge and mesa grasslands and (2) valley bottom and depression grasslands.

Grasslands on ridges and mesas develop on thin soils with continuous rock outcrops or underlying rock within 20 to 30 centimeters of the soil surface. Lithic ustropepts, Lithic ustorthents, and Tipic ustropepts (inceptisols) support these vegetation formations. Shrub species such as *Byrsonima crassifolia, Curatella americana, Pithecelobium dulce, Gliricidia sepium, Libidivia coriacea, Cochlospermum vitifolium, Guazuma ulmifolia, Psidium sartorianum, Miconia argenta, Alibertia eduis*, and *Smillax malis* dominate, mixed with a large variety of acacias, cactus, terrestrial bromeliads, and spiny palms. *Pinus caribea* and *P. oocarpa* are important components of these savannas in Guatemala, Honduras, and northern El Salvador (Calderón and Stanley, 1941; Johannessen, 1963). In southern Costa Rica, along the banks of the Terraba River between the towns of Buenos Aires and Cajon, savanna occupies an undulating topography, with deep, acidic residual soils that have high clay content (Ustoxic palehumult). *Byrsonima crassifolia* and *Curatella americana* dominate this formation.

The other type of Pacific savanna is found in valley bottoms and small inundated depressions, where deep black or grey vertisols occur (typic pellusterts and vertic ustropepts). In Guatemala, Honduras, and El Salvador,

Crescentia alata dominates, mixed with *Calycophyllum candidisimum, Bactris subglobosa*, and *Blepharodendrom mucraratum.* This pattern is not typical in Costa Rica (Vargas, 1987a,c), nor in Nicaragua (Terán and Incer, 1964), where monospecific formations of *Crecentia alata* dominate, popularly called *jicarales.*

Savannas are depicted on vegetation maps of Panama (Tejeira, 1975), but phytogeographic, ecological, and floristic studies of Panama's dry regions are scarce. Tossi (1971) suggests that plant specimens in herbariums and studies conducted in Panama have focused on more humid areas.

The second grassland type found in Central America occurs in protected valleys and highlands. It is found on the Caribbean side of the isthmus in areas of lower precipitation. Zacapa and El Progreso, with 470.9 millimeters of rainfall; Manacal, with 1,151 millimeters; and Coyoles, with 912 millimeters are examples. Aridity at these sites is due to their location in deep valleys that lie in rain shadows. They are in the middle watersheds of the Motagua, Ulua, and Aguan Rivers and in the upper watershed of the Patuca River. The ecologically limiting factors are low precipitation and its uneven distribution during the year. Shrubs such as *Curatella americana, Byrsonima crassifolia, Erythroxylum fiscalense, Asclepias latifolia, Juliana adstringens, Pisonia aculeata,* and *Phyllestylom brasilensis* dominate these formations. Acacias, cactus, and bromeliads are abundant. The environment is very arid, and defoliation during the dry season gives these landscapes a desolate appearance (Piñeda, 1984).

The third type of savanna is found in the Peten and Belize and on the Mosquito coast. The lowlands of the Peten and Belize receive more than 2,000 millimeters of precipitation annually (according to Flores, a meteorological station in Belice). The dry season is short or nonexistent. However, excessively drained calcareous soils on ridges and mesas create edaphic aridity. On these sites, shrub vegetation made up of *Curatella americana, Byrsonima crassifolia, Pinus caribea, Randia truncada, Brosinum alicastrum, Lysiloma latisilqum, Caesalpinia gaumesi, Acacia pennatuda, A. collinssi, A. gaumini*, and *Quercus oleoides* dominates. The latter is also present in all of the Pacific lowland grasslands to northern Costa Rica. It is associated with poor soils with deficient moisture retention (Montoya, 1966). Soils in microdepressions are saturated, and *Crescentia alata* clearly dominates.

Precipitation on the Mosquito coast is 2,741 millimeters annually, and there is no dry season. April, with 82 millimeters of rainfall, has the lowest rainfall. Despite the high precipitation regime, the sandy, excessively drained soils create edaphic aridity. The grasses *Trachypogon montufari, Panicum maximum, Tripsacum laxum*, and *Arundinella deppeana* are the most notable in the herbaceous cover. These are sturdy grasses that resist prolonged water deficits well, but they do not support livestock. Many Cyperaceae (*Stenophyllus paradoxus*) and geophytes (*Bulbostylis paradoxus*) are present. *Pinus oocarpa* accompanies this herbaceous cover, mixed with shrub species typical of arid sites, such as *Curatella americana, Byrsonima crassifolia,* and *Calliandra houstoniana* (Parsons, 1955).

Spiny Shrub Formations (Thickets)

The spiny shrub formations represent an evolution of grasslands toward a higher vegetation complex. Abandonment of cultivated parcels, especially pastures, produces this evolutionary stage in Central America. Release from the pressures of grazing and burning favor succession from grassland to the next vegetation complex, that is, shrubs, which include plants from 50 centimeters to 7 meters tall. Variation in ground cover is high. Here, we consider any plant from 50 centimeters to 7 meters tall that branches from the base of the trunk to be a shrub.

The most frequent floristic components are *Curatella americana* and *Byrsonima crassifolia*. They are commonly associated with 23 cactus species and 13 acacia species, including *Acacia cornigera, A. hindissi, A. sphaerocephala, A. ferruginea, A. collinssi*, and *A. farnesiana*, all well described by D. H. Janzen (1966). Other species present in these formations are *Asclepias curassavica, Erythoxylum fiscalense, Pithecelobium dulce, Suprechtia deamii, Guazuma ulmifolia, Jatropha urens, Bix orellana, Calliandra houstoniana, Karwiskia calderonii, Chomelia spinosa, Randia spinosa, Albertia edulis, Croton* spp., *Smillax mallis, Cupania glabra, Acalypha* spp., *Dyphisa robinioides, Pisonia aculeata, Zizyphulus guatemalensis, Ximenia americana, Phyllostylon brasilensis, Prosopis juliflora, Acrocomia* sp., *Guetarda ramuliflora, Apeiba tibourbau, Crataeva tapia*, and *Cleoma spinosa.*

Most of the area in thickets is found in El Progreso, Zapaca, and Chiquimula in the middle watershed of the Motagua River in Guatemala; in the interior valleys of the Comayagua and Aguan rivers, the upper watershed of the Choluteca River, and the areas around Tegucigalpa in Honduras; and to the north of Lakes Managua and Nicaragua and on the Matagalpa and Esteli mesa in Nicaragua. When the savannas of the Mosquito coast in Honduras and Nicaragua are protected from fire and other anthropogenic factors, they evolve toward a thick-leafed shrub formation. This thicket formation is very dense, dominated by such species as *Curatella americana, Byrsonima crassifolia, Miconia argentea, Quercus oleoices*, and the palm *Arcoelorraphe*.

Dry Deciduous Forest

Dry deciduous forest was the climax vegetation formation that existed in the Pacific lowlands of Mesoamerica in the pre-Hispanic era. When the Spanish reached Mesoamerica, peoples with an intimate knowledge of the dry forest occupied these regions. Theirs was pragmatic knowledge, including a thorough understanding of the biological role of the vegetation type. Today, this region has been transformed into savannas, thickets, croplands, and degraded secondary forests. Dry tropical forests cover only 2 percent of the Central American land area today, about 11,304 square kilometers. Small patches of this forest occur north of Santa Ana and around Lake Guija in El Salvador. In Nicaragua, a significant expanse of dry forest occurs west of Matagalpa, and there are smaller expanses in the Pacific lowlands (Terán and Incer, 1964). Costa Rica has the largest remaining area in dry forest in the isthmus. This expanse of forest is protected in a series of national parks (Santa Rosa, Barra Honda, and Palo Verde), forest reserves (Volcán Orosí), biological reserves (Carara), and wildlife refuges (Palo Verde). A small area of tropical dry forest occurs in Panama on the Azuero Peninsula. These areas are truly relicts of the once extensive tropical dry forests of Mesoamerica.

Tropical dry forests develop in areas where precipitation ranges from 1,000 to 2,000 millimeters annually, with a five- to six-month dry season and temperatures of 24° to 32°C. The upper story reaches a height of 20 to 25 meters. *Enterolobium cyclocarpum, Anarcadium exelsum, Ceiba pentandra, Lysiloma seemannii, Hymenaea courbaril, Lonchocarpus minimifolius, Cassia grandis, Dalbergia retusa, Bombacopsis quinatum, Tabebuia pentaphylla, Bursera simaruba, Brosimum costaricanum, Cochlospermun vitifolium, Pithecelobium saman, Bixa orellana, Spondias mombim, Sterculia apetala*, and *Swietenia macrophylla* are among the principal species of dry forests.

The Origins of the Dryland Grass and Shrub Vegetation Formations

The dryland grass and shrub formations of Central America are a secondary vegetation type, caused by anthropogenic transformations of the dry and very dry forest types. There are both historical and ecological explanations of their origin. Historical explanations focus on the role of humankind in their development.

Many authors agree that savannas and thickets in the worldwide intertropical belt were created by human action. The same may be true for Central America. Montoya (1966) states that agriculture by indigenous peoples caused the savannas. Lundell (1945) explains that the savannas and thickets of the Peten and Belize were caused by the agricultural practices of the Maya. B. L. Turner (1974, 1976) believes that the Maya caused large-scale transformations in the environment through intensive agricultural systems that employed terracing and irrigation. Wagner (1964), however, attributes the formation of the savannas to the Spanish conquest, which led to burning and the development of adverse edaphic conditions. Barrera (1976, 1977) also argues that the Maya adapted to the equilibrium of the tropical forest by practicing selective exploitation and allowing regeneration, using polyculture systems in regrowth areas for agricultural production.

In short, any historical explanation of the development of savannas creates polemic arguments. Detailed historical, anthropological, and geohistorical studies into the degree and types of interventions by both indigenous groups and the Spanish in the 17th, 18th, and 19th centuries are needed. G. Vargas (1983, 1987b), for example, has attempted to explain the origin of the savannas in Guanacaste Province in northwestern Costa Rica. Using historical sources, he concluded that indigenous groups were not responsible for the development of this savanna. Rather, the savanna developed during and after the Spanish conquest and was encouraged in the nineteenth and twentieth centuries when pastures were introduced to support extensive livestock production systems. The repeated use of fire to control weeds and stimulate regrowth of pasture grasses was an important component in developing and maintaining grasslands. Fire is selective; it favors some species and works against the regeneration of forest species that cannot tolerate burning. Fire, along with the hydraulic deficits that exist in soils derived from igneous rocks, definitely favors the development of a herbaceous and dry shrub flora. The very woody, spiny, and fire-resistant vegetation that forms the principal component of Costa Rica's grasslands and thickets is a classic example of the resulting vegetation type.

Whereas humankind is definitely one factor in the origin of the savannas, other ecological factors have also played a key role. Edaphic, climatological, and topographic factors determine where these vegetation types will develop and how extensive they will be. Succession is the replacement of one vegetation type by another in a determined time and space. It is a process that involves structural substitutions, from herb to shrub to tree, for example, and changes in floristic composition. Succession can work in two directions. It is a progressive evolution when an area goes from bare soil to the stable and complex vegetation type called a climax formation. However, evolution can also progress from a climax formation to a bare soil.

From this perspective, the grasslands and thickets are a secondary vegetation succession type, originating in a regressive evolution from dry or very dry tropical forest. This regressive evolution was provoked by humans (indigenous groups or the Spanish), and the secondary vegetation type is maintained by annual burning, which prevents succession backward to the climax vegetation type. Edaphic, climatological, and topographic factors played a major role in the formation of the grasslands. However, grasslands can evolve backward to the original vegetation type when fire and grazing are eliminated.

Many studies of grassland dynamics have shown a progressive evolution toward shrubs and trees in areas protected from fire and grazing for periods of 6 to 10 years. For example, San José and Farinas (1978) have verified this evolutionary process in the Venezuelan grasslands, which evolve toward a tall thicket called *matas*. The *matas* consist of shrub species such as *Cochlospermun vitifoliun, Curatella americana, Bowdichia virgilioides*, and *Godmania macrocarpa*. These species reach heights of 8 to 10 meters. Koechlin (1978) made similar observations in the Brazilian Mato Grosso. Here the shrub vegetation type is known as *cerrado*. The *cerrado* has the same structural characteristics as the Venezuelan *matas* and contains the same species, as well as *Terminalia argentera, Jacaranda* spp., *Cassia ferruginea*, and *Cariniana exelsa*. Schnell (1976) analyzed the evolution of the savanna toward open forest in the Guinea and Congo regions of Africa.

In these three cases, typical dry forest tree species were not observed, perhaps because of the almost complete lack of dry forest remnants in the regions under study. The contrary is the case in Costa Rica, where several typical dry forest species were found in parcels in Santa Rosa National Park that were protected from fire for more than seven years (Vargas, 1987a,c). All of the species mentioned in the Brazil and Venezuela studies were found, as well as significant regeneration of dry deciduous forest species, such as *Enterolobium cyclocarpun, Dalbergia retusa, Cassia grandis, Pithecelobium saman, Astronim graveolens, Hymenaea courbaril, Bombacopsis quinatum, Sterculia apetala*, and *Luehea candida*. This regeneration is a product of the patches of forest that remain in this Costa Rican savanna. The three regions share one other important characteristic, the presence of many spiny species.

Based on these studies, I conclude that the savannas and thickets are not a climax vegetation type. Rather, they are one stage of vegetation succession and may further evolve toward dry secondary forest or high thickets when protected. Both types of evolution may occur in Central America, depending on the presence of remnant patches of dry forest and the degree of soil degradation.

Source of the Dry Grassland and Shrubland Flora

Two great nuclei of dispersion for dry herbaceous and shrub flora can serve as the source of Central America's dryland vegetation. One is the xeric region of the Tehuantepec isthmus (Tabasco, Chiapas, Oaxaca, and Veracruz) in North America. The other is the South American xeric region of the Venezuelan plains and northeastern Brazil (the cerrados and caatinga). The species that originated in each of these regions were identified and quantified. In the Tehuantepec isthmus 260 species were found, and 245 were found in the South American nucleus. Later, by consulting the registry of Central American flora, it was determined that 52 percent of the northern xeric region reach as far south as Costa Rica, and 34 percent reach their southern limit in the Panama Canal Zone. Only 3 percent of the southern xeric region species reach a northern limit in Costa Rica. Five factors explain this marked domination of northern xeric flora from the Tehuantepec isthmus in Central America.

The first is the tectonic and morphological history of Central America. Central America consists of two geological provinces, nuclear or northern Central America and southern Central America (Dengo, 1968). Structurally, nuclear or northern Central America is a prolongation of the North American continental platform. Its southern limit is approximately at 12°30' north latitude, in Nicaragua. This platform began to develop in the late Paleozoic (230–250 million years B.P.), and its development continued until the late Cretaceous (60–70 million years B.P.). South of the platform lay the Central American canal, an interoceanic passage that prevented terrestrial contact with the South American continent. Southern Central America developed over this canal. Nonetheless, it was not until the late Pliocene (1.8–5 million years B.P.) that vulcanism and uplifting closed the interoceanic canal to form a thin isthmus, which permitted passage from north to south and vice versa (Lloyd, 1963). Thus, northern Central America has experienced a long period in which a continuous land mass extended from North America to 12°30' north latitude. The southern limit of this historic land mass also defines the southern limit of the northern floristic unit.

Once the isthmus formed, a biological bridge came into existence, consisting of a coastal plain and interior hills characterized by relatively uniform precipitation, temperature, and humidity. As a result, dryland species from the northern xeric nucleus began to disperse toward the south. This north-to-south dispersion has been verified in several studies. Janzen and Martin (1982) cite a total of 36 dry tropical forest species dispersed by large fauna during the Tertiary because of their large, hard fruit and flowers. Among these species are *Spondias mombim, Crescentia alata, Byrsonima crassifolia, Acacia farnesiana, Andira inermis, Caesalpinia coriaria, Enterolobium cyclocarpum, Himenaea courbaril, Pithecelobium samam, Prosopis juliflora, Bunchosia biocellata, Brosimum alicastrum, Chlorophora tinctoria, Acrocomia vinifera, Bactris guinensis, Alibertia edulis, Genipa americana, Apeiba tibourbou*, and *Bromelia pinguin*. Horn (1985), in his study of Quaternary sediments along the coast of the Nicoya Peninsula, found several dry genera: *Bursera, Guazuma, Luehea, Brosimum, Piper, Byronima, Cordia, Spon-*

dias, Ficus, Eugenia, and *Psidium*. All are typical of the Tehuantepec xeric flora.

The second factor that explains the dominance of northern flora is the Wisconsinan glaciation (10,000 years B.P.), when the sea level fell from 60 to 90 meters (Derrau, 1980). On both Central American coasts, the continental land mass increased, possibly forming a level, dry plain with uniform temperatures and geologically young soils. This uniform corridor permitted a north-to-south migration of dry herbaceous and shrub species from the xeric Tehuantepec nucleus on both coasts. It is equally possible that a south-to-north migration occurred from the Venezuelan savannas and Brazilian cerrado that may have reached Panama and southern Costa Rica (P. C. Stanley, 1968).

The third factor is the climatic conditions during the Pleistocene. Gómez (1986) has extrapolated climatic conditions for Costa Rica from paleobotanic and paleoclimatic studies. He describes the Pacific Coast as having temperatures of 20°–23°C, low precipitation, and aridity in the lowlands. After the Pleistocene glaciation (10,000 years B.P.), climatic conditions were more humid. However, for the last 5 to 6 million years, the climate of tropical America, including the Central American isthmus, has varied little (Frakes, 1979). Based on the principle that the most tolerant flora and fauna will be the most prevalent in a region, it is logical to suppose that the dry vegetation of Central America dispersed under climatic conditions of irregular precipitation regimes with periods of prolonged drought and high hydraulic deficits. This factor interacts with the physiographic characteristics of the region, which was dominated by plains and hills that did not present any great obstacle to north-south dispersion from Tehuantepec.

Fourth, the very humid region of the Darièn plays the role of an ecological filter, forming a barrier to the migration of dry flora from the Venezuelan plains and northeastern Brazil toward the north. A small number of these dry South American species did overcome this obstacle and are now found in Chiriquí and the Azuero Peninsula. Nonetheless, even here species originating in Tehuantepec dominate those from South America. However, it is significant that the xeric florae of Tehuantepec decreased by 17 percent in Chiriquí and 34 percent in Azuero, whereas in northeast Costa Rica they form 52 percent of the dry vegetation. This is explained by the presence of another very humid region in southern Costa Rica, with precipitation of more than 4,000 millimeters per year. The very wet tropical forest in this region plays the same role of ecological filter as does the Darien forest.

Finally, the Pacific lowlands of Central America have been the most densely populated region of Central America historically. Here, human impacts have been decisive in the transformation of forest into savannas and thickets, and human actions have maintained these vegetation types. Fire and soil degradation are critical. The former favors the dispersion of fire-resistant dry herbaceous and shrub flora along the Pacific flank of the isthmus. This vegetation is more competitive and adapts to a greater ecological range of conditions than do forest flora.

Conclusion

The areas occupied by dry herbaceous and shrub vegetation in Central America are increasing steadily because they coincide with the areas of greatest population, poverty, and pressure on the land. Large-scale land ownership also characterizes these areas. Under these conditions, the scant remaining forested areas are subject to extreme pressure, cut down to develop cropland or simply for use as firewood.

These areas are also affected by droughts, which are becoming more prolonged. These dry periods hurt the large landowners, who produce rice, sorghum, sugar cane, tobacco, and cotton, and cause smallholders to enter into poverty and indebtedness and lose their land. Upon selling their land to the large landowners, they become part of the mass of unemployed workers.

These conditions present a dichotomy: natural resource conservation versus human need and rural development. However, sustainable rural development requires knowing and understanding the resources available in a region. This chapter is an initial attempt to define the resources of these heavily populated Central American regions. However, detailed studies for each country, including field reconnaissance, are needed before more of the region's biological wealth is lost.

11

Soils of Mesoamerica

Alexis Vásquez-Morera

During the last four decades, the development of the agricultural sector in Central America has focused on increasing production, resulting in an irrational expansion of the land base that is devoted to agriculture. This irrational expansion has serious consequences for the resources that sustain development, especially on soils, forests, and water. A development strategy focused on expanding production has caused a major misallocation of soil resources in Mesoamerica because current land uses are not based on a consideration of land use capability. Severe erosion is the end result of a soil management strategy focused on increasing production and leading to overuse of the soil resource in many areas. Compounding the problem, the increased use of chemicals in more intensive agricultural production systems causes further degradation of the soil and the environment. Although land resources are abundant in the region, unequal distribution of ownership and differences in soil quality dictate very different potential uses in different parts of each country.

The lack of a regional plan in the agricultural sector, one based on reliable information, is one of the main reasons that there is such a lack of awareness of these problems in the region. Lack of planning and poor information result in poor decision making. The importance of gathering more reliable information about the natural resources of Mesoamerica is becoming critial for three main reasons: (1) the expansion of the agricultural frontier in all of the Central American countries, (2) the intensification of production on lands already in agricultural use, and (3) the growing need for sustainable development. Better information will make it possible to determine the most appropriate ways to develop and use the region's natural resources. A better understanding of the region's soils is especially critical in achieving greater success in regional economic development.

Soils in Mesoamerica have been described for individual countries and on a regional basis at different scales and pedologic levels. In some cases, there is confusion in describing land types and land use. Nevertheless, despite the difficulties, a general description of Mesoamerican soils is presented in this chapter, based on studies carried out by many different researchers. I also indicate some of the major land use limitations or potentials for each of the major soil groups in the region.

The Soils of Mesoamerica

Table 11.1 shows an approximate equivalence between the soil taxonomy classification system used by FAO-UNESCO (1976), the *World Reference Base for Soil Resources* (ISSS, ISRIC, FAO, 1994), U.S. Agency for International Development (1992), and USDA (1975). For some of these relationships, references were taken from regional or national soil surveys developed by different agencies and authors (INIAP, 1988; Parsons Corporation, 1971; Vásquez, 1989). The FAO-UNESCO terminology is used here, which permits us to classify Central American soils by geomorphologic regions.

Folded Calcareous Chains and Lowlands Adjacent to the Peten

Central America's folded calcareous chains are parallel chains arranged in the shape of a large arc. They are formed by hard, light-colored crystalline calcareous rocks. These geoshapes are most frequently found in southern Mexico, Guatemala, and Belize. The prevalent soils that occupy the uplands of the chains are lithic eutric cambisols, lithic chromic cambisols, rendzinas, and lithosols. In the lowlands adjacent to the Peten, these soils are replaced by pellic and chromic vertisols, eutric and humic gleysols, orthic and gleyic luvisols, and some planosols.

Nonvolcanic Central American Highlands

The nonvolcanic Central American plateau is a strongly dissected highland plateau, starting in Guatemala, getting wider in Honduras, and ending at the line about

Table 11.1 Preliminary Soil Taxonomy Classification Equivalences

Ferric acrisols	Ultisols and oxic subgroups of alfisols
Humic acrisols	Ultisols (humults, udults)
Haplic acrisols	Ultisols (haplohumults)
Plinthic acrisols	Ultisols (plinthic subgroups of humults, udults, and ustults)
Gleyic acrisols	Ultisols and alfisols (aquic suborders, great groups, and subgroups)
Orthic lithic acrisols (Lithic Leptosols)	Lithic subgroups of ustalfs
Humic lithic acrisols (Umbric Leptosols)	Lithic subgroups of humults and ustoxs)
Pachic andosols	Andisols (humic and pachic subgroups from udands and ustands)
Eutric andosols	Andisols (ustands, udands)
Vitric andosols	Vitrands, vitric subgroups of ustands and udands
Lithic chromic cambisols (lithic leptosols)	Lithic subgroups of tropepts and orthents
Lithic eutric cambisols (lithic leptosols)	Lithic subgroups of tropepts and orthents
Lithic dystric cambisols (lithic leptosols)	Lithic subgroups of tropepts and orthents
Lithic humic cambisols (umbric leptosols)	Lithic subgroups of tropepts and orthents
Dystric cambisols	Inceptisols (dystropepts)
Mollic cambisols	Mollisols (haplustolls, hapludolls)
Calcaric cambisols	Inceptisols (lithic ustropepts and lithic eutropepts)
Rendzic leptosols	Mollisols (rendolls)
Lithic leptosols	Entisols (lithic subgroups of orthents)
Haplic ferralsols	Oxisols (haplustox)
Gleysols	Inceptisols (aquepts), entisols (aquents) and oxisols (aquox), aquic great groups and subgroups in those soil orders
Eutric gleysols	Inceptisols (aquetps), entisols (aquents), aquic great groups and subgroups in those soil orders
Humic gleysols (umbric gleysols)	Inceptisols (umbraquepts) and ultisols (umbraquults)
Fluvisols	Mollisols (hapludolls, haplustolls), fluventic subgroups from tropepts chromic reddish luvisols: alfisols (paleustalfs)
Orthic lithic luvisols (haplic luvisols)	Alfisols (lithic haplustalfs)
Ferric luvisols	Alfisols (rhodustalfs)
Gleyic luvisols	Aqualfs, aquic subgroups from udalfs and ustalfs
Plinthic luvisols	Alfisols (plinthic subgroups from aqualfs, udalfs and ustalfs)
Orthic luvisols (haplic luvisols)	Alfisols (haplustalfs)
Dystric nitisols	Alfisols and ultisols (pale great groups)
Eutric nitisols	Alfisols (paleustalfs)
Histosols	Histosols
Thionic histosols	Histosols (sulfic subgroups from saprist and hemist)
Regosols	Entisols (typic subgroups from troporthents and ustipsamments), vitrands
Vertisols	Vertisols
Pellic vertisols (haplic vertisols)	Vertisols (hapluderts and haplusterts)
Chromic vertisols	Vertisols (chromic subgroups from uderts and usterts)
Planosoles	Alfisols and vertisols with abrupt textural changes
Dystric planosols (dystric leptosols)	(Typic subgroups from albaqualfs and albaquults)

Source: Contributions from Dr. Alfredo Alvarado are gratefully acknowledged in the formulation of this table.

700 meters above sea level in Nicaragua. It is formed by Tertiary ignimbrites and by rhyolithic tuffs. This region is strongly dissected by numerous rivers and streams, whose waters flow into the Caribbean Sea. The soils are mainly lithosols and dystic cambisols. However, there are lithic eutric cambisols and some calcic cambisols in the intermontane valleys, and to the south, in Nicaragua, there are dystric nitosols and dystric lithic cambisols. In this region, the valley floors are occupied by fluvisols, planosols, and gleysols, all of which are very important for agricultural production.

Lowlands of the Isthmus and the Pacific Littoral

The lowlands of the isthmus and the Pacific littoral extend from the Darien, in Panama, through the central part of Costa Rica, where it adjoins the southern end of the Central American volcanic highlands. The most common formations are tuffs from the Tertiary age. However, there are also sedimentary marine formations from the Miocene. This is a relatively narrow central mountain chain with few flat areas and low coastal lands, basically littoral inlets. The mountainous landscape is covered by lithic orthic and humic lithic acrisols, as well as dystric nitosols. The Darien region in Panama has some interior valleys with gleysols, fluvisols, and gleyic acrisols, and the low coastal Pacific lands include planosols, vertisols, and regosols.

Volcanic Highlands and Lowlands Adjoining the Pacific Littoral

The subregion of volcanic highlands and lowlands that adjoins the Pacific littoral is a relatively narrow longitudinal zone of the Central American isthmus. It is bounded by a mountain chain on the Pacific side, from the southern border of Mexico through Guatemala and El Salvador to Nicaragua. It continues from there to the central mountain chain of Panama. In Costa Rica and Panama, the lower volcanic lands are surrounded by basic and acidic tuffs from the Tertiary, whereas Quaternary vulcanism characterizes the rest of the subregion. Despite the prevalence of craggy hills in the Pacific coastal chains, the thick layer of volcanic ash that covers these hills is free of erosion. This uneroded condition will continue as long as the forest cover is maintained.

The principal soils are humic and mollic andosols, although vitric andosols are the most common soils on the lowlands. Some of the oldest layers of volcanic ash give rise to chromic reddish luvisols and vertisols. Regosols occupy the highlands, where the ejecta are thick pumice stone, ash, or scoria. On the Pacific side, eutric lithic cambisols, orthic lithic luvisols, and lithosols are found in places where erosion is very active. In addition to the vertisols and luvisols, there are very limited but important gleysols and regosols in the Pacific coastal lowlands.

Caribbean Lowlands

The Caribbean lowlands follow the Antillean coast from southern Belize to Panama. They form a narrow coastal belt, which is wider on the eastern coast of Honduras, Nicaragua, and Costa Rica and narrows again along the Panamanian coast. It includes a littoral with shallow coastal water, intermittent coastal mangrove swamps, lagoons and sand belts. The Caribbean lowlands also cover salt marsh zones, low coastal terraces, and a mass of spurs of rhyolites and basic volcanic tuff from the Tertiary, as well as sandstone and schistose clay that gradually merge into the Central American highlands. The soils in the north are ferric acrisols and ferric and orthic luvisols, and in the south they are orthic, ferric, and plinthic acrisols with dystric planosols and thionic histosols.

Plateaus of the Maya Mountains (Belize)

The plateaus of the Maya Mountains in Belize consist of a mass of quartzite from the Jurassic, overlain by schistose phylithic clay and sandstone. The granitic batholithic nucleus is broken into blocks, which form craggy hills. The soils that have been identified in this region include humic acrisols and humic cambisols, as well as plinthic and gleyic luvisols. In the hills it is possible to find dystric lithic cambisols, humic lithic cambisols, orthic lithic acrisols, and lithosols, as well as rendzinas.

The Suitability of Major Soil Groups for Agriculture

Soils of the Pacific coastal plains are enriched periodically by the contribution of fine volcanic ash and sediments from the uplands. As a result, these soils are generally very fertile and productive and they are the most exploited in the region for agricultural purposes. Despite steep slopes, the periodic accumulation of volcanic ash on the Pacific coastal mountain soils prevents the formation of rocky outcrops under natural conditions. The rocky outcrops that are present are therefore considered a new and isolated phenomenon in the region, caused by deforestation and inappropriate agricultural land use. While such inappropriate agricultural use will cause erosion, these areas are classified as suitable for forestry or for restricted agricultural use, mainly for perennial crops.

The nonvolcanic highlands, the highlands of the isthmus, the Maya Mountains, and the highlands in northern Mesoamerica are characterized by shallow or highly lithic soils. Consequently, these soils are marginal for agricultural development. Although much of this land is suitable for forestry, a considerable percentage should be devoted to environmental protection and conservation. The soils with the highest agricultural potential in these subregions of Mesoamerica are the intermontane valleys that occupy the highlands and the alluvial valleys on the coastal plains. They can produce coffee, plantation crops, fruits, vegetables, and field crops.

The principal constraints for many soils of the Caribbean coastal lowlands are related to drainage. Nevertheless, many of the soils of alluvial origin have

Table 11.2 Hilly Lands in the Central American Region

Country	Total area (km²)	Percentage of hilly and highlands	Population (1986)
Guatemala	108,889	82.0	8,600,000
Belize	22,965	32.0	159,000
El Salvador	20,877	95.0	5,100,000
Honduras	112,088	82,0	4,600,000
Nicaragua	140,746	75.0	3,300,000
Costa Rica	50,800	73.0	2,700,000
Panama	77,060	76.0	2,200,000
TOTAL	533,425	75.5	25,250,000

Source: H. J. Leonard, 1986. *Recursos naturales y desarrollo en América Central: Un perfil ambiental regional* (trad. del Inglés por G. Budowski y T. Maldonado, 1987) (San José: CATIE).

demonstrated a high agricultural potential when they are adequately drained. This is the case in Honduras, Costa Rica, and Panama, where these soils are used for growing bananas, field crops, cacao, coconut, citrus, tubers, and other crops. Other soils of the coastal lowlands have high clay content, are very acidic, or are not very fertile. Generally located among low hills, these soils are very fragile because they erode easily when deforested, making them unsuitable for agricultural use.

In summary, the principal constraints associated with agricultural use of Central America's soils are slope, soil depth, and climate (Leonard, 1986). Table 11.2 shows the distribution of hilly lands for the entire region, according to Posner et al. (1983).

Land Use Capability

Land use capability refers to the most intensive land use that can be productively maintained over time without causing damage to the resource base. Land use capability has not been evaluated sufficiently in Central America. In many cases, the data used to determine it correspond to actual or current land use, not to a true evaluation of land use capability (Leonard, 1986). Moreover, the methodology used differs from one country to the other and sometimes between regions in the same country. Nonetheless, preliminary data are provided in table 11.3. Overall, data for actual land use show a decrease in forested area in the region, largely the result of an increase in lands used for grazing.

Table 11.3 Land Capability in Central America (percentage of land)

	Agriculture			Forests		
Country	Total Area (km²)	Intensive Annual Crops	Limited Annual Crops, Perennials, Grazing	Perennial Associated Crops and Timberland	Wood Production	Forests for Protection
Guatemala	108,889	4.0	22.0	21.0	37	14
Belize	22,965	16.0	23.0	15.0	27	19
El Salvador	20,877	24.0	8.0	30.0	28	28
Honduras	112,088	11.0	9.0	13.0	66	—
Nicaragua	140,746	4.0	9.0	35.0	52	—
Costa Rica	50,800	19.0	9.0	16.0	32	24
Panama	77,060	9.0	20.0	6.0	43	18
Weighted average	533,425	9.1	14.2	21.9	54.8	

Source: H. J. Leonard, 1986. *Recursos naturales y desarrollo en América Central: Un perfil ambiental regional* (trad. del Inglés por G. Budowski y T. Maldonado) (San José: CATIE).

Table 11.4 Changes in Use of Land in Central America (1960, 1980, 1990)

Country	Agriculture			Grazing			Forests			Others		
	1960	1980	1990	1960	1980	1990	1960	1980	1990	1960	1980	1990
Guatemala	4	17	17.23	10	8	12.73	77	42	36.06	N/A	33	33.98
Belize	N/A	3	2.46	N/A	2	2.10	N/A	44	44.39	N/A	51	51.05
El Salvador	32	35	33.38	29	29	29.44	1	7	5.02	N/A	29	30.16
Honduras	13	16	16.02	18	30	22.70	63	36	30.57	N/A	18	30.71
Nicaragua	10	13	10.76	14	29	44.63	54	38	30.32	N/A	20	14.36
Costa Rica	9	10	10.32	19	31	45.24	56	36	32.12	N/A	23	12.32
Panama	7	8	7.58	12	15	20.23	N/A	N/A	N/A	N/A	N/A	N/A

Source: H. J. Leonard, 1986; L. Pedroni and J. Flores Rodas, 1992.

Tables 11.3 and 11.4 show that 23.3 percent of the land in Central America is suitable for agricultural use. However, actual land use in 1990 indicated that 40 percent of the region's lands was used for agriculture. Similarly, according to land use capability classification, 76.7 percent of Mesoamerica's land should be covered by forest. However, only 34 percent was forested in 1990. These data clearly show an inappropriate level of deforestation in most areas of the region. The large increase in land devoted to grazing has occurred at the expense of land available for both crops and forestry. Furthermore, actual land use studies include a category called "other use," accounting for 26 percent of the land area. How these lands are used is not specified, confounding attempts for a rigorous analysis of actual land use in Central America.

These data demonstrate that the information about current land use is very contradictory and confusing, varying according to different sources. For example, for Nicaragua, FAO (1990b) indicated that 6.11 percent of the land was used to raise crops for export and internal consumption. Leonard (1986), however, showed that 10.76 percent of Nicaragua's land was in crop production. And Vargas (cited by Pasos et al., 1994) indicated a much higher area in crop production than either of these sources: crops, 33 percent; grazing lands, 25 percent; and forest, 42 percent.

It is also important that the lands with the highest potential for agricultural use in Central America are, in fact, often underexploited because they are used as grazing lands. Moreover, the lands with the lowest potential are overexploited, mainly for the production of crops for internal consumption by small farmers. Reforestation replaces only 7 percent of the area that is deforested annually in the region.

Conclusions

One of the most important challenges for Central American countries is to improve agricultural production in order to satisfy increasing internal demands for food. However, this must be achieved while simultaneously reducing the environmental damage caused by overexploitation of lands and poor land management. Current knowledge about soils and their suitability for different land uses in the region is very poor. In every country in Mesoamerica, different methodologies and criteria have been used. Sometimes, these methodologies result in incorrect estimates of the land that is suitable for agriculture, grazing, timber, or protective uses. In many places, basic soil studies to support these estimates do not even exist.

Soil surveys at a scale of 1:50,000 are needed throughout the Central American region to classify land use capability in detail. This classification would serve as the base for defining and implementing socioeconomic development strategies. The effort could begin at a scale of 1:200,000, as a first approximation, useful for decision making at the national level. However, in areas dedicated to intensive agricultural use, much more detailed surveys, at a scale of 1:25,000 or greater, are necessary. It is also very important to standardize the methodology and criteria used to determine land use capability throughout the region.

Soil surveys and land use capability studies are also indispensable for establishing strategies for land use planning. These strategies deal with the analysis and strategic distribution of the land use for different geographic areas in a country or region. They take into consideration the potential of natural resources; land use capability, especially for areas that require special attention; population distribution; and human activities. Land use planning is an efficient tool to determine how different kinds of human uses of the resource base—urban, agricultural, industrial, recreational, and forestry—can expand and still remain in harmony with the natural environment.

The importance of land use planning has been recognized by different institutions. In *Nuestra Propia Agenda* (*Our Own Agenda*) BID-PNUMA (1991) acknowledges that such planning is a fundamental tool for achieving sustainable development. The authors urge governments to orient their development and investment policies, in conjunction with private initiatives, to contribute to this goal.

12

Aquatic Ecosystem Deterioration in Latin America and the Caribbean

Catherine M. Pringle
Frederick N. Scatena

> Of all of our natural resources, water has become the most precious. . . . In an age when man has forgotten his origins and is blind even to his most essential needs for survival, water along with other resources has become the victim of his indifference.
>
> Rachel Carson

On a global scale, the most serious aquatic pollution problems are found in rapidly developing countries (Meybeck, Chapman, and Helmer, 1989). Here, pollutants that gradually appeared in industrialized regions over the passage of a century or more are rapidly building up in the compressed time span of decades (Boon, 1992). Pollution has emerged as a significant and alarming feature of many water bodies in Latin America and the Caribbean during the last 30 years. The control of water pollution is consequently a major challenge for water management in many areas of the region (UN ECLAC, 1990c).

Rapid increase in population growth, particularly in urban areas, is one of the most significant factors in this increase in pollution. Population growth is typically not accompanied by the development of waste treatment facilities, land use planning, and pollution control. It is estimated that almost 50 percent of the world's total population will live in the humid tropics by the end of the century (Gladwell and Bonell, 1990). Therefore, methods to deal with water quality and quantity problems are especially needed for this region and similar areas throughout the world.

Compared with those in temperate regions, there is very little baseline information on the biology of undisturbed tropical aquatic systems, making it difficult to assess the effects of human activities. Understanding the nature of hydrologic flow paths in tropical environments, how they link up different ecosystems, and the potential impact of anthropogenic alterations to these flows is very important for maintaining the ecosystem's integrity and the sustainability of water resources. A stream integrates processes that occur in the terrestrial environment and airshed that it drains. For instance, a headwater stream that is draining montane cloud forest is physically and biologically linked to the coastal zone. Factors such as the quality of lowland drinking water and the health and biological integrity of sea grass and coral reef communities can reflect land use activities in the mountains. Conversely, human activities in lowland stream reaches, such as water abstraction, can affect the biological integrity of highland streams by interfering with the life cycles of migratory aquatic organisms (see Pringle and Scatena, chapter 12).

Anthropogenic effects can operate over a range of scales. First, within-stream effects produce organic and inorganic pollution. Second, land use changes result in catchment-level effects. Third, flow regulation such as dams, reservoirs and water diversions also create catchment-level effects. Fourth, interbasin transfers and acid deposition are examples of supracatchment effects. Finally, global changes such as global warming affect the entire hydrologic cycle.

In this chapter, we emphasize the importance of aquatic ecosystems to whole-ecosystem integrity in Latin America and the Caribbean. We briefly review the history of human use of aquatic resources in the region, factors that affect aquatic resource deterioration, and strategies to mitigate this deterioration. In another chapter in this volume (chapter 13), we focus on the Commonwealth of Puerto Rico and the country of Costa Rica, discussing how geography, climate, ecology, human population growth, and political and economic factors such as globalization have interacted to shape water resource issues and problems in these two areas. In chapter 13 two case studies from specific regions of Puerto Rico and Costa Rica examine the ecological effects of these factors on aquatic resources and ecosystem sustainability.

Human Use of Water Resources

Knowledge of the spatial distribution of water resources and the human use of this resource is fundamental in understanding water management and sustainable development. Some of this baseline information is provided by the United Nations Economic Commission for Latin America and the Caribbean (ECLAC) (UNECLAC 1990a,b). The major hydrographic divisions of Latin America and the Caribbean are shown in figure 12.1. Table 12.1 illustrates the general distribution of human activity by major hydrographic region.

Latin America and the Caribbean have relatively abundant water resources, with an average annual rainfall of 1,500 millimeters, 50 percent higher than the world average. The region also contributes almost one-third of the total worldwide land drainage entering the ocean (Inter-American Development Bank, 1984). However, although Latin America is among the most humid regions of the world, it also contains some of the most arid areas, such as the Atacama Desert in northern Chile.

Differences in the physical availability of water, coupled with variations in human population densities and activities, produce strongly contrasting patterns of use and transformation of water resources within the region (UNECLAC, 1990a). Vast areas of Latin America are still considered relatively undisturbed. Almost half is classified as forest and woodland, although much of this is secondary forest. However, industrial regions that have large urban populations also exist. As human populations expand, human-dominated ecosystems are exerting an increasing effect on these relatively undisturbed systems.

Making generalizations about water use in Latin America and the Caribbean is difficult. The overall pattern of human water use has been described as "spatially sporadic and highly concentrated," with the greatest use concentrated in coastal areas (UNECLAC, 1990a). Changes in patterns of land use have been significant since the time of earliest human settlements and are probably the most important means by which human activities affect hydrographic patterns in the region.

Large dams were not constructed in Latin America until late in the colonial period. The earliest existing large dam was built for irrigation on the Saucillo River in Mexico in 1750. Irrigation has remained the major purpose of dam construction. It predates the arrival of the Spanish in many areas of Latin America, including the Caribbean, Mexico, Peru, and the central valley of Chile and adjacent Argentina. Since then, the proportion of area cultivated under irrigation has increased dramatically. Mexico has the greatest amount of irrigated land, but over the last two decades the largest increases have occurred in parts of central and southern Brazil, Central America, and Cuba (UNECLAC, 1990a).

Hydroelectricity has become a significant secondary purpose of dams with increasing industrialization. The rate of dam building has remained high since 1950. Though regulation was initially restricted to smaller streams, large river systems that drain into the Atlantic Ocean in the Plate basin have been dammed (figure 12.1). The water held in reservoirs has increased more than twentyfold since 1945 (UNECLAC, 1985). Nonetheless, most large river systems in Latin America remain unregulated (UNECLAC, 1990a).

The gross domestic product of Latin America has grown at an average annual rate of almost 5 percent over much of the last four decades. Resulting changes in the internal economies of the region, such as the increasing importance of the manufacturing sector, have affected water use. Latin America is the one region of the world, however, where the agricultural frontier continues to expand, and many hydrographic regions remain predominantly in agriculture. Industrial and urban activities are heavily concentrated in a few catchments, where they have a highly significant but localized influence on flow patterns and quality of surface waters. For example, the highest concentrations of human population and economic activity occur in only 3 of the 26 major hydrographic divisions: the Gulf of Mexico, South Atlantic, and Plate basins (figure 12.1; table 12.1). These three basins contain almost 40 percent of the human population and account for 52 percent of the total gross domestic product of the region (UNECLAC, 1990a). Many river basins remain only sparsely populated.

Although Latin America and the Caribbean may enjoy the most abundant freshwater resources in the world on a per capita basis (Witt and Reiff, 1991), regional water shortages frequently occur. For instance, Rice (1986) considers the problem of regional water shortages in the Caribbean, which received new emphasis when Exxon Company tankers were caught filling up with Hudson River water for export to their refineries in Aruba.

Major Contributors to Water Problems and Effects on Human Populations

There has been no systematic regional evaluation of the evolution of water pollution and/or its economic consequences in Latin America and the Caribbean. Further-

Table 12.1 Latin America and the Caribbean Distribution of Human Activity by Major Hydrographic System

Hydrographic system	Area km^2	Population 1960	Population 1970	Population 1980	Density 1980	Annual avg. percentage increase (1960–1980)	(millions of 1980 US dollars) GDP 1980	GDP 1985	% GDP Agriculture	Industry	Service
Central America and the Caribbean											
California	471,473	2,820,032	4,132,946	5,739,565	12.17	3.62	15,100.8	16,620.4	16.2	20.9	63.0
Caribbean	646,213	16,840,767	22,384,393	28,497,374	44.10	2.66	32,857.4	36,332.2	22.5	24.8	52.6
Caribbean Islands	230,789	20,328,660	24,430,255	30,044,411	130.18	1.97	55,517.0	63,565.5	12.7	33.7	53.6
Gulf of Mexico	474,552	19,724,062	28,113,599	39,783,373	83.83	3.57	109,288.7	120,005.2	8.5	35.4	56.1
North Pacific	328,406	8,121,876	10,618,230	13,820,162	42.08	2.69	26,095.6	28,673.9	19.6	21.8	58.6
Northern Endorheic, Mexico	140,840	699,272	919,139	1,143,122	8.12	2.49	2,807.6	3,090.2	19.5	27.6	52.8
Rio Bravo	214,096	1,621,683	2,243,483	3,112,607	14.54	3.31	10,672.9	11,747.0	8.5	44.0	47.5
Southern Endorheic, Mexico	236,637	2,257,220	2,829,615	3,727,082	15.75	2.54	8,493.3	9,348.0	35.4	23.9	40.7
Yucatan	141,523	832,437	1,098,061	1,710,271	12.08	3.67	3,518.4	3,872.4	17.0	26.6	56.4
South America											
Amazonas	6,157,253	10,414,471	13,684,801	18,416,972	2.99	2.89	19,242.7	19,924.1	22.9	22.1	55.0
Central Chile	116,002	5,466,021	6,809,934	8,324,396	71.76	2.13	18,245.3	18,022.3	7.6	36.8	55.6
Central Venezuela	142,419	4,356,758	6,346,751	8,661,137	60.81	3.50	30,562.5	28,189.1	7.0	29.8	63.2
Guayanas	468,235	895,431	1,149,533	1,186,631	2.53	1.42	1,702.0	1,655.0	14.8	35.3	49.9
Endorheic, Argentina	706,869	3,923,597	4,405,305	5,333,450	7.55	1.55	11,556.8	10,127.7	19.4	27.8	52.8
Maracaibo	101,688	2,450,814	3,296,334	4,114,182	40.46	2.62	12,062.3	11,265.3	8.2	29.3	62.5
Northeast Brazil	881,361	12,415,643	15,647,407	19,175,848	21.76	2.20	13,284.2	14,466.3	18.2	16.0	65.8
Orinoco	1,116,599	3,490,445	4,475,929	5,646,580	5.06	2.43	12,278.3	12,069.5	19.0	27.1	53.9
Pacific: Arid Climate	590,419	6,630,031	9,575,465	12,795,372	21.67	3.34	19,070.3	18,977.5	10.7	41.5	47.8
Pacific: Tropical Climate	348,495	12,840,420	17,273,856	21,506,679	61.71	2.61	26,547.9	28,283.2	28.8	22.9	48.3

Pampa	621,207	2,347,099	2,887,705	3,666,148	5.90	2.25	9,896.1	8,672.4	14.9	45.8	39.3
Patagonia	487,645	201,851	285,909	402,784	0.83	3.51	1,691.3	1,482.1	14.5	49.6	35.9
Plate	3,878,926	42,987,903	56,231,529	70,061,929	18.06	2.47	204,115.6	207,430.0	11.3	37.1	51.6
Sao Francisco	617,778	10,758,761	13,051,100	15,607,537	25.26	1.88	20,445.7	22,265.0	17.6	21.7	60.7
South Atlantic	795,875	16,634,026	20,799,476	30,500,661	38.32	3.08	73,943.9	80,471.3	9.5	26.9	63.6
South Pacific	343,471	1,405,130	1,564,822	1,736,796	5.06	1.07	2,698.6	2,664.7	27.0	18.0	55.0
Titicaca	112,501	753,130	912,488	1,089,602	9.69	1.86	671.3	619.0	17.7	34.6	47.7
Total Latin America	20,371,272	211,217,540	275,168,065	355,804,671	17.47	2.64	742,366.5	779,839.3			

Sources: GDP 1980 and 1985: ECLAC, Statistics Division, national accounts (computer printout), 1988.
GDP Caribbean Islands and Guayanas: United Nations, *National Accounts Statistics: Analysis of Main Aggregates*, 1982 (ST/ESA/STAT/SER.X/2), New York, 1985, Sales No. E.85.XVII.4. Population and Area: National censuses and *United Nations Statistical Yearbook*. Structure of GDP: ECLAC and national reports.
GDP by sector: Economic Commission for Latin America (ECLA), *Distribución regional del producto interno bruto sectorial en los países de América Latina*, Cuadernos Estadísticos de la CEPAL series, No. 6 (E/CEPAL/G.1115), Santiago, Chile, 1981.

Notes: 1. Year for the structure of GDP by country: Argentina 1968/Brazil 1980/Chile 1980/Colombia 1975/Ecuador 1965/Panama 1968/Peru 1980/Mexico 1980/Uruguay 1961.
2. The regional distribution of GDP for Bolivia, Costa Rica, Guatemala, Honduras, Nicaragua, and Venezuela has been calculated in proportion to the distribution of population in 1980 by administrative region and the base of national GDP.

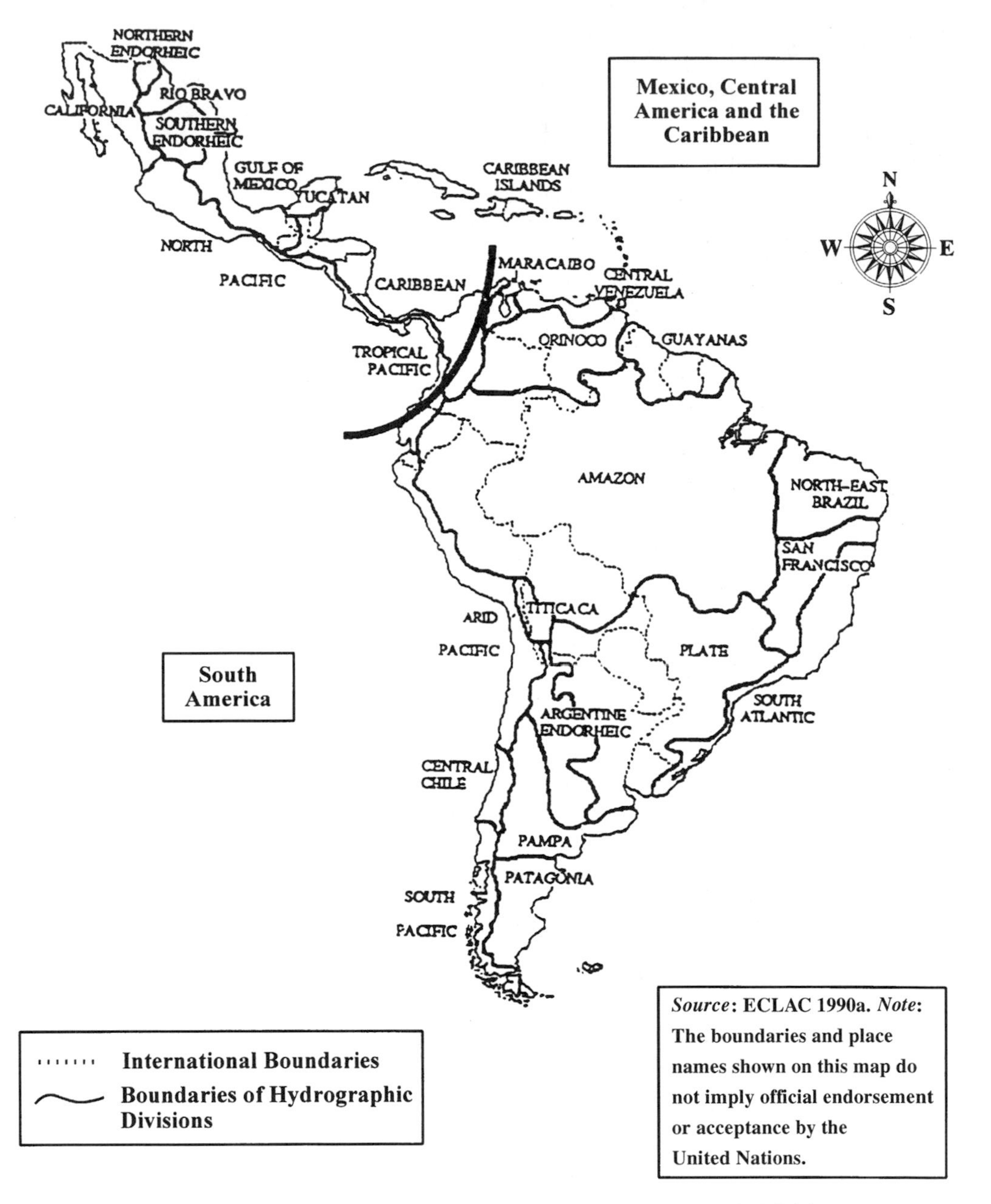

Figure 12.1 Latin America and the Caribbean: Main Hydrographic Divisions.

more, the overall magnitude of aquatic resource deterioration is not known. A report that synthesizes existing reports and available information on water quality and its regulation concludes that although some progress has been made toward solutions, water quality continues to decline steadily in the region and control efforts are weak (UNECLAC, 1990c).

Many anthropogenic factors affect tropical and subtropical ecosystems of Latin America, including deforestation, agriculture, mining, urban and industrial development, river diversions, damming, impoundments, and the destruction of wetlands and riparian zones (Gladwell and Bonell, 1990). The discharge of untreated and/or inadequately treated domestic and industrial wastewater in combination with nonpoint source pollution such as agricultural runoff are major factors that contribute to the rapid decline in water quality. Several rivers that have very large flow volumes, for example the Cauca and Magdalena Rivers of Colombia and the Mantaro River in Peru, show the severity of the problem (UNECLAC, 1990c). In Latin America, sewer systems are accessible to only 41 percent of the urban population, and more than 90 percent of collected wastewater is discharged completely untreated into rivers (Nash, 1993).

Polluted drinking water causes increasing human health problems from water-related diseases (see Reiff, 1993). For instance, epidemic cholera, one of more than 20 serious water-related diseases, struck Latin America in 1991 for the first time in this century. Epidemics of cholera occurred in Peru, Ecuador, and Colombia, with additional outbreaks in Mexico, Guatemala, Chile, Bolivia, El Salvador, Panama, and Brazil (Witt and Reiff, 1991). These epidemics and outbreaks have been tied to rapidly expanding human populations and untreated municipal wastewater (PAHO, 1990a). Many agree that the high rate of infant mortality and the incidence of various intestinal infectious diseases in Latin America and the Caribbean, compared with developed countries, is at least partially attributable to the biological pollution of surface water by human wastes (UNECLAC, 1990c).

The most serious nonpoint source pollution is runoff of sediment, nutrients, and pesticides from deforested land and storm water flows from urban areas. Increased amounts of sediment can result in substantial economic losses downstream by affecting irrigation, navigation, hydroelectricity, and potable water supplies (see Pringle and Scatena, chapter 12). As one of many examples, deforestation in the watershed behind the El Cajón dam, constructed in Honduras in the early 1980s, resulted in a rapid buildup of sediment in the reservoir, seriously compromising hydroelectric power production. By 1994, over half the 25,000-acre reservoir system was a giant mud flat. Consequently, cities and regions throughout Honduras were allocated power in four- and six-hour units on an irregular schedule. In 1994 alone, Honduras will spend $60 million on imported energy, and it is estimated that power shortages alone are costing the country $20 million a month in lost industrial production (Gollin, 1994). Almost $1 billion was spent on construction of the El Cajón dam, but virtually nothing was spent to protect the watershed. As part of a belated effort to stop deforestation in the El Cajón River watershed, the Honduran Forestry Development Corporation shut down 6 sawmills that were operating near the reservoir. However, another 20 mills are still in operation. In September 1994, the Inter-American Development Bank and the government of Honduras agreed on a $20.4 million emergency loan to reforest and protect the watershed (Gollin, 1994).

Irrigation water is often a major source of suspended solids, sediments, pesticides, dissolved salts, and fertilizers. The dissolution of naturally occurring salts in the soil by irrigation is causing water salinization, which renders the water unsuitable for other uses. Also, as in other parts of the world, the reuse of drainage water for irrigation accelerates the process of soil salinization, increasingly affecting many areas in South America, Central America, and Mexico. For example, on the coast of Peru, an estimated 34 percent of the land is estimated to suffer from salinization and drainage problems (Alva et al., 1976). Desertification claims 2,250 square kilometers of farmland in Mexico each year (Grainger, 1990). In two-thirds of all Mexico's arable lands, the water supply is the main factor in limiting agricultural productivity (Cummings, 1989). Dwindling water supplies coupled with soil erosion are responsible for abandonment of at least 1,000 square kilometers of farmland each year (M. Redclift, 1988).

Consumption of fertilizers in Latin America and the Caribbean increased by approximately 97 percent between 1973 and 1985. This is a much smaller volume than that used in industrialized countries. However, fertilizer runoff contributes substantially to the eutrophication of freshwater systems, resulting in increased biological oxygen demand, algal growth, and human health hazards (UNECLAC, 1990c).

Agricultural pollution of freshwater by herbicides and pesticides is also a growing problem in Latin America and the Caribbean. Although the total volume of pesticide consumption is not known, pesticide imports increased by almost half between 1971 and 1973 and 1983 and 1985 (Postel, 1987). Chemical substances are often employed that are restricted or no longer permitted in countries with more stringent environmental legislation (UNECLAC, 1990c). For example, the pesticide dibromochloropropane (DBCP) is classified by the World Health Organization as extremely hazardous, and it is banned in many countries. However, it is used in Ecuador, Honduras, and possibly Colombia and Panama (Chetley, 1987). The pesticide DDT (dichlorodiphenyl-trichloroethane) is still widely used in several countries of Central America. One contributing factor to this dangerous use of pesticides is that there is no centralized authority to regulate pesticide trade, use, and application in many countries (UNECLAC, 1990c).

Water diversions from rivers for irrigation and drinking water can also threaten the ecosystem's integrity and cause water quality problems. A compelling example is provided by the River Guayas in Ecuador. Water abstraction has resulted in saltwater intrusion to such an extent that it precludes using the river as a source of potable water for the city of Guayaquil during periods of low flow (UNECLAC, 1990c). Water abstraction from rivers also reduces the water available for the dilution and transport of domestic and industrial sewage, increasing the concentration of pollution. This can be particularly problematic when peak periods of sugar and coffee processing in tropical countries coincide with both low flows and maximum water demand (UNECLAC, 1990c).

Groundwater pollution is of growing concern in the region, particularly in cities that rely on springs and wells for drinking water and irrigation (e.g., Mexico City and Havana). The increase in nitrate content in groundwater associated with the use of sewage for irrigation and with other agricultural activities is an increasing problem in some South American countries such as Chile (Pena-Torrealba, 1993). However, contamination of groundwater by fertilizers and toxic agrochemicals in Latin America and the Caribbean does not yet represent as common a problem as it does in more industrialized countries. This is due to the relatively lower use of these materials in agriculture in the region than in most industrialized countries and possibly to climatic and soil characteristics (UNECLAC, 1990c). Certain soils in tropical climates have a lower risk of nitrate leaching than do similar soils under temperate conditions (Golubev, 1984). The growing exploitation of groundwater for irrigation and other purposes has led to contamination of freshwater coastal aquifers by saltwater in many regions, including the Caribbean, Argentina, Brazil, El Salvador, and Mexico (UNECLAC, 1990c). This is a particular problem in coastal areas underlain by limestone.

Pollution from acid rain and dry deposition is becoming an increasing concern in Latin America because of both local sources and long-range transport from northern areas. However, little information exists on the relationship between air and water pollution in the region (see Rodhe and Herrera, 1988, and UNECLAC, 1990c, for discussions of this problem).

Effects of Water Resource Deterioration on the Biological Integrity of Aquatic Ecosystems

Overall, very little scientific information has been collected about the direct effects of human activities on the "nonhuman" ecology of tropical aquatic ecosystems in Latin America. Environmental effects of human activities on the ecology of aquatic systems are often overlooked, even though the maintenance of the biological integrity of freshwater systems results in important ecosystem services to human populations and is crucial to ecosystem sustainability. The little we do know is often confined to the ecology of coastal environments, particularly coral reefs (Connell and Hawker, 1992; Goenaga, 1991; Guzmán and Jiménez, 1992; Seeliger, deLacerda, and Patchineelam, 1988). For instance, high levels of common pesticides and herbicides, including DDT and DDT derivatives, contributed to the mortality of 50 to 80 percent of the coral on Chiriquí reefs off the coast of Panama between January and April 1983 (Glynn et al., 1984). Herbicides are regularly sprayed on agricultural fields in adjacent coastal areas, and there have been unofficial reports that crop dusters release surplus chemicals directly into the Gulf of Chiriquí (Glynn et al., 1984). Similarly, heavy metal contamination from sewage discharge, oil spills, and the misuse of agricultural chemicals and fertilizers are problems throughout coastal areas of Central America (Guzmán and Jiménez, 1992). Rivers that drain the Caribbean coast of Costa Rica and Panama are increasingly loaded with suspended sediments because of deforestation. These rivers carry heavy metals many kilometers from the source to the sea, where they are polluting coral reefs (Guzmán and Jiménez, 1992).

Determining direct cause-and-effect relationships in highly dynamic riverine ecosystems is difficult (Petts et al., 1989), and this contributes to our lack of information about the human impact on freshwater biota and overall freshwater ecology in Latin America. However, the sheer magnitude of freshwater deterioration is dramatically affecting the ecological integrity of many regions in Latin America. These effects are consequently receiving more attention (FAO, 1993a), particularly when they have an economic impact. For example, decreases in riverine fish catches and in general abundance have been related to the increase of toxic substances used in agriculture and industry in La Plata, a regulated river-floodplain system in Argentina (R. Quirós, 1993). Fruit- and seed-eating fish in the genera *Colossoma* and *Brycon* and the big catfish *Paulicea lutkenii* have virtually disappeared from the commercial catch in the lower Paraná, La Plata, and Uruguay Rivers (Fuentes and Quirós, 1988; R. Quirós, 1993). Migratory fish species, such as *Basilichthys* and *Lycengraulis* spp., which usually move upstream from the estuary in the winter, have almost disappeared from commercial catches in the middle Paraná. Commercial catches of the predaceous fish *Salminus maxillosus* have been decreasing since the late 1940s in the lower basin (R. Quirós, 1993).

Strategies to Address Aquatic Ecosystem Deterioration in Latin America and the Caribbean

Protecting aquatic and riparian resources requires a complex mix of land use and water supply management. Protection is difficult even in countries such as the United States, where many federal and local laws apply to water supplies and riparian zones. The lack of enforcement and the lack of a legal focus on riparian zones and nonpoint source pollution have limited the effectiveness of these regulations (Adler, Landman, and Cameron, 1993; Lamb and Lord, 1992). A recent review of the management of tropical freshwater systems concluded that ecological integrity and sustainability have not been achieved by focusing solely on aquatic resources, ignoring the terrestrial linkages, or by focusing solely on civil structures (Dudgeon et al., 1994).

Multiscale management strategies that consider the longitudinal and lateral dimensions of aquatic habitats and the role of riparian vegetation were recommended (Dudgeon, 1994).

The river basin concept has not been widely applied in water management in Latin America and the Caribbean. Although isolated application of the concept has occurred in some countries in the past, there are few contemporary examples (Lee, 1991). Integrated river basin management—management efforts that observe watershed boundaries and consider the entire river basin—is not commonly used, even at the planning stage. Brazil and Chile, however, have received recognition for their recent innovations in water management policy. These innovations may eventually lead to the creation of national water management systems based on the concept of integrated river basin management (Lee, 1991). One of the first steps in preventing pollution is to assess the health of water bodies. This requires the development of monitoring programs that assess pollution levels in major water bodies. The importance of national water quality programs and their absence throughout much of Latin America and the Caribbean is discussed elsewhere (Ongley, 1993).

In the United States, freshwater conservation efforts increasingly focus on river basin management and maintenance of the integrity of the entire ecosystem (Doppelt et al., 1993; Naiman et al., 1995). Management and conservation efforts have moved away from a focus on commercial fishing and recreational interests toward issues of biodiversity and restoration (Dewberry and Pringle, 1994; Pringle and Aumen, 1993). However, political and socioeconomic factors in the United States currently threaten basic environmental legislation, such as the Clean Water Act of 1972, that underpin this current conservation focus in the United States. For example, over the last two decades, the Clean Water Act has reduced annual discharges of toxic chemicals and raw sewage into U.S. lakes and rivers by about a billion pounds and 900 million tons, respectively (Adler et al., 1993). Because of this statute, 66 percent of U.S. waters are now classified as safe for swimming and fishing, up from 36 percent before the enactment. Consequently, conservation efforts in the United States have recently concentrated on protecting and improving this threatened legislation.

In contrast, in developing tropical countries, the maintenance of water quality is being driven primarily by human health concerns (PAHO, 1992). About 80 percent of tropical diseases are water related and can be attributed to poor or nonexistent sewage treatment and lack of safe drinking water (Gladwell and Bonell, 1990). Large segments of the population do not have access to potable water supplies. Many surface water problems in Latin America, such as low oxygen levels that result from high organic loading, high levels of toxins, and fecal coliforms, resemble problems that were common in the United States more than two decades ago, before the Clean Water Act led to massive water cleanup and sewage treatment programs.

Most Latin American and Caribbean countries have begun to develop a body of law on pollution control, including legislation that empowers public agencies to take steps to control water pollution (UNECLAC, 1990c). A general discussion of laws aimed at controlling water pollution in Latin America can be found elsewhere (UNECLAC, 1990c).

Public understanding and involvement are also critical components of any water quality improvement effort, but they have frequently been overlooked (Nash, 1993). Environmental surveys conducted in Costa Rica of three socioeconomic groups indicated that only 52 percent of those surveyed ($n = 300$) perceived any link between population growth and environmental quality, and of those, 33 percent could not describe the link (Holl, Daly, and Ehrlich, 1995). These findings underline the importance of environmental outreach and education. An important step in solving water quality problems in Latin American countries and the Caribbean will be the identification and organization of user groups. American Rivers, the Pacific Rivers Council, the Nature Conservancy, the Sierra Club, Trout Unlimited, and other nongovernment organizations (NGOs) have stimulated and carried out many programs in freshwater conservation in the United States (Pringle and Aumen, 1993). Lawrence (1993) also emphasizes the importance of active community involvement and participation for effective protection of Jamaica's deteriorating water resources. However, these groups do not exist or are in the very early stages of development in much of Latin America and the Caribbean.

Developing countries have the opportunity to learn from costly environmental mistakes made elsewhere. For instance, given the increasing demand for hydroelectric power in Latin America, dams and hydroelectric facilities can be designed to avoid mistakes made in developing countries, as described by Rosenberg, Bodaly, and Usher (1995).

Freshwater aquatic ecosystems have received increased attention and protection in some areas of Latin America over the last decade. For instance, in 1985 the president of Brazil declared the Pantanal Matograossense a priority area for research and conservation as part of the opening ceremonies of a worldwide campaign directed at the conservation of wetlands (Alho, Lacher, and Goncalves, 1988). The Pantanal is the world's largest wetland. It is one of the largest breeding grounds for waterfowl and an important habitat for many of Brazil's threatened or endangered mammals, including jaguars, giant anteaters, and swamp deer. This wetland remains largely unstudied. It is currently threatened by deforestation, expanding agriculture, illegal hunting and fishing, and pollution of the water with herbicides and pesticides (Alho et al., 1988).

Environmental decline of water resources demands urgent attention. Problems can be addressed through remedial and preventive measures in many cases. To do this, however, will require addressing a spectrum of interests, including (1) the innovation and dissemination of less polluting technology; (2) concentrated efforts by local, national, and international agents to clean up wastes and improve waste disposal (Bartone and Salas, 1984; Inter-American Development Bank, 1984; PAHO, 1990a,b, 1991, 1992); (3) conservation measures to protect marginal and common lands, to promote reforestation, and to protect water quality and fisheries; (4) clarification of land tenure and adoption of agrarian reform; and (5) measures to address poverty (Lee, 1988; UNFPA, 1991; Witt, 1984). Options for policy responses in the domain of environmental pollution and population growth are described elsewhere (UNFPA, 1991).

The Lack of Baseline Information on Tropical Aquatic Systems

Baseline information on the ecology of tropical aquatic ecosystems is an urgent need in both the neotropics and Asia for the development of ecologically sound water management strategies (Dudgeon et al., 1994). Much of the information available has been generated from studies in temperate systems (U.S. Forest Service, 1984). Those few studies that address the management of tropical systems (Krishna, 1990; Santiago-Rivera, 1992) often do not consider the unique ecological characteristics of tropical ecosystems and/or requirements of tropical aquatic biota (see Pringle and Scatena, chapter 12).

For example, very little is known about migration patterns, food supplies, or breeding grounds for even commercially important fish species in tropical Asia (Dudgeon et al., 1994) and the neotropics. Also, the lack of long-term data on stream discharge throughout the neotropics reduces the ability to design effective water-intake systems for municipal water supplies and other uses. Because of these gaps in knowledge, in-stream flow needs of aquatic species and river ecosystems are also not known. In-stream incremental flow models have shown promise in Puerto Rico (Scatena, 1995) and Hawaii, but there is insufficient information for widespread use in other tropical areas.

Protection of riparian zone vegetation along the margins of streams and wetlands is an important conservation strategy that is based on the premise that riparian zone function and structure have a controlling influence on aquatic habitats (see Boon, Calow, and Petts, 1992; Naiman, 1992). Although there is little information about the structure and function of riparian zones in tropical ecosystems (Scatena, 1990), their protection is particularly important for detrital-based aquatic food webs, which are typical of many tropical streams. Much more information is needed on riparian zone–stream interactions in tropical areas.

The ecological consequences of dam construction are complex and are seldom studied in tropical regions (Goldman, 1976; Obeng, 1981). There is a critical need for such studies in Latin America and the Caribbean (see Pringle and Scatena, chapter 12). For example, in the last decade researchers have discovered that fish are becoming contaminated by mercury in new reservoirs created in the temperate zone (Hecky et al., 1991; Rosenberg et al., 1995). The methyl mercury accumulating in fish is apparently microbially transformed from ambient natural mercury sources in newly flooded reservoirs characterized by high amounts of decomposing organic debris. Reservoirs are now recognized as a leading cause of this contamination. However, very few studies have examined this phenomenon in tropical areas (see Yingcharoen and Bodaly, 1993, for examples).

Developing new methods of monitoring water quality for tropical aquatic systems is another critical need. The biological and chemical indexes of water quality developed for temperate zone systems are often not appropriate for tropical systems. For instance, coliforms are the standard indicator of recent fecal contamination in the United States. However, recent research in Puerto Rico and elsewhere reveals that tropical ambient temperatures and humidity can promote high coliform counts, including *Escherichia coli*, in primary forest undisturbed by human activities (Carrillo, Estrada, and Hazen, 1985; Hazen et al., 1987).

Summary

Latin America and the Caribbean are generally characterized by relatively abundant water resources. However, rapid deterioration in water quality because of population growth, land use changes, and associated increases in demands for water has emerged as a major challenge in many areas of the region in the last few decades. Recent freshwater conservation efforts in the United States have focused on river basin management and maintenance of the integrity of the entire ecosystem. In developing tropical countries, however, concern for water quality is driven primarily by human health concerns. Large segments of the population do not have access to potable water supplies, and approximately 80 percent of tropical diseases are water-related and can be attributed to poor or nonexistent sewage treatment. The importance of the problem is evidenced by an epidemic of cholera that struck many areas of Latin America in 1991 for the first time in this century. Water resource protection and management is crucial to both the health and the economy of the region. Strategies to mitigate aquatic ecosystem deterioration include (1) the development of integrated river basin management strategies, (2) the establishment of water quality and quantity monitoring stations, (3) biological assessments, (4) land use planning, and (5) the establishment of minimum flow standards and riparian zone protection areas.

Baseline information on the ecology of tropical aquatic ecosystems is needed for the development of ecologically sound water management strategies in Latin America and the Caribbean. Clearly, much of the information available to resource managers of tropical aquatic systems has been generated from studies in temperate systems. Those few studies that address the management of tropical systems often have not considered the unique ecological characteristics of tropical ecosystems. More information is particularly needed for tropical areas on the life history and ecological requirements of freshwater biota; riparian zone–stream interactions; and the ecological consequences of stream regulation, dam construction, and water abstraction. Also, given the differences between temperate and tropical systems, indexes of water quality developed for temperate zone systems are often not appropriate for tropical systems (e.g., fecal coliform counts). Consequently, the need to develop new and effective biological and chemical methods of monitoring water quality is urgent.

Acknowledgments National Science Foundation grant DEB-95–28434, to the University of Georgia, and BSR-88–11902, to the International Institute of Tropical Forestry (Southern Forest Experimental Station) and the Center for Energy and Environmental Research (University of Puerto Rico), partially supported the writing of this chapter. We thank J. Benstead, T. Laidlow, and A. Raminez for their comments on the manuscript.

13

Freshwater Resource Development

Case Studies from Puerto Rico and Costa Rica

Catherine M. Pringle
Frederick N. Scatena

The history of freshwater resource development and associated environmental problems in both Puerto Rico and Costa Rica illustrates how geography, climate, ecology, human population growth, and political and economic factors like globalization have interacted to shape water resource issues. The effects of these factors on aquatic resources and the ecosystem's sustainability are examined for specific regions of each country in two separate case studies. The first deals with conflicts between water use and the retention of biological integrity in the Caribbean National Forest of Puerto Rico. The second documents changing land use and aquatic deterioration in the Caribbean lowlands of Costa Rica. The circumstances presented in these two case studies show how increases in human population and land use activities are threatening forested watersheds and associated water resources throughout Latin America and the Caribbean.

History of Freshwater Resource Development and Problems in Puerto Rico

Puerto Rico is the smallest island (8,895 square kilometers) in the Greater Antilles. Properly called the Commonwealth of Puerto Rico, it is a territory of the United States. A population census showed that there were 3.5 million persons, with 396 inhabitants per square kilometers, in 1990 (U.S. Department of Commerce Bureau of Census, 1992), a density that is higher than the most densely populated countries in the European Union. In contrast to other countries in Latin America, population pressure in Puerto Rico has traditionally been relieved by immigration to the United States. Approximately 40 percent of the island is made up of mountains, 35 percent of hills, and 25 percent of coastal plain (Pico, 1974). The island has approximately 1,300 streams, 17 of which are large enough to be classified as rivers. There are no natural lakes, and locations for reservoirs are few. The contamination of groundwater from accidental spills, industrial wastes, and leaking septic tanks has limited the groundwater and surface water that can be developed without treatment in Puerto Rico (Hunter and Arbona, 1995; Zack and Larsen, 1993).

Puerto Rico is one of the wettest islands in the Caribbean and has a global airshed that receives precipitation from at least five major weather systems: the northeast trade winds, tropical depressions and hurricanes, northern cold fronts, and systems originating from the Pacific and the Amazon basin (Lugo and Scatena, 1992). The total annual rainfall over the island is approximately 1,930 millimeters per year (Quiñones, 1989). However, rainfall is not evenly distributed, and it ranges from nearly 5,000 millimeters per year on mountain peaks that face the trade winds to areas with subtropical, dry life zone conditions on the windward side of the island. Of the total precipitation that falls on Puerto Rico, 65 percent is lost to evapotranspiration and 33 percent to surface runoff and ground water recharge (Quiñones, 1989). Only 1.4 percent of the total rainfall is consumed. Of total stream flow, approximately 70 percent flows out to sea during storms. Municipal uses of this water are impossible because of the large short-term fluctuations in stream flow and because even the largest reservoirs on the island are small and can be filled several times during one large storm (Morris, 1994).

Europeans first visited Puerto Rico in 1494, during Columbus's second trip to the New World. Early accounts document the abundance and high quality of water, and the island was often called the "land of many rivers." The rate of landscape transformation by Europeans was initially very rapid but slowed after 1530 with the discovery and exploitation of the continental New World (Scatena, 1989). Settlements and agriculture were generally restricted to coastal plains and intermountain valleys during the first three centuries of colonization. The most rapid change to the island's landscape began in 1815, when the Spanish Crown removed trade restrictions and opened Puerto Rico to immigration and world trade. By 1893, water quality problems were such that the colonial government recommended protected zones along the margins and headwaters of streams (F. H. Wadsworth, 1949). By the middle of the 20th century, more than 90 percent of the island's forests had been cut (Birdsey and Weaver, 1987).

The island's economy was based primarily on agriculture from the time of European discovery to World War II. The economic base has changed to a manufacturing and service-based economy over the last several decades. Beginning in the 1940s, government programs of tax exemption successfully stimulated the manufacturing industry. Reforestation of abandoned agricultural areas coincided with this economic change (Birdsey and Weaver, 1987; Thomlinson et al., in press), as well as a switch from agricultural to urban and industrial pollution of aquatic systems. This rapid industrialization, combined with a lag in infrastructural development and laxity of pollution control, has resulted in severe water problems, including a landfill crisis and a heritage of toxic dumps (Hunter and Arbona, 1995). Land use changed from small-scale agriculture or large-scale sugar cane plantations to pasture and, subsequently, to forest or urban landscapes on much of the island. Forest islands and riparian corridors have served as nuclei for forest regeneration of abandoned agricultural lands during this transition (Birdsey and Weaver, 1987; Thomlinson et al., in press).

Throughout the present century, the island's central government has had a virtual monopoly on the production and distribution of water. The government's first major involvement with water resources began with the creation of the Puerto Rican Irrigation Service in 1908. This agency was created to meet the demand for irrigation caused by planting sugar cane in the coastal plain. The agency was also responsible for constructing the first reservoirs on the island during the years 1914–1919 (Pico, 1974). The legislature created the Bureau for the Utilization of Water Resources in 1925 in response to demands for electrical power. In 1941, the legislature joined the various water management agencies in the Water Resources Authority, which was later divided into the Electrical Power Authority and the Water and Sewage Authority.

No fewer than 20 dams were constructed, at an average rate of 1 dam every 17.2 months, between 1928 and 1956 (Hunter and Arbona, 1995). By the early 1950s, virtually every stream on the island was used somewhere. Nearly all urban areas were served by aqueducts from surface reservoirs, and there was an extensive system of hydroelectric dams (Pico, 1974). During the 1960s and 1970s, water management efforts focused on the channelization of rivers for flood mitigation. Many large rivers in urban areas that had previously supplied water were channelized during this period. This channelization of rivers to reduce flooding can also reduce groundwater recharge and affect the quantity of water that enters the estuaries (Quiñones, 1989).

By the early 1980s, increased demand and poor maintenance of water supply, treatment, and distribution facilities began to plague the island. At times, water rationing, low pressure, or intermittent service affected 40 percent of the resident population (PRASA, 1995a). By the 1990s, problems became most acute in urban areas of Puerto Rico's north coast. No new water supplies had been developed in this area since 1972, despite almost a doubling of the population over the same period.

The Puerto Rico Water and Sewage Authority was also debt-ridden by the 1990s and was unable to meet increasing demands for water and local and federal environmental regulations. For example, in 1990, 82 percent of the sewage treatment plants were under court jurisdiction for failing to meet water quality standards (Hunter and Arbona, 1995). Epidemic outbreaks of gastroenteritis that affected more than 9,000 people at a time were common, as were fish kills that resulted from runoff from poultry farms, sewage treatment plants, and pesticide-laden agricultural areas (Hunter and Arbona, 1995; Pagan and Austin, 1970). In addition, broken pipes, leaks, water theft, and faulty meters were so prevalent that 40 percent of the water distributed by the Water and Sewage Authority was unaccounted for (or lost) (EIS, 1995). Losses from comparable distribution systems in the United States and Europe range from 15 to 25 percent (EIS, 1995).

Besides severe water quality problems, the quantity of water that can be obtained from surface and groundwater supplies has also become problematic. Excessive withdrawals, saltwater intrusion, and contamination have diminished groundwater resources (Hunter and Arbona, 1995; Zack and Larsen, 1993). Surface water supplies have also diminished because many of the island's principal reservoirs have become partially filled with sediment since their construction, curtailing their effectiveness in providing water and reducing flood peaks. For example, the reservoir that is the principal water supply for the San Juan metropolitan area has lost 47 percent of its initial capacity since it was constructed 40 years ago (PRASA, 1995b). The remaining usable life of the reservoir is about 20 years under optimal conditions, and the cost of restoration and maintenance dredging is estimated at $100 million over a 20-year period. In addition, the dredging has environmental costs associated with using up to 13.5 acres of wetlands as disposal sites (PRASA, 1995b).

The present public water supply in Puerto Rico cannot face prolonged droughts due to diminishing supplies, reduced reservoir storage capacity, inadequate infrastructure, and excessive losses (Morris, 1994). In 1994, during the worst drought in Puerto Rico's history, water rationing affected 1.4 million people. This drought, which has an estimated recurrence period of 30 to 40 years, was so severe that the entire island was declared an agricultural disaster area by the U.S. federal government. The water supply in the metropolitan San Juan area was cut off from 12 to 36 hours at a time for several months. The Puerto Rico planning board estimated that the total economic loss from the drought was $165 million, of which $92 million was from the agricultural sector. The Water and Sewage Authority alone suffered losses of about $20 million in revenues and spent $15 million in emergency measures.

In summary, the current crisis in water quality and quantity in Puerto Rico is the result of rapid economic growth and unplanned development, followed by infrastructral lags and laxity in pollution control (Hunter and Arbona, 1995). Historically, water crises on the island have been managed by developing large new reservoirs and by raising water prices to pay for the infrastructure (Morris, 1994). However, this approach is limited, given the lack of additional sources of large quantities of and public dissatisfaction with the failure of past rate increases. The current strategy to deal with these problems is to privatize daily operations and maintenance, increase the number of small water sources, and work toward an integrated supply and distribution network.

Conflicts between Water Use and the Retention of Biological Integrity in the Caribbean National Forest

Increased water demand by the expanding human population in Puerto Rico, combined with recent droughts, have placed severe pressure on streams that drain the Caribbean National Forest (CNF). The CNF, in the northeastern corner of the island (figure 13.1), is one of the largest remaining tracts of old growth forest (11,269 hectares) in the Caribbean. The CNF is located near the San Juan metropolitan area, where one-third of the island's inhabitants live (figure 13.1). The CNF is an important area for recreation and tourism and is a major site for tropical research. The Man and the Biosphere Programme of UNESCO declared it a Biosphere Reserve in 1976. The U.S. National Science Foundation designated it as a site for long-term ecological research (LTER) in 1988. Scientists from the International Institute of Tropical Forestry, the University of Puerto Rico, and other universities and institutions throughout the world conduct research in this highland area. The CNF is host to 61 species of trees endemic to Puerto Rico, of which 26 are endemic to the Luquillo Mountains (Little, 1970). It also supports the only remaining wild population of the endangered Puerto Rican parrot, *Amazona vittata*, an endemic species threatened by extinction.

A water use budget recently developed for the CNF (Naumann, 1994) shows that significant stream dewatering or water abstraction is occurring in order to supply municipalities with potable water. Twenty-one water intakes are operating within the forest boundaries, and there are nine large intakes outside the forest in lower reaches of streams that drain the forest. On an average day, more than 50 percent of riverine water that is draining the forest is diverted into municipal water supplies via water intakes before it reaches the ocean. Several rivers have no water below their water intakes for much of the year (Naumann, 1994).

The impact of massive water abstraction and diversions on integrity of the stream ecosystem remains largely unknown in Puerto Rico and elsewhere. As Pringle and Scatena (chapter 12) point out, much of the information available to resource managers of tropical aquatic systems has been generated from studies of temperate systems, which do not consider the ecological requirements of tropical aquatic biota. For example, streams that drain the CNF have a relatively simple food chain, typical of oceanic islands, characterized by migratory shrimp and fish species (Lugo, 1985). Almost all fish and shrimp species must spend some part of their life in the estuary to complete their life cycle. Their migration along the stream continuum forms a dynamic linkage between stream headwaters and their estuaries. Newly hatched larvae of amphidromous shrimps migrate downstream; complete their larval stage in the estuary; and migrate upstream, where they live as adults, often dominating the macrobiotic assemblage of headwater streams (Lugo, 1985).

Massive water withdrawals can seriously threaten the biotic integrity of stream ecosystems in the CNF by affecting the population dynamics of migratory shrimps and fishes, which are part of estuarine and coastal ecosystems (Benstead, March, Pringle, and Scatena, in press; March, Benstead, and Pringle, 1996). It is therefore important to understand the effects of engineering and civil structures like dams and associated water withdrawals on stream and coastal ecosystem processes, such as the distribution and migratory patterns of aquatic biota and longitudinal gradients in stream salinity. Such data can be used to develop ecosystem management plans for coastal tropical streams and rivers, not only in Puerto Rico, but also elsewhere throughout the Caribbean. These data are essential to resource managers in their attempts to balance water use and biological integrity.

Initial scientific studies conducted during the non-drought year of 1995 indicated that water abstraction and damming in the lower stretches of one of the main river drainages in the CNF significantly affected shrimp recruitment (Benstead et al., in press; March et al., 1996). Recruitment decreased because of the direct mortality

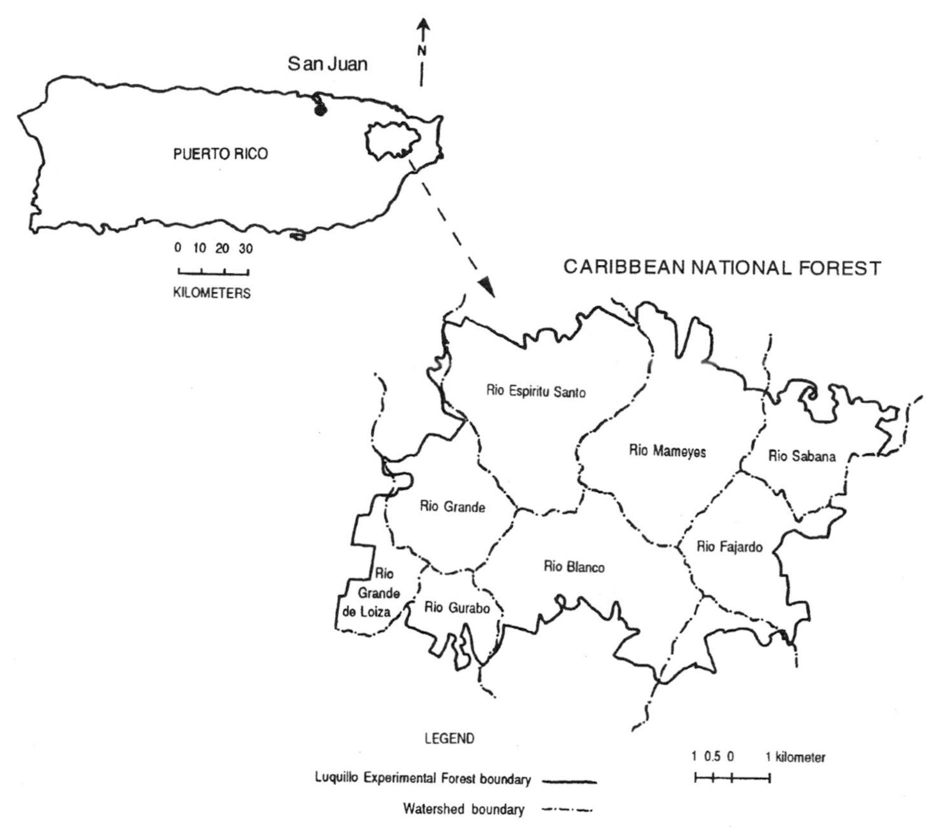

Figure 13.1 The Caribbean National Forest in Northeastern Puerto Rico with Its Nine Major River Drainages

of more than 50 percent of migrating larvae pulled into water intakes for municipal water supplies. Also, below-dam discharge was insufficient to prevent saltwater intrusion during periods of low flow, leading to a sharp physicochemical discontinuity at the dam site several kilometers inland. Furthermore, the dam functioned as a predation gauntlet. Juvenile shrimps returning upstream from the estuary faced severe predation by marine and freshwater fishes below the dam. High mortality of migratory larval and juvenile shrimps can affect the biotic integrity of aquatic systems in the CNF because shrimps have important effects on ecosystem-level processes, such as primary production, organic matter processing, sedimentation, and algal and insect community composition (Covich et al., 1991; Pringle, 1996; Pringle and Blake, 1994; Pringle, Blake, et al., 1993).

The most recent studies predict that the demand for water in the municipalities along the northern border of the CNF will increase from 28.3 million gallons per day (MGD) to 36.1 MGD between the years 1990 and 2040 (U.S. Army Corps of Engineers, 1993). Most of this increased demand will occur before the year 2000. Today, all except one of the nine major streams in the CNF have dams and associated water withdrawals. A proposal is currently being considered to dam the last remaining undammed river, the Río Mameyes. The current and future water withdrawals from the CNF will clearly create increasing conflict with other important functions of the forest, such as recreation, scientific research, and the maintenance of the original biodiversity of the island.

Water resource problems faced by the expanding population of Puerto Rico represent the tip of the iceberg and are indicators of what many rapidly developing neotropical areas are or will be experiencing in the near future. Costa Rica is a case in point.

History of Freshwater Resource Development and Problems in Costa Rica

Costa Rica (51,900 square kilometers) has three major geographic regions—the Caribbean flat and open coast lowlands, the interior highlands, and the hilly Pacific coast—and includes a diversity of habitats in 12 ecological life zones (Holdridge et al., 1971). The orientation of mountain ranges along the longitudinal axis of the country contributes to the high number of watersheds characterized by steep gradients in their headwaters (see Vargas, chapter 10). Costa Rica has more than 100 river basins and has been subdivided into 34 major watersheds (Hartshorn et al., 1982).

The country is a democracy, with a population of 3,029,700 in 1992 and a population growth rate of 2.6 percent (Mata and Blanco, 1994). The Central Valley (6 percent of the nation) is the most densely populated portion of the country. It is inhabited by more than 1.6 million people, which translates to a population density of 300 persons per square kilometers (Arcia, Merino, and Mata, 1991). In contrast, the Pacific lowlands have a population density of 20 persons per square kilometers, and the Atlantic lowlands have a density of 15 persons per square kilometers.

Before the arrival of the Spanish, Costa Rica was inhabited by scattered Indian populations that practiced shifting agriculture (WRI, 1991). The native population of Costa Rica did not recover for 300 years after the arrival of the Spanish. The population density was one person per square kilometer, with about 100,000 mature trees per person, at the time of independence in 1822 (Tosi, 1974). Agricultural settlement began in the Central Valley and in Nicoya, primarily to produce sugar cane, tobacco, and coffee. Coffee exports started by 1825 and quickly grew to dominate the economy in subsequent decades. Permanent crops replaced the forests of the Central Valley by the end of the century. Farms were stable and prosperous, and less than 10 percent of the national territory was developed for agriculture and pasture, with natural forest dominating the rest.

The end of the nineteenth century was marked by a wave of colonization toward the coasts, particularly the Pacific Coast (Sandner, 1972). Extensive rangelands were created for horses and cattle, and the forests of the Puriscal, Candelaria, and Dota Mountains and those on the western slopes of the Tilaran and Guanacaste Mountains were cleared. Deforestation has increased exponentially since 1922 because of official expansionist policies, liberal land tenure laws, and high population growth rates (WRI, 1991). A cattle subculture dominated the 1950s, encouraged by policies designed to increase beef exports and supported by ample credit programs funded almost entirely by international agencies. Subsidies for certain land uses, especially cattle ranching, were fundamental in the massive conversion of Costa Rica's forests into pasture and induced massive speculation for unclaimed land (Annis, 1990; WRI, 1991). Landless *campesinos* rapidly claimed unoccupied national territory, much of which was best suited for forestry because of its steep terrain, high rainfall, and poor soils. Typically, the land was deforested, with lumber cut and burned or left to decompose, to make the minimum "improvement" needed to assure possession. Forest clearing has been the traditional method of claiming land. Under Costa Rican law at that time, clearing was considered to be an improvement that raised land value (Hartshorn et al., 1982). *Campesinos* would then sell the land illegally as "improved" land to wealthy buyers with bank connections, who then converted it into cattle farms (WRI, 1991).

Expansion of the cattle-ranching frontier has clearly been at the expense of forests and water resources. Costa Rica lost 50 percent of its forest cover between 1940 and 1986, resulting in significant problems in soil erosion, reservoir siltation, and water quality. In the 1980s, Costa Rica had one of the highest annual rates of deforestation in the world, peaking at 100,000 acres per year.

The annual deforestation rate has now fallen to about 20,000 acres because much of the remaining primary forest is within national park boundaries. However, pressure on national parks and forest reserves is increasing because of increasing poverty levels (Hansen-Kuhn, 1993). Although no extractive or exploitive practices are allowed inside the parks, there is severe squatter pressure in some areas, particularly in protection zones and forest and wildlife reserves, which are less well protected than national parks (Garita, 1989; Hartshorn et al., 1982).

The banana industry in Costa Rica and throughout Central America has also had serious negative effects on inland and coastal water resources. The entire ecosystem is disrupted to provide optimal growing conditions for bananas. Thousands of hectares are cleared and relandscaped, with streams often channelized and rerouted. Streams receive high sediment loads, along with pesticide and fertilizer contamination. As R. J. Vargas (1995) points out, it is likely that the Costa Rican government would be less supportive of the banana industry if environmental impacts were considered in traditional cost-benefit analyses.

The history of banana development in Costa Rica has paralleled the development of the railroad system. Large international banana companies worked closely with the construction of the Atlantic Railroad to form a monopoly of the banana export and transportation business. The term *banana republic* was applied to Costa Rica and other countries in Central America because of the power that United Fruit Company had over the governments. By 1930, the United Fruit Company owned 1,409 hectares in the Atlantic region of Costa Rica. These lands were eventually abandoned because of soil deterioration, and production moved to the Pacific side of the country. The move resulted in economic collapse in Limón Province, and the environment still bears the scars. Since then, the banana industry has gone through peaks and lows. Today, Costa Rica is the second-largest banana exporter in the world (Ecuador is first), with more than 50,000 hectares in banana plantations (CORBANA, 1994).

Although Costa Rica has long been acknowledged as one of the wealthiest and most stable countries in Central America, it entered a period of economic difficulty in the late 1980s. Export-driven structural adjustment programs introduced by the International Monetary Fund (IMF) and the World Bank to stem this economic decline have apparently exacerbated the situation (Hansen-Kuhn, 1993). Costa Rica is now more heavily in debt than before, poverty is increasing, and there have been severe costs to the environment (Hansen-Kuhn, 1993). For example, production of new export crops in Costa Rica typically involves massive use of agrochemicals because many nontraditional crops are not native to the country and are highly vulnerable to pests and diseases. An interview with a representative of the Costa Rican nongovernment organization CEDADE (Centro de Capacitacion Para el Desarrollo), indicated that a study of melon producers revealed heavy use of such agrochemicals as Tamaron, Paraquat, and Lannate (Hansen-Kuhn, 1993). Fifty-eight percent of the melon farmers reported that water supplies were poisoned by agrochemicals, and more than 70 percent of them reported witnessing the death of domestic or wild animals after spraying. Furthermore, 75 percent of the farmers reported health problems that they attributed to insecticides and fungicides (Hansen-Kuhn, 1993).

At least 1,500 farm workers per year seek medical attention for pesticide poisoning, and government officials say that this number seriously underestimates the actual cases (Hansen-Kuhn, 1993). Thousands of banana plantation workers, exposed to DBCP (dibromochloroprepane) and other chemicals banned in industrial countries for years, report continuing health problems, including sterility (Collier, 1991).

Two Costa Rican laws deal with watershed planning. The Forestry Law of 1986 includes basic management guidelines. It prohibits clearing land in a 10–20 meter strip (depending on the inclination) on either side of streams, 100 meters from the shoreline of lakes, and in a 200-meter radius around natural springs. The Water Law of 1942 (see Vargas, chapter 7) also restricts some activities. In 1995 an Executive Decree on Effluents and Water Contamination mandated that commercial operations must have treatment facilities that will prevent all solid and liquid waste and residual waters from causing harm to the environment or public health (R. J. Vargas, 1995).

Laws often remain unenforced, however (see de la Rosa, chapter 14). Lack of enforcement is shown by open dumps near streams, expansion of urbanization into areas close to rivers, unregulated waste disposal into streams and rivers, and increasing pollution of groundwater supplies. There is a serious need for regional environmental planning from a watershed perspective. An example of the magnitude of the problem can be found in two of the most polluted rivers in the country, the Grande de Tárcoles and the Reventazón. Four main sources contribute to pollution in these basins. (1) The agricultural industry contributes solid and liquid wastes from coffee-processing plants; 45 percent of these industries do not have any type of wastewater treatment. (2) Industrial wastewaters from food-processing, tannery, textile, chemical, and cardboard products industries are a main source of pollution. (3) Agricultural wastes include fertilizer and pesticide runoff from coffee, sugar cane, and banana plantations. (4) Domestic sewage is also a major source of pollution.

Virtually every major watershed in Costa Rica is undergoing degradation because of deforestation and inappropriate land use. Sustainable and rational exploitation of water resources is contingent on the protection and restoration of watersheds, particularly forest cover on steep mountain slopes. Deforestation can cause severe erosion and hydrological changes under conditions of steep topography and high precipitation. An estimated 400–800 tons of soil per hectare per year are eroded in steep areas of the country that have been converted to pasture (Hartshorn et al., 1982). Topsoil is permanently lost to the ecosystem, causing high turbidity in lowland rivers and ruining drinking water, fisheries, and other resources.

Changing Land Use and Aquatic Resource Deterioration in the Caribbean Lowlands of Costa Rica

Despite rainfall of almost 4 meters per year, the human populations in some lowland regions of Costa Rica's Caribbean foothills are having water quality and quantity problems. For example, the town of Puerto Viejo de Sarapiquí and adjacent areas (figure 13.2), at the juncture of the Sarapiquí and Puerto Viejo Rivers, face severe water resource problems. This is due, in part, to the rapid population increase of Sarapiquí County, from 20,000 to more than 46,000 people in just three years, because of the recent and extensive development of banana plantations. Significant banana production by foreign-owned companies such as COBAL (Compania Bananera Atlantica LTDA), BANACOL (Bananeros de Colombia), GEEST (Caribbean Corporation), and BANDECO (Banana Development Corporation) began in Sarapiquí in 1991, resulting in an influx of immigrants to the area, many of whom have become land squatters. This has become a major social and economic problem in Sarapiquí, seriously exacerbating water problems in the region.

Although the lowland area is drained by numerous streams and rivers, local surface waters and groundwaters, previously tapped for potable water supplies, are contaminated with fecal coliform, introduced by cattle and people. The development of banana plantations and the intensive use of pesticides and nematicides are also negatively affecting the water quality of the area (R. J. Vargas, 1995). Aquifers in the area are surficial and therefore vulnerable to contamination by agricul-

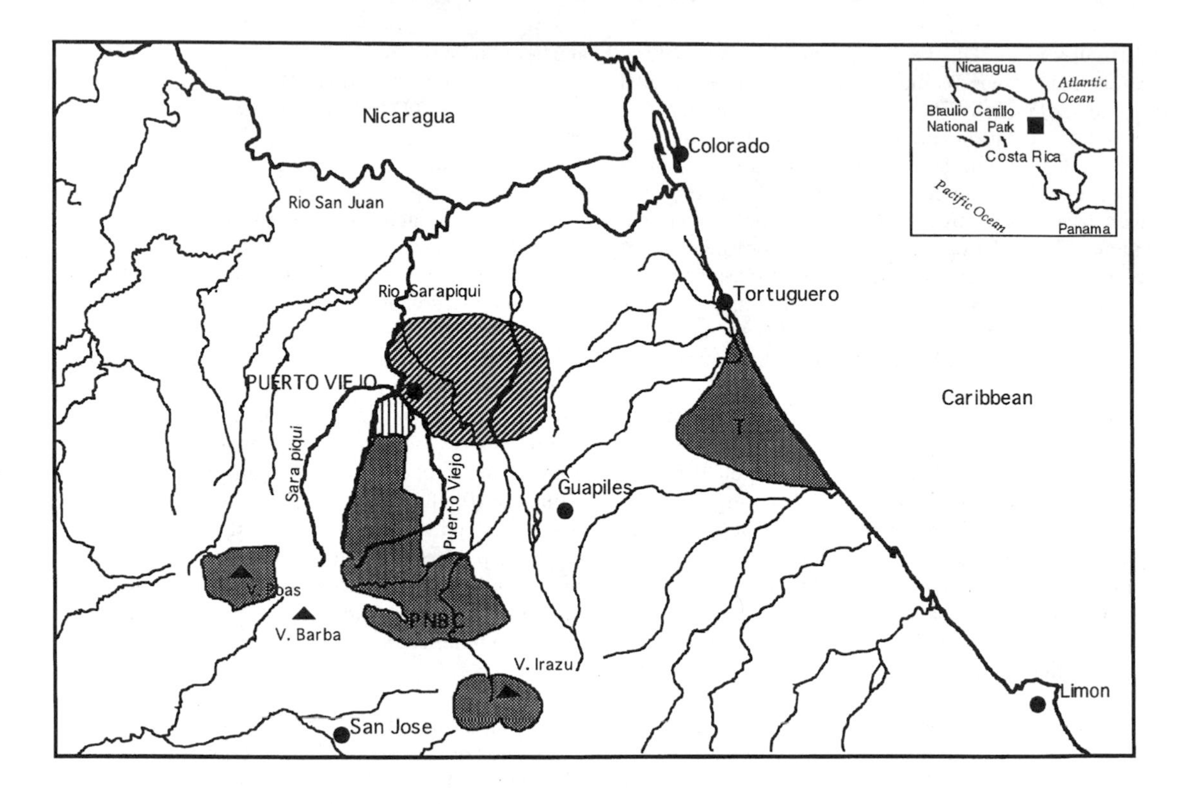

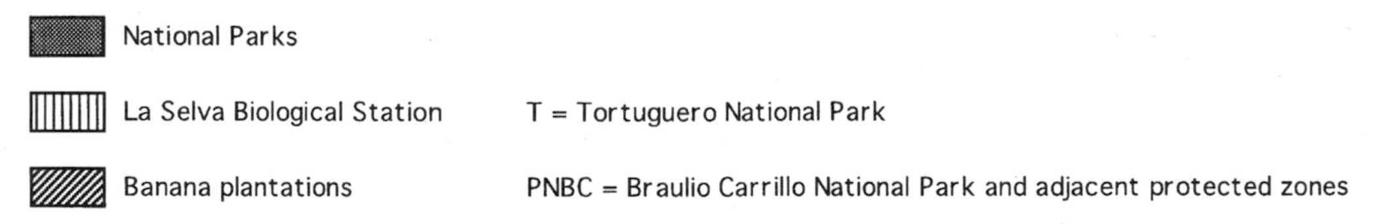

Figure 13.2 The Town of Puerto Viejo de Sarapiquí with Stream Drainages, National Parks, and Banana Plantations on Costa Rica's Caribbean Slope

tural or urban activities. Recharge of aquifers is from precipitation runoff and infiltration by the lateral flow of the Sarapiquí, Sucio, and Puerto Viejo Rivers.

The upper watersheds of the Sarapiquí and Puerto Viejo Rivers are currently protected because they lie within the 47,077-hectare Braulio Carrillo National Park (figure 13.2). Most other areas in the watersheds of these rivers remain unprotected and have been cleared for agriculture or urbanization. This deforestation has resulted in greater runoff, decreased infiltration rate and aquifer recharge, and increased erosion and sedimentation in rivers.

The watersheds within Braulio Carrillo National Park are the source of potable water for several communities, including the three districts of Sarapiquí County (about 46,000 people). Braulio Carrillo National Park is the last intact tract of primary rainforest that spans extremes in elevation, from near sea level to 10,000 feet, on Central America's Caribbean slope (Pringle, 1988). The high infiltration capacity of forested watersheds in the park helps to regulate surface water in the Sarapiquí and Puerto Viejo Rivers. Changes in land use in these protected areas would have detrimental effects, such as increased flooding and decreased water supply and quality, on lowland communities. A recent study of forested areas adjacent to the northeastern portion of Braulio Carrillo National Park shows that recharge of the aquifer occurs at an annual rate 2.6 times greater than in pasture areas. Also, annual runoff in pasture is 9.3 times greater than in forested areas (reported in R. J. Vargas, 1995).

In the lowland community of Puerto Viejo de Sarapiquí (population about 20,000) and nearby *barrios*, contamination of local surface waters and wells has resulted in the town's recent dependence on springs that originate at the northern end of Braulio Carrillo National Park (figure 13.3). In response to water shortages and quality problems, the Aqueduct and Sewage Authority of Costa Rica (AYA) installed plastic pipe with a 15-centimeter diameter. It transports water from the spring

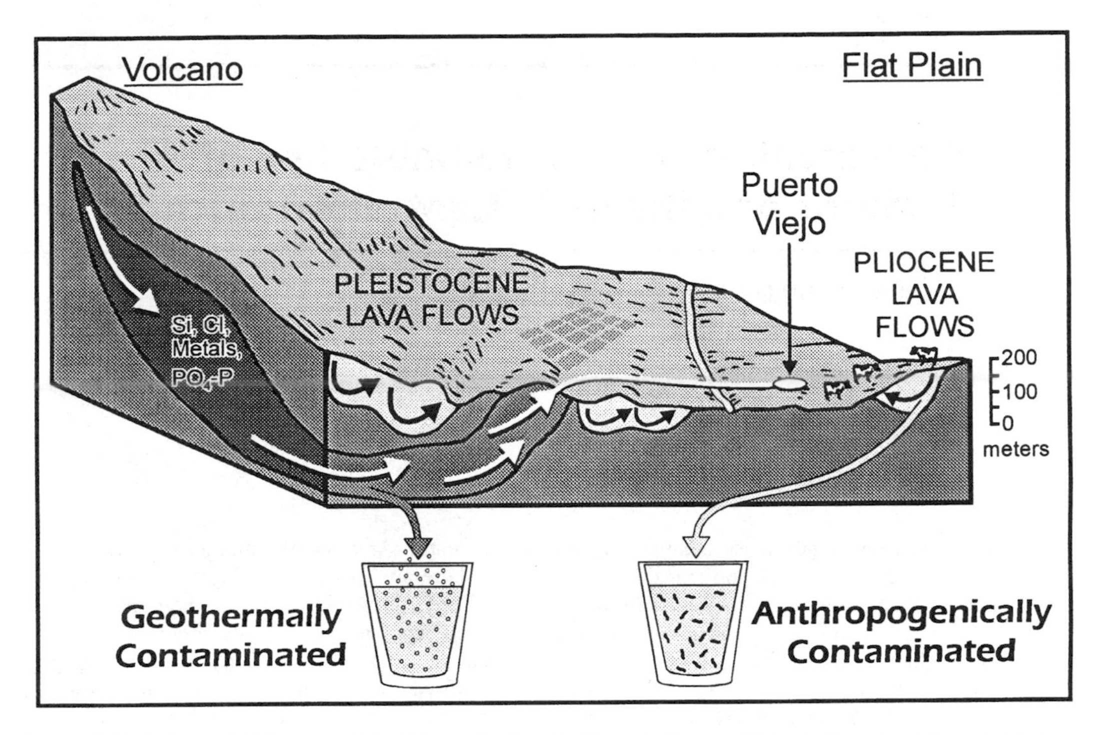

Figure 13.3 Schematic Diagram of the Water Quality Problems in Puerto Viejo de Sarapiquí, Located between a Dormant Volcano and the Flat Coastal Plain

approximately 8 kilometers to Puerto Viejo, where it is distributed to 95 percent of the downtown population. The quantity of potable water from this source is currently insufficient to serve other areas of Puerto Viejo and nearby *barrios*.

Although this new drinking water supply is currently uncontaminated by fecal coliform, the water has a very high mineral content. Studies show that it is geothermally contaminated and can be classified as sodium-chloride-bicarbonate (Na-Cl-HCO_3) water (Pringle, 1991; Pringle, Rowe, et al., 1993; Pringle and Triska, 1991), based on a classification of geothermal water types that is widely used by geochemists. Many residents of Puerto Viejo do not like the taste and complain of black residues in their cookware. Here, however, the mineral content of the water is not so high as to render it unpotable. Geothermally contaminated waters are quite common along the volcanic spine of Central America (Pringle, Rowe, et al., 1993) and are often much higher in mineral content, rendering them unsuitable as sources of potable water. For instance, Lake Managua in Nicaragua is seriously contaminated from volcanic sources. The occurrence of geothermally contaminated waters in volcanic areas of Central and South America will undoubtedly act as an additional constraint for rapid development in these regions.

Socioeconomic changes that result from the rapid development of the banana industry in Sarapiquí are placing increasing pressure on the water resources of the area. Projected increases in the human population of Sarapiquí over the next few decades show that there will be increasing pressure on the relatively undisturbed watersheds of Braulio Carrillo National Park as a source of water for domestic and agricultural use, underlying the importance of watershed protection.

Conclusion

These case studies from Puerto Rico and Costa Rica illustrate the seriousness of water resource deterioration in Latin America and the Caribbean. Both quantity and quality of water are critical issues. The need for adequate baseline information is urgent. Formulation of adequate regulations and laws and, even more important, enforcement of the laws and regulations that already exist are key to maintaining water resources for future generations.

Acknowledgments National Science Foundation grant DEB-95–28434, to the University of Georgia, and BSR-88–11902, to the International Institute of Tropical Forestry (Southern Forest Experimental Station) and the Centre for Energy and Environmental Research (University of Puerto Rico), partially supported the writing of this chapter. We thank J. Benstead, T. Laidlaw, and A. Raminez for their comments on the manuscript.

14

Conservation and Sustainable Use of Streams and Rivers in Central America

Carlos L. de la Rosa

Streams and rivers are among the most endangered ecosystems in Central America. Population growth, urbanization, and conversion of natural ecosystems into agricultural land in Latin America advance much more rapidly than our knowledge of the ecosystems or our efforts to conserve them. Even in countries like Costa Rica, which has led conservation efforts in Central America (Boza, 1993; Calvo, 1990; Gámez, 1991), the environmental problems from deforestation, aquatic pollution, air pollution, and general environmental deterioration are serious. Protected areas, such as national parks and wildlife refuges, are under the constant threat of invasion, insularization, habitat deterioration, or other destructive processes. Many of these problems stem from the traditional, unsustainable form of development in which natural resource exploitation occurs without regard for renewability or sustainable management. These are not only conservation issues but also socioeconomic issues, and they require solutions that incorporate socioeconomic elements.

The development of methods and techniques to evaluate the integrity and general health of aquatic ecosystems and the incorporation of this knowledge into overall development schemes remain key issues in aquatic resource conservation in Central America. Biological monitoring techniques and evaluation have been developing for many years in temperate regions (R. O. Gilbert, 1987; Ward, Loftis, and McBride, 1990). Recent efforts to establish bridges between the scientific community and organized citizen groups have led to a series of conservation initiatives based on solid scientific knowledge, particularly in the monitoring of aquatic environments through their biological components (Courtemanch, 1994). The merging of a well-established body of scientific knowledge with the growing public interest in environmental quality has produced environmental legislation that allows participation of all sectors of society to achieve better environmental quality standards.

In Central America, the picture is quite different. Here, there are still large, relatively intact tracts of natural environments along with their associated undisturbed ecosystems. Here, also, there are enormous pressures to improve the quality of life of the population, very high birthrates and population growth, severe poverty problems, and a prevalence of subsistence economies. All these factors lead toward the unsustainable use of natural resources. As a result, we observe high deforestation rates, severe air and water pollution, solid and liquid waste management problems along with their associated health problems, and an overall lack of planning in how to confront and resolve these problems.

At the same time, Central America has been the world's arena for experimenting with conservation strategies for the last 15 years. Enormous resources have been and are being invested in conservation and sustainable development projects, especially in forestry, biodiversity, health, and water quality. Costa Rica, a small Central American country (51,000 square kilometers), has led many of these efforts because of its privileged geographical location, its social stability, and the large area of protected lands within its territory. Costa Rica has been the "world's laboratory for tropical conservation" (Boza, Jukofsky, and Wille, 1995) for the last two decades, receiving aid and assistance from many countries for these efforts. Between 1987 and 1989, the United States alone invested more conservation and research resources in Costa Rica than in any other country in the world (Boza et al., 1995; Hambleton, 1994). Many other countries, particularly Canada, Germany, Norway, Holland, Sweden, Spain, England, and Japan, continue to establish ties and projects in Costa Rica. Other Central American countries that are receiving special attention are Nicaragua, Panama, and Guatemala. These efforts have produced a growing body of knowledge about Central American natural systems and particularly about Costa Rica's natural environments.

How is this knowledge used in the conservation and sustainable use of natural resources? Can we see the results of the accumulation of this scientific knowledge in effective legislation, better use of aquatic resources, and greater citizen participation in the decision-making processes that affect environmental quality? If not, what elements are missing that would allow these countries to begin and to consolidate this process?

In this chapter, I explore the apparent contradiction of Costa Rica's conservation efforts, particularly in the aquatic ecosystems, analyzing the causes of the present state of streams and rivers in this country. The core of the discussion is the interaction among science, government, and the general population and how these three elements will determine the integrity and future use of the aquatic ecosystems in Costa Rica and Central America. I offer one case study that exemplifies the difficulties encountered in bringing scientific information and conservation into the public consciousness. Another shows the interplay among science, government, and industry (see end of this chapter).

The State of Scientific Knowledge about Costa Rica's Aquatic Ecosystems

Ecological research in aquatic systems is a recent field in Costa Rica and in Central America. Much aquatic research has concentrated on species of health interest to humans, such as disease vectors or reservoirs, or on commercially important species, such as fish and shrimp. New World freshwater ecology is really a northern phenomenon. Most of the theory and practice of aquatic ecology, including its applications for environmental monitoring, regulation, and restoration, were developed in temperate regions.

Taxonomic knowledge of the groups, even at the family or genera level, is essential for the development of effective monitoring tools. Taxonomic knowledge of aquatic organisms in Costa Rica is very good for some groups, such as fish (Bussing, 1987, and references), Odonata (de la Rosa and Ramírez, 1995; A. Ramírez, 1992, 1994, and in press; Zloty, Pritchard, and Esquivel, 1993), Ephemeroptera (Edmunds, 1982; Flowers, 1992; Flowers and de la Rosa, in press), and some Diptera such as Simuliidae, Culicidae, Ceratopogonidae, and Trichoptera (Holzenthal, 1985, 1995). However, some groups are known as adult insects and are therefore of limited utility to aquatic ecologists who need to identify immature stages. Very few keys and identification guides are available at the generic or specific level for aquatic macroinvertebrates. Additionally, the distillation of this taxonomic and ecological knowledge into monitoring and evaluation tools for scientists and nonscientists is missing. Indexes of community integrity and pollution tolerance for macroinvertebrates are still too new and too preliminary to be used effectively in the field. Much of the research has been conducted by a few local scientists in coordination with foreign scientists. Aquatic biologists as a group are barely organized in Costa Rica.

Ecological work on freshwater communities is also scattered and scarce. Recent attempts to bring together current knowledge have produced important compilations, such as the December 1988 and March 1995 issues of the *Journal of the North American Benthological Society*. A review of current ecological knowledge of Middle American streams and rivers (de la Rosa, 1995c) concluded that "research in the region's streams is urgent given the recent history of land conversion to agriculture and urbanization." Biological diversity as a whole is likely to have been reduced because of deforestation and land use changes. Key paradigms, such as the river continuum concept (Vannote et al., 1980), have generated abundant discussion and testing in temperate regions and New Zealand, but little of it has been applied to tropical streams (de la Rosa, 1995c). Finally, pollution tolerance levels have not been developed for any species in the region. Only recently is some work beginning to produce data about pollution tolerance, although still unpublished (Yamileth Astorga and Luisa Castillo, Centro de Investigaciones de Ciencias Ambientales, Pesticide Program, Universidad Nacional, Costa Rica, personal communication).

These limitations create a dilemma. The conservation community cannot wait for science to come up with the necessary knowledge. However, neither can this community effectively carry out its conservation role without it.

Water Legislation and Institutional Capabilities

There is no shortage of environmental laws in Costa Rica (see Vargas, chapter 7). The Ministry of the Environment and Energy (MINAE), formerly the Ministry of Natural Resources, Energy and Mines (MIRENEM), has an impressive array of laws and regulations, some of them dating from the last century. Costa Rica subscribes to the Convention on International Trade in Endangered Species (CITES) and to other international agreements that regulate the traffic of species and the use of common resources like marine fishes and other sea products. Recently, the country has developed a solid image of conservation and support for global and local environmental concerns. Costa Rican environmental laws and regulations are continually evolving and becoming more specific and freer of loopholes.

Nonetheless, there remain two major concerns about the development of Costa Rica's legislative package. First, environmental laws need to be supported by scientific knowledge; otherwise, some laws or regulations make no sense. For example, the General Health Law and the Water Law both contain several articles that specifically prohibit all pollution or deterioration of rivers, streams, and their watersheds, especially those that provide drinking water for human settlements. However, no guidelines exist to establish what is pollution or deterioration; users of soil or water within watersheds have no responsibilities assigned to them, nor do any serious sanctions apply to those who break the law (Salazar, 1993; Zeledón, 1994). Second, laws need to be enforced. The population often looks the other way and breaks laws when there is no law enforcement in sight. Government officials and the population at large commonly lack knowledge of the existing legislation.

Local groups have begun the arduous tasks of interpreting environmental legislation, proposing changes in accordance with specific conditions in Costa Rica, and promoting knowledge and application of the law among the population. One example is the Centro de Derecho Ambiental y de los Recursos Humanos (Center for Environmental Law and Human Resources). However, river conservation in developing countries is still predominantly a question of socioeconomics (McNeely, 1988, cited in Seghers, 1992). Costa Rica's economy, like Trinidad's (Seghers, 1992), is now based in great part on ecotourism (Boza et al., 1995). Clean rivers and a healthy watershed are essential to this new income-generating industry.

Society's Perception of Water Resources and Environmental Quality

A recent study sheds some light on the perceptions of Costa Ricans about the environment, population, and biodiversity (Hall, Daily, and Ehrlich, 1995). In their conclusions, the authors emphasize the urgent need for more effective environmental education, including adequate treatment of specific topics such as the social and economic cost of environmental degradation. This educational effort should make topics relevant and useful to individuals rather than emphasizing global aspects of environmental degradation that are, as a rule, beyond the grasp of most rural and urban populations.

Environmental education efforts about freshwater ecosystems in Costa Rica have been few and only partially effective. It is easy to perceive that environmental degradation is "bad" when it causes a person to loose something in the process. In northern Costa Rica, road construction and maintenance have affected many rivers and streams in the region during the last 15 years. Many communities have felt and voiced their concerns about this destruction (see Case Study I). When interviewed, many people complain about the destruction and describe how much nicer it was before the intervention (de la Rosa, 1993a,b). However, most people in Costa Rica feel helpless to do anything about the current tide of degradation. Local governments are generally very weak, and they themselves are often the cause of environmental degradation. Powerful economic interests, outdated government policies, inadequate legislation, institutional weakness, and lack of enforcement of existing legislation make it very difficult for the common citizen to get involved in the improvement or maintenance of environmental quality.

Volunteer and Citizens' Groups in Costa Rica

The monitoring of water quality and ecosystem integrity by citizens' groups is a viable alternative for the development of a database that can be used by government and society to maintain or improve the quality of their aquatic (and other) environments. Volunteer or citizens' monitoring programs have been operating in the United States for almost a century (Firehock and West, 1995). In the United States there are several very well organized nongovernment organizations (NGOs) that are very active in stream monitoring and water quality research. Among them are (1) Adopt-a-Stream Foundation, (2) Izaak Walton League of America Save Our Streams Program, (3) Aquatic Project WILD, (4) Volunteer Monitor, (5) Soil and Water Conservation Society, (6) River Network, (7) American Rivers, and (8) Pacific Rivers Council (also see Pringle and Aumen, 1993). Even government organizations such as the U.S. Environmental Protection Agency (EPA) have recognized the usefulness of these groups and have developed specific materials for them (e.g., EPA, 1995). Some of these organizations have attempted to form chapters in Costa Rica, particularly the Global Rivers Environmental Education Network (GREEN), from the University of Michigan, but without much success. There are a few local initiatives, such as the Asociación Pro Conservación Acuática de Costa Rica (Association for Aquatic Conservation), which publishes bulletins and other communications that deal with water conservation and management issues.

Most successful volunteer monitoring programs depend on identifying selected macroinvertebrates that are divided into groups and defined according to their tolerance to pollution. Several indexes and groupings have been proposed and tested, and their usefulness for interested groups has been examined (for a review of methods and issues related to data collection and its quality, see Penrose and Call, 1995, and references). The success of these volunteer groups has been ascribed to four main factors (Firehock and West, 1995): (1) improved methods and training for volunteers and improved quality of the data they obtain; (2) government recognition of the value of getting the public involved, which is critical; (3) the combination of more stringent government regulations for water quality and reduced budgets for local governments, requiring more efficient local monitoring, which in turn increases the reliance on volunteer groups; and (4) public awareness of the problems, which is the result of many years of environmental education. All these require a scientific knowledge base that presently does not exist in Central America, as well as effective channels of communication among science, government, and the public.

Assessing and Minimizing Environmental Impact on Streams and Rivers

It is safe to assume that there are no pristine or virgin streams in the world. No matter how remote a river may be, human activities have had an effect on them (Seghers, 1992). Careful analysis reveals that the human presence has left its mark in even the best of cases, such as headwater streams on the apparently undisturbed forested slopes of some Costa Rican volcanoes. Hunting has been a traditional activity in these areas. Some large mammals known for their use of streams, for example, tapirs (*Tapirus bairdii*, Tapiridae) and wild pigs (*Tayassu pecari* and *Dicotyles tajacu*, Tayassuidae), have all but disappeared from most forests. Also, the surrounding agricultural lands affect even the remotest areas. Pesticides have been recorded in aquatic insects, trees, and sediments from some "pristine" streams in northwestern Costa Rica (Stanley and Sweeney, 1995), most likely carried there by winds. Finally, most lowland rivers in Costa Rica, where human activities are concentrated, have suffered extensive deforestation and habitat manipulation, leaving few rivers undisturbed or largely unaffected. In northern Costa Rica, for example, more than 90 percent of the original vegetation cover has disappeared because of unrestricted logging and infrastructure development, mainly roads and human settlements (de la Rosa, 1995a,b). Most of the rivers that are within protected areas, such as national parks or wildlife refuges, are not necessarily representative of rivers outside the parks. Costa Rica has many life zones (Holdridge, 1967) and the original vegetation and habitats have all but disappeared.

How does one assess environmental impact when the basic biological and ecological knowledge needed to characterize aquatic communities is missing? In the case of Costa Rica's northern zone, we adopted a strategy similar to that proposed by Plafkin et al. (1989), who developed and refined rapid bioassessment protocols for streams and rivers. A series of physicochemical and biological characteristics were chosen to serve as the basis for the assessment. "Least disturbed" sites were then selected on a variety of streams and rivers. These sites acted as references for evaluating the integrity and degree of disturbance of other sites. As knowledge of the rivers increases, more refinements can be added to the database.

One needs to select the set of measures to be used in the study. These measures are a compromise between the quality of information gathered and the cost of obtaining it. We utilized a large database, gathered during a land use and water quality study, for the environmental impact studies conducted in northern Costa Rica. Data from 25 sites on 16 streams and rivers were examined (de la Rosa and Barbee, 1994). A set of physicochemical and biological parameters was defined from these data and used in the environmental impact study (see Case Study II).

The methods chosen for these studies are not necessarily the most accepted for scientific research. There has been much discussion in recent literature about the development of appropriate tools for stream monitoring, pollution detection, and basic database development. On the one hand, some aquatic biologists have criticized the use of "pollution kits," or simplified data collection methods that reduce the credibility and usefulness of the data obtained. On the other hand, the growing number of environmental groups, including nongovernment agencies, high schools, and citizen groups, need quantitative tools for monitoring environmental quality. However, they do not have the training, the resources, or even the need to use highly sophisticated methods (Penrose and Call, 1995). Efforts to create bridges between the scientific community and these organized citizen groups should continue so that we can increase the quality and usefulness of the information gathered and improve the methods that can be used by nonprofessionals (Courtemanch, 1994; Penrose and Call, 1995, and references).

Minimizing the environmental impact of development activities is largely based on faith in Central America, if it is done at all. The tools used by biologists are different and much weaker than those used by industry and government in those areas or countries where there is a mandate to minimize the impact from such activities. Engineers who design and implement construction projects such as roads, buildings, and other infrastructures seldom have even the minimal knowledge of how they affect the environment with their activities. A road engineer, for example, sees a stream as a source of materials for the road or as an obstacle to cross with a bridge. He or she may also see it as a source of water for washing machinery, making materials, and discharging road runoff. The important issues for the engineer are economic ones—how to get the job done at the lowest possible cost. Unfortunately, most of the population sees streams and rivers similarly. They may also see them as sources of food and domestic water and as recreation sites. When road building impinges on these other uses by the public, conflicts may occur. Often, however, these alterations are seen as the price that must be paid for "development." Biologists have serious problems in convincing engineers that streams and rivers are ecosystems and that engineering activities may seriously impair the survival of these ecosystems.

Developing tropical countries are still far behind their developed counterparts in making ecological knowledge part of society's mental picture. Few scientists and citizen groups deal with these issues in Central America. The basic knowledge needed for the development of environmental tools is lacking. The pressing need in Central America is to develop a set of tools adequate to meet two goals—social action to resolve environmental problems and scientific research to advance the basic knowledge necessary for a more educated society.

Putting It All Together: Strategies for Achieving Sustainable Use of Aquatic Resources in Central America

Several key elements are missing in the overall plan for strengthening conservation and adequate management of rivers in Central America and in other developing countries throughout the world. Boon (1992) lists the following among the needs: (1) the application of theoretical ecology to river conservation; (2) increased research efforts, including habitat requirements or riverine biota, taxonomic work, scientific publication of the results, and better international coordination of research efforts between industrialized and developing countries; (3) improved procedures for environmental assessment; (4) long-term monitoring efforts; and (5) public education and participation.

The interplay among scientific knowledge, appropriate environmental education, and citizens' actions cannot be overemphasized. Science provides the knowledge and the tools for effective environmental monitoring; government provides the legal framework and the enforcement tools for making laws stick; and the population participates in various degrees, from demanding law enforcement to becoming activists themselves in cooperation with government agencies. All three must work in coordination, feeding one another and advancing in the achievement of the conservation and management goals.

Research efforts in Central America need to be strengthened in both applied and "pure" science. Recent efforts to integrate North American researchers with their Latin American counterparts, such as the "America's Program" of the North American Benthological Society, are efforts in the right direction, although the results so far have been discouraging. Very few Latin American researchers are active in the society, and interest in increasing participation and integrating research groups is limited. These efforts are further hindered by a common phenomenon in Latin America of which few North Americans are aware. Most scientists in developing countries have to wear many hats. Most Latin American scientists also hold teaching and administrative positions and are often pulled away from intensive research by these obligations, as well as by the government. Furthermore, funding for research is extremely limited, access to current literature is difficult, and attending international meetings and buying equipment and supplies are difficult and expensive. Latin American freshwater scientists in particular are extremely frustrated by these constraints, and progress in their fields is slow.

Government should depend more on the scientific community to help develop sound and practical legislation. Without the participation of this community, the development of legislation becomes a hit-and-miss game, in which lawmakers attempt to respond to higher government directives with law packages that lack scientific foundations. For example, the most recent environmental law passed by Costa Rica's congress was approved and published before most of the scientific community had a chance to read it, much less comment on it. It is likely that major revisions in this law will be required almost immediately. Latin American governments and the institutions in charge of environmental regulation are further constrained by limited budgets and low technical capacity. Salary schedules for field personnel are very low, and so is the training they receive. This leads to high rates of corruption among local public officials, especially when they interact with industries that generate high revenues, such as coffee, timber, and wildlife commerce. As a result, most laws are ignored or even broken, with official participation at the local level.

Volunteer groups in Latin America suffer from the same deficiencies in knowledge and technical capacity as government agencies. They are, however, much more flexible and motivated to acquire knowledge through informal channels. Volunteer groups in Costa Rica are highly motivated, but short-lived, because their actions seldom carry the punch that solid knowledge brings. Participation by the general populace is also weak, and private economic interests consistently prevail over the public good.

We should definitely start with science if we had to set priorities for actions to resolve the weaknesses just described. This is perhaps the easiest problem to solve. Latin America is likely to continue to be the tropical laboratory, not only for conservation science, but also for scientific inquiry. There is a largely untapped potential in the development of the freshwater sciences in the neotropics, but its development must be congruent with the social and economic realities of the region.

Scientific development requires more participation by the international community in the practical needs of Latin American societies. Scientists from industrialized countries working in the tropics need to increase and redirect some of their efforts toward issues that would help these countries develop while still preserving part of the incredible biological diversity in their territories. In other words, more scientists need to invest in conservation issues than in pure science. Otherwise, their natural field laboratories will continue to degrade and disappear at the current alarming rate. Strengthening government and citizen participation can occur more effectively from the inside. Knowledge, as both formal and informal environmental education, will encourage initiatives and participation.

Case Study I: Environmental Education about Aquatic Ecosystems and Citizens' Action Groups—Bringing Science to the Masses

There have been several attempts in Costa Rica to bring water quality and environmental monitoring tools to the public, particularly to conservation groups and schoolchildren. Tina Laidlaw, while a graduate student at the University of Georgia, set up a local version of "Adopte una Quebrada" (Adopt-a-Stream), using materials derived and adapted from the original Adopt-a-Stream Program (Yates, 1988), the Georgia Department of Natural Resources (1994), and Rapid Bioassessment Protocols for Tropical Streams (de la Rosa and Barbee, 1994). The program had four major goals: (1) to teach children and educators in the Puerto Viejo de Sarapiquí region (see Pringle and Scatena, chapter 13) basic elements of aquatic ecology, (2) to stimulate their interest in studying local streams, (3) to use the information collected to evaluate water quality in local streams, and (4) to help rehabilitate degraded areas and protect areas still in good condition (Laidlaw, 1996).

Laidlaw reports that the Puerto Viejo group completed sampling for an entire school year in November 1995. Their data from the Quebrada Grande, a second-order stream that flows through pasture and then through the town, shows that dissolved oxygen levels in the stream were normally 7–8 milligrams per liter. High nutrient concentrations were occasionally detected. Although they observed nonpoint source pollutants that were affecting the system, the nutrient tests were somewhat subjective, based on colorimetric equipment, and should be verified with more accurate testing methods.

> Student interest in the program resulted in the group expanding their monitoring effort to include a more pristine, but geothermally influenced stream in the nearby La Selva Biological Station. The group also sampled the Quebrada Grande at a site upstream of the town, although results showed few differences in chemical concentrations, and monitored a more degraded stream in Río Frío, a neighboring town, where they found obvious signs of wastewater discharge directly into the stream. (Laidlaw, personal communication)

The students and teachers offered suggestions for expanding and improving the program. Most of the group enjoyed the program and wanted to see it continued. Based on the program's success at building awareness about water resource issues and teaching students about science, La Selva Biological Station decided to maintain the program and is looking for additional funding sources. Laidlaw feels that too many of the programs she has evaluated focus on giving groups chemical kits and leaving it at that. Once funding dries up and the kits need to be refilled, the program dies. The cost of the programs should be a major consideration in Costa Rica. "These programs must be available to all groups," stressed Laidlaw.

Other recent efforts include the Environmental Education Program of the Environmental Management Office of the U.S. Agency for International Development, now a private foundation called FIREMA. It published and distributed various materials related to stream and river biology and conservation (de la Rosa, 1992; de la Rosa and Barbee, 1993; OMA, 1993–1995) and field guides for identifying aquatic organisms (de la Rosa, in press; Flowers and de la Rosa, in press; A. Ramírez, in press). These materials have been distributed to over 1,000 schools in Costa Rica, and more than 800 educators have participated in activities related to stream biology conservation.

Both programs were designed and carried out by individuals with formal training in aquatic biology. They are pilot projects or experiences that need to be fine-tuned and replicated to become effective means for social and behavioral change. However, these and similar programs must be incorporated into the regular official curricula of the schools. The enlistment of government educational institutions is necessary to create the broad coverage that environmental education requires. However, even though these programs should ideally begin with public officials to date, they have responded weakly and inconsistently.

Based on these experiences, we draw two major conclusions. First, environmental education efforts in Costa Rica, especially those related to streams and rivers, water quality and environmental impact, are few and very localized. They are mostly pilot programs that need to be consolidated and described in print so that the experiences can be replicated elsewhere. Second, the active participation of the educational community, from teachers and students in government officials, is essential in achieving the coverage that environmental education requires to be effective.

Case Study II: Environmental Impact Studies and Sustainable Use of Aquatic Resources—The Case of the Road Maintenance Association, Northern Costa Rica

An environmental impact study (EIS) is a tool for detecting the effects that a given activity may have on the environment. In Costa Rica, EIS also includes steps that must be taken to minimize the impact, a restoration plan for the affected site, and a series of checks and controls over the process. The final goal is to adopt whatever advantages the environment offers and that someone wants to exploit without destroying or degrading the resource unnecessarily. Many activities require an EIS, among them are tourist facility development, construction of ports and shipyards, lumber mills, any industrial development, and extraction of materials from rivers.

Impact studies for extracting materials from rivers (quarrying) are done by a certified (college degree) geologist, the only professional allowed by law to sign the EIS. The geologist can confer with biologists and other specialists during the study, but it is the geologist who assumes the responsibility for the results and recommendations. Unfortunately, geologists in Costa Rica do not receive any formal training in environmental impact, biology, or ecology. To assess potential impacts adequately, they should either seek appropriate training or consult biologists and ecologists. Often, neither is done.

An EIS became a prerequisite for river rock extractions in 1992. Since then a series of modifications and refinements to the laws and requisites have been made. In 1993, a national Environmental Impact Commission (EIC) was created to organize the information, develop lists of requirements for the different types of EIS, review all applications and studies, and determine their validity. There were seven members in the commission, none of them with a background in biology. The commission quickly prepared lists of requirements and information that should be included on each type of EIS, establishing a format in which to prepare and present them. The information was extremely detailed, perhaps too detailed, elevating the cost of the EIS, sometimes to astronomical sums. As a result, some subjects were treated superficially to keep the cost of the studies reasonable.

The U.S. Agency for International Development (USAID) promoted the formation of a Road Maintenance Association (AMV) in Upala in northern Costa Rica in 1990. The AMV was responsible for maintaining the large network of rural roads built as part of a previous USAID effort in the region. The AMV had qualified road engineers and equipment paid for or purchased with USAID funds. To reduce the impact of quarrying on local streams, USAID commissioned its own Environmental Management Office (OMA) to monitor activities and provide assistance to AMV regarding all aspects of existing Costa Rican environmental legislation and environmentally sound quarrying techniques. One large component of this technical assistance was the elaboration of EIS.

A large database gathered during a land use and water quality project, also financed by USAID, was used by OMA for the EIS in northern Costa Rica. Twenty-five sites on 16 rivers and streams were examined during this study (de la Rosa and Barbee, 1994). Physicochemical and biological parameters from these data were defined and used in the EIS. A set of tools developed over many years of trial and error and basic scientific research were applied in the EIS (e.g., Calvo and de la Rosa, 1993; Calvo, de la Rosa, and Norman, 1994). Using these guidelines, OMA developed or adapted the tools to evaluate several issues: (1) the initial conditions of the body of water at the proposed extraction site; (2) the possible effects the proposed extraction would have on the integrity of the ecosystem; (3) the possible effects the environment could have on the proposed quarrying; (4) corrective or mitigative actions that would help reduce the impact of extraction; and (5) a plan for the recuperation of the site.

These efforts were partially effective in avoiding undue damage to streams where quarrying took place. However, these environmental evaluation tools were applied only as long as USAID funds were available. When USAID's project ended in March 1995, the AMV went back to its former practices, for three main reasons. The first was economics. These studies are too expensive and complex for industry to apply because of the lack of low-cost, standardized techniques. The second was regulatory. Although there is legislation that requires an EIS, the legislation is not enforced and the laws are broken. Finally, education plays a role. Environmental concepts are usually too esoteric for engineers and geologists to apply because environmental impacts and basic ecology are not part of their current technical knowledge.

This case exemplifies one drawback of attempting to apply scientific methods in industry, especially an industry that is not prepared or ready to incorporate the expenses involved. We draw several conclusions.

First, river quarrying is an industrial activity that is likely to increase in the future in Costa Rica and in other Central American nations. Road building and maintenance will require a steady flow of road-building materials. Often these materials will come from rivers.

Second, governments like Costa Rica's are trying to regulate the quarrying in their rivers to reduce the impact of these activities on the ecosystems and on general aquatic environmental quality. However, they will have a hard time achieving effectiveness until better monitoring and enforcement are in place.

Third, an EIS is required in Costa Rica before a permit is issued to quarry a stream. The information required is detailed and very specific. However, the tools available for geologists, biologists, and other professionals in charge of these studies are insufficient to truly assess the environmental impact.

Fourth, governments also have a low capacity to monitor and regulate the extractions or to evaluate the EIS. Most extractions are finally approved, regardless of whether or not they will severely affect the system.

Fifth, there is no national plan for regulating the quarrying of streams and rivers. The decision to approve one permit is made in isolation, without considering all permit requests. This allows, for example, the development of a series of contiguous extraction sites for miles along a river, in effect making the entire river into a quarry.

Finally, there is no preselected set of streams and rivers that can be exploited, leaving some intact or untouched. All rivers of the country are potentially subject to extraction except for those inside protected areas.

Acknowledgments I would like to thank the U.S. Agency for International Development in Costa Rica, particularly Mr. Arturo Villalobos, for his support during the years some of these projects were implemented in Costa Rica's northern zone. Thanks are also due to Tina Laidlaw, Alonso Ramírez, Nicole Barbie, Thelma Ledezma, William Chavarría, Rigoberto Argüello, Gabriela Calvo, R. Wills Flowers, Daniel H. Janzen, Mario Boza, David Norman, and to the directors of the Asociación de Mantenimiento Vial (AMV) for their contributions, discussions, and efforts in advancing some of the topics covered in their paper. I would also like to thank Drs. U. Hatch and M. E. Swisher for their invitation to contribute to this volume and their comments to the several drafts of the manuscript.

PART III

Food, Fuel, and Fiber Production: Are They Sustainable in a Globalized Economy?

15

Traditional Farming Systems

Panacea or Problem?

H. L. Popenoe
M. E. Swisher

Smallholder agricultural production in Mesoamerica faces a crisis. The region confronts increasing problems of population growth, impoverishment of rural populations, and environmental deterioration. It is clear that past policies, technologies, and practices have been unable to raise the vast majority of the rural population out of poverty and provide the region's millions of smallholders with a sound footing in sustainable, commercially viable agricultural production. In this chapter, we examine the roots of this crisis, review the history of agricultural technology in the region, and suggest some new paradigms for agricultural development to meet the needs of small farmers.

One of the driving forces in the current agricultural crisis in Mesoamerica is the interaction between population growth and inadequate land tenure structures. Population growth rates in the region are some of the highest in the world. As early as the 1960s, political leaders inside and outside the region clearly recognized the potential conflict and unrest that would result from the combination of increased population, decreased landholding size, and enhanced expectations on the part of rural peoples.

Land Reform and Technology Transfer: Failed Strategies

One major solution posed was land reform. Every nation in Mesoamerica experimented with one or more forms of land reform between 1965 and 1985. Mexico and Costa Rica provide two examples of the different approaches that were tried. In Mexico, the *ejido* approach was used: landless or near landless farmers, largely of primarily indigenous ethnic background, were given lands that were held in common by the local community. The justification for this approach was that the *ejido* represented a modification of the traditional commercial systems of land tenure that the pre-Columbian cultures practiced in the region. In Costa Rica, individual land ownership remained the norm. The landless were provided with plots of varying sizes, and the people were organized into groups of settlements. Early efforts at land reform focused on opening up frontier areas, which are often in remote regions of the country with little social or physical infrastructure. The justification used was twofold: the areas were underutilized, and settling the regions would solve both the land tenure issue and help develop the nation by intensifying agricultural production.

In retrospect, it is clear for many reasons that land reform alone could not solve the problems of population growth and inadequate parcel size. First, the solution was, at best, a temporary one. Land redistribution, without increased productivity per unit area and without reductions in population growth rates, works at most for one or two generations. Moreover, the increase in the number of farms from the division of large estates requires more infrastructure and services (Popenoe, 1984). Governments were unwilling or unable to make this necessary investment. We find, therefore, that the percentage of rural landless or near landless in the region has not changed significantly over the past three decades. Land reform may have bought some time, but it has proven to be an impermanent solution to the problem.

Second, land reform contributed directly to deforestation, environmental degradation, and loss of biodiversity. Forested lands and other natural ecosystems were viewed until recently—and probably still are to a large degree—as "underutilized" lands. Settlers, as in the case of the recent expansion of population into the Atlantic coastal zone of Costa Rica, were encouraged to move into these frontier regions, clear the land, and make it "productive." The result is a loss of natural ecosystems and biodiversity. Again, retrospection reveals serious flaws in the settlement strategy. Many, if not most, of the frontier regions of Mesoamerica were unpopulated because they are poorly suited to permanent agriculture. One of the tragedies of land reform is that many areas were cleared and put into agricultural production, only to provide a poor living for a few people for a few years—at the cost of enormous losses in valuable natural resources, ecosystems, and biodiversity.

Third, most land reform measures never addressed two basic problems. The first was the unequal ownership of prime agricultural lands. In many nations—El Salvador is a good example—prime agricultural lands are in the hands of a few wealthy landowners. Unwilling to tackle these powerful groups, most Mesoamerican governments were faced with trying to relocate small farmers onto marginal lands, increasing the potential for environmental degradation by small farmers. The second problem was that smallholders in virtually every country lacked the fiscal and technological resources needed to participate in industrial agriculture. Given their resources, including poor land in many cases, they simply could not compete with international or national producers that could utilize industrial models of production. Most nations attempted to bypass this problem by instituting trade barriers to protect national producers from international competition or by subsidizing national producers, either through input subsidies or maintaining artificially high prices for selected commodities, such as rice and corn, or a combination of both processes.

By the 1960s, the importance of technical assistance was also recognized, and most development programs from the 1960s onward focused on technology transfer, using many models, most of which were based on the experience of the industrialized nations. Many farmers have adopted "modern" technologies, although it is unclear whether planned development programs or the penetration of international agribusiness firms has been the major factor in causing the changes in production. It is arguable, for example, that companies such as Dow Chemical and Ciba-Geigy have been more effective in getting farmers to adopt "their" technology, such as agrochemicals, than national and international development programs have been in promoting the adoption of such technologies as integrated pest management (Barfield and Swisher, 1994).

The changes, however, have not permitted most small farmers to enhance productivity, earn greater income, and thus enjoy a higher standard of living. Today, with structural adjustment, loss of favored nation status, and increasingly unregulated international trade, millions of small farmers are paying the price for these policies. They are no better, or perhaps less, able to compete against industrial production of basic grains, like corn and wheat, today than they were in the 1960s, despite major improvements in transportation.

Also, most limited resource farmers, particularly the corn producers, did not participate much in the benefits of the green revolution. In many areas, average corn yields are still not much more than 1,000 kilograms per hectare, about the same as at the time of Columbus's arrival in the New World. Although improved corn varieties and technologies are available, much corn production has been displaced to more marginal soils by other, higher-income crops.

Pre-Columbian Farming Systems in Mesoamerica

To understand why agriculture in Mesoamerica has not succeeded in achieving the same kinds of productivity advances that have occurred in western Europe, the United States, and Southeast Asia, it is important to understand how the systems that we see today have evolved. Although a range of production systems was used by pre-Columbian populations in Mesoamerica, in general they can be divided into two types: those that relied on extensive land use and those that relied on intensive labor inputs. Slash-and-burn agriculture is an example of the former, and the raised-bed systems of Belize or the Chinampas in Mexico are examples of the latter. Other labor-intensive systems included home gardens in the Guatemalan highlands; irrigated systems such as the *tablon* system of Panajachel (Mathewson, 1984) in the Oaxaca Valley of Mexico; and terraced hillside systems like Cayo, Belize, and Sololá in Guatemala (Donkin, 1979). Many of these systems, whether based on extensive land use or labor, required high levels of management and relied on high germ plasm diversity, especially compared to European systems of agriculture in the same period.

Traditional slash-and-burn agriculture is perhaps one of the earliest forms of food production in the New World. As practiced by modern descendants of the Maya, it involves cutting a plot of forest or second growth during the dry season. The site is burned immediately before the start of the rains, and corn is planted at intervals with a dibble. Occasionally beans and squash may be planted at the same time or shortly thereafter. The crop may be weeded once or twice before it matures and is harvested. The plot or part of it is then abandoned to forest fallow, and the farmer will clear a new area for the following crop season. The duration of the fallow depends on soil fertility, the presence or absence of grass, and the availability of land. After a period of at least 2 and up to 20 years, the farmer will return to the same site and start the process over again. The system is quite efficient in terms of labor inputs if enough land is available. As human population density increases, however, fallows are shortened and productivity declines.

The pre-Columbian farmers of Mesoamerica also practiced intensive forms of horticulture, particularly on slopes and in wetland areas. The *chinampas* (the famed floating gardens of Xochimilco) of Mexico are well known, and Palerm (cited in G. C. Wilkin, 1971) has characterized them as "possibly one of the world's most stable, intensive, and productive cultivation systems." *Tablones*, a variant of *chinampas* (Mathewson, 1984), are quite common in the Guatemalan highlands, especially around Panajachel. These raised beds vary from 20 to 60 centimeters in height and are usually rectangular in shape. They are separated by small irrigation ditches from which water is dipped and splashed on the beds with a long wooden scoop. The heavily fertilized beds are planted with a mixture of vegetable crops. Since production is mainly market-oriented, the area devoted to this practice has expanded in recent years. Undoubtedly, the present-day system has not changed much since the conquest.

A large variety of crops was available to pre-Columbian farmers in Mesoamerica and was comparable to the large inventory maintained by their counterparts in the Andes. Harlan (1971) proposed that agriculture had originated in three primary centers: the Near East, North China, and Mesoamerica. He considered South America to be a secondary center. Earlier, Vavilov (Harlan, 1971) had proposed eight centers of plant domestication, of which Mesoamerica was one. The traditional triad of staple crops—corn, beans, and squash—originated in Mesoamerica. Amaranth, chiles, avocados, *Annona* sp, and New World cotton also originated in the same region. Other local crops have a South American origin but were domesticated, or at least introduced, in Mesoamerica long before the Spanish conquest (Ford, 1984). Well-known examples are cassava, tomatoes, sweet potatoes, peanuts, pineapple, cacao, and maybe *Xanthosoma* sp.

These different crops allowed systems suitable for a variety of ecological niches. Thus, by the time of the conquest, the inhabitants had evolved not only agricultural systems but also a set of crops that were adapted to most of the environments of Mesoamerica. From this base civilizations developed that appeared to be in harmony with the environment.

In general, these agricultural systems were well adapted to the landscapes of Mesoamerica. Relying largely on increasing knowledge, labor, and local resources, farmers evolved systems over the centuries that provided stable food supplies for the societies of which they were a part. The development of these agricultural systems allowed major population centers to evolve at such widely separated sites, as varied in geography and ecology, as the Central Valley of Mexico, the Tehuacán Valley, Oaxaca, Chiapas, Yucatán, Belize, Peten, Guatemalan highlands, the coastal areas of Costa Rica and Honduras, and the volcanic uplands of Salvador and Nicaragua. Many of the lowland sites have not been amenable to modern agricultural systems until the last few decades. Did the original inhabitants have crop production strategies that were better adapted to seemingly marginal environments than we have today? Were these ancient systems more sustainable than those currently in use?

We do know that these ancient systems supported major ceremonial centers, such as Teotihuacán, Chichén Itzá, Palenque, Tikal, Kaminal Juyu, and Copán, that survived for many centuries. Some of these places are no longer suitable for agriculture that uses current technologies. The decline or collapse of some indigenous systems may have been related to environmental degradation (Brenner, 1994), climatic changes (Hodell, Curtis, and Brenner, 1995), and associated social upheavals. Obviously, the traditional agricultural systems were not completely benign in their effects on the landscape, but one might argue that the impacts were less than those caused by the conquest or modern technology.

The European Conquest

With European settlement in Mesoamerica, agricultural production systems underwent enormous changes. The agricultural landscape that we see today is, in many ways, two landscapes, one superimposed on the other. The pre-Columbian components consist of indigenous crops. The most notable in the landscape today are corn, squash, and beans, which are still produced on most small farms in the region. Other remnants of the pre-Columbian agricultural landscape include isolated systems, such as the *chinampas* of Mexico, and home gardens, which continue to rely heavily on traditional, pre-Columbian germ plasm, management, and knowledge. Over this landscape, we find European crops, such as wheat and, most important, livestock like cattle and sheep. One of the most striking features of the pre-Columbian landscape is the absence of large animals. Domesticated livestock was a relatively small component in pre-Columbian agriculture, although ducks, turkeys, and dogs were present. There were, however, no large ruminants.

It is not surprising that one of the first major alterations to the natural and agricultural ecosystems of Mesoamerica that the Europeans imposed was to clear trees and plant grasses. European agriculture, at the time of the conquest and today, was highly dependent on animals. Animal protein was critical in the European diet, and animal traction was important in agricultural production in that animals were the basis of the European transportation system. The large expanses of grasslands that we see today in Mesoamerica are virtually all created landscapes. Natural grass landscapes in the region are basically limited to two types of sites. The first type is wetlands; the second, highlands, particularly dry highlands. The large natural grassland ecosystems of South America, the United States, and Africa simply did not exist in Mesoamerica.

The easiest natural ecosystem for the Europeans to convert to grass-based agriculture was the seasonally dry forest on the Pacific side of the Mesoamerican region. Today, we find that these forests, more than any others, are in danger of disappearing. Only scattered remnants remain in Mesoamerica. While some of these lands have been converted to crop production, many are still dominated by introduced grasses and livestock, primarily beef cattle. One of the most striking features, then, of the modern Mesoamerican agroecosystem is the large expanse of introduced grasses devoted to livestock production. It is such a prevalent feature of the landscape that most take it for granted, rarely stopping to think about the enormity of this transformation or the rapidity with which it occurred.

Europeans also brought their food crops with them. By and large, the staple crops introduced by the Europeans have only competed moderately well with indigenous staple foods in Mesoamerica. Wheat and rice are important crops, but corn remains the staple of the diet for much of the region's population. However, crop germ plasm, introduced by Europeans, although often not of European origin, has had an enormous impact on the Mesoamerican landscape in other ways. One major transformation has been the introduction of plantation agricultural crops. Oil palm, coffee, sugar cane, and rubber are all introduced crops. The land area devoted to plantation agriculture in Mesoamerica fluctuates over time, but plantations remain striking, highly visible, superimposed features on the landscape.

More recently, the introduction of germ plasm in the form of nontraditional vegetable crops has become important. Throughout Mesoamerica today, crops preferred by European and North America palates proliferate. They include, for example, broccoli, cauliflower, snow peas, lettuce, and green beans. Until recently, these crops were produced mostly on a limited scale, largely for sale to local, urban populations. With structural adjustment and an increasing emphasis on production for export, these crops have come to occupy larger portions of the landscape. Small farmers in some areas, such as the highlands around Guatemala City and the Central Valley of Costa Rica, are now producing the crops for international export, as well as local consumption.

Finally, the Europeans brought to Mesoamerica an entirely different approach to agricultural production. The European system stresses monocultural production. Unlike many pre-Columbian systems, which were often based on polycultural plantings of a few to many species, European systems focus on molding the environment to fit the production of a single crop. Although Europeans practiced crop rotation and strip farming at the time of the conquest, they had never developed the intricate systems of crop associations typical of pre-Columbian Mesoamerican agriculture and based on the enormous genetic diversity of crop plants in the region. Today, even where pre-Columbian crops remain an important component in the agroecosystem, we are much more likely to see them planted in large monocultural fields.

As agriculture became more and more industrially based in Europe and the United States, the industrial model of production was also introduced in Mesoamerica. Inputs of chemical fertilizers, pesticides, and energy in the form of mechanical traction and processing became the norm. While never adopted by all farmers, most Mesoamerican farmers today do utilize industrial inputs.

In summary, the agricultural landscape that we see today in Mesoamerica is an uneasy mix. It contains remnants of pre-Columbian systems with a strong superimposition of European germ plasm, technology, and management practices. The degree to which pre-Columbian agricultural systems continue to be important varies. Home gardens are still a feature of many Mesoamerican households. Although likely to be overlooked because of the small area they occupy, they are often important in the stability and diet of rural families. In a few regions, pre-Columbian polyculture systems, based almost entirely on crop, not animal, production, remain. In other areas, European systems throughly dominate the landscape. Costa Rica's Guanacaste Province, with its large-scale cattle ranches, and the Honduran northern Caribbean coast, with its huge banana and oil palm plantations, are two examples.

The conversion of the Mesoamerican landscape to a mix of European and pre-Columbian patterns has often resulted in environmental deterioration and loss of biodiversity. It is, of course, impossible to say how the pre-Columbian system would have evolved had the change to European agriculture never occurred. It might have produced the same environmental degradation and destruction of natural ecosystems that we see now.

It is clear that European production systems have not been able to successfully replace pre-Columbian systems in Mesoamerica for small farmers. The attempt to replace traditional subsistence systems, based on genetic diversity, traditional knowledge, and intensive labor and management inputs, has not been fruitful. Small farmers in Mesoamerica, even when they try to adopt industrial technologies fully, rarely succeed.

A New Paradigm?

There are many opportunities to enhance the value and use of indigenous crops and cropping systems in Mesoamerica. Not only do local farmers have a comparative advantage in the production of these crops and associated practices (pest control, postharvest storage, marketing), but also their cultivars and practices may be better adapted to local ecological niches and thus require fewer industrial inputs (pesticides, chemical fertilizers, mechanization); the end result is that they are more environmentally benign.

Small landholders in Mesoamerica have already shown that they can adopt to a highly commercialized agriculture for the international marketplace by their success with the production of nontraditional crops for export, such as snowpeas, broccoli, and asparagus. The expansion of this type of production in recent years has been one of the main engines of growth for the small farmer. Unfortunately, this rapid growth to satisfy foreign markets has caused increased soil degradation, environmental (and human) contamination with agricultural chemicals, and a loss of the genetic diversity of indigenous crops.

Recently, the international marketplace has become more receptive to exotic crops like kiwi fruits, carambola (star fruit), and pomelos. The book *Lost Crops of the Incas* by the U.S. National Research Council (1989) has increased the popularity of such Andean crops as quinoa, cherimoya, tamarillo, and pepino. Many of these are now marketed in Europe and the United States.

Some Mesoamerican exotic crops such as chayote, mamey sapote, amaranth, and pitaya are slowly finding their way into the international marketplace. Could not this trickle be increased to a flood by choosing good crop candidates, selecting good varieties, popularizing many of their attributes, and upgrading traditional production practices? Many of the small landholders might find that they have a comparative advantage in the production of indigenous crops. The use of traditional methods of soil fertility maintenance, integrated pest management techniques, and water management might be more sustainable and less harmful to the environment than the heavy use of the pesticides, chemical fertilizers, and machinery that is often characteristic of nontraditional crop production. The associated preservation of local crop germ plasm would be an additional benefit and prevent the further erosion of traditional genetic resources. Promising examples of plants for promotion include leafy greens such as chaya (*Cnidosculus chayamansa*), chipilin (*Crotalaria longirostrata*), and hierbamora (*Solanum nigrescens*); the flowers of loroco (*Fernaldia pandurata*); and the fruits pitaya (*Hylocereus undatus*) and Chico spapote (*Manilkara achras*).

Such a process of retrieving some of the traditional crops and management practices has already begun in Andean America. A concerted effort in Mesoamerica might be an important step in retrieving some of the important features of farming knowledge that have withstood the test of time. Efforts to improve rural welfare could focus on traditional crops, indigenous cropping systems, and the physical and cultural environment in which they are produced. The intent would be threefold: (1) to popularize little-known traditional crops for nontraditional markets, which would provide new income to native farmers; these new markets could be domestic or international; (2) to provide data on nutritional value, crop production, and pest management options for native species and cultivars, which would provide the basis for improved nutrition and food supply of the native farming population; and (3) to develop training materials for rural extension activities.

Not only would this approach be a way of preserving unique cultivars through commercialization, instead of in situ or ex situ conservation, and of slowing environmental deterioration, but it would also be a way of creating more employment in the counryside through labor-intensive activities.

16

The *Frijol Tapado* Agroecosystem

The Survival and Contribution of a Managed Fallow System to Modern Costa Rican Agriculture

M. Rosemeyer
K. Schlather
J. Kettler

Many agricultural development strategies have failed because local indigenous knowledge of agricultural techniques has not been taken into account. The undervaluing of these traditional methods has often led to the transposition of inappropriate technologies onto socioeconomic and ecological conditions for which they are not adapted (Moock and Rhoades, 1992). These large technological jumps have not been sustained once the promoters of the nonindigenous technology have left the area. The term *sustainable* implies not only ecological soundness but also suitability for local use over time.

There are two main approaches to the development of modern, sustainable agricultural systems—one is to develop a new system from what we know about how modern systems work, and the other is to modify traditional systems, usually by increasing production. Researchers such as Ewel and Haggar, in their experiments at La Selva Organization for Tropical Studies (OTS) field station, have followed the first approach (Haggar, 1994; Haggar and Ewel, 1994). We have chosen to take the second with experiments that modify the traditional *frijol tapado* system at Finca Loma Linda, located near the OTS Las Cruces Biological Station. This effort has included the continuous collaboration of a farmer, Darryl Cole, and his employees.

The *frijol tapado*, or slash mulch bean, system is a pre-Hispanic system of bean production still in use today in Central America and northern South America (Thurston, 1994), especially in high rainfall areas. The system is important in Costa Rica, where presently 30–40 percent of the country's dry beans are produced by this system (Rosemeyer, 1995).

The key characteristic of the *frijol tapado* system is a mulch layer of fallow secondary vegetation. It is similar to the slash-and-burn system, except that the slashed vegetation is left to form a decomposing mulch layer over the soil instead of being burned. In the last month of the rainy season, passageways are cut in one- to two-year-old second-growth vegetation and seeds are broadcast into it. The secondary vegetation is chopped down with a machete and then rechopped and spread evenly on the ground. The bean seed germinates within the 2.5–10-centimeter mulch layer and emerges from the mulch in a few days. The crop is left to develop with virtually no cultural treatment until harvest during the dry season. Before the first bean crop, the land is fallowed for one to two years; for the second bean crop, the land stays fallow for nine months to produce vegetation for the mulch layer (Rosemeyer, 1995). At present, in many areas of Costa Rica, population increases have put pressure on the productive capacity of the land, and fallow years have been eliminated. This leaves only the nine months between bean-growing seasons to produce the vegetation for the mulch. In Acosta-Puriscal of Costa Rica, farmers say that they have produced beans continually in this manner for over a hundred years (Arias, 1995). The fallow is composed of native tree, bush, and herbaceous species, some of which resprout after the beans are harvested. Since trees are a part of this system's fallow, it is denoted as an agroforestry system.

The *frijol tapado* system is practiced by medium-sized landholders on their own lands or by small landholders on rented or sharecropped lands of medium or large farmers of Coto Brus (Conway et al., 1991). Small farmers do not have the available land for this extensive system, and large farmers are often cattle ranchers, who are less interested in bean production. This is also true in other areas of Costa Rica such as Sarapiquí (Schelhas, in press).

The following five questions present challenges for all traditional hillside agroecosystems that are surviving in the 20th century and are vying for a place in the modern arena of global competition and privatization of the agricultural sector. The *frijol tapado* system is used as an example of how some of these challenges are being met in the development of sustainable agricultural systems for the tropics.

Is It Possible to Achieve Sustainability If Agricultural Inputs Are Used to Increase Yields?

Agricultural sustainability in its broadest sense means the ability of a cropping system to maintain yields without degrading the resource base. However, population increases and expectations of a rising standard of living cannot be ignored. It is obvious that not only do we need a sustainable food supply (figure 16.1), but we also need an increasing food supply. This is a double challenge.

To maintain productivity in a sustainable manner, nutrient cycles must be closed. Therefore plant and animal residues, the byproducts of food processing, as well as human sewage (and technically human bodies after death), would need to be returned to food-producing lands. If residues are not applied, exportation of nutrients in the form of food from these plots will continue to mine the soil resource without replenishment. The application of nutrients in some form will be necessary to sustain productivity when nutrient cycles are not closed. Since the complete closure of nutrient cycles is unlikely, when sustainability is a goal applied inputs should be (1) of a renewable source, (2) low in quantity, and (3) efficiently used by the crop plant.

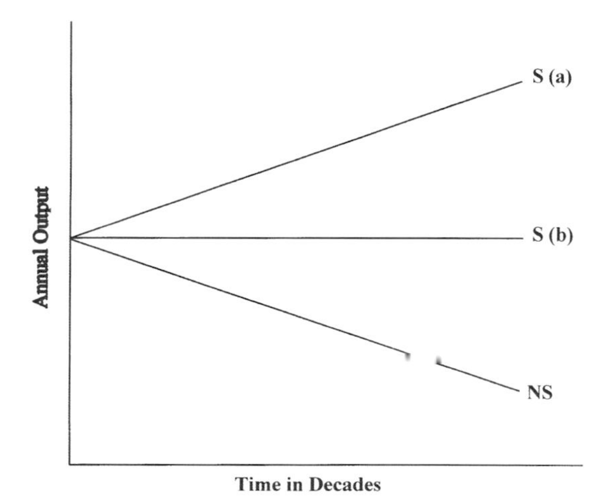

Figure 16.1 Annual Productivity over Time Challenged by the Need to Be Both Sustainable and Increasingly Productive as Populations Grow and Expectations Rise

To increase long-term productivity in a sustainable manner, nutrients need to cycle in a relatively closed manner and to cycle more rapidly, or more nutrients need to be put into the cycle. Increased rapidity of cycling has been attained, along with the decentralization of food production systems, in small, tropical, diverse home gardens, where composting of plant residues takes place. Apparently, large Mayan populations were supported by this agricultural system (Stavrakis, 1978), as are high population densities in Indonesia today (Michon and Mary, 1985). Modern standards of rural sanitation for human sewage dictate the use of latrines, and nutrients may be maintained in the food production loop by frequently rotating latrine sites, where fruit trees are later planted (Gliessman, 1988). Alternatively, urban organic wastes and sewage could be composted, shipped, and applied to agricultural lands at a greater distance from urban centers, as has been done by "honey wagons" in China (Cox and Atkins, 1979). If we view all nutrients moving in and out of humans as part of a human component, we can increase this component either by putting new acreage under production or by intensifying existing agricultural acreage with applications of biological (and produced on-farm) or agrochemical (produced externally) inputs. Since there are few new arable lands to be brought under agricultural production, augmenting the quantity of the nutrients through external inputs has been the easier alternative.

The renewability of the input itself is important. Whether the nutrient input is renewable within the time frame that sustainability is being defined will determine whether the input application itself is sustainable. Renewable inputs like nitrogen (N) from symbiotic nitrogen fixation or solar-powered N-fixation are less problematic than an input such as phosphorus (P), which is nonrenewable over the human lifetime. Symbiotic N-fixation is dependent on reasonable levels of soil P—the needs for these two nutrient inputs cannot be entirely separated (Isreal, 1987).

The *frijol tapado* system can be functionally defined as sustainable since it has been practiced for at least 500 years in the same geographic areas without apparent environmental degradation. Therefore, by starting with the premise that this system is sustainable, experiments were initiated to determine whether it could be modified to respond to current needs for increased production while maintaining its characteristics of sustainability. Efficient use of external inputs such as fertilizers and increased internal nutrient cycling by fallow enrichment are to be key factors in modifying the system for greater productivity and sustainability.

In these experiments, the *tapado* system, both in its traditional and modified states, is compared to the soil-based *espequeado* production system. In the latter, beans are planted in the soil with a pointed stick, fertilizer is band-applied, pesticides and herbicides are used as needed, and there is no mulch layer. Many reports of farm yields have indicated that the *espequeado* system has resulted in higher yields than the *tapado* system on commercial hectarages (CIAT, 1986; Jiménez, 1978).

The experimental site is located at 1,200 meters in altitude on an Andosol soil and receives an average of 3,600 millimeters of precipitation. The experimental design is a randomized complete block with five or six repetitions (depending on the experiment). Our results show that the unfertilized *frijol tapado* system has consistently produced yields equivalent to the agrochemical intensive, nonmulched *espequeado* system that uses the recommended rate of 325 kiligrams per hectare 10-30-10 fertilizer. This was not expected since most of the non-experimental literature reports that yields of the *espequeado* system are approximately twice that of the *tapado*, though there is one report in which they are equivalent (Araya and González, 1986). In these reports, previous land use and slope of the land used for these two systems are often different, so that published yield differentials from farmers' fields cannot be directly attributed to the system but rather to the fertility of the land.

Of nitrogen (N), phosphorus (P), or potassium (K), P was found to be the limiting nutrient in the slash mulch, or *frijol tapado*, production system, as it is in other bean production systems in Latin America (Schwartz and Gálvez, 1980), and yields have increased with P application (Rosemeyer and Gliessman, 1992). When fertilizer is applied to the *frijol tapado* system at two-thirds of the rate applied to the *frijol espequeado* (216 vs. 325 kiligrams per hectare 10-30-10), the yield of the former is significantly greater in the first year (Rosemeyer, 1994). In the second through the fifth years, adding only one-third of the fertilizer recommended for *espequeado* results in significantly greater bean production than does the *espequeado* with the full complement of fertilizer over a period of five years on the same site; the second year's higher yield is probably due to the residual fertilizer contribution from the first year. In the first three years, when the bean harvest was unaffected by drought, yields of both *frijol tapado* without nutrient application and *espequeado* with 325 kiligrams per hectare of 10-30-10 averaged 900 kiligrams per hectare; with one-third application (108 kiligrams per hectare 10-30-10), yields averaged 1,100 kiligrams per hectare; and with two-thirds and three-thirds (216 and 325 kiligrams per hectare 10-30-10), bean yields averaged 1,700 kiligrams per hectare. The yield of the *tapado* system with the two-thirds fertilizer treatment (216 kiligrams per hectare 10-30-10) was almost twice the yield of the *espequeado* system with the full fertilizer treatment.

The two systems are also different with respect to the efficiency of applied nutrient use by the bean plant. Nutrient use efficiency of the yield (NUEY) is defined as the yield of kiligrams per hectare with fertilizer minus the yield without fertilizer, divided by the nutrient (in this case, P) applied (kiligrams per hectare of elemental form). During the first three years of this experiment, NUEYs were greater at the one-third and two-thirds rates of application in the *frijol tapado* system than in the *espequeado* system. The NUEYs of the highest levels of fertilizer application were not significantly different from that of the *espequeado* (table 16.1).

Fertilizer for the *tapado* system may not be needed every year. There are substantial residual effects of the three years of P application in the *tapado* system in comparison to the *espequeado* system. On the one hand, when fertilizer applications are eliminated after the third year, yields in the *espequeado* system (where plant densities were the same) drop a statistically significant 50 percent in comparison to *espequeado* plots that received fertilizer for the fourth year. On the other hand, yields in the *tapado* system remain at the same levels, and these yields were 40 percent higher than the fertilized *espequeado*, although not statistically different. The same pattern appeared in the fifth year as well, although the yields of the two lowest fertilizer applications were significantly lower than their split-plot halves, where fertilizer continued to be applied.

As yield increases, the net exportation of P from the land increases. The net drain of P from *frijol tapado* systems—seed and plant residues were not returned to the land because of the risk of disease—is about 5–10 kiligrams per hectare when yields are 1,000–2,000 kiligrams per hectare (Flor, 1985). Net release of P from the decomposing mulch layer over a bean-growing season is about 5–6 kiligrams per hectare (Rosemeyer, 1990). In Andosol soils with a total P (mostly unavailable) of 1,000–3,000 kiligrams per hectare (Fassbender, 1969) and vegetative P of 100 kiligrams per hectare (Gessel et al., 1977), theoretical soil depletion could take place in about 110 to 620 years. On other soil types, this may be more problematic—for example, on Amazonian Oxisols, total soil P levels are only 336 ppm (parts per million), of P in a layer of 0–15 centimeters (Benavides, 1963) and 70 kilograms of P per hectare in the vegetation (Klinge, 1976), where P exhaustion of the soil layer could take place in about 40 to 80 years. If there were no closure of nutrient cycles through the application of bean residues and sewage, the application of P would be especially important in these soils, even if there were rapid P cycling.

Table 16.1 Nutrient Use Efficiency of the Yield (NUEY) over Three years, in Coto Brus, Costa Rica

System	P applied kg/ha	1990–1991	1991–1992	1992–1993	1990–1992
Tapado	14.58	15.50	20.64	36.60	24.25*a*
Tapado	29.16	23.88	21.24	39.88	28.33*a*
Tapado	43.86	12.58	13.92	23.96	16.82*b*
Espequeado (30 x 30 cm spacing)	43.86	10.05	10.06	27.96	16.02*b*
Espequeado (30 x 50 cm spacing)	43.86	8.58	5.22	25.15	12.98*b*
F over 3 years					$p = .0004$
Ortogonal contrast of the *tapado* vs. *espequeado* systems = $p < .0001$					

Source: Rosemeyer, M. E., 1995b. Eficiencia de aplicationes de fósforo en los sistemas frijol tapado y espequeado a treves de tresaños. In R. Araya and D. Beck (eds.), Taller Internacional Sobre. Bajo Fósforo en Frijol Comun. CIAT (Centro de Investigación de Agricultura Tropical), San José, Costa Rica.

Note: Nutrient use efficiency of the yield (NUEY) = the yield with fertilizer application minus the yield without fertilizer application divided by kg/ha of nutrient (in this case phosphorus) applied in elemental form. Numbers followed by the same letter (*a*, *b*) are not significantly different, according to Duncan's Multiple Range Test.

Can a Land-extensive Traditional Agricultural System Be Intensified?

Intensification in an agronomic sense means (1) less time when the land is allowed to remain in fallow and (2) the application of more inputs, usually both capital and labor in developing countries, where the ultimate goal is to produce more food per land area per unit of time.

We intensified the system by reducing the fallow time to nine months and, in addition, tested two approaches in applying inputs: fertilization, as mentioned above, and fallow enrichment, using nitrogen-fixing trees. The fallow enrichment strategy adds nutrients from on-farm sources in a low-input manner available to small farmers. This also remedies the inevitable reduction in biomass available for mulching that results from the elimination of the fallow period. Over the course of five years, where the fallow was not enriched through the addition of N-fixing trees or fertilizers, the secondary vegetation biomass had decreased from an estimated 30 tons per hectare to 15 tons per hectare in our experimental plots.

Therefore, experiments on the enrichment of fallow with N-fixing trees were begun to test whether sufficient biomass could be produced by the N-fixing trees and, consequently, P from the tree biomass to maintain bean yields through time (Kettler, 1995). A strategy of using internal sources of nutrients should result in a reduction of off-farm expenditures for inputs; however, over time, the mining of farm nutrients might result.

In experiments at the same farm, presently in their fifth year, four species of native N-fixing leguminous trees were planted singly and in mixture to enrich the fallow. The trees were planted every 0.5 meters in rows 4 meters apart. Every year at the time of bean sowing, the trees are cut to approximately 1 meter in height, and the biomass is chopped up with the secondary vegetation. In the first year, biomass production was significantly higher in all fallow enrichment treatments than in the *frijol tapado* system without trees. Bio-available nutrients were also significantly higher in the fallow enrichment treatments. Mulch from the *Inga* single-species treatment (only one type of tree planted in the alleys) and the *Inga* mixed-species treatments (*Inga* planted with other N-fixing tree species in the alleys) provided greater amounts of inorganic P and N (Kettler, 1995).

The trends in bean yield seem to follow overall mulch biomass production and the availability of soil nutrients; however, bean yields were not significantly different from the *tapado* control the first year of implemented treatments (second year of the experiment). In the third and fourth years where tree-mulches were applied, the two trees that produced significant biomass increased bean yield (Rosemeyer and Kettler, in press).

Agricultural intensification can also be defined as increasing annual economic returns to the land (Boserup, 1965). Gross economic returns from intensification of the *frijol tapado* system with fertilization are greater than the *frijol tapado* system controls in estimations based on our experimental plots (table 16.2). The *tapado*

Table 16.2 Economic Return to Investment of the *Tapado* and *Espequeado* Systems

Year	*Tapado* without Fertilizer[a]	*Espequeado* + 325 kg/ha 10-30-10[b]	*Tapado* + 216 kg/ha 10-30-10[b]
1990–1991	c 62.000	c 49.000	c 115.000
1991–1992	c 62.000	c 31.000	c 101.000
1992–1993	c 95.000	c 62.000	c 173.000

Source: M. E. Rosemeyer (in press). Avances en investigación de frijol tapado con énfasis en fuentes alternativas de fósforo. CEDECO, San José.

Note: The tapado system used no fertilizers or had 216 kg, 10-30-10/ha, and the *espequeado* used 325 kg/ha, 10-30-10, in colones/ha for the years 1990–1991, 1991–1992, and 1992–1993. Based on data from experimental plots at Finca Loma Linda, Coto Brus, Costa Rica.
[a]Calculations used a return of c 4000/46 kg of beans. Kg/ha yield /46 kg × c 4000 = return. Cost of seed and labor is not included.
[b]Calculations used c 4000/46 kg of beans and c 43.4/kg of 10-30-10 fertilizer. Rendimiento/46 kg/qq × c 4000/qq – cost of fertilizer = return. Cost of seed, fungicide, herbicide, and labor to apply them is not included for the *espequado* system.

system without fertilizer also had greater gross economic returns than the *espequeado* with fertilizer during the three years reported here. If the weeding or herbicide costs had been included in this analysis, the relative economic advantage of the *tapado* system would have been even greater.

Should Agriculture Be Practiced on Fragile Hillside Ecosystems?

This is a question that is being debated in intellectual and policymaking circles in Costa Rica at present. A coalition of the Ministry of Agriculture and Food and the Agricultural Organization of the United Nations (MAG-FAO) has (1) developed a land use capability classification based on the USDA capability classes (MAG-MIRENEM, 1995) and (2) classified Costa Rican lands at 1:50,000 according to Geographic Information System (GIS) techniques (Lucke, 1993). The objective is to bring land use into accord with its "capability" by using a combination of regulations and incentives.

One application of the capability classification is the targeting of lands for soil conservation projects. Areas where actual use and recommended use are most at variance can be identified. One GIS, using overflight data from 1980, has identified 500 hectares of problem areas in Coto Brus (Laurent, 1992). In general, the land use most at variance is pasture on steep slopes, where perennial or forest cover is recommended.

Although the land use capability classification in its present form is useful, one major limitation of the USDA system and its tropical modifications (Lal, 1995a; MAG-MIRENEM, 1995) is that it was developed for an environment (the United States) where agriculture means mechanized agriculture. This therefore excludes many agricultural systems that use hand labor, as is common in developing tropical countries. The percentage slope is one of the key factors in this classification system because of both the ease of mechanization and the potential for erosion. Agroecosystems, which use either hand labor and/or soil conservation techniques such as mulching, are subsequently banned from slopes, along with mechanized systems of conventional tillage.

The effect of mulches and plant residues in sharply reducing the sheet erosion of soils has been reported in Africa, Asia, and Latin America (Lal, 1995b). The *frijol tapado* system caused 6.5 times less erosion than the nonmulched *espequeado* system (Bellows, 1992). Soybean residue mulch on a Brazilian Oxisol reduced sheet erosion from 20.2 metric ton per hectare where residues were burned to 6.5 metric ton per hectare (Bertoni and Lombardi, 1985); soil erosion was reduced from 26.8 metric ton per hectare under conventional tillage to 1.4 metric ton per hectare under no tillage, where rainfall intensity was 60 millimeters per hour (Barker and Wunsche, 1977). Mulching reduced soil erosion 86 percent in comparison to conventional tillage under soybeans of a sandy soil on a 15 percent slope in Korea (Im et al., 1977). Minimum tillage of cassava on slopes was significantly equivalent to cassava planted with conventional tillage on flat areas, although it produced 9 times less erosion (Reining, 1992).

Mass erosion can also be a severe problem in hillside agriculture, especially on Ultisols and Andosols (Pla, 1995). Living barriers of trees, as in alley-cropping, are considered an effective way to combat slippages since the roots of the trees generally penetrate more deeply into the soil (Hamilton, 1994).

An estimated seventy percent of land in beans and 65 percent of corn are planted on hillsides in Central America in 1990 (table 16.3). Also, perennial crops such as coffee and much of the grazing land in Central America are on hillsides (Lindarte and Benito, 1991a). This may be due to the fact that (1) the majority of land is sloping hillside, (2) the best land for growing beans is on the hillside, or (3) the expansion of high-value export crops on flatland has marginalized to hillsides the small farmers who plant crops for the local markets and for subsistence.

In Costa Rica, only 23.8 percent of the land is considered flat and 18 percent suitable for annual crop production (TSC, 1982). Land that is flat is usually unavailable to small farmers, who produce food crops for local

Table 16.3 Percentage of Crop Area on Hillsides in Central America

	Guatemala	El Salvador	Honduras	Nicaragua	Costa Rica	Panama
Corn	30	70	42	80	80	90
Wheat	75					
Rice	25			10	0	60
Sorghum	20	70	90	25	0	0
Beans	65	60	60	95		
Pasture	65	80	40	59	70	70
Vegetables	8	10	60		80	30
Fruits	63	80				15
Roots and tubers	80		46		50	95
Coffee	60	95	90	80	75	95
Plantain		3				
Dual-purpose cattle	40	60	45	60	30	70
Dairy cattle	90	0.5			50	95
Meat cattle	25		55		30	70
Small animals	98	50				

Source: Lindarte and Benito, 1991a.

consumption. In Costa Rica, the flat, highly productive coffee land in the Central Valley has been converted to urban uses. Flat alluvial land on the Atlantic Coast near Limón has been in banana production and is presently expanding almost to Tortuguero National Park. Alluvial land on the Pacific Coast near Golfito was in bananas until 1984, when the United Fruit Company replanted it with African oil palm. The northern plains of Costa Rica are classified as dry or moist and mainly dedicated to cattle grazing for export and internal consumption. In flat, northern San Carlos near Nicaragua, citrus for export is being planted, with semimechanized beans between rows of citrus.

In summary, there is little flatland in Costa Rica, and most of it is dedicated to export crops, with the exception of the semimechanized beans noted above. Since the 1960s, the Costa Rican government has promoted the expansion of farmers into cattle production for export, and in the 1980s into nontraditional export crops such as flowers, pineapples, and ornamentals (traditional export crops in the 1980s were coffee, bananas, meat, and sugar) (Lehman, 1992). Although boosting exports was designed to obtain foreign exchange that Costa Rica has needed since the oil crisis of 1980, the government support of the export sector has resulted in one of the largest per capita debts in the world (Korten, 1995).

Beans, though a major portion of the Costa Rican diet, have not been a national priority for production in the past seven years because of the emphasis on export cropping. In the mid-1980s, CIAT (Centro Internacional de Agricultura Tropical), with the Ministry of Agriculture (MAG) and the Consejo Nacional de le Producción (CNP), aggressively increased bean production for national self-sufficiency through the promotion of green revolution technologies—high-yielding bean varieties grown with fertilizers and pesticides under the *espequeado* system (CIAT, 1986).

Although Costa Rica's bean self-sufficiency was achieved in 1985 (CIAT, 1986), the momentum has not been maintained and at present Costa Rica imports 40 percent of its beans from Nicaragua and occasionally China (Petritz, 1996). Adoption of this technology was greatest in Pérez Zeledón, where CIAT, MAG, and the CNP focused their work. However, in many other parts of Costa Rica where the *tapado* system continues, farmers say they have tried using the new varieties but that the *criollo*, or traditional varieties, produce more yield. In Coto Brus, approximately 95 percent of the bean production is under the *tapado* system using traditional varieties (M. Quirós, 1994). With the expansion of export crops on flatlands, small farmers are displaced to more marginal, hilly areas (Durham, 1979; Seligson, 1984). Small farmers produce the country's basic grains such as corn and beans, in these marginal, hillside areas.

Presently about 30 to 40 percent of the beans grown in Costa Rica are cultivated under the *frijol tapado* system, and approximately the same hectarage has been cultivated in this manner since the 1970s (Rosemeyer, 1995). Farmers consider hillside land best for beans grown by this system since water runs off quickly, preventing high humidity and disease. East-facing slopes are sought because the early morning sun dries the leaves and prevents disease (Rosemeyer, 1995).

Although it may not be desirable to use hillside land because of the potential for erosion, unless there is an attractive, income-producing alternative, it seems unlikely that the situation will change. In the Caribbean and El Salvador, the natural regeneration of rural land substantially increased because of urban migration

(Lugo, 1988), prompted by massive government welfare programs. If an aggressive incentive program is funded, it might be possible to remove farmers from the hillsides, but the government is grappling with privatization (see below) and the pressure to reduce incentives. The development of an expanded industrial sector and *maquila* industries (tax-free industries located in developing countries to take advantage of cheap labor) as has occurred in Honduras, has been promoted by the government since the 1980s as part of structural adjustment (Korten, 1995); however, in the last 2 years the Costa Rican *maquila* and other industries (Brennan, 1996) are relocating elsewhere as the government increases taxes in its search for revenue and as sources of cheaper labor become available.

If We Increase the Productivity of the *Frijol Tapado* System, Which Is Being Practiced on Hillsides, Will We Promote a Greater Use of Hillsides for Agriculture?

Understanding the motivation of farmers who are using the *frijol tapado* system is key to answering this question. If a farmer practices *frijol tapado* as a subsistence strategy, one would not expect to see an expansion unless subsistence farming is increasing. If a farmer plants for commercial production, one would expect to see an expansion of *frijol tapado* lands if suitable lands are available and there is a market for the crop. In 1992, farmers estimated that of the beans they produce under the *frijol tapado* system, about 40 percent are produced for subsistence and 60 percent for sale (CEDECO, 1992). Thus, if production methods are improved, we might expect an increase in commercial production; growth in subsistence production would be largely based on growth of a low-income population.

What Will Be the Effect of Privatization on Bean Production under *Frijol Tapado*?

In recent years, Costa Rica has begun the task of privatizing various sectors of the economy, as described by Szott (chapter 21). Reductions in subsidies and price supports to farmers are part of the requirements of the International Monetary Fund (IMF) for eligibility for new loans (Korten, 1995). Until 1995, the price of beans had been supported by the Costa Rican government through purchase from farmers at guaranteed prices. Since then, prices to the producer have decreased, and consequently, so has bean production; for example, in Coto Brus, the hectares under production have decreased from an estimated 2,600 in 1993 to 1,600 in 1995 (E. Navarro, 1993, 1995). Prices to the consumer have not decreased because middlemen buy the beans at low prices from the farmers but do not pass on the savings to the retailer. The new IMF loans include money to help form farmers' marketing cooperatives to commercialize beans (Barquero, 1995). Perhaps this will give them a better price and an incentive to produce beans, though at the end of 1996 this had still not been implemented.

What system will farmers choose if commercial production does expand in the future? Privatization could cause an increase in *frijol tapado* over *espequeado* hectarage since fewer inputs are needed for the former. One would expect farmers to adopt more internally derived inputs, such as alley cropping with N-fixing trees to increase production where they own their own land. The use of trees in the *frijol tapado* system could also decrease mass erosion on the slopes. Where farmers rent land, one might expect that such inputs as fertilizer would be more attractive than investments, like planting trees.

Conclusion

If productivity of traditional agroecosystems is to be increased in a relatively sustainable manner ensuring their mainstream survival into the twenty-first century, inputs are critical. The type of inputs—renewable or internally or externally derived—and whether they are efficiently used by the crop plant are important criteria. Traditional systems can be modified for increasing production with externally- and internally-derived inputs, as seen by the *frijol tapado* system. Erosion rates of hillside land under mulched systems can be equal to those in flat areas. Documentation of erosion control in traditional systems will be useful for the adaptation of USDA-type land capability classification to socioeconomic and ecological conditions in the tropics. Whether production will be expanded under the *frijol tapado* system will depend on a variety of factors, including prices and markets for beans and the degree of marginalization of the small producer. Future experimentation by farmers and researchers will determine the extent to which the principles involved in the *frijol tapado* agroecosystem are relevant to the production of other Mesoamerican crops.

17

Food Crop Production Systems in Central America

Raúl A. Moreno

The Central American isthmus is often treated as a single analytical unit from a geopolitical perspective. Nevertheless, from an ecological perspective, the region is so diverse that most generalizations lead to some degree of error. Several factors contribute to environmental variability in the region. Most important, 70 percent of the area in agricultural production consists of highlands of different altitudes and relief. Both the Atlantic and Pacific Oceans influence the climate of this mountainous agricultural region, another source of variability. The isthmus may well have the richest flora and fauna in the world because it serves as a bridge between the North and South American land masses. Because of this extremely large environmental diversity, any global analysis of the food production systems of the Central American isthmus is an exercise in generalization and is, therefore, imprecise.

From a social and political point of view, the unequal access to the resources needed for agricultural production is similar in the region to that in the rest of Latin America (Gunder, 1970). Large farms produce traditional export crops, such as bananas, African palms, sugar cane, or meat, based on extensive livestock production systems. Smaller farms produce the food consumed by the region's population. These smaller farms themselves exhibit great socioeconomic diversity. Some are very nearly subsistence production units, while others are completely oriented toward production for the market.

The family management that typifies these small production units makes it necessary to consider both ecological and socioeconomic factors. Crop production and other farm activities are highly interrelated. Therefore, any analysis of small farms remains incomplete unless the interactions between crops and other components on the farm, such as the production of livestock or perennial plants, are taken into consideration. Other activities, like off-farm employment and the management of community lands, also form a necessary part of the analytical framework.

Although agricultural production illustrates many common features throughout Mesoamerica, the agriculture of two Central American nations differs significantly from that of other countries in the region. The most important production systems in Costa Rica have evolved differently from those of other nations for political and economic reasons. Panama's agriculture differs because service industries are so important in the economy (R. Moreno, 1980).

The purpose of this chapter is to describe briefly some important food crop production systems in Mesoamerica and to draw attention to the changes that have occurred in these systems because of the necessity to intensify production. The chapter also discusses some of the interactions between food crop production and the natural environment, especially from the point of view of the sustainability of production. Finally, the potential effects of macroeconomic policies on food crop production in the region are briefly examined.

Basic grain production systems, based on maize and beans or maize and sorghum, are the focus of much of this analysis. These systems are typical of small farms that occupy steeply sloping lands in Central America. Extensive dryland rice production, which occurs on flatlands, is another important grain production system in the region. Root and tuber crops, produced both for internal consumption and for export, also occupy important acreages in the isthmus. Horticultural crops produced for both internal consumption and export are prevalent in higher-altitude zones, where they form part of a group of agricultural systems that includes plants and animals typical of the middle latitudes. Horticultural crops, including potatoes, are less extensive in the land area devoted to their cultivation than the crops mentioned previously. However, the value accrued from them is high, making them economically important.

Production Systems Based on Maize and Beans

Both maize (*Zea mays*) and the common bean (*Phaseolus vulgaris*) are native to Mesoamerica. These crops have been grown in association in the isthmus since humankind became sedentary in the region (Harrison and Turner, 1978). This system of production in association extends into other tropical parts of Latin America as well (Patiño, 1965). Besides being complementary nutritionally, the physical attributes of the two crops contributed historically to the common practice of raising them in association. In the past, climbing-bean types were more common in Mesoamerica than bush types, and maize forms a natural support for climbing beans.

This production system takes many different forms throughout Mesoamerica. In high altitudes, tall maize cultivars and a large grain bean are grown in association. The maize cultivars grow slowly because of the low temperatures of high altitudes. Bean cultivars with a long vegetative growth period are the preferred types in highland regions. In midaltitude areas, associations or relay cropping systems of tropical maize cultivars and determinate (bush) bean cultivars are common. In lower altitudes, the lima bean (*P. lunatus*), the cowpea (*Vigna unguiculata*), or the tepary bean (*P. acutifolius*) replaces the common bean, depending on environmental conditions.

The maize and bean association is practiced in areas where rainfall is seasonal, with a bimodal distribution. A brief dry season, called the *canicula*, or "little summer," separates the two rainfall peaks in zones under the climatic influence of the Pacific Ocean. Since irrigation is not practiced, plant growth and development depend greatly on the availability of water and nutrients in the soil. One of the most important characteristics of the precipitation regime in the maize and bean production areas is the high variability in total rainfall in different areas and from year to year. The lower the total rainfall, the less predictable the date when the rains will begin, the date when they will end, and the probability that the *canicula* will occur (Hargreaves, Hancock, and Hill, 1975).

Maize is usually planted at the beginning of the first rainy season of the year. Beans are planted between the maize stalks shortly after the maize has reached physiological maturity. This chronological arrangement between the two species is the most common in the isthmus. In many regards, it resembles relay cropping more than cropping in association. The maize crop develops rapidly with the first rains of the year. Maize yields are closely related to the availability of water in the soil during the critical period of grain fill. This normally occurs before the *canicula*. Development of the bean crop depends on the soil moisture provided by the second, shorter rainy season of the year. Bean yields are closely related to two factors: (1) the availability of water in the soil profile toward the end of the second rainy season and (2) the absence of rainfall in the period immediately after grain fill and during the harvest season.

The interactions between this production system and animal production are limited. Crop residues are used as feed for cattle, and most of the small animals on the farm depend on maize production for feed.

Maize and Sorghum Production Systems

This combination of maize and sorghum species, planted sequentially or in relay, is common in the driest regions of Central America. The common unimproved sorghum varieties, called *maicillo* or *millon* by farmers, are photoperiod-sensitive, tall, and of low yield potential. These cultivars probably derive from kafir sorghum types introduced into Mesoamerica less than 350 years ago (Wall and Ross, 1975). Some regard this cropping system as a risk-aversion mechanism that farmers use to reduce the risk from lack of precipitation. Sorghum grows reasonably well under conditions of higher soil moisture tension (lower moisture availability) in the soil profile. The system predominates on thin, low-fertility soils.

Many combinations of these two crops are possible. The most common practice is to plant sorghum between the maize stalks when the latter reach knee height. After emergence, the sorghum continues its vegetative growth phase until the days shorten during the months that correspond to winter in the northern hemisphere. The short days induce florescence, followed by grain fill.

This system may continue uninterrupted for many years. However, under conditions of low plant population and very limited use of chemical inputs, maize-sorghum intercrops have also been a component in agricultural systems that involve long rotation cycles and multiple land uses. For example, the maize-sorghum association is part of one rotation of more than 20 years. In this system, trees of the *Mimosa* genus (*carbonales*) are planted, maize is grown in monoculture, and pasture (*Hyparrhenia rufa*) is planted as the final component in the rotation (Mateo, Díaz, and Nolasco, 1981). Between these two extremes, many other systems are possible. Species other than maize give different combinations, including such associations as sorghum and the common bean or sorghum and the tepary bean. Sorghum is also planted in monoculture (Hawkins, 1985).

Usually, maize-sorghum associations are intimately related to cattle production. Sorghum grain is a food for both humans and animals. Crop residues also have many uses, principally as feed for animals. After the grain harvest, the residues almost always provide standing forage for animals. Farmers who do not themselves have animals usually rent the land that contains maize and/or sorghum residues to farmers who do have livestock (Sinclair and Romero, 1990).

Rice Production Systems

Although it originated outside the region, rice is one of the most important components in the modern diet of Central Americans, especially in cities. Governments in the region have intervened strongly in the production, processing, and national and international marketing of rice because it is very important in the diet of the urban working class.

Two forms of rice production are common. Small farmers raise rice principally to meet their own consumption needs. Traditional rice production systems form an integral component of their complex farming systems. Tall cultivars with good culinary qualities but low yields per unit area of land typify these traditional production systems. Traditional rice production varies from migratory to more modern systems. In the latter, rice is planted on a leveled field, but only manual implements are used. Larger-scale commercial production systems provide rice for the majority of the population. Commercial production is in the hands of farmers with good access to production inputs, especially land and capital.

Panama and Costa Rica produce most of Central America's rice on large farms, using modern technology. More than 60 percent of all Central American production occurs in a "rice band" in these two nations. It includes the Chorotega, Central Pacific, and Brunca regions of Costa Rica and extends into Panama, passing through Veraguas, Cocle, and the Province of Panama itself. El Salvador, Honduras, and Guatemala produce the remainder of the rice crop in the region. Production in these nations has increased considerably in recent years.

Rice yields are higher in Costa Rica than in Panama because Costa Rica's agricultural sector is stronger and because Costa Rican farmers have had better access to technology and capital. These factors have led to an even greater concentration of production on large farms in Costa Rica. In Panama, where the service sector dominates the economy, rice is produced on smaller farms, often on state cooperatives, and modern technology is applied less intensively. The area that is cultivated by using more traditional technology, and achieving lower yields, is considerable in Panama. It is estimated that approximately 50,000 hectares are planted by using the traditional technology. Only 13,000 hectares are planted by using the more modern technology—mechanical soil preparation and harvest—although rice planted on these lands achieves yields almost three times greater than those of the acreage planted with traditional technology (Panama, 1988–1990).

Highland Production Systems

The altiplano, an extensive highland region in Central America, is an important zone in terms of the spatial distribution of crop production systems in the isthmus. In almost all of the Central American countries, these higher-elevation zones with lower temperatures are used to produce crops and raise livestock breeds commonly found in temperate nations. Land values have traditionally been high in these regions because of the high productivity of the soil. As a result, intensive production systems are the norm. Several vegetable crops, especially the potato, are produced primarily in high-altitude zones. Milk production, based on crosses with European breeds, sometimes occurs in the same regions. In some high-altitude areas, farmers are also trying to produce temperate fruit crops, with little success in terms of quality. Temperate fruits are now grown in some areas of Cerro Punta in Panama, part of Cartago Province in Costa Rica, the highest zones of Las Segovias in Nicaragua, La Esperanza in Honduras, and most of the altiplano of Guatemala, to name only a few.

The principal characteristics of horticultural crop production systems are the large investments required and the high risks incurred because of volatile market prices and a high potential for pest damage. Potato production, perhaps the most important of these systems, clearly exhibits these characteristics. The investment required is large, mostly because of the cost of seed, which accounts for more than 50 percent of the total cost of production in some years. The technology used can include high levels of inputs because of the large initial investment and the high potential for profits (personal communication, MAG-Cartago, Costa Rica, 1995). Potato production has expanded more aggressively in Central American than in the rest of Latin America in recent years (Horton, 1987). Nonetheless, production is insignificant worldwide. Mean production by country does not exceed 30,000 tons annually.

Tropical Root Crop Production Systems

Manioc, cocoyam, and the aroids are not important crops in Mesoamerica from the point of view of total consumption. Nonetheless, they are important in several areas, including significant acreage in the Province of Veraguas in Panama, the Atlantic lowlands of Costa Rica, southern Nicaragua (especially Rivas), and the regions of Honduras where the *garifuna* people live. Manioc is grown both in monoculture and in association with maize, whereas cocoyam and the aroids are grown principally in monoculture (Moreno and Hart, 1978). Most of this production occurs in small parcels with high labor inputs. Marketing is the principal problem for farmers who produce tropical root crops. The high cost of seed is a secondary problem for cocoyam and the aroids.

Production in the Atlantic lowlands of Costa Rica provides one of the most interesting examples of tropical root crop production. Manioc, cocoyam, and aroids have become a significant source of agricultural export earnings for the country. These systems are closely tied to a marketing chain, controlled principally by foreigners, that begins at the farm and extends to the supermarkets of Miami and New York (H. Monge, 1994; personal communication, CNP, Costa Rica, 1994).

Intensifying Food Crop Production

The maize-bean intercropping system undoubtedly constituted the food base of the first inhabitants of Mesoamerica. The maize-sorghum system has a more recent and less important role in the region's history. Both systems are now having significant changes.

The necessity to intensify production, because of population growth, has caused many modifications of the maize-bean intercropping system. The first of these was to replace common climbing-bean cultivars with determinate bush types. Separating the vines of climbing-bean cultivars from the maize at harvest is time consuming and difficult. Bush types therefore permit a greater efficiency in labor use, especially for harvesting, drying, and shelling. Another consequence of intensifying the maize-bean system is to separate the maize and bean components both in space and in time. The result is larger acreages in both maize and bean monoculture. Monoculture production simplifies the tasks involved in maize production, increasing labor productivity.

Maize

Intensified maize production, including monocultural production, is most observable in three areas in Central America. These regions, called the "maize band," include the parts of Guatemala and El Salvador closest to the Pacific Ocean, the "maize arch" that extends through Honduras and Nicaragua, and the "maize island" that stands alone in Costa Rica. These areas are historically important maize and bean production zones, where the two species have been grown in association for many years. However, spatial specialization in land use is now observable. Monocultural maize production in these three important subregions is growing. This does not mean that beans are not grown today in these areas but rather that more area is devoted to maize.

The maize band occupies fertile soils of volcanic origin that have received additional sedimentary deposits washed from nearby mountains. The band begins in the western part of the altiplano of Guatemala. It includes the central southeastern part of Guatemala and extends along the Pacific coast of El Salvador to a point just short of the Gulf of Fonseca. In the part of the maize band that lies in El Salvador, 70 to 80 percent of the maize crop is planted without an associated bean crop. Improved maize varieties, among them high-yielding hybrids, are the most common. Machinery and agricultural chemicals are used in most cases. Only 20 percent of the hybrid maize is planted in association with beans. In contrast, unimproved (*criollo*) maize varieties, still grown in this region of El Salvador, are planted in about equal amounts in monoculture (51,000 hectares) and in association with beans (47,000 hectares).

Low labor productivity has been one of the most outstanding characteristics of small farm agriculture in Central America. Labor use has decreased in the maize band because of the intensification of monocultural maize production. The maize-bean intercrop requires about 100 days of labor per hectare, whereas monoculture maize production requires only about 50 or 60 days. Temporary employment has increased at the expense of permanent employment, especially in the maize-producing regions of Guatemala. Labor productivity for maize production has increased principally because the number of days required per hectare of maize is lower. Reduced labor for weeding is especially significant. The increased labor productivity is not due primarily to an increase in yield. Monocultural maize yields are higher only by 90 to 130 kilograms per hectare than yields obtained when maize is planted in association with beans.

The most significant characteristic of maize production in the maize band is the high level of production for the market. More than 60 percent of the production enters the national market. Another important characteristic is the superior control of postharvest grain losses, which have decreased significantly since maize has become a commercial crop.

The maize arch is another region of concentrated maize production, extending across the frontier between Nicaragua and Honduras. Most of this arch occupies intermontane areas. One extreme lies in the north and northwestern region of Honduras and the other in Leon and Chinandega in Nicaragua. The central part of this arch is in Olancho, Honduras, extending into Nicaragua, where it includes the Departments of Jinotega and Matagalpa. In Honduras, because of the intensification of drought and soil degradation, maize production has moved from Choluteca and Valle toward the north and northwest. Today, approximately 50 percent of Honduras's total maize crop comes from the north and northwest.

Maize yields are lower in this maize arch than in the maize band described previously. Yields in north and northwestern Honduras are between 1.5 and 2 tons per hectare. Currently, yields are even lower in Nicaragua than in Honduras because of the economic destabilization caused by the war. In 1987–1988, farmers in

Jinotega and Matagalpa obtained yields of only 1.2 tons per hectare of maize.

A tendency toward specialization in monocultural maize production is evident in the Honduran part of the arch. In contrast, the tendency in Nicaragua appears to be the maintainance of the maize-bean cropping system in which beans are planted in succession after the maize crop. These differences may exist because maize production is an older tradition in Jinotega and Matagalpa than in the north and northwest of Honduras. The latter are areas of recent settlement and lower population density.

The last area of concentrated maize production in Central America is the "maize island" in the Brunca and Huetar Atlantica regions of Costa Rica. Its importance, both in the area planted and the total production, is less than that of the areas described previously. According to 1985 data, maize yields in this area reach 2 tons per hectare on the average. Precipitation is high for maize production in the island. The Huetar Atlantic zone has characteristics typical of the humid tropical lowlands, making maize a crop that is ecologically out of equilibrium with the environment. As a result, losses in the field and after harvest are high. Many soils of the Brunca region are highly acidic, which also affects maize yields negatively (Santana, n.d.).

The technology used to intensify monocultural maize production has been limited principally to the use of improved or commercial hybrid maize varieties (Córdoba, Barreto, and Crossa, 1993). Technologies very similar to those used in temperate nations are applied to monocultural production. These include mechanical soil preparation, seed disinfection, atrazine application for weed control, and sometimes fertilization with urea or with formulas recommended by agricultural extension services or commercial fertilizer dealers. Overall, yields are highly reflective of the level of technology applied. Profitability, however, does not necessarily reflect the technology used (PRIAG-Universidad Libre de Amsterdam, 1993).

Beans

The other component of the maize-bean intercrop system has also been slowly separated out of the association. The common bean is planted and managed as a monocultural crop more and more frequently. However, describing bean monoculture systems is more difficult than describing those for maize. Statistics about bean production are not extensive and are often very imprecise. Part of the problem in statistical reporting relates to how different enumerators classify common bean production systems. Planting beans immediately after the maize harvest, rather than before the harvest, is becoming more common. Many data collectors consider this a monocultural bean production system and report it as such. Others call this a relay planting system. Yet others consider it to be a form of maize-bean association. Therefore, the data about the area planted in beans in association, in relay, or in monoculture depend greatly on the criteria applied by the enumerator. Although the data are scarce and imprecise, intensive bean cultivation generally occurs in the same areas as intensive maize production.

Beans are a more profitable crop for the farmer than maize. Therefore, the best soils are planted in beans. Sloping lands are preferable because the species are intolerant to an excess of water in the soil and to a lack of oxygen. Farmers sell a larger portion of the bean crop than the maize crop, compared with the amount of the crop consumed by the farm family, because of the high profitability of beans. However, bean parcels are smaller than maize parcels on the average. Beans are the crop that provides the majority of income for many farmers (Díaz, 1982).

Beans are a relatively profitable crop, and prices can be quite high in some years. As a result, bean production tends to occur in areas of recent settlement and very high precipitation. Under these conditions, beans become a high-risk crop because of disease and pest problems. Furthermore, they provide very little soil cover and therefore pose a risk of accelerated soil erosion. Bean production in the Caribbean region of Honduras is a good example of the expansion of production into areas of high precipitation and higher temperatures than are normal for this species. Bean production in this region reflects high prices in the national market at the time that beans are harvested in the Caribbean area of Honduras (CIAT, 1993).

Other interesting bean production systems are used by itinerant farmers. Many farmers migrate from the hillsides or intermontane valleys of the central zone of Central America during the dry season. They move into the more humid tropical regions of the Mesoamerican nations to plant beans, thereby obtaining an additional harvest during the year. Another itinerant system occurs in mountainous areas. Monocultural bean production is a typical hillside farming system in the mountainous zones of Central America. Most of the beans grown in monoculture are planted on rented lands. Bean production alternates with a fallow that allows natural vegetation to reoccupy the plot. Many farmers who own small farms also own or rent parcels of land outside the boundaries of the parcels that they occupy. When they own the distant parcel, they often raise coffee. However, when it is rented land, it is more common for them to raise beans for a few years and then move to another parcel. This itinerant system of bean production is also typical of other parts of Latin America.

A very unusual system of monocultural bean production, called the *frijol tapado* system, occurs in Costa Rica. Historically, Costa Rica is the Central American country that has had the highest labor cost. *Frijol tapado,* derived from migratory agriculture, is primarily a way to maximize returns per unit of labor invested. The basic technique is to broadcast seeds on the surface of a parcel of land where certain types of weeds grow. Later, this natural vegetation is chopped down with a machete, and the bean seedlings emerge through the thick, loose cap of decaying organic material formed by the vegetation. No other care is given to the crop until harvest (J. Monge, 1985). Scientists, principally from other countries, have described and studied this mode of production in detail (Rosemeyer, Schlather, and Kettler, chapter 16; Thurston et al., 1995). In the past, in some cases until recently, maize and other species have been produced by production systems similar to *frijol tapado* (Patiño, 1965). Clear indications of the intensification of this production system are not notable, although some farmers have attempted to use improved varieties and broadcast application of phosphorus fertilizer (Alfaro, 1994).

The technologies applied to intensified maize production—planting on a large scale, using machinery and modern technology, and planting on flatlands—are virtually never used for bean production in Central America. Traditional varieties continue to be the most common, despite intensive efforts by national and international institutions to encourage the use of improved bean varieties. Very few improved bean varieties are in common use at the current time, and the technology used to intensify production remains very traditional. Nonspecific herbicides such as paraquat and, sometimes, fertilizer constitute the most common technology in monocultural bean production. Further, pest and disease problems affect beans more than maize (Schwartz and Gálvez, 1980).

Maize-Sorghum Intercrops

Modifications in the maize-sorghum crop association have also taken place. Hybrid maize cultivars have replaced traditional maize cultivars in areas of high population density, such as some parts of El Salvador. Improved sorghum varieties have also gained acceptance in these areas. The improved varieties are planted immediately after or almost simultaneously with the maize harvest. Fertilizer is applied during the succession, principally to the maize component. Herbicide use is also becoming more frequent for the maize crop. Because of the separation of the two components in the maize-sorghum association, most farmers use the sum of the technologies recommended for each crop when produced in monoculture in the new production system.

Criollo, or unimproved sorghum varieties of the kafir type, have been slowly displaced in the zones of concentrated maize production. These cultivars, occasionally planted in association with maize, have moved into even drier areas, onto even thinner soils, and into the hands of even poorer farmers. Medium-sized farms and some that formerly produced cotton are now raising hybrid sorghum for agroindustrial uses. Overall, the separation of the components of the maize-sorghum association has resulted in the production of hybrid sorghum in intensive agricultural systems. Nevertheless, this region produces 50 percent less sorghum now than it produced in the early 1980s. Price changes and increased importation of other sources of animal feed by companies that produce concentrates have resulted in reduced production.

Rice

Commercial rice production has always been very intensive. Rice production in several countries has been sufficient to satisfy national demand in many years. National research institutions dedicate a goodly part of their resources to the genetic improvement of rice, frequently in collaboration with international institutions. Most of the technology needed to increase dryland rice production is available. Whether or not to apply the technology is an economic decision that the farmer must make, based on the prices of rice and of inputs. Improved cultivars, herbicides, harvest equipment, fertilizer recommendations, and other technological inputs have been tested repeatedly.

High-latitude Horticultural Crops

A process of intensification and specialization characterizes the production of highland horticultural crops. These vegetable production systems in their current form are the result of an evolutionary process. Originally, the areas were devoted to basic grain production. Later, these systems evolved to associations or rotations of basic grains and horticultural crops. Today, they are specialized systems of monocultural production of potatoes or other horticultural crops oriented completely toward sale in the commercial market. The better the road infrastructure and market information flow, the greater the specialization in highland horticultural production systems. For example, in 1950, 58 percent of the agricultural land in Guatemala was used for basic food crop production. Today basic food crops occupy only 37 percent (von Braun, Hotchkiss, and Immink, 1989).

Except for potato production, the technology available for highland horticultural crop production has been imported. Because of the high profitability of these crops, there is a tendency to use more technology than is necessary. There is better institutional coordination for research and technology transfer for potato production. As a result, the existing technology is more appropriate than that available for the production of other horticultural crops. However, the components of current technology rely strongly on the use of fertilizers, herbicides, and chemical products to control pathogens and insects in all cases (Kaimowitz, 1992).

The necessity to generate export earnings in some countries, Guatemala, for example, has led to the production of such horticultural crops as snow peas, broccoli, parsley, and other crops destined completely for the export market. The insertion of farmers into the export market has caused several changes. The new crops have in part replaced more traditional crops, and labor use in the region has increased. The new systems have generated some export earnings but have also exposed farmers to the high risks characteristic of the international market, especially to the variability in international prices for export products (von Braun et al., 1989). The export firms also recommend the technology used on farms that produce for export.

Tropical Root and Tuber Crops

The tropical root crops have undergone a similar process of intensification. Nevertheless, modern technology is applied to these crops, either in production or in postharvest processing, only when they are produced for export. Costa Rica is the most characteristic case (H. Monge, 1994).

Food Crop Production and Sustainability

Central America confronts the challenge of increasing food crop production to satisfy the growing demand for food that the continually increasing population creates. At the same time, farmers are under pressure to ensure that increased production occurs under a new regime of environmental management that ensures the sustainability of production. For the agricultural sector, this challenge means producing more without negatively affecting the resource base on which agriculture depends, including soils and water and vegetable and animal life.

Small farmers produce on hillside lands a significant part of the basic grain supply for these nations. About 50 percent of the remaining forests in Central America persist on these same hillsides, making them an important place for diversity of flora and fauna. Therefore, an important interaction between agriculture based on basic grains and regional biodiversity occurs in this habitat. Hillside farming leads to soil degradation from erosion, requiring a constant search for new, fertile lands for production. As a result, biodiversity and therefore ecological equilibrium are threatened in these zones. Nonetheless, the original deforestation on hillsides is often attributable to other economic interests and not always to small-scale grain producers (Utting, 1993).

Approximately 50 percent of the land in Central America is in pasture and forests, principally on hillsides, that have no clearly identifiable owner. This abundance of uncontrolled lands permits farmers with little or no land to produce basic grains, using very high discount rates for the future costs and benefits associated with producing in this habitat. The necessity to survive puts the availability of land for future generations in danger (Lindarte and Benito, 1991b).

Under hillside conditions, the yearly production depends greatly on the quantity and distribution of precipitation, leaving few fundamental management decisions under the farmer's control. Lack of precipitation and poor precipitation distribution have been the principal causes of low yields in recent years. For example, in the 1994–1995 production cycle, losses reached more than $160 million for crops planted in the first rainy season alone. The five Central American nations lost 651,000 tons of maize, 37,000 tons of beans, and 42,000 tons of rice because of drought (Consejo Agropecuario Centroamericano, 1994). Yield variability, therefore, depends largely on phenomena that the farmer cannot control.

Uncertainty about the dates on which the rainy season will start and end, as well as doubts about whether the interpluvial dry season will occur, force farmers to make management decisions to ensure production, even when their actions may incur a certain degree of environmental degradation. Under these conditions, when farmers are pushed by necessity to protect themselves from a "false start" of the rainy season, they wait until several rainfalls have occurred to plant. These precipitation events are generally highly erosive because the soil is barren in preparation for planting. This set of events, year after year, contributes to a constant process of soil degradation that characterizes these regions of hillside production.

Many national institutions are developing technologies to avoid this problem of soil erosion on Central American hillsides. A traditional management practice used by small farmers who produce grain, burning crop residues, is frequently cited as a phenomenon that contributes to soil degradation. However, from a scientific point of view, and considering only the practice of burning crop residues and herbaceous weeds, sufficient evidence to attribute soil degradation directly to the practice of burning does not exist. Normally, less than 10 percent of the total crop area is directly affected by burning when crop residues alone are burned (CIAT, 1993). The loss of nutrients because of burning, principally nitrogen, may be a factor that should be considered. Some bean cultivars planted in Costa Rica have been shown to fix almost 25 percent of the nitrogen that they use (Cervantes, personal communication, 1995). This would represent a considerable loss of nitrogen when bean residues are burned. However, when bean residues are used as animal feed, this nitrogen loss could be much less.

Many other management practices used by farmers who produce basic grains are arguable from the point of view of natural resource management. Nonetheless, all have pressing economic justifications. Plant density is an example. Plant density for both maize and beans could theoretically be increased in maize-bean intercrops, resulting in increased yields. Higher plant density would also provide greater protection to the soil. However, the potential for competition for available soil moisture between the two crops leads farmers to use lower than optimal plant densities. The spatial arrangements in which maize and beans are planted are another example. There are spatial arrangements other than those now used that would provide better soil cover. However, it is better to view the existing spatial arrangement as a decision taken to maximize the efficiency of labor for weeding. Protection from erosion is a secondary consideration.

A series of cultural practices also provide available technologies that could contribute significantly to the sustainability of traditional production systems in Central America. They include using mulches; incorporating legumes, principally velvet beans, into the system (a favored research topic by many international institutions); constructing barriers to prevent soil movement; and other technologies. All these practices have been adopted to some degree, depending on the circumstances facing the farmer. However, only time, adequate technical policies, and the resources invested in extending these technologies to farmers will decide whether these practices are more widely adopted (Buckles et al., 1994; Calderón et al., 1990; M. Flores, 1995).

Dryland rice production presents other challenges. Loss of genetic diversity is due to both extensive and intensive cultivation of a group of cultivars based on a narrow genetic base. This is the principal reason that rice in Central America is highly susceptible to *Rhizoctonia*, *Piricularia*, and some insect pests that seriously affect production. Genetic uniformity and the uniformity of cultural practices are the causes of growing problems of weed management. Weeds highly specific to rice are increasing, leading to greater application of herbicides and resulting water contamination (NAS, 1972).

The best-known problem in highland horticultural production is the high application rates of pesticides, often applied incorrectly. Exaggerated pesticide application works against ecological equilibrium and threatens the health of people living in these zones. This problem is especially characteristic of potato production. High initial investments, due to the price of seed and the potential profitability of these crops, leads farmers to apply chemical products more to protect the capital that they have invested than to protect the plants from insects and diseases. The resulting altered ecological equilibrium favors the appearance of ever more pathogenic variants of pests, leading to yet greater pesticide application. The result is a vicious cycle that is repeated year after year. Integrated pest management has not achieved desirable rates of adoption in these regions. The lack of appropriate technology for highland production, mentioned earlier, is another ongoing problem (Kaimowitz, 1992).

Structural Adjustment and Food Crop Production

Structural adjustment programs have characterized economic activities in the Central American region since the 1980s. Environmental variability makes it impossible to make many generalizations about the interactions between agroecosystems and the environment in Central America. Similarly, structural adjustment programs have developed differently in the countries of the region. It is reasonable to presume that they will affect farmers in the region differently, depending on the degree to which they are integrated into the market economy. The effects of structural adjustment on farmers will depend on the form in which they are applied in each country and the policies developed for the agricultural sector. In general, the Central American agricultural sector faces reduced economic importance, reduced employment, and a general reduction in investment because of the application of structural adjustment programs in the isthmus (Kaimowitz, 1994). The most generally accepted thesis for the impact of structural adjustment programs on basic grain producers is that weaker farmers will be more negatively affected while more powerful farmers will consolidate their economic power and control over the land.

At the macroeconomic level, economic stabilization, which is the goal of structural adjustment, is won by devaluing the currency, increasing interest rates, and reducing real salaries for the population as a whole. In this setting, the demand for basic products should decrease, because of a decrease in the population's ability to buy, especially in the urban sector. The amount of product offered should decrease as well because of restrictions on credit and high interest rates. This combination of factors will affect prices according to the elasticity of demand for each product, which tends to be positive. Farmers who are highly dependent on credit should be the most seriously affected by the restrictions on credit. On the average, no more than 10 to 15 percent of all credit in the agricultural sector goes to basic grain production in Central America. Rice is the crop that receives the greatest part of the agricultural credit. Therefore, a reduction in rice offered for sale is expected in those countries that carry out complete programs of structural adjustment.

Structural adjustment will affect small and mid-sized farmers who produce basic grains, like maize and beans, to the degree that they participate in the market. Without doubt, those mid-sized farmers who, after much effort, have become creditors will be the most affected. The effect will be significantly less for small farmers who are not now creditors or who do not need credit. Reductions in state intervention in credit programs will greatly affect small farmers who are residing in isolated areas. In these areas, any possibility of credit will disappear, as will state purchase of grains, because state institutions involved in price stabilization will withdraw under structural adjustment.

Devaluation will affect production costs negatively. As a result, farmers who use imported inputs will suffer reduced profitability, although their competitiveness in the export market will increase if international prices are favorable. The higher relative cost of technology based on external inputs could lead farmers to use technological components that are more environmentally friendly. These components might include green manures, soil erosion barriers, mulches to control weeds, and increased efficiency of fertilizers, for example.

Economic flexibility is another important goal of structural adjustment. This goal can be achieved, in theory, by opening new markets and by eliminating subsidies and market protection. Market liberalization will have a positive affect on farmers who are competitive in the international market, such as those who produce horticultural crops for export. Those who produce basic grains efficiently may be stimulated to turn to export crop production, which would decrease the availability of basic grains. The impact of structural adjustment on farmers who produce grains for national consumption will depend on international grain prices. On the one hand, if prices are lower than the national cost of production, grain imports will reduce grain prices internally and reduce the profitability of their systems. Market liberalization of grains among the nations of the region, on the other hand, will produce greater uniformity in grain prices in Central America. Theoretically, this could make it easier to import grains from the United States or Europe, which would reduce or stabilize grain prices.

In practice, the degree to which structural adjustment policies are applied in Central America varies. Today, adjustment is scarcely the topic. Rather, policies are aimed at fiscal stabilization, a step that follows from the process of monetary stabilization. The original purposes of these policies in the agricultural sector were to (1) strengthen agriculture's role as a generator of revenues through export earnings, (2) improve the terms of trade between rural and urban areas, (3) liberalize the agricultural market, and (4) generally decrease state intervention in the agricultural sector. The impacts of these policies have been different in the various Central American nations. Production costs have increased in some countries and decreased in others. The same has been true for commodity prices. Initially, input prices increased rapidly, whereas labor and land costs increased more slowly. Later, this tendency reversed itself. Analyses conducted to date have not revealed any clear tendency for the observed impacts of structural adjustment to conform to the desired results (Wattel and Ruben, 1992).

Conclusions

Food crop production in Central America is being affected by globalization through its concomitant policies of stabilization and structural adjustment and through increasing market opportunities. Although maize associations with beans and sorghum on moderately elevated hillsides, horticultural crops at higher elevations, and extensive rice production in flat areas still represent important food production systems, several changes in cropping systems are becoming apparent. These include more monoculture, more high-valued horticulture export crops, more use of fragile hillsides for traditional crops, less investment in food crops, and increased market opportunities and risk.

Food crops traditionally produced in association with maize are increasingly monocropped because of the need to increase production and improve labor efficiency. Monocropped maize production systems that use high-yielding varieties (HYVs) are increasingly concentrated in specific zones of several Central American countries. Concurrently, beans are also monocropped more frequently.

The need and opportunity to gain foreign exchange has resulted in the promotion and expansion of high-valued, high-elevation horticultural crops for export to North American and European markets. Traditional food production systems have changed dramatically in several countries of the region, particularly in the high plateau of Guatemala.

Because of land ownership patterns, small farmers typically produce their food crops in zones of erratic rainfall and thin soils. These conditions significantly increase the probability of dramatic harvest losses, leading to a decline in the economic and social sustainability of these areas. From an ecological sustainability perspective, small-farm crop production is typically accomplished on fragile lands and hillsides, where attempts to reduce erosion through cultivation practices can involve investment that many small farmers are not able to make.

Although it is difficult to generalize the effects of the recent policies of structural adjustment and stabilization, because of the diverse way these policies have been implemented in the different countries of the region, it appears that agricultural investment for food crops has declined. This decline is largely due to the apparent reductions in real income among the labor class. These laborers and their families are an important component of the demand for traditional food crops.

Market opportunities are expanding, especially for export crops. Subsistence farmers are not likely to benefit from the increased orientation to overseas markets that typically do not focus on traditional food crops. Market opportunities are more likely to be available to larger farmers based on volume produced, access to credit and other production inputs, and the ability to inventory products to take advantage of price fluctuations. Small- and medium-sized commercial farmers will often not be able to afford the increasing interest rates that have generally been associated with stabilization programs. In addition, many subsidies to agriculture have also been removed. Future investigations should be undertaken to document in detail the effects of these programs and potential initiatives to mitigate their negative effects.

18

Organic Farming in Central America

Gabriela Soto Muñoz

The organic farming movement has grown steadily since the mid-1960s, especially in Europe and the United States. Organic farming is becoming a well-known new production strategy in Central America, pushed by international buyers and local growers. However, while the organic movement is promoted in Costa Rica, it is common for older farmers to say, "I have been an organic farmer all my life, as my father and his father were." What is organic farming? Is it simply a traditional farming system with a new name? Is it a modern, more marketable farming system? Or is it a philosophy that tries to reset the natural balance?

This chapter provides a brief overview of the organic farming movement and the politics associated with it in Central America. Basic organic farming techniques are discussed, and some examples of organic farming systems in Costa Rica are described.

Definition of Organic Farming

Several different philosophies, techniques, and agricultural practices are needed to arrive at a comprehensive definition of organic farming. Some movements, such as biodynamic farming, permaculture, and natural farming, stress a holistic perception of the farm. Other groups, such as low-input farming or ecological farming, focus on the efficiency of the system and its ability to reduce input costs and outside dependancy (Castañeda, 1995). The different organic movements around the world are joined in the International Federation of Organic Agriculture Movements (IFOAM), based in Germany, which created a well-defined and comprehensive list of the principal goals of organic agriculture at its General Assembly in São Paulo, Brazil, in 1992. This list (IFOAM, 1992) describes the common principles of all the philosophies that are comprised in the organic movement:

- To produce food of high nutritional quality in sufficient quantity
- To interact in a constructive and life-enhancing way with all natural systems and cycles
- To encourage and enhance biological cycles within the farming system, involving microorganisms, soil flora and fauna, plants, and animals
- To maintain and increase long-term fertility of soils
- To use, as much as possible, renewable resources in locally organized agricultural systems
- To work, as much as possible, with materials and substances that can be reused and recycled, either on the farm or elsewhere
- To give all livestock living conditions that allow them to engage in their natural behavior
- To minimize all forms of pollution that may result from agricultural practices
- To maintain the genetic diversity of the agricultural system and its surroundings, including the protection of plant and wildlife habitats
- To allow agricultural producers a living, according to the UN's code of human rights, to cover their basic needs and to obtain an adequate return and satisfaction from their work, including a safe working environment
- To consider the wider social and ecological impact of the farming system.

According to this set of principles, organic farming is not just a system in which synthetic agrochemicals are replaced with natural pesticides. It is a production system that helps establish and maintain the ecological balance within the farm. In contrast to conventional farming systems, organic farming's main goal is not to obtain the highest productivity per hectare but to produce a sufficient quantity of healthy food over the long run.

The first aim of IFOAM, to produce food of high nutritional quality in sufficient quantity, is a critical one. A common argument used by conventional farmers is that feeding the world's population through organic farming is impossible. In fact, world hunger may be more a problem of food distribution than of food production. Perhaps more important, productivity on organic farms can be the same as in some conventional systems. For example, a well-established organic coffee farm can produce as much or more coffee than conventional farms (Boyce et al., 1994). However, most organic farmers do say that the initial productivity of an organic farm varies greatly, depending on how drastic the transition is from the conventional system previously used. Productivity can equal that on conventional farms after the organic farming system is well established. Organic farmers also argue that conventional farmers need to modify their farming systems by increasing biodiversity and improving soil conditions to maintain both production and equilibrium.

Nature Farming, from Japan, has similar goals. This movement's (Nature Farming, 1990) goal is to ensure the harmony and prosperity of human beings and all other life by conserving the ecosystem. They stress "conforming to the laws of nature and respecting the soil." According to the conventional farming perspective, the soil is a support and nutrient supplier for plants. For organic farmers, the soil is a source of life, a system that requires equilibrium and protection to function properly.

The organic certification agencies, mainly active in the United States and Europe, also play a role in defining organic farming by guaranteeing that a product is organic for marketing purposes and issuing a "certified organic seal." They establish lists of specific standards that certified organic products must follow. Meeting certification standards is important because farmers can usually sell certified products at a higher price than either conventionally produced or noncertified products.

For some, the role of certification agencies in defining organic farming is a source of concern because mere compliance with specific standards does not necessarily ensure that a farming system is sustainable. For example, a monoculture plantation can be certified organic if pesticides have not been used for three years and if there is no risk of contamination by pesticides from neighboring farms. However, this does not mean that other, broader goals, such as enhancing biodiversity and protecting the soil, are fulfilled. Nonetheless, some flexibility in standards (not including those that deal with the use of synthetic agrochemicals) has helped many farmers start to farm organically. Without this flexibility, some farmers might never move to organic production.

Some consider organic farming not only a way to produce food but also a philosophy of life. These members of the organic movement would like to see social issues included in the certification standards (García and Najera, 1995). For example, some argue that underpayment or other mistreatment of labor should prevent certification. Members of the IFOAM (1992) General Assembly in Sao Paulo instructed the IFOAM Standards Committee to develop standards for social rights and fair trade. No regulations have been approved yet, but the Standards Committee has submitted suggestions to the membership for discussion. A decision will probably be taken by the IFOAM General Assembly in 1996.

Organic Farming in Central America

If organic farming means a system that does not use agrochemicals and that will maintain constant productivity over decades with little or no damage to the environment, farmers in Central America have been farming organically for several centuries, although their produce has never been certified (Castañeda, 1995; Gonzálvez, 1995; Mojica, 1995; Quijada, 1995; and Trejos, 1995). Traditional farming systems have provided a stable lifestyle for many farmers until recently (Gonzálvez, 1995). However, most traditional systems rely on rotation and long fallow periods (Gonzálvez, 1995; Popenoe and Swisher, chapter 15). In the last 20 years, population growth and decreased land availability have constrained this type of shifting agriculture, often relegating it to hillsides (S. E. Carter, 1991). Land scarcity prevents farmers from rotating plots or using long fallows (see Rosemeyer, Schlather, and Kettler, chapter 16). The resulting systems cause serious environmental deterioration. We therefore must develop new systems that combine the valuable aspects of traditional farming with technological advances that allow higher production on smaller areas. Organic farming in Central America focuses on creating new systems that use the best of both traditional and organic systems to produce an abundant, healthy product.

Organic farming has some characteristics that make it ideal for small farmers. Labor, not cash, is the main initial investment necessary for organic production (van Bemmelen, 1993). The market for organic products is increasing rapidly. Many consumers are willing to pay higher prices for products produced on small farms or under socially sound conditions. Some will accept the variability in product quality typical of much small-farm production (van Bemmelen, 1995).

One main goal of organic farming is to eliminate the dependency of farmers in less developed countries on large agribusiness companies in developed countries. Many Central American agricultural systems have depended for years on seeds, fertilizers, and pesticides produced by large companies in Europe and the United States (Gonzálvez, 1995). The organic movement promotes the use of local varieties and locally produced renewable resources for soil fertilization and pesticides. Some groups, such as those who advocate ecological farming and low-input farming, even discourage the use of natural pesticides if they are produced by large companies.

Nongovernment organizations (NGOs) involved in small-farm development are responsible for the initial growth of the Central American organic movement (Castañeda, 1995; Gonzálvez, 1995; Mojica, 1995; Quijada, 1995; Trejos, 1995). Most are funded by international agencies. The NGOs working for habitat protection in national parks or reserves, for example, have turned to organic farming as an environmentally sound alternative (Soto, 1993).

Until recently, the governments of most Central American countries have been involved very little or not at all in the organic movement. However, involvement has increased in recent years. In Costa Rica, the Ministry of Agriculture formed a National Organic Farming Committee in 1995. It includes representatives of NGOs involved in organic farming, organic farming associations, universities, certification agencies, the Commerce Council, and the Costa Rican government (La Gaceta, 1995; Organic Crop Improvement Association, 1995). The government, in collaboration with organic farmers' associations and universities, developed the organic certification law, published in February 1997. Pressure to establish government regulations comes from the European Union's requirement that national certification agencies be regulated by the government (Brul and van Elzaker, 1995).

There has been no government support for the organic movement in Guatemala, Honduras, or El Salvador, although the movement has grown steadily because of the work of NGOs. In Panama, the organic movement is in its initial stages. For most farmers it is still an unknown term. However, NGOs and the Panamanian government have started promoting the movement, financed by international institutions from Germany and Japan, for example (Quijada, 1995).

Nicaragua is the only Central American country with a national certification agency. Perhaps more important for marketing, it has developed its own standards. Most Latin American certification agencies, to be able to export to Europe, the United States, and Canada, have followed the standards established by foreign agencies.

The role of the universities, as well as research overall, is still very small in organic farming in Central America. Very few universities have organic farming in their curriculum. However, this attitude is changing quickly because sustainability has become one of the best funded areas by international agencies.

In the Central American Symposium of Organic Farming in 1995, Castañeda summarized some general limitations for the development of the organic farming movement in Guatemala. However, these limitations apply to the whole region because the situation is similar in most other countries:

- Lack of adequate government support policies for the development of the organic movement
- Lack of appropriate technical assistance
- Lack of written information on organic farming techniques developed in local tropical conditions
- Lack of adequate credit policies
- Lack of coordination between NGOs and persons promoting organic farming
- Lack of a well-developed local market
- Lack of local companies working in the development of biological inputs for farm use
- Lack of legal status for inspection and certification of organic products
- Lack of a well-structured research program on organic farming at public and private research institutions.

Marketing Organic Products

Central America's internal market for organic products is very limited, probably because of the higher prices demanded for organic products[1] and the lack of consumer education about their potential advantages. At this point, the international market is still the best option for the organic farmer, although organic products are sold in Mesoamerica in some natural food stores, conventional grocery stores, and farmers' markets. In Costa Rica, for example, one supermarket chain sells organic vegetables. However, consumer demand has not increased in two years, and only a couple of branch stores are still marketing them.

A main principle of organic farming is first to supply healthy food to farmers and local communities and to export only excess production. However, higher prices in the international market make exportation more attractive than local sale, and most organic production from Central American is exported internationally. For example, Costa Ricans do not have access to the country's organic coffee, although it has been exported to Europe and the United States for several years. Certified organic products exported from Central America to Europe, Japan, and the United States include coffee, cotton, ginger, vanilla, cacao, bananas, ajonjolí, lemon grass, and other products. Table 18.1 provides examples of products exported from Nicaragua.

Organic Farming Techniques

The Federal Organic Foods Production Act of 1990, the National Organic Standard Board of the United States, the European Union, and IFOAM have established a minimum set of common standards that must be met for a product to be certified organic. These standards are flexible, and decisions are made on a case-by-case basis. Nonetheless, some general rules are common to all of the certification programs, and several of the most important rules follow:

Table 18.1 Organic Certified Products Exported from Nicaragua, 1994–1995

Product	Amount (metric tons)	Production Location	Type of Farmers
Coffee	245	San Juan, Rio Coco, Carazo, Mombacho	Small farmers in 14 cooperatives
Cotton	100	Chinandega, León	35 small farmers
Sesame (ajonjolí)	136	La Paz Centro, León	
Red and black beans	680	León, Somotillo	Small farmers

Source: V. Gonzálvez, La agricultura orgánica en Nicaragua, in *Simposio centroamericano sobre agricultura orgánica, marzo*, J. Garcia and J. Najera (eds.). (San José: Universidad Estatal a Distancia, 1995).

- *No use of synthetic agrochemicals.* Almost every agency has a different list of generic products allowed, prohibited, and/or restricted (Nature Farming, 1992; OCIA, 1995). Based on these lists, farmers decide which products they can use. Usually the lists are generic and open to some interpretation, and they can be reviewed and modified by each agency's standards committee.
- *No use of synthetic fertilizers for at least the last two or three years.* European regulations require two years, but most North American agencies require three. Some agencies require an initial soil analysis. However, a written confirmation that no synthetic fertilizers have been used for the last two or three years is usually sufficient.
- *No risk of contamination from neighboring operations.* Some agencies establish a minimum distance between a farm where synthetic agrochemicals are used and the organic farm (often 20 meters). They also normally recommend planting living fences between conventional and organic farms and building diversion canals if the conventional farm is uphill from the organic operation. There is still no established minimum distance from farms using aerial applications of synthetic agrochemicals. The type of synthetic agrochemical used, wind direction, and forest cover are some variables considered in these cases.
- *Soil-building program.* The farmer must show a soil improvement plan because the soil is the most important component in the organic production system. Among the practices normally used are (1) compost, cover crops, crop rotation, or intercropping; (2) leguminous crops to improve soil nitrogen levels; and (3) soil conservation practices to eliminate erosion. Mineral fertilizers such as calcium carbonate or dolomite and rock phosphate are allowed. Algae and fish emulsion are commonly used to supply nitrogen and calcium to the soils by farmers who can afford it. Biofertilizers such as *Rhizobium* and *Azospirillum* are also commonly used.
- *Weed management.* Weed management is usually manual and incorporates other traditional cultural practices such as cover crops, shading and pruning, and crop rotation. Very few natural herbicides are sold in Central America.
- *Insect pest and disease management.* Insect pests and diseases are seen as symptoms of poor management, not as problems in and of themselves (Chaboussou, 1995). Diseases and pests are not major production problems on many well-managed organic vegetable farms, even when natural pesticides are used. Again, crop rotation, cover crops, and intercropping are some cultural practices used to control pests and diseases. Biological control is also common.
- *Audit trail.* Good records of all farm activities are required for certification, a problem when farmers do not know how to read and write. An individual must be appointed to maintain records and other paperwork for associations of farmers who want to be certified.
- *Choice of crops and varieties.* Farmers must consider local soil and climate conditions when selecting the varieties and species to be used on the farm.

Types of Organic Farms

There are several general types of organic farms in Central America. Some have been traditionally managed, without the use of synthetic agrochemicals. Some traditional corn and bean production systems and some coffee, banana, and cacao plantations fit this category (Castañeda, 1995). In other cases, farms were managed conventionally, including the use of synthetic agrochemicals, but have been abandoned for several years, allowing them to meet organic requirements. The number of abandoned farms has increased greatly in regions of guerrilla warfare in El Salvador and Nicaragua, for example (Gonzálvez, 1995; Trejos, 1995). In Costa Rica, some farmers abandoned their farms because of disease problems, high production costs, and low international market prices. Finally, there are farmers who have farmed conventionally but are now consciously attempting to convert to an organic system. Coffee, banana, and cacao production in Costa Rica provide examples of one or more of these types of farms.

Coffee

Costa Rica has grown and exported coffee since the previous century. Synthetic agrochemicals were not used in coffee production until the middle of this century. Fungicides and herbicides came into use in the mid-1950s. However, coffee plantations still exhibited high biodiversity; for example, many species of trees provided shade. It was not until the mid-1970s that shade trees were removed and intensive Caturra (an unshaded coffee variety) plantations were established throughout the country. Production increased from 35 *fanegas* per hectare in 1965 to 78 *fanegas* per hectare (1 *fanega* = 20 hectoliters) in 1989 with the adoption of this new, unshaded system (ICAFE, 1992). Costa Rica's yield became one of the highest in the world.

The intensive, unshaded system allowed higher production in a shorter time than traditional, shade-grown coffee production systems. However, plant demands for nutrients were also much higher. As a result, the use of fertilizer increased. Herbicide use also increased because large blocks of trees were pruned in the new system, exposing the soil for extended periods (Boyce et al., 1994), whereas trees were pruned selectively in the traditional system. Soil exposure increased erosion, and new root pests appeared (Montealegre, personal communication). Profit per hectare increased but so did the investment required.

Promoted by the Ministry of Agriculture, the shade-free production system was established on most large coffee plantations in Costa Rica. However, some traditional, older farmers and the ones who could not afford the high input costs for the new system maintained the traditional, shade-grown system. These farmers made up the first group of modern organic coffee growers in Costa Rica.

The traditional coffee production system relies on abundant shade trees, most commonly citrus, musaceas, and nitrogen-fixing trees. Old coffee varieties such as Híbrido or Nacional, which are more tolerant to low nutrient levels and the major common diseases, are used. Trees are pruned selectively, so that the soil is always covered. Erosion is low, and weeds are controlled manually. The average production under this system is about 15–25 *fanegas* per hectare, much lower than the national average in 1989. However, cash inputs required to maintain production are also minimal.

In the mid-1970s, it seemed probable that these small farmers would disappear because of competition from the conventional, unshaded production system. However, new opportunities arose for them with the increased demand for organic coffee, at the higher prices it commands. Coffee-processing plants started marketing organic coffee. They sought out these small, traditional farmers for the international market. Other factors have also encouraged farmers to become more interested in organic production systems. High prices for synthetic agrochemicals and low coffee prices on the international market are important, but so, too, is an increasing awareness of the damage to the environment caused by the excess use of agrochemicals.

As a result, other farmers who did adopt more intensive production systems are also interested in organic coffee production today. However, research in this production is still very limited. Farmers themselves test new production techniques, looking for ways to maintain intensive production but using only those products allowed by the certification agencies. This is not always easy. Switching from conventional to organic production by replacing only the products used, without converting the entire production system, can be very costly. Nonetheless, conventional farmers who began experimenting with organic production long ago are now achieving production levels that equal those of their conventional systems (Jorge Issac Mendez, personal communication). On these farms, some shade trees are being introduced, and chicken manure, lime, and blood meals are commonly used to improve soil conditions, as is compost in a few cases. Diseases are controlled by Bordeaux mixes, pruning and shade management, and weeds are controlled manually (Boyce et al., 1994).

Cacao and Bananas

Both cacao and banana production were largely abandoned in the South Atlantic region of Costa Rica during this century. Internationally based banana companies abandoned their plantations in the beginning of the century because of devastating infestations of "Mal de Panama" (*Fusarium oxysporium*). Much later, about 20 years ago, Costa Rican farmers abandoned their cacao plantations because of another disease, Monilia. As a result, to earn a living many farmers turned to working for banana companies that had reestablished their corporate plantations, using resistant varieties. Others simply lived on very limited incomes earned by harvesting the highly reduced production they could get from their plantings and selling it for generally low prices to local buyers.

Recently, small farmers' associations in the area supported by NGOs, began to look for organic buyers for their products. Today, associations of organic farmers work together to export organic cacao and bananas. About 800 farmers organized in farmers' associations are selling organic cacao and bananas in the European and North American markets. They are backed by educational programs that focus on organic farming, postharvest management, and organic certification. A reliable market, better prices, and the support of the associations have motivated farmers to improve farm management.

Old cacao and banana plantings may appear at first glance to be secondary growth forest. Crop density is approximately 400 plants per hectare for each crop. Shade trees are abundant and diverse. This production system has remained unchanged for the last 20 years. However, since the new market for organic products has appeared, farmers have begun to increase plant density, incorporate pruning for disease management, eliminate unhealthy plants, practice manual weed control, and manage shade. Now, farmers who can do so are applying lime, following recommendations based on soil analyses financed by farmers' associations. Productivity is increasing and new income is improving families' welfare.

Future Challenges

The organic movement in Central America was, to a large degree, introduced by NGOs that were working with small farmers. Today, there is an international market for organic products that has made organic farming very attractive for some large farming operations as well. Growing without synthetic agrochemicals is not the only goal of organic agriculture. Other goals are to produce healthy food, to protect the environment, to encourage fair trade, and to provide adequate social rights for both farmers and consumers. These issues are especially important in regions like Central America, where land is not distributed equally and the agricultural structure does not support some farmers, especially small farmers, well.

Note

1. Farmers charge a higher price for organic products than conventional ones to compensate for the predicted lower productivity and the higher labor costs characteristic of organic farming. Farmers frequently inflate the price for organic products, even when one or both drawbacks are not present, a major incentive for farmers to convert to organic production. Nonetheless, a large sector of the organic community opposes overpricing because it limits the consumption of organic products to those who can afford to pay. For example, the Association of Organic Farmers (AAO) of São Paulo, Brazil, opposes overcharging for their organic products. They have organized the largest farmers' market for organic products in Brazil, and higher prices are not charged.

19

La Pacífica, 40 Years of Farm Ecology

A Case Study

Werner Hagnauer

Hacienda La Pacífica is one of the few farms in the New World tropics that has been managed by using ecological principles over a long period of time and for which adequate records about the management experience and lessons learned have been maintained.[1] This midsized cattle ranch in Guanacaste Province is located 4 kilometers west of Cañas between the Río Blano, Río Tenorio, and Río Corobicí (figure 19.1) in the semiarid zone originally covered by tropical dry forest (Holdridge, 1967; see also Vargas, chapter 10). It has served as a research site for two generations of agricultural, biological, and social scientists. This chapter describes in some detail the background and strategy for the farm's management from 1970 to 1985 and the conceptual framework on which it was based. In many ways this experience represents an example of agroecology in real life.

Farm History

On a 1945 map, today's Hacienda La Pacífica appears as a 1,332-hectare parcel in a larger, 7,611-hectare farm, Hacienda Paso Hondo. An older source, United Fruit Company Farm Map NR 96, dated 1925, shows the same farm with an area of 8,867 units (presumably hectares or *manzanas*, a local measurement), of which 8,200 were forest, 456 pasture, and 8.5 bananas. Despite the confusion over measurement units, this map provides an overview of the ecological context in 1925. Approximately 95 percent of the land on the farm was covered by forest. Six years later, in March 1931, Hacienda Paso Hondo was sold, apparently in an abandoned condition, to Maximiliano Soto, who planted 1,060 *manzanas* (about 700 hectares) of improved pasture, Jaragua (*Hyparrhenia rufa*), and probably some Guinea grass (*Panicum maximum*). He had about 1,500 head of cattle.

The northern part of Hacienda Paso Hondo—La Pacífica—owes its name to Pacífica Fernández de Castro, who designed the flag of Costa Rica at the beginning of this century. With the construction of the Pan-American Highway in 1942, La Pacífica was cut off from the famous old Paso Hondo farm. It was sold again to Florentino Castro in 1945, and after that it changed hands several times. The northern part of La Pacífica eventually ended up in the hands of farmers and loggers from the Central Valley, and they sold it to the CIBA Corporation in New York, a subsidiary of CIBA Basel (Switzerland), in October 1955. By then, the property was once again abandoned and in poor condition. CIBA wanted to establish plantations of medicinal plants such as Dioscorea (*Dioscorea floribunda*) and Rauwolfia. Dioscorea was thought to be a promising source for Cortisone, and it was believed that Rauwolfia could supply the raw material for Reserpin, a heart regulator.

After seven years of experimenting, this project was abandoned and the farm manager, the author of this chapter, leased the farm as a tenant for another seven years, until 1970. A merger of two big chemical companies, CIBA and Geigy, in 1969 brought more change. The farm was sold to a new company, Hacienda La Pacífica S.A., founded by the family of the author and a group of Swiss, German, and Dutch colleagues. None of the new partners was familiar with farming. All of them were business people, and their interest in buying shares was simply to invest in a safe operation. In today's modern terms, they wanted "sustainable development" for the farm, and the concept of sustainability became the guideline for future farm management.

After 15 years of successful farming, Hacienda La Pacífica was sold in 1985. The new owner, who was heavily involved in sustainable development (Schmidheiny, 1992), agreed that the farm would serve as a base for research, education, extension, and ecotourism in applied agroecology, with the cooperation of the Asociación Centro Ecológico. The owner also agreed to continue the existing management strategy and to con-

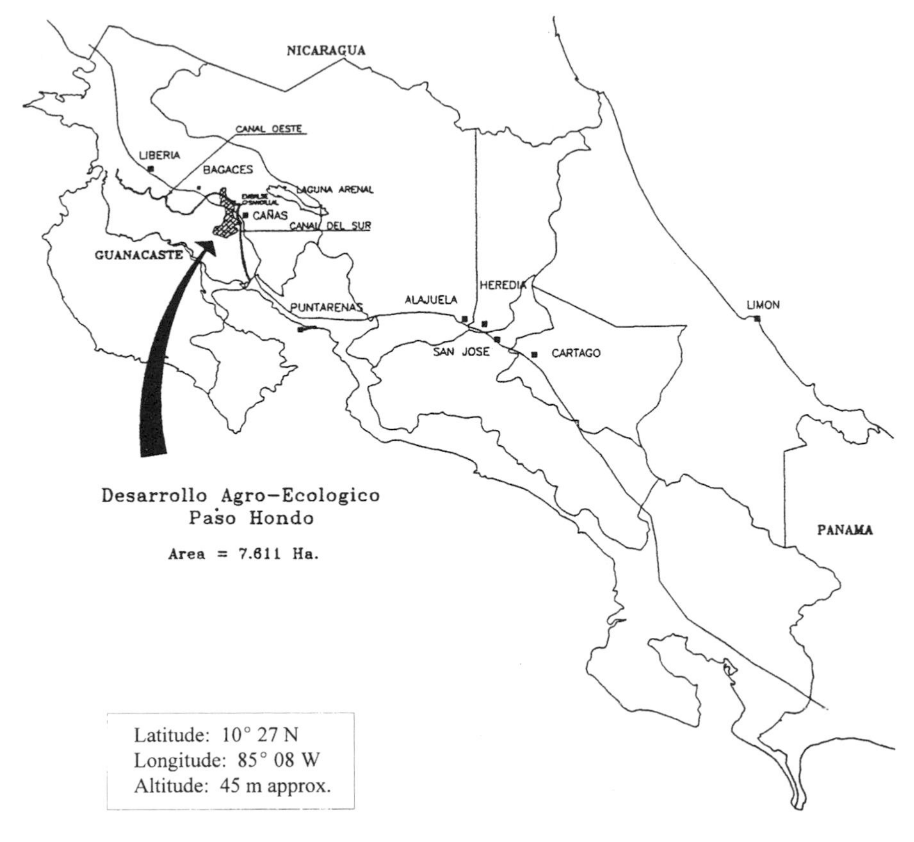

Figure 19.1 Location of Paso Hondo Farm with Costa Rican Boundaries, May 1945

duct long-term research projects that could not be carried out by short-lived political institutions.

In the 1940s, the southern part of the original Paso Hondo farm was sold to Don Carlos Pasos, a Nicaraguan citizen known as General Pasos, who planted the first cotton in the area. In 1952, the first cotton was exported from a new ginnery in Cañas. Then the property was split into various parts and changed hands again. In the late 1940s, an irrigation ditch with a capacity of about 350 liters per second was constructed, using only a small Caterpillar D2 and hand labor. In 1977, this part of Paso Hondo became the first section of today's Arenal/Tempisque Irrigation District. The very different management strategies and fates of these two parts of Paso Honda are discussed near the end of this chapter.

The Agroecological Setting

Precipitation

An examination of rainfall patterns between 1925 and 1995 in Cañas (figure 19.2) gives a good picture of the most important factor for producing a crop in the tropical dry forest zone. Although the average precipitation of about 1,500 millimeters per year appears, at first glance, to be adequate for many crops, the reality for a farmer is different. He or she must consider the extreme variability in rainfall, not simply base decisions on average conditions. The difference between the driest year, 1980 with 800 millimeters, and the wettest year, 1933 with 2,880 millimeters, is an astonishing 2,000 millimeters. The standard deviation is 450 millimeters, and the mean is 1,500 millimeters. This interannual variability is large in all regions affected by the El Niño Southern Oscillation, such as the Guanacaste region where La Pacífica is situated (Borchert, 1997).

Rainfall Intensity

A normal *aguacero*, or rainshower, in the tropical dry forest zone can produce 1 liter of water per minute per square meter (Hagnauer, 1980). Herrick (1993) reports a maximum intensity for any 15-minute period in 1991 on La Pacífica of 111 liters per hour (1.85 liters per second). He studied the infiltration rate in the soil in a Jaragua pasture and found an equilibrium infiltration rate of 40 millimeters per hour. On any cropland not properly contoured, such heavy rains mean erosion,

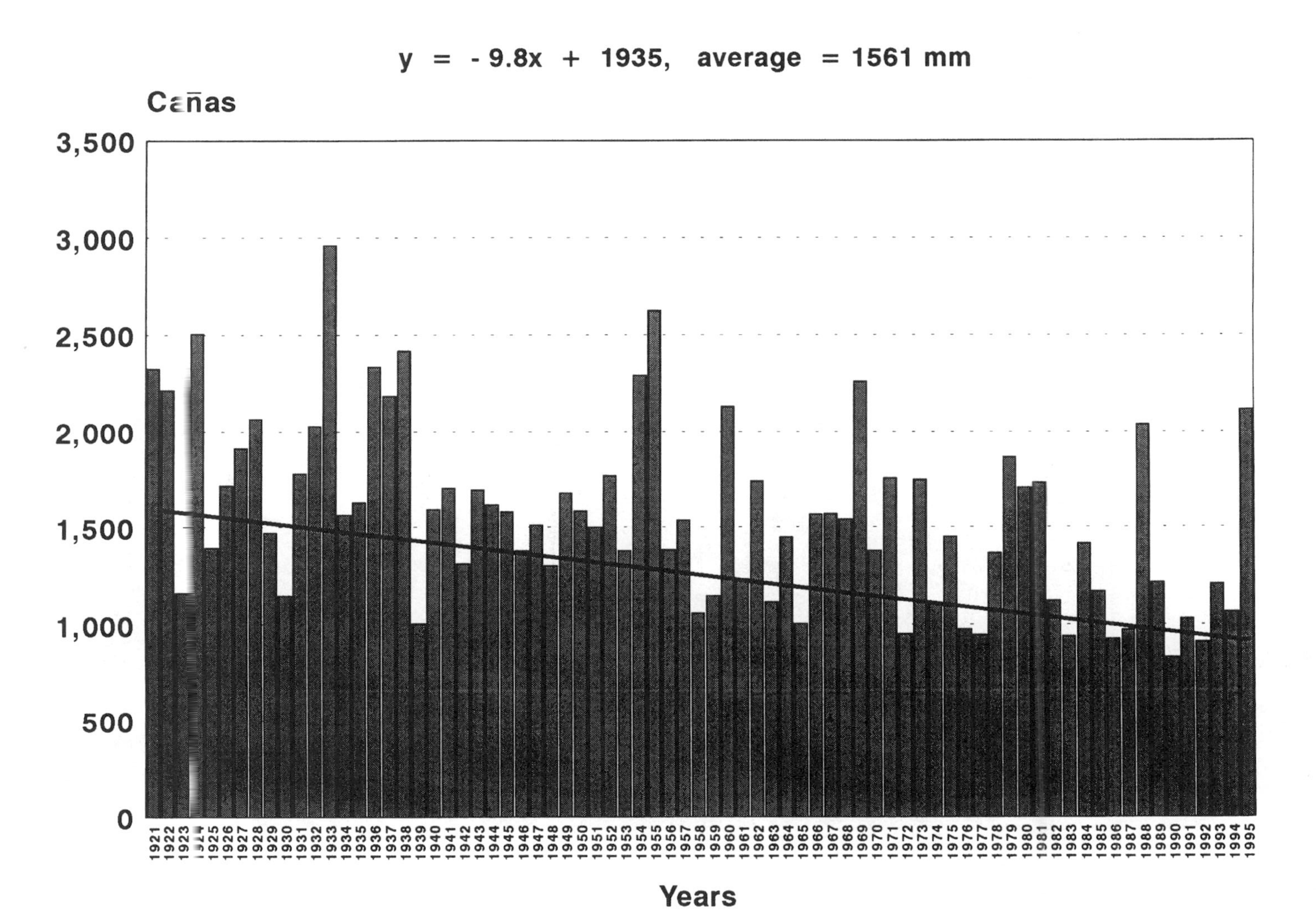

Figure 19.2 Precipitation in Cañas, Costa Rica, 1921–1995

mud, problems for machinery, and serious soil compaction by machinery or high stocking rates on pastures. A consistent soil cover becomes, therefore, a high priority in farm management.

Wind and Evapotranspiration

Another important climatic factor that affects crops and livestock in the dry tropics is the wind. Very few multiyear systematic studies have been published on this topic. Wind speed was measured for a short time in the 1960s, and peaks of 80 to 100 kilometers in February and March were not unusual. Herrick (1993) recorded an average of 38 kilometers per hour for the highest month, February, and 11 kilometers per hour for the lowest, October. Thomas (1992) observed in March 1990 a windspeed of 21 kilometers per hour in a pasture protected by windbreaks on La Pacífica; 54 kilometers per hour over the canopy of the forest; and 3.6 kilometers per hour inside the forest, which was 2 meters above ground level.

High winds produce severe soil drying. Daubenmire (1972) reported dry soils down to 1 meter five to six weeks after the end of the rainy season in a high precipitation year (2,265 millimeters of precipitation), February 1970. Reich and Borchert (1982) observed a drop in soil humidity of 71 percent in December 1979, at a depth of 30 centimeters, only three weeks after the last rain, and a drop of 15 percent down to 1 meter, after six weeks. This was after an accumulated rainfall of 1,868 millimeters in the previous rainy season. Because of high wind speeds, the relative humidity often falls to 25 to 30 percent at noon, and evapotranspiration increases to a level where the growth of crops and pasture, even under irrigation, may be seriously slowed or completely interrupted. The consequences of this rapid soil drying, even during the rainy season, include the loss of the first cotton crop or low yields for late-planted sorghum or corn, for example (Hagnauer, 1980).

During his research, Herrick (1993) registered a moisture deficit of more than 60 millimeters for every month during the year 1991. These findings are widely confirmed by farmers' experience and visual observations of crops and pasture. They also explain why irrigation in the dry tropics, although very helpful, is not the complete solution to increased productivity as is sometimes hoped.

Soil Temperature

In open, unshaded Jaragua pasture, soil surface temperatures may reach 50°–80°C. In a detailed site study, Herrick (1993) found the maximum average midday temperature for a single day to be 67.2°C, which agrees well with Daubenmire's (1972) findings. Thomas (1992) recorded 40°C as a maximum inside the dry forest on the same farm. Daubenmire called these extremely high temperatures to the attention of both farmers and extension professionals. It was a relatively unknown factor, one not taken into serious consideration when new crops or production systems were recommended. Such conditions may affect germination and the growth of crops, especially in large, unshaded fields. They also restrict the choice of possible crops, varieties, or animal breeds and can turn crop farming into a high-risk venture if overlooked. Livestock without shade in open pastures suffer, milk production and weight gain remain low, and feed concentrates are not used efficiently.

Soils

In 1968, Stewart Harris of Kansas State University classified the soils of the farm as alfisols and lithosols, and he identified six different soil series. The surprising result of this short study was a complex mosaic of colors on a map that covers about 60 to 70 hectares of crops and pastures spread over nine different fields on La Pacífica. A visual survey by airplane of the crops, pasture, and weeds in the fields confirmed Harris's findings and raised the question of how fertilization, weed control, and cultivation can be applied efficiently under conditions of such high variation in soil types over short distances.

Farmers must therefore organize their farms to survive under these conditions without destroying the production base—the land and the soil. They must deal with extreme values and not with statistical averages. A first step in mitigating these adverse factors is to improve the local microclimate at the farm level by increasing the biomass. This means reforesting cleared land.

Farm Management through the Years

In 1925, forests covered about 90 percent of La Pacífica's land. Mahogany, balsa, pochote, and others, all growing on humid soils and not typically native to the dry tropical forest zone, were extracted and exported to England. In 1931, Maximiliano Soto planted Jaragua grass as an improved pasture for the first time. About 80 hectares were planted, fenced, and kept clean of weeds and shrubs by burning. As far as we know, in doing so he introduced to the region the first planned cattle management system based on an improved pasture.

Early in the 1950s this part of the farm was leased to a progressive and experienced cattle rancher from Tempisque. He took care of the Jaragua, maintained the fences, and controlled the weeds. The unused center part of the farm was leased to a traditional cattleman. He grazed and browsed his herd—cows, calves, and steers all together—in an unfenced area of about 600 hectares of open and heavily logged and damaged dry forest, commonly called the *sitios*. Animals had to walk about 2 kilometers to reach watering places on the river. The progressive farmer canceled his contract, alleging that he was losing too many animals for unknown reasons in the Jaragua pastures. Don Genaro, the traditional farmer, commented in 1960 at the time his contract was withdrawn, "Sorry, Machito, this was such a damned good place for my animals." These two comments express two different ecological experiences, unwillingly made by a progressive and a traditional farmer. We will refer to their experiences again later in this chapter.

Crops: An Unsuccessful Strategy for Sustained Production

To produce quick cash flow to cover the expenses of experimentation and administration, CIBA introduced intensive, modern agriculture, planting such crops as cotton, rice, and sorghum. The land was cleared by bulldozer and then plowed in straight furrows running up and down the slopes. This caused serious soil erosion in the first year of production (1960), when 2,250 millimeters of rainfall occurred. Soil protection became a first-order problem for resolution. After this experience, La Pacífica was one of the first and very few farms that established terraces and introduced contour planting for all crops and pasture.

The planting of cotton for seven years gave another unwanted lesson in agroecology. In 1960, the first cotton crop was sprayed twice to control insects, using only a farm tractor. In the second year, airplanes had to be hired. Strong winds caused dangerous flying conditions and problems in applying the insecticides properly to avoid spray drift. Windbreaks out of the surrounding bush and shrubland were built to reduce these problems, with a maximum spacing of about 150 to 200 meters between windbreaks. In later years, this "loss" of farmland was considered to be a benefit for both crops and livestock, although it had not been planned as a tool for ecological farm management or to improve biodiversity and the microclimate.

Despite the availability of technical assistance and credit for cotton production, and an emphasis on increased and diversified farm production for the export market, cotton was not a sustainable crop for La Pacífica. In the course of a few years, cotton pests, such as boll weevil (*Anthonomus grandis*), bollworms (*Heliothis* sp.) and white flies, built up and the number of spray applications increased.[2] In 1965, 19 aerial applications were made, but the cotton crop was still lost. Artificial fibers were introduced, cotton prices dropped on the world market, and it became clear that cotton, in the long run, was a high-risk crop that was not sustainable.

Many other crops were tried over the years. Table 19.1 lists several alternative crops tried between 1960 and 1985 to diversify production. Most of them were recommended by official institutions, such as the Ministry of Agriculture, or extension services, private companies, or other knowledgeable sources. Many of these crops did survive in local gardens for home consumption, but they could never contribute significantly to the commercial income stream. In fact, they lost money, but since the acreage planted was never high, the deficits they produced could be covered by other enterprises on the farm.

Table 19.1 Alternative Enterprises Tested at La Pacífica, 1960–1985

Corn	Vegetables	Hot pepper
Soybeans	Cucumbers	Sweet pepper
Sunflowers	Limes	Cassava (yuca)
Kenaf (fiber)	Lemons	Yam (tiquisque)
Castor beans	Oranges	Sweet potato
Beans (frijoles)	Soursop (guanábana)	Cashew nut
Cowpeas	Mangoes	Star fruit (carambola)
Peanuts	Coconuts	Grapes
Velvet bean	*Vinca* sp. (seeds)	

Source: Records in personal farm notebook, unpublished.

Despite their poor economic performance, crops still have an important role to play in ecological farm management for La Pacífica. Crops are needed for rotation with pastures to reduce the inevitable soil compaction that results from the cattle. Such a rotation may include from 7 to 10 years of pasture and 2 to 3 years of crops, giving a cultivation factor (R) of about 35–40, as suggested by A. Young (1989) for the semiarid zone. Pasture, with careful herd management, replaces the fallow necessary for fragile tropical soils and guarantees the sustainability of a cropping system in the dry tropics.

Despite government and private sector technical and financial assistance in the 1960s, the financial status of many farms that depended on crop production in Guanacaste became untenable after three to seven years in operation. Soils were often exhausted, more fertilizers and pesticides were needed, and bank loans came due. Local and international banks were willing to extend additional loans, but farmers were rapidly accruing large mortgages and increasing financial risk.

In July 1969, Arenal Volcano erupted and covered a large part of the province with ash. The volcano was both a culprit and a savior for farmers. The banks offered settlements for the accumulated debts over a term of 20 years with interest rates as low as 4 percent. In reality, this was nothing more than a superficial economic cure laid over a developing ecological catastrophe brought on by disregard of the climatic and edaphic conditions already described. Many farmers did not understand the underlying problems, but eventually many quit farming, sold their farms, started other businesses, or simply abandoned their farms. Others, however, turned back to traditional cattle ranching. La Pacífica adopted the latter approach in 1970 because its new shareholders felt that ranching offered the lowest risk on their investments.

Cattle: An Ecologically Sound Alternative

The CIBA Corporation started experiments with medicinal plants in 1956 but used only a tiny portion, about 0.5 to 1 hectare, of the 1,332 hectares of land on La Pacífica. The rest ran wild. Large amounts of grass, dry shrubs, and trees grew up, and in the dry season in 1961 a bush fire burned almost half the farm. This accident was the starting point for thinking about the importance of livestock on the farm. The farm manager began to build a properly managed cattle herd, not only to control bush fires, but also as a long-term and stable activity that could serve as a financial backstop for experimentation and speculative crops.

Starting with 100 Brahman heifers, a Brown Swiss bull, and 20 Guernsey and Brahman crossbreeds, the herd grew slowly without much attention in the remaining dry forest or *sitios*. Although some wondered how the herd could be healthy and in good condition, the experience of the traditional cattle rancher mentioned earlier was repeated and confirmed. The cattle served several functions. They kept the fuel for the inevitable dry season bush fires low, they cleaned up the cotton left on the ground when the strong winds started blowing during harvest, and they grazed the stubble of the sorghum and rice to prepare the fields for new plantings. Although these functions have never been economically evaluated, they were important economic inputs for La Pacífica.

In 1967, a record system was introduced for the herd. In 1972 artificial insemination was introduced, and all data back to 1965 were collected and recorded at the Computing Center at the Federal Institute of Technology (ETH) in Zürich, Switzerland. These data were evaluated periodically by students and researchers between 1972 and 1985 (Aragón, 1981; Bachmann, 1983; Hagnauer, 1978, 1992; Kropf, 1978; Kropf et al., 1983; Moreno, 1985). The discussion that follows is based largely on these publications, combined with the practical conclusions and observations on the farm.

When artificial insemination was introduced, the shy and nervous Brahman cows, used to large pasture and dry forest areas, had to be tamed. This was done by milking in the traditional way, tying the calf to the front leg of the mother. Initially the milk not used for home and farm consumption was sold locally as a byproduct of the taming process, but it quickly became a profitable commercial product. A Brown Swiss bull was already in use to upgrade the Brahman cows. A second milk breed, the Simmental, of Swiss origin with high milk production, was also introduced through artificial insemination. The objective was to have a better milk supply for people and calves and to supply the highly deficient and fast-growing national milk market.[3]

The local milk cooperative (Cooperativa de Productores de Leche) offered to buy milk from La Pacífica in 1975. With an assured market, a modern dairy plant with a cooling tank equipped for energy recycling was installed,[4] although setting up an ultramodern dairy operation was not the intent. Rather, La Pacífica sought to improve the traditional local system by using modern tools and relying on the available resources of cows, forage, and labor.

Producing maximum yield with sophisticated technology and high inputs of concentrate, fertilizer, and other agricultural chemicals was rejected from the start. However, despite the fact that 85 percent of the milk was of premium quality, acquiring the initial financing for buildings and equipment was a major problem. Credit was not available for such a low-input, moderate-yield system. The World Bank and National Bank credit lines were designed for high-input and high-technology systems. Nonetheless, because of the agroecological conditions of the farm, bank loans for setting up a high-performance operation like those that had already failed on other farms in the region would have been foolish.

The bottlenecks for any kind of animal husbandry in the tropics are the protein content and the digestibility of the available forage. Assisted by ETH in Zurich, we estimated that tropical forages suitable to conditions at La Pacífica could eventually produce about 4 to 6 kilograms of milk per day, or a weight gain for steers of roughly 300 to 500 grams per day as a yearly average (Kropf, 1978). In fact, crossing Brahman cows with European dual-purpose bulls by artificial insemination highly improved both milk and beef production, with a low additional economic input. About 5 to 6 kilograms of milk per day were obtained, allowing about 1.5 to 2 kilograms for the calf and 3 to 5 kilograms for the market. No feed concentrate was supplied, except sugar molasses and chicken manure during the dry season. Urea and synthetic phosphoric acid were also used to correct the phosphorus deficiency in the diet because supplying the phosphorus directly to the cow is less expensive than spreading it on the pasture. All the farm manure was recycled to the pasture (Swisher, 1982).

Bachmann (1983) analyzed 1,150 pairs of cows and calves between 1975 and 1982 for milk production, weight gain, calving intervals, influence of the breeding season, and other factors. Some of his findings are presented in figure 19.3, slightly adapted and summarized. These data show that crossbreeds in this production system produce 160 grams of beef, the calf's weight gain, for each kilogram of milk produced during the year, or 85 grams of meat per day of lactation. Milk and meat production is economically at a break-even point

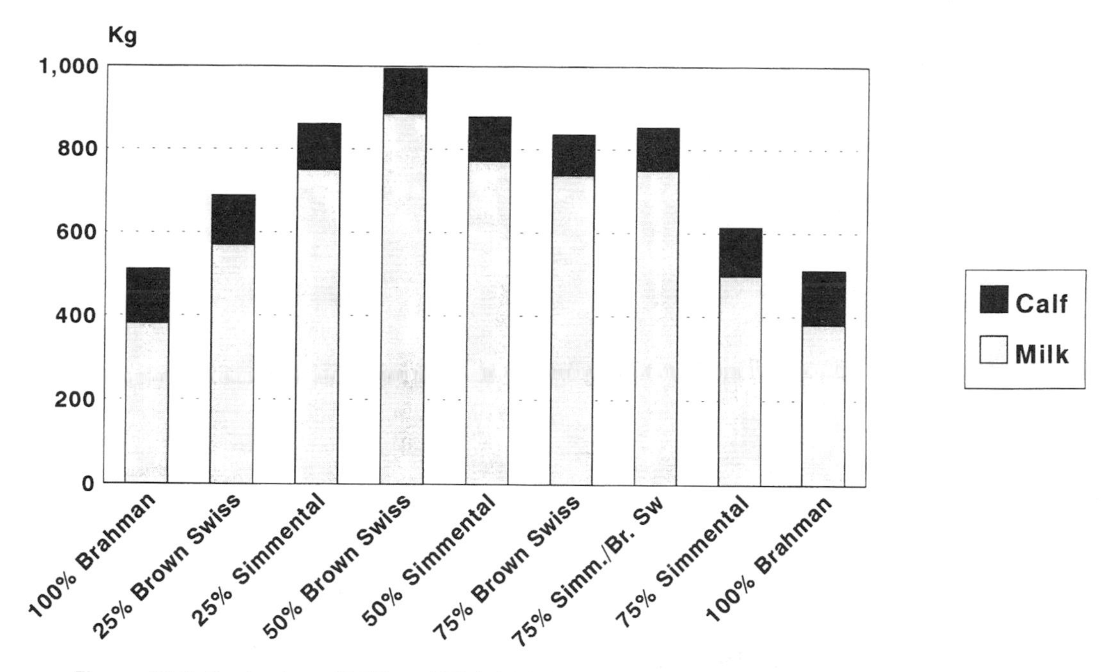

Figure 19.3 Production of Milk and Weight Gain of the Calf for Different European Crossbreeds

with a milk-to-beef price ratio of 1 to 4 (1 kilogram on the hoof should have the same price as four kilograms of milk).

The system produces for each kilogram of milk 85 grams of meat per day of lactation. This is roughly a daily food ration for two people. This balanced production will not create surpluses of either of the two products on the market and does not require subsidies because it is a low-input system. The data also show that the traditional crossbreeding to produce animals with about 50 percent improved genetic material and 50 percent local genetic material was successful. This tactic rapidly produced resistant animals that gave a sustainable and balanced yield of milk and meat produced on locally available forage resources. Bachmann (1983) analyzed the influence of the year, season, breed, and other factors on herd performance, and Moreno (1985) described the structure of the system in detail.

The herd performance obtained probably represents the natural ceiling for this environment. The overall standard performance of the herd, shown in table 19.2, became very stable over the years. Having five products for sale diversified cash income for the farm. Milk provided a weekly paycheck for the farm workers. Fat steers went to the packing house for export. Young, one-half blood breeding bulls from selected mother cows were sold to small and large farms. Young one-half blood heifers were also sold as breeding stock. Old, culled cows of excellent slaughter quality and high weight (450 kilograms) went to the national market.

These results were obtained by matching the animals' performance to the natural environment and by rejecting maximum yields as an objective. This is the opposite of the strategy of the green revolution, which emphasizes maximum crop yields through high inputs of fertilizer and pesticides, usually financed from external, international resources.

Keys to Economically Viable, Ecologically Sound Cattle Management in the Dry Tropics

Pasture Management

In 1961 two researchers from the Interamerican Institute for Agricultural Science (IICA) in Turrialba started trial plots at La Pacífica to observe and compare the behavior of different providences of Jaragua and native legumes from different parts of Costa Rica. Although this promising project was short of funds and was canceled, Pohl (1980) did identify a total of 42 different grasses on the farm. However, no studies were conducted for the phenology or agronomic importance of these native species.

During the 1960s, the Ministry of Agriculture and other institutions encouraged the establishment of improved pastures, such as African star grass (*Cynodon nfluensis*) and pangola grass (*Digitaria decumbens*). About 70 hectares of pangola and some African stargrass were planted at La Pacífica. However, these pastures needed high levels of routine fertilization and were

Table 19.2 Standard Performance of Crossbred Herd at La Pacífica

Milk production per lactation	800–1200 kg
Calving interval	13.5 months
Weaning weight at 8 months	160–180 kg
Slaughtering age of steers	33 months
Delivering weight on the hoof	420 kg
Insemination performance	1.3–1.5 inseminations per pregnant cow
Optimum stocking rate	1.5 animal units per hectare

Source: Bachmann, F., 1983, and personal farm records.

susceptible to pests. As a result, no additional areas were planted and more attention was devoted to legumes. Opler (1969) recorded a total of 47 species of native legumes on the farm. We classified them according to growth characteristic and acceptability to cattle, as shown in table 19.3 (Aragón, Hagnauer, and Kropf, 1976). This large number of legumes perhaps grew on the ranch because the farm never used herbicides extensively (because of the neighboring cotton plantings) and stopped applying all synthetic chemicals in 1969.

In 1972 the farm, in collaboration with the University of Florida, took part in a program to improve pastures by incorporating legumes. About 15 hectares of *Siratro*, *Stylosanthus*, and *Desmodium* greenleaf and silverleaf were established, with excellent results. However, after three years, these dense stands slowly disappeared and the native species or cultivars took over. As a result, we learned to appreciate native legumes as a natural resource available on the spot and free of cost.

In all the years through 1985, weed control was achieved by rotating the herd through different pastures and mechanically chopping weeds, using small 35- and 65-horsepower tractors with a bush hog. The phenological development of weeds was observed, and chopping was strictly timed to increase weed destruction. Driving over wet or irrigated pasture was avoided to help keep soil compaction low but could not be avoided entirely. The rotation of pasture into cropland helped solve the compaction problem, one of the key roles for crops in cattle ranching.

Table 19.3 Growth Characteristic and Acceptability to Cattle of Native Legumes on La Pacífica

	Herb	Shrub	Trees	Total
Total	15	26	6	47
Accepted	14	15	3	32
Percentage accepted	93%	53%	50%	68%

Source: A. Aragón, W. Hagnauer, and W. Kropf, 1976.

Stocking Rate

Different types of pasture permitted higher or lower stocking rates. From 1968 to 1975, records were kept of the number of animals grazing in each pasture. Table 19.4 shows the average stocking rates that were effective for each type of pasture. One animal unit (AU) is defined as a cow of 400 kilograms live weight grazing for one day. A calf is calculated as 0.2 AU and a heifer as 0.8, according to their metabolic weight.

Over the years the higher-yielding species such as *Paspalum*, *Cynodon*, and *Digitaria* disappeared. Only Jaragua, growing in combination with the native legumes mentioned earlier, persisted without serious problems and with little attention. Specialists usually rate Jaragua very low because protein content and digestibility drop quickly as it matures. However, with frequent rotation and grazing to maintain a height of 7 to 15 centimeters, Jaragua grows in polyculture with legumes, producing an overall protein content as high as that of other grasses. This grass-legume mix can serve as the base for an economically and ecologically viable livestock operation.[5] Research at the University of Minnesota in grassland plots with mixed species goes in the same direction, showing a better pasture performance in stands with up to six species (Schneider, 1996).

Jaragua pastures were allowed to set seed every three or four years and were then bailed for hay. This hay contained 3 to 4 percent protein with low digestibility (about 45 percent or less) and, combined with sugar molasses (48 percent sucrose) and urea, made an excellent and cheap supplement for pregnant cows and young stock during the dry season. At La Pacífica, chicken manure was another source of feed supplement, a by-product of the henhouse with up to 8,000 layers. We started to use it in the 1960s without any processing in quantities of 30 to 50 tons per dry season.[6]

Table 19.4 Average Stocking Rates for Different Types of Pasture on La Pacífica

	Animal Units per Hectare
Paspalum notatum, irrigated	2.33
Digitaria decumbens, irrigated, fertilized	2.13
Cynodon nfluensis, unirrigated	1.71
Hyparrhenia rufa (Jaragua), irrigated	1.62
Hyparrhenia rufa (Jaragua), unirrigated	1.28
Dry forest/savanna	0.52
Farm overall (1965–1985)	1.52

Source: Records in personal farm notebook, unpublished.

Irrigation

In 1969, a neighbor suggested taking water out of the Corobicí River to supply drinking water for the livestock on both farms, although he did not want to irrigate. Rights to 600 liters per second of water and the construction costs for the main ditch were split between the two farms.[7] The neighboring farm could not complete construction as rapidly as La Pacífica. As a result, all the water was available to La Pacífica. We decided to let the excess water overflow into the windbreaks and into the uncleared forest patches. In the dry season of 1970, the farm had, for the first time, the benefit of living, green windbreaks and drinking water for the herd. The irrigation of windbreaks and woodlands provided a great but unanticipated improvement to the farm environment and microclimate.

During the next seven years, La Pacífica developed an irrigated perimeter of about 300 hectares for pasture and crops. The main emphasis was to supply drinking water for cattle and to irrigate pastures in the dry season. Supplementary irrigation for wet season crops in rainfall-deficient years was a secondary emphasis. This strategy permitted optimum efficiency in using the available water. The water supply from the river was as high as 1,200 liters per second at the beginning of the dry season but dropped to 300 to 400 liters at the end of the season. The flow was too unreliable and too low to justify the high investments required to level land for flooded rice production.

Over the years, and based on accumulated experience, the importance of the different water uses changed. On the one hand, environmental improvement, a low priority when the irrigation system was first developed, grew in importance and by 1985 this was the most important use of water. On the other hand, drinking water for cattle and pasture irrigation, originally the highest priority uses, decreased in importance. The ranking of importance in the use of water was as follows:

Importance in 1968	Purpose	Importance in 1985
1	Drinking water for cattle	4
2	Pasture irrigation	2
3	Crop irrigation	3
4	Improvement of environment	1

Besides the ditches and terraces, 12 ponds were constructed as reserves for drinking water for cattle. Vegetation along irrigation and drainage ditches was protected, permitting it to grow rapidly. The ponds quickly became a refuge for waterfowl never seen before on the farm, and thus they eventually became an important attraction for tourists and photographers.[8] These changes, taken together, produced a remarkable improvement in the microclimate of La Pacífica for both animals and vegetation. The subterranean water table rose as much as 4 meters. The roots of many trees could reach this depth, even in the dry season (Borchert, 1997; Hagnauer, 1993).

The Role of the Forest

In 1962 Gerardo Budowski, a forester and ecologist, with a group of students made the first list of trees found on La Pacífica. In the following years many researchers contributed to this list, and by 1985 a total of 112 species was registered. In 1984, a forest inventory of representative plots to record volumes, abundance, and frequencies of the different species was under preparation, but the new management edited only a revised list with 116 species. Shrubs and herbs in the forest have never been systematically recorded, and therefore no clear picture of the overall biodiversity of the dry forest on the farm is available. Unfortunately, the opportunity to compare an untouched, protected area and a browsed area over a period of 35 years was lost. The fight against bush fires in one of the protected areas (approximately 120 hectares) was continuous but not always successful. In 1984 we abandoned the idea of a private reserve and put cattle in the area to use the browse, as proposed by Conklin (1987).

Special attention was always paid to protecting old trees for seed production and natural reforestation. D. Janzen (1973) had published papers on this issue, and we were aware that natural reforestation occurs over a long period. We exerted no preference for any species, and no tree was cut without the express order of the farm manager. The most important trees for use on the farm reproduced naturally in surprisingly high numbers. Table 19.5 lists the species of major interest. By 1985, we believed that the future timber supply for the farm for buildings and fencing was assured, but no commercial sale of timber was planned.

Browse for Animals

Forest patches, windbreaks, river borders, dry creeks, and similar habitats are excellent forage reserves for the dry season. Conklin (1987) mentioned two of the most advantageous aspects of many deciduous trees in Guanacaste's dry land forest:

1. They drop their sugar, oil, or nitrogen-rich fruit during the later half of the dry season, providing feed to herbivores when there is very little else to eat. Sixty-six percent of the trees mature their fruit in the dry season, with April being the peak of fruit drop. Edible fruit compromises about 50 percent of this fruit drop and the other 50 percent is wind dispersed.

Table 19.5 Tree Species of Major Interest to Farm Operations on La Pacífica

For Construction	For Fencing and Corrals
Enterolobium cyclocarpum	*Gliricina glauca*
Pithecellobium mangense	*Lysiloma divaricata*
Bombacopala quinatum	*Tabebuia rosea*
Swietenia humilis (mahagony)	*Mora tinctoria*
Cordia aliadora	*Cassia bicapsularis*
Tabebuia ochracea	

Source: W. Hagnauer, unpublished list.

2. They tend to leaf out in anticipation of the next rainy seasons and become available for browsing before the grass begins to grow. Seventy percent of the tree's leaf flush is in the last ten days of April, weeks before available soil moisture has been replenished.

She also suggested that an additional factor could be exploited by allowing deciduous species to become "evergreen" in riparian environments. Leaf drop occurs on schedule according to the normal phenology of the species, but the leaves flush out again within a few weeks when year-round water is present. Such a suggestion may be reprehensible for foresters or National Park Service personnel but not for a farmer who needs to combine natural cycles and management to provide forage for his herd. This principle could be extended to all the new irrigation ditches.

With careful management, valuable feed resources in "unused" or "useless" land can be saved and transformed into protein for human food (figure 19.4). Cattle should graze or browse scrub and forest areas at the beginning of the dry season to remove and trample the fuel for bush fires. They should remain in these areas until they have eaten the fruit and dry leaves on the ground and be removed before leaf flush begins (D. Janzen, personal communication, 1982). The stocking rate should be lower than 0.5 AU per hectare. Table 19.6 (from Conklin, 1987) shows only a few data that demonstrate the high value of some very common browse species. The low digestibility ratios are not a concern because animals adapt to wide ranges under natural conditions.

Fencing

There is no livestock management in the tropics without fences. In the first 10 years (1960–1970), about 125 kilometers of three-wire fences were built on La Pacífica. With a spacing of 2.5 meters between posts the use of 4-by-4-inch posts approximately 2 meters long, this fence required 9 to 10 cubic meters of timber per kilometer. Wild and shy cattle need closer post spacing, and unnecessary rough treatment will exacerbate the problem. Under some conditions, stronger posts and more wires will be needed. The timber used in 1 kilometer of fence may rise quickly to 20–35 cubic meters. The size of each pasture is an additional problem. A 1-hectare plot of 100 by 100 meters, recommended for intensive grazing, requires 400 posts. A 10-hectare parcel of 333 by 333 meters needs a more extensive operation of only 133 posts per hectare, or three times fewer posts. Corrals are another, often luxurious use of timber. A 60-by-60-meter corral with four pens easily demands 25 to 30 cubic meters of hardwood, corresponding to the timber production of 25 to 30 hectares of dry forest.

The type of pasture and herd management possible depend greatly on the timber resources available to a farm. This was never seen as a problem in the past. The timber came from clearing the land and was never carefully used. Today, this issue is never properly and clearly explained to farmers and is completely ignored by technical advisers and bankers. It is especially serious in the dry tropical forest zone where overall forest yields may be little more than 1 cubic meter per hectare and where the planting of living fences is critically dependent on the rainfall. Careful forest management must, therefore, go together with animal husbandry to provide adequate forest reserves for the farm. The pride of any farmer, a nice, strong fence, is not only a costly hobby but also one that directly affects forest resources.

Biodiversity on the Farm

Many believe that farms destroy biodiversity. This may be generally true for any monoculture, including coffee, bananas, oranges, sugarcane, cattle ranches, macadamia, and teak or other forest plantations. However, this need not necessarily be the rule. A wide range of plant and animal species were recorded on La Pacífica (table 19.7), although an inventory of all species was never completed. Numerous studies of various species of bats, opossums, and other nocturnal mammals and insects, carried out from 1964 to 1990, confirm the species diversity (for a list of mammals sighted on the farm, see Clarke, 1982). A long-term study started in 1970 by Glander (1980, 1992) and continuing through the present, provides a wealth of information on the population of howler monkeys (*Alouatta palliata*) that inhabit La Pacífica. A systematic census of the forest windbreak areas was carried out by Clarke and colleagues in 1984 and again in 1991, and it was compared with records from capture and mark sessions from 1974 to 1976. The population of howlers remained relatively constant over this 18-year period with approximately 350 animals living in 18 social groups (Clarke, Zucker, and Scott, 1986). Glander (1996) estimates a total of

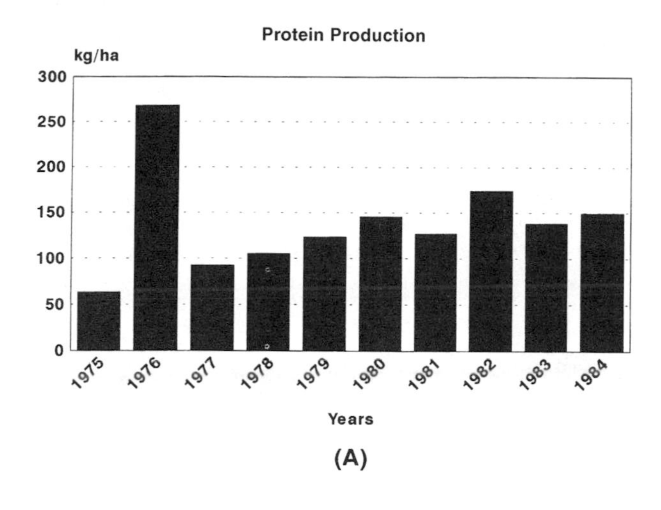

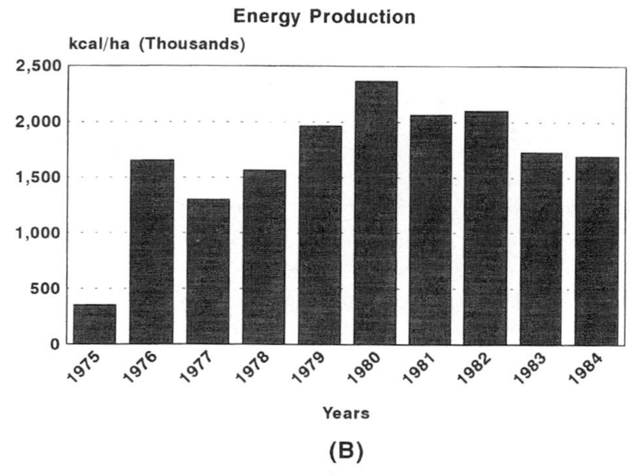

Figure 19.4 Protein and Energy Production per Hectare, 1975–1984

550 howlers in 30 groups when an area added to the farm in 1992 is included.

Biodiversity was an important factor in farm management on La Pacífica. For example, predator insects may play a decisive and underestimated role in pest control. Pesticides were never applied after 1969, but caterpillar attacks on sorghum never caused damage and were controlled by predators. After the disastrous cotton harvest in 1969, the situation on La Pacífica resembled that described by Rachel Carson (1962). However, because of the forests and windbreaks on the farm, the situation became normal again only four or five years after we stopped applying pesticides. Installation of the irrigation system also enhanced biodiversity. Apparently a reserve of insects, birds, reptiles, and mammals recovered rapidly, probably a main factor in the success of the farm. Forests and shrubland patches, riverbanks, and irrigation ditches should have a high priority as sources of biodiversity in programs that focus on sustainable agriculture.

The Improvement of Microclimate

The change in microclimate due to windbreaks, irrigation, and artificial ponds has never been systematically monitored. However, we recognized these important changes by looking at the animals and observing when and where they rested and grazed. The old part of Paso Hondo, where land reform, large-scale irrigation, and intensive agriculture erased all the native vegetation, illustrated the difference, allowing the wind to blow at full speed over kilometer-long fields. Similar differences can be observed between the old earth ditches on Paso Hondo (more than 50 years old) and La Pacífica (25 years old) and the new, concrete-lined irrigation ditches. The former are covered with vegetation, but the new ditches are traps for domestic and wild animals and have no vegetation at all. Unfortunately, however, forest, shrubs, and trees are still considered an obstacle to drive around with tractors and combines by many farmers and engineers.

Table 19.6 Crude Protein and In Vitro Digestibility of Forest Species

	Percentage Crude Protein	Percentage In Vitro Digestibility
Acacia spp.	25.5	65.0
Pithecellobium mangense	23.8	41.2
Enterolobium cyclocarpum	22.8	36.0
Tabebuia ochracea	22.2	40.6
Gliricidia sepium	20.7	51.8
Cajanus cajan	20.5	39.3
Cordia aliadora	19.6	36.9

Source: N. Conklin, 1987.

Economic Results (1965–1985)

Figure 19.5 shows land use on La Pacífica from 1965 to 1984.[9] By 1985, the forest reserve stabilized at 35 percent, including windbreaks, borders of rivers, irrigation ditches, and steep land unfit for pasture or crops. Because of excellent beef and milk prices in the early 1980s, cropland was reduced to 10 to 12 percent, although a good rotation would be six to eight years of pasture followed by two to three years of crops. Pasture was grazed longer than usual; the necessary plowing to reduce soil compaction and weed problems was delayed in view of the attractive prices in the livestock sector and the broad diversification of farm activities.

Figure 19.6 shows the gross income of the farm in colones and in dollars. The decrease in dollar values in 1980–1981 reflects the influence on the internal farm economy of a monetarist-guided administration from 1978 to 1982, a trend repeated today with tendencies toward free markets and deregulation. Export products such as beef, bananas, and coffee are much more sensitive to these fluctuations than products sold on the local market, such as milk, rice, fruits, and vegetables. Diversification and production for the national market are therefore not only ecological but also basic economic issues in farm management. Today, both national and export markets are changing rapidly and climatic conditions are uncertain. A farm needs at least five or six products to be strong enough financially to survive occasional climatic or market fluctuations.

In figure 19.7, the total gross margin in U.S. dollars is plotted for the different enterprises. Gross margin is defined as income minus all costs directly related to any activity, including the costs of hired labor and purchased materials. The positive impacts of the dual purpose cattle operation (beef and milk) are evident. This enterprise was without doubt the sustaining and stabilizing element over all years. Crops, however, were risky and did not play an important role.

Figure 19.8 shows both gross income and gross margin for the two different management strategies: high-input crop farming from 1965 to 1970 and low-input cattle ranching, based on local resources and crop rotation, from 1971 to 1984. Both are expressed in dollars to remove distortions due to devaluation of the local currency. Gross margin is income available to repay loans, pay interest, pay overhead, make farm improvements, and provide capital rent on investment. Expressing it as a percentage of income is useful. With a high-percentage share, the farmer retains more cash. Intensive production systems tend to have low-share ratios. As a result, although farm income seems high, little is left over after paying all production costs. In the period 1960–1970 the share was 28 percent, and in the period 1971–1984 it was 55 percent. This means that in 1965–1971, 2 hectares of land were required to produce the same amount of gross margin as 1 hectare produced in the 1972–1984 period.

In figure 19.9, the gross margin (in U.S. dollars) of La Pacífica is calculated by activities on a per hectare per year basis as an average for the 20 years from 1965 to 1985. This is the long-term result that should be used for decision making. Livestock provided modest gross margins but never resulted in losses. Rice showed higher gross margins than livestock but was risky. Cotton produced marginal results. Sorghum produced a low cash flow but served as an excellent forage reserve for the dry season. The other enterprises tested usually

Table 19.7 Number of Species, by Groups, at La Pacífica

Trees	116
Shrubs	No record
Grasses	42
Legumes (herbs)	16
Other plants	Unknown
Birds	120
Reptiles	15
Fish	Unknown
Mammals	30

Source: W. Hagnauer, unpublished list.

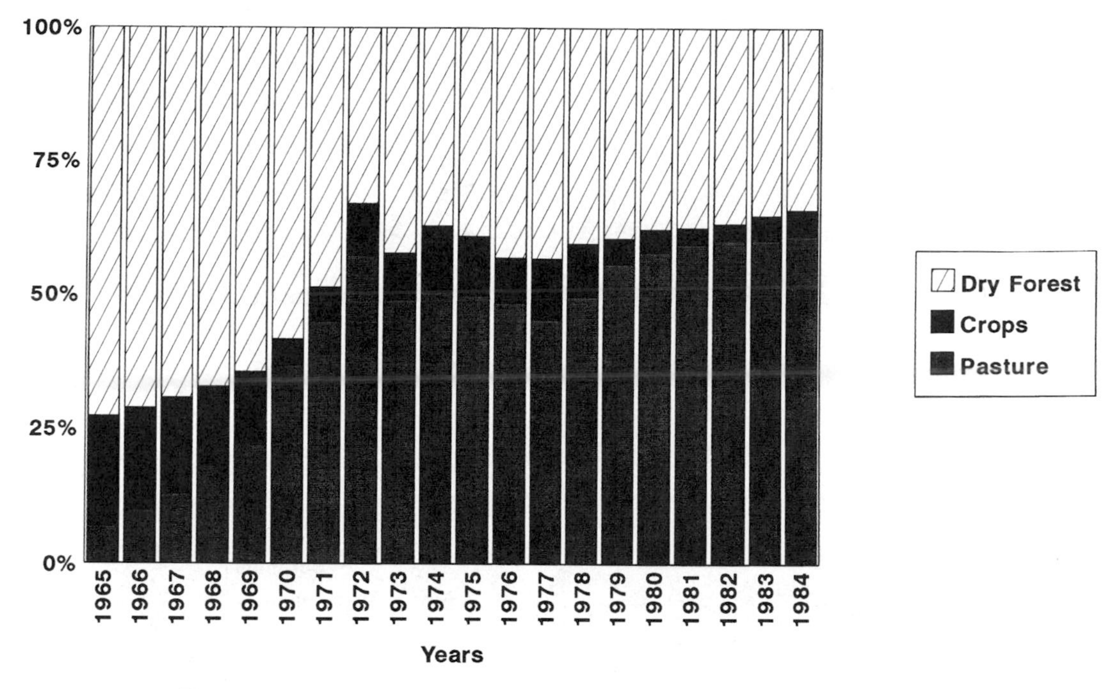

Figure 19.5 Land Use in Percentage, 1965–1984; Farm Size, 1332 Hectares.

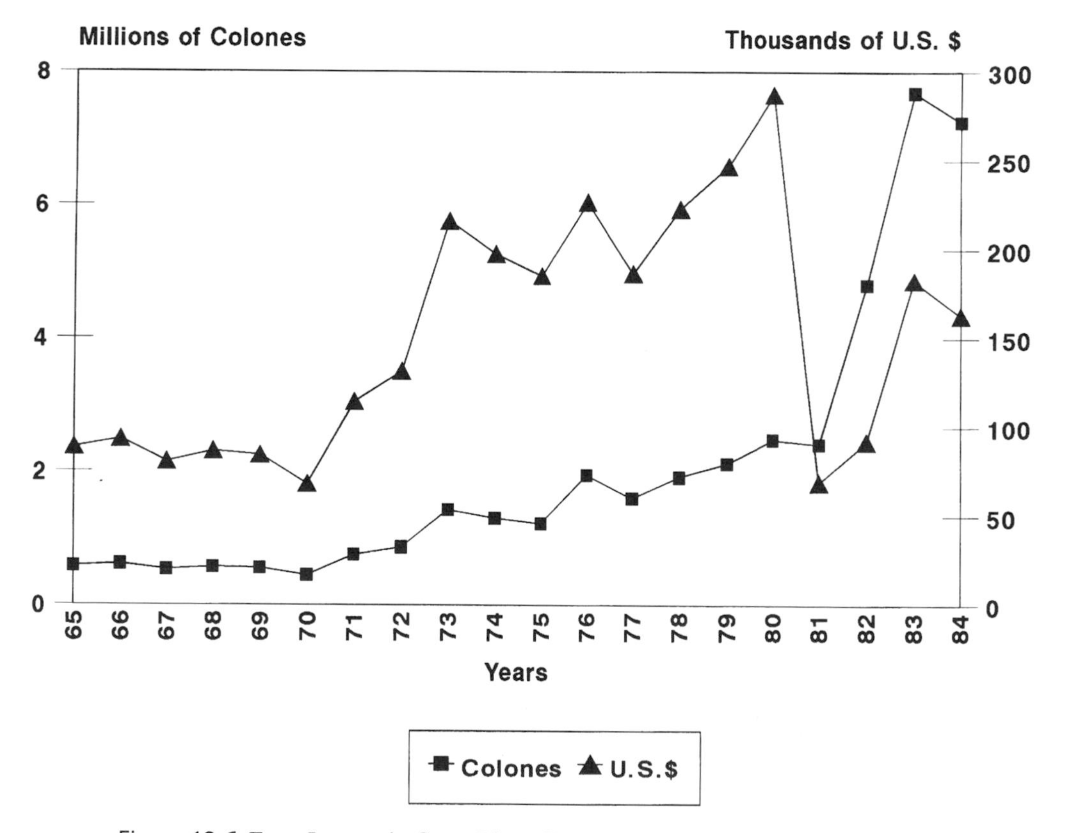

Figure 19.6 Farm Income in Costa Rican Colones and in U.S. Dollars, 1965–1984

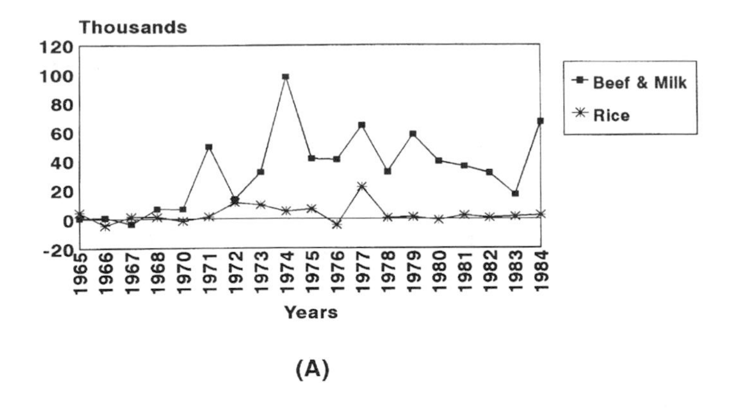

(A)

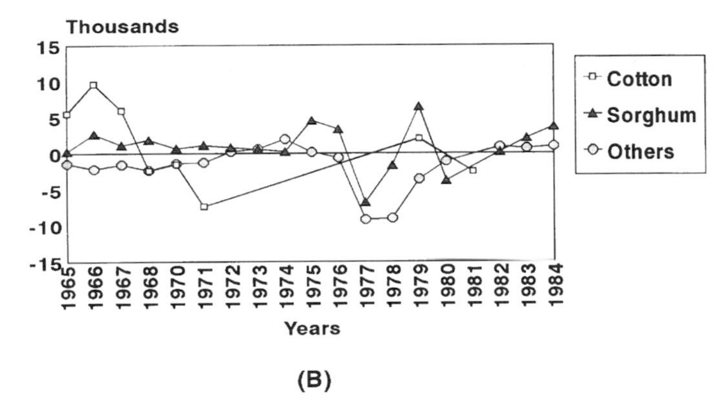

(B)

Figure 19.7 Gross Margin in U.S. Dollars by Main Activities

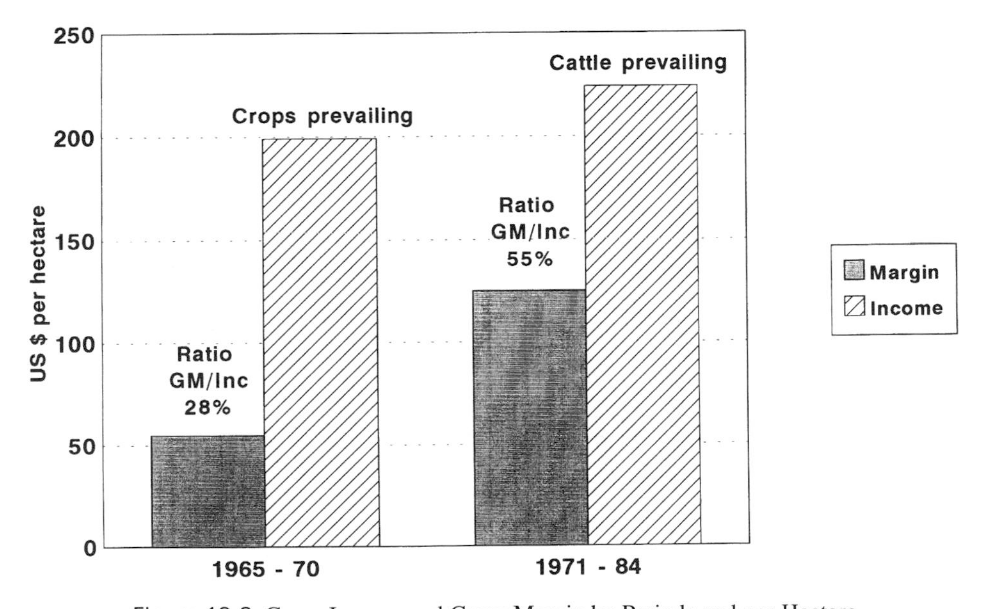

Figure 19.8 Gross Income and Gross Margin by Periods and per Hectare

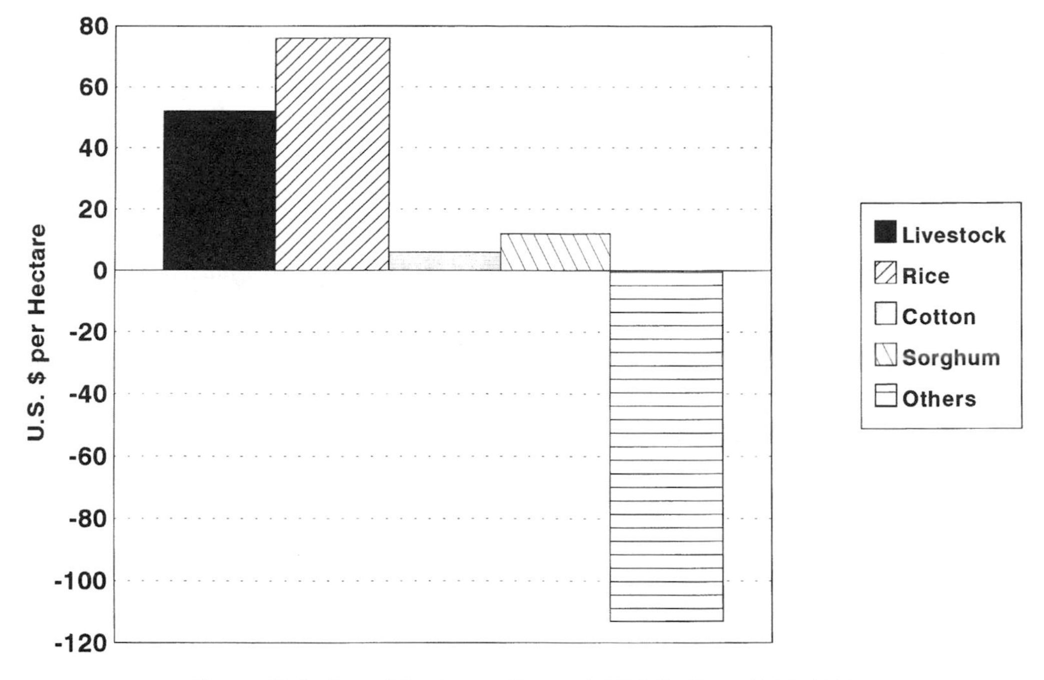

Figure 19.9 Gross Margin per Hectare in U.S. Dollars, 1965–1984

turned out to be expensive trials and errors, often recommended by institutions and specialists.[10] At the end of the intensive, high-input period in 1970, La Pacífica was in debt to banks and suppliers for five times the value of its yearly production. Changing the strategy of the farm management to maximum use of locally available resources reduced debt to zero in 15 years (figure 19.10). In the same 15 years, American agriculture piled up a debt of $206 billion (U.S. National Research Council, 1989) by using high-input technology. The American farm credit system lost $4.6 billion in 1985–1986, and many farmers lost their farms in this crisis.

The Use of Labor

Many also believe that cattle ranches use little labor and therefore provide few employment opportunities. Figure 19.11 compares the labor (in hours) used in cattle, rice, cotton, and sorghum per hectare over 20 years on La Pacífica. A more detailed study can be seen in table 19.8. In these years, our combine was equipped to handle grains in bags. Today harvesters handle grains in bulk and crops are usually sprayed by airplanes, reducing labor demand for grain crops even further. The labor per hectare required to produce rice may well drop to 10 hours or less. Further, a ranch laborer has a full-time, year-round job, whereas agricultural labor for mechanized production is highly seasonal. The social consequences of full-time and part-time employment have strong social implications for local communities.

Energy Budget

La Pacífica's energy efficiency was a concern for management for several years. In 1985, the Organization for Tropical Studies (OTS) agroecology course compiled

Table 19.8 Labor Demand for Cattle and Rice Production, 1981–1984

	Labor Hours per Hectare	
Year	Cattle[a]	Rice[b]
1981	62	20
1982	60	20
1983	61	21
1984	60	20
Ratio	3	1

Source: W. Hagnauer, farm records.

[a]Includes manual labor—herd care, electric milking, fencing, irrigation—and labor by tractor—mechanical weeding.

[b]Includes mechanized labor—plowing, planting, pest control, harvest.

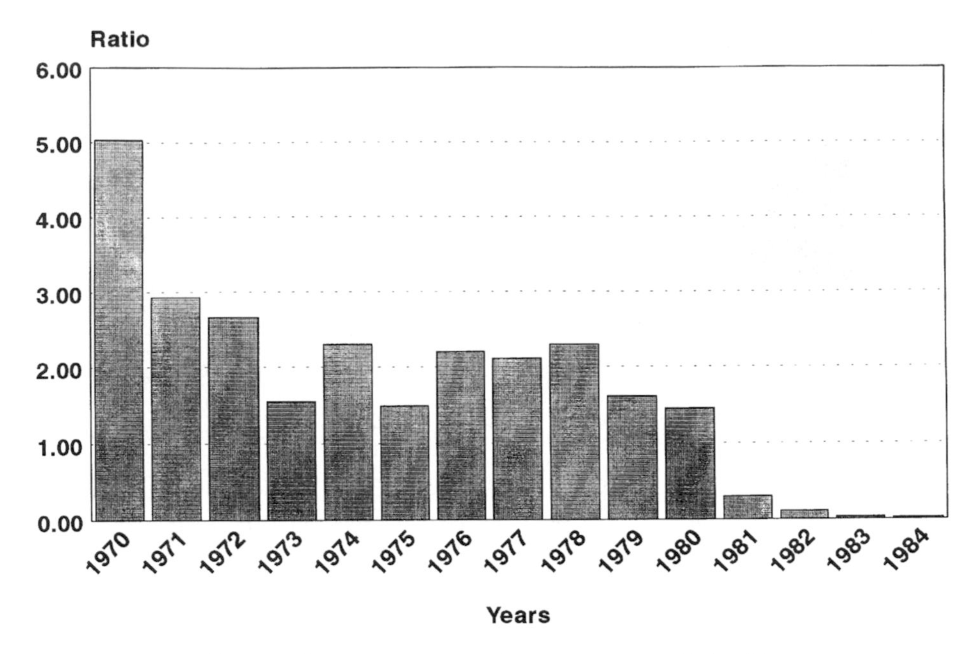

Figure 19.10 Ratio of Farm Debt to Income, 1970–1984

an energy budget for the farm. Table 19.9 shows some of the key results. Energy inputs and outputs are about the same for livestock production on the farm, and the energy ratio almost achieves the break-even value of 1. About half the energy inputs for livestock comes from sugar molasses, a product purchased from a nearby sugar mill. Without this input, livestock production achieves a positive energy ratio of 1.91 to 1.00.

Further, "forage" in this analysis includes land preparation, planting, chopping king grass (*Panisetum* hybrid) or sugarcane, and transporting the chopped fodder to the barn. This practice is useful in overcoming the shortage of green fodder during the dry season. However, it is expensive not only in calories but also monetarily and should not be recommended as a general practice for the dry tropics. Calculations for silage production, a practice never introduced at La Pacífica, give negative results for both the gross margin and energy efficiency. If the input and output of energy does not balance, the economics of the enterprise will hardly be positive, and as a result farm subsidies will be requested (Hagnauer, 1993).

The Lessons of La Pacífica

Principles of Farm Management

In 60 years of working in agriculture, 36 years of which were spent in Costa Rica, I learned that agriculture in the tropics is much more complicated and complex than agriculture in temperate climates. The reasons for this include high climatic variability, fragile soils, and a lack of or disregard for accumulated agricultural wisdom and traditions. The experience gained at La Pacífica shows that sustainability, even under highly variable conditions, is possible. Some basic lessons include the following:

Table 19.9 Energy Balance for Crops and Livestock

	Total Kilocalories				
System	Energy Input	Energy Output	Ratio Out/In	Area (ha)	Output (kcal/ha)
Crops	54,374,745	156,710,000	2.88	50	3,134,200
Livestock	558,181,705	550,994,000	0.99	810	680,240
Farm total	612,556,450	707,704,000	1.16	860	822,912
Livestock without molasses	283,181,705	550,994,000	1.95		

Source: Swisher, M. E., C. Vázquez, B. Bergmann, R. Linder, N. Rank, and M. Rosemeyer, 1985.

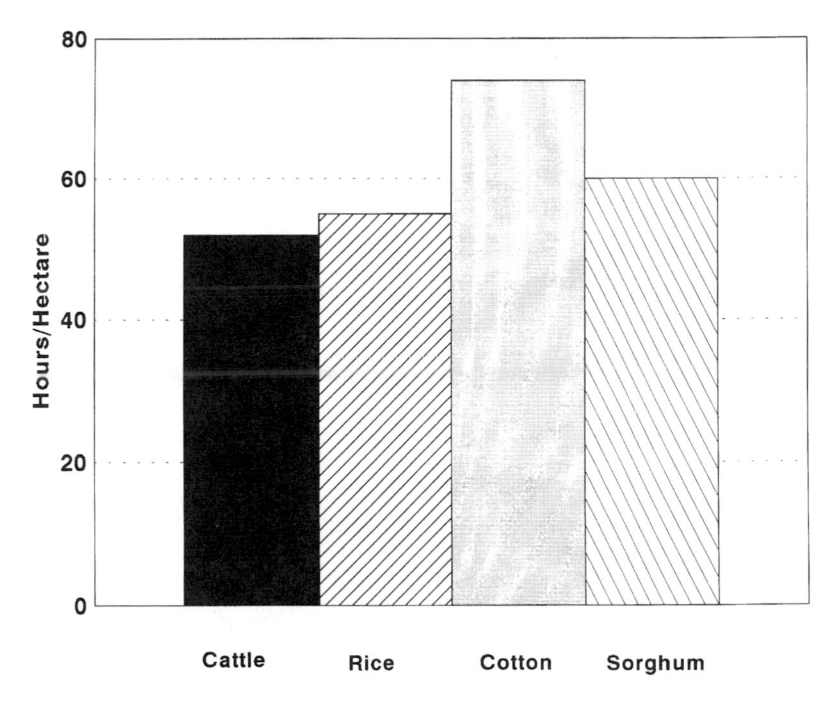

Figure 19.11 Labor Hours per Hectare in Different Activities, 1965–1984

1. Maximum yields in animal husbandry and crops should not be the top priority.
2. Yield goals should be adapted to fit the conditions and limits imposed by the local environment.
3. Boosting yields by using fertilizer, hormones, and pesticides, as well as the excessive use of heavy farm machinery, should be avoided.
4. Adequate diversification between crops and livestock is critical.
5. Integration of forestry into farm management is a key to success.
6. A minimum farm size of about 80 to 100 hectares is necessary.
7. A low-input philosophy should dominate. This implies maximum use of available on-farm resources, including the recycling of all stubble, browse, by-products, and waste materials.
8. Inputs of outside investments should be restricted to an absolute minimum. New bank loans should be incurred only in case of an emergency.
9. Bookkeeping, with cost accounting, is essential, and an energy budget is very helpful.

The Problem of Farm Ownership

Acquiring and selling farms occurs frequently in Costa Rica, in sharp contrast with the European tradition. This is shown by the case of La Pacífica, which was sold several times in less than 70 years. In fact, this has been the rule, not the exception, in the farm's history. However, establishing a sustainable farming system is hardly possible under such conditions. Frequent changes in farm ownership, seen repeatedly in the agricultural "development" of the past decades, result in a trial-and-error approach to farm management. New owners tried new crops, new varieties, new breeds, and better formulas for fertilizers and pesticides, often supported by the narrow insights of specialists and backed by bankers and agroindustry. Basic biological research was not regarded as essential and was never taken into consideration. Farm economics was virtually nonexistent, limited to bookkeeping for tax purposes or to justify new loans. Severe economic and financial difficulties were repeatedly resolved by political lobbying without a sound analysis of their origin. The case of La Pacífica, while breaking this pattern in some regards, also typifies this experience in many ways.

Social Factors

Farmers and peasants today want their share of the wealth in affluent societies. At the same time, the industrialized countries want to develop markets. Citizens in both developing and industrialized nations are alike in wanting to improve their standard of living. The question is, Who pays what and at what price? Who takes the better share? Figure 19.8 shows that a low-input system based on livestock leaves a better share of the gross income to the farmer. A smaller part of his or her income goes to suppliers and banks. In the extreme case, such as an extensive cattle ranch, a very high percentage of the income remains on the farm.

Another social issue in low-input systems is the use of the labor. During the last years of managing La Pacífica, the ratio between labor costs and expenses for materials remained constant at about three to one. For every three colones paid as wages to the workers, only one colon was spent on materials and the services of third parties. The value aggregated to the product was high. This ratio is usually inverted in a high-input, fully mechanized system. More money is spent for materials and services than for labor. This money goes to dealers and suppliers, leaving the risk of an intensive farming system to the farmer and providing little income for local labor.

The Problem of the Farm's Size

Small farms created by land reform, such as that carried out by the Instituto de Desarrolo Agrario (IDA), will face problems in achieving sustainability and economic survival (Hagnauer, 1978). Large-scale irrigation projects planned for the region around La Pacífica, which provide water to dozens of small and many large farms, will meet equal obstacles. To live well on a few hectares, the farmer has to intensify the operation. Intensification in the tropics brings a farm to the edge of mining the soil (Aragon and Hagnauer, 1980). Small farms, although popular politically, should not and cannot be evaluated purely from a political or social perspective. The first argument for larger farm size is the necessity to diversify into livestock (dual purpose), a critical component in survival on a farm. The second argument is the necessity for windbreaks and forest patches, land uses that cannot occur on a small farm. Finally, small farms do not provide their operators with a liveable income. If we compare the cost-benefit data in table 19.10, we see that a rice grower, according to SENARA (Servicio Nacional de Riego y Avenamiento) needs a minimum of 6 hectares to make the same minimum legal wage of a farm laborer. In the more conservative appraisal of the Banco Nacional, the grower would need as much as 14 hectares to break even with a farm laborer.

The practical results of the first 15 years of the irrigation project Paso Hondo show that many recipients of land-reform parcels are now leasing their land to others. Many have also sold their land under various terms or conditions or failed to repay debts (Hagnauer, 1993). These "small" farms, therefore, start to become bigger. Today, a second step seems to be underway; they start to raise cattle. It is hoped that over time they will provide shade for the animals and plant windbreaks and forest, and under these ecological conditions, an evolution occurs that involves larger parcels, the addition of cattle, and eventually setting aside land for windbreaks and small forest patches.

Splitting viable farms into pieces does not make economic sense and is the wrong approach to solving a social problem. A comparison of La Pacífica's history and its status in 1985 with that of the southern part of the original Paso Hondo farm is illustrative.

The Paso Hondo Farm

Like La Pacífica, the southern part of the old Paso Hondo farm changed hands, but it took a different development and management path. In the 1940s, one of the owners built an irrigation ditch with a capacity of 350 liters per second and irrigated sugarcane, rice, and pasture, as was done years later at La Pacífica. In this period, it was a typically extensive but diversified farm.

In the 1970s, Paso Hondo became part of the national irrigation program, SENARA. In 1977, as part of land reform, the 4,300 hectares of Paso Hondo were divided into 122 parcels ranging from 7 to 90 hectares. (However, this is a misleading figure because many of the smaller farms were leased or even sold to neighbors who were operating larger parcels, up to 300 hectares or more.) Paso Hondo was completely cleared to plant

Table 19.10 Comparing Three Different Data Sources in Rice Production

	Rice Crop 1994			Pacífica
	Bancio	Senara	Pacífica	Overall[b]
Gross income[a]	199,756	223,135	42,780	48,690
Total cost	177,495	171,070	30,802	21,424
Gross margin (bruto per ha)	22,261	52,065	11,978	27,267
Ratio GM/Inc (%)	11%	23%	28%	56%
Breakeven, bags of 46 kg[a]	44	43	8	5
Hectares to produce minimum salary	14	6	27	12

Source: SENARA, 1996 (Farm accounts 1960–1985).

[a]Price col. 4000 per bag.
[b]Diversified operation.

Table 19.11 Yields, Paso Hondo, Etapal, 1994

1994	Hectares	%	Colones Income	Colones Costs	Colones Yield	Gross Margin (%)	Income per Activity (%)
Rice	4,442	58	991,166	759,893	231,273	23	48.9
Sugar cane	2,616	34	797,017	573,375	223,642	28	47.3
Milk	121	2	8,976	4,430	4,546	51	1.0
Beef	421	6	31,996	18,978	13,025	41	2.8
Overall	7,600	100	1,829,155	1,356,676	472,486	26	100.0
Per hectare			240,678	178,510	62,169		
U.S.$/hectare			1,528	1,133	395		

Source: SENARA, 1996.

rice and sugarcane. No trees were left for windbreaks or soil protection. General policy during the reform was to convert traditional farms into a high-yielding, modern agriculture that could afford to pay for the high investment in ditches and infrastructure associated with large-scale irrigation projects. Livestock was excluded for political reasons.

The cost of the main and secondary irrigation ditches, with gates to the farms, was $19 million, or $3,163 per hectare of irrigated land, which contrasts greatly with the $21.15 needed to build the irrigation ditch at La Pacífica only 10 years earlier. Another $700 to $800 per hectare is needed to build ditches and level land inside the farm. The total cost for infrastructure comes to approximately $4,000 per hectare of irrigated land.

Table 19.11 reflects the situation in 1994 of the whole district (not only the Paso Hondo farm). Fifty-eight percent of the area was planted in rice and 34 percent in sugarcane, but only 6 percent was devoted to livestock. Crops such as melons, sweet or hot peppers, watermelons, coconuts, and other fruit are not mentioned, although they played an important role in the justification of the irrigation project (BEL, 1976).

Paso Hondo changed from a traditional, extensive, and fairly diversified farm to an expensive, high-technology, and high-input system. Today this system depends only on two crops—rice (49 percent of the income) and sugar (47 percent). Only one or two varieties of each are produced, and both are economically and ecologically fragile. If we compare the gross margins (table 19.12), 1.4 hectares of milk and beef operation would produce the same gross margin as 1 hectare of rice or 0.6 hectare of sugarcane, and a diversified farm would avoid all the economic and ecological risk of these monocultures and have the same gross margin.

Conclusion

What is, then, the minimum size of a farm necessary to survive economically in a sustainable way without mining the natural resources, and what are the large farms about? Based on the data from our own farm, I believe that a minimum farm size of 100 hectares is needed. This size allows diversification and stabilization of production through crop rotation and livestock. It permits the farmer to introduce enough biomass (windbreaks, forest patches, etc.) to improve the environment and its microclimate and build up a long-term timber and wood supply. A country also needs larger farms, both as pioneers for testing innovations (even ecological ones) and to produce basic crops in large quantities to meet local demands for food and fiber. Ecoreserves should be a part of both larger and smaller farms, perhaps based on a progressive percentage of the farm size, starting with 10 to 15 percent in smaller farms and increasing to 30 to 40 percent in larger properties. This would depend

Table 19.12 Gross Margins and Their Relations, 1994

	Cost	Yield	Gross Margin		Relation
Rice	171,070	223,135	52,065	23%	1.00
Sugar cane	219,180	304,670	85,490	28%	0.61
Milk	36,610	74,181	37,571	51%	1.39
Beef	56,482	95,226	38,765	41%	1.34

Source: W. Hagnauer, farm records.

on the soil and topographic conditions. Today this is still a factor in older, intact farms in our region and gives the necessary chance for biodiversity to recover from the many errors of the past.

In all these years more questions were raised than answered at La Pacífica, but one thing became clear: *farming is applied ecology*. Applied ecology on a farm covers a broad spectrum of problems and includes the economic and social, as well as the biological and agronomic, aspects of agriculture.

Notes

1. All that is written here could not have been done and analyzed without the presence and strong influence of researchers and students from OTS in San José, CATIE (Centro Agronómico Tropical de Investigación y Enseñanza) in Turrialba, the University of Costa Rica in San José, and ETH in Zurich, who stayed on or passed through the farm from 1962 through 1985 (Fleming, 1988). Many of them first came as students in courses and then came back after the courses to work on specific projects. Their worries, concerns, and daily questions provided many suggestions for day-to-day management and general farm strategy. They supplied new books, reprints, and photocopies to fill the farm library and the long evenings. Between 1962 (the first CATIE course in forestry) and 1985 (the first agroecology course), more than 170 different papers were published about La Pacífica and its environs (D. Janzen, personal communication, 1982).

2. Entomologists, specialized in these pests, argued that the windbreaks caused the pest problem and recommended special insecticide applications to the edges of the windbreaks to kill the pests in their refuges. We felt then, and still do, that the windbreaks in fact served as refuges for natural predators. The presence of these zones of high biodiversity may be one important factor in making it possible for La Pacífica to recover a natural equilibrium of pests and predators relatively rapidly, once pesticide applications on the farm were abandoned in 1969.

3. Although La Pacífica never intended to create a new beef breed, the Simmental-Brahman crossbreed later became very popular for meat production in the southern part of the United States under the name of Simbra.

4. A "2-by-4 herringbone" design was used, in which two milking machines, each one milking four cows, are accommodated in the form of a herringbone.

5. The latest trend in pasture management in Switzerland follows this same pattern; in 1995, 63 of 90 farmers preferred this "new" system (Thomet and Hadorn, UFA Revue March 3/96, Bern, ISSN 1420–5106).

6. Processed chicken manure is now a market product in Costa Rica.

7. A topographer laid out the ditch. The farm's own machinery built the main ditch of 2.5 kilometers, finished in 90 days with a cost per hectare of potential perimeter of $21.15. The same topographer trained the farm personnel, including the manager, to build irrigation terraces, water gates, and dividers.

8. In the course of the construction of the new concrete-lined ditches by Servicio Nacional de Riego y Avenamiento (SENARA), and because of changes in farm management, almost all of the refuges were eliminated and lost.

9. The data used in this section refer strictly to the farm activities related to land use. To avoid complicating the issue, we excluded the workshop, the poultry, and the tourist business, although they played an important role in the farm's diversification.

10. Most farmers do not keep books or practice cost accounting. Nonetheless, many have a good feeling for these relationships. Their resulting tendency to reject untested alternatives gives them a reputation for being antiquated, conservative, or unconcerned about progress. Their experiential understanding is also the reason many of them would like to return to cattle ranching. Unfortunately, some cannot because of the inflexible institutional structure of the irrigation project now in place.

20

The Sustainability of Milk Production in Mesoamerica

Carlos A. Jiménez-Crespo

The role that cattle ranching has played and continues to play in deforestation is discussed elsewhere in this volume (Boza, chapter 6; Fisher, chapter 23; Mayne, chapter 9; and others). The attention devoted to the role that ranching plays in environmental degradation in Mesoamerica leads almost inevitably to the conclusion that cows are indeed the culprits in many of Central America's environmental problems. However, cattle ranching to produce beef is only one way that cattle are exploited in Central America. They also provide animal traction in many nations, and milk is an important source of animal protein in the diets of many of the people.

Milk production systems differ greatly from extensive ranching operations. Few farmers clear land to raise cattle for milk production, for example, which is largely more intensively managed than beef production. Often, small farmers, rather than large-scale ranchers, play an important role in providing milk to a nation's population.

However, despite the many differences between milk and beef production systems, it is not clear that the former are inherently sustainable in tropical Mesoamerica. In this chapter, I discuss some factors that will determine whether milk production in Central America, as it is practiced today, is sustainable. I describe the major milk production systems and evaluate the future of this agricultural enterprise in the region.

Two groups of factors are important for the sustainability of milk production: (1) economic and policy factors and (2) biological and ecological factors. The first group includes the policies and laws that affect, or may affect, the economic framework for milk production. The cost of imported inputs and the price received for milk products will determine whether milk production is economically sustainable in the region. The current use of corn in the diet of dairy cows is a good example of the relationship between input costs and product price. Only a few years ago, the use of this grain in balanced feed mixes for dairy cattle was prohibitively expensive. The PL 480 grain program of the United States radically changed the situation. This program was designed to open markets in the Caribbean basin to U.S. grain producers. As a result, large quantities of grain are now imported at international, competitive prices. This example shows how market liberalization can affect milk production systems. Future changes in world trade agreements, such as GATT (General Agreement on Tariffs and Trade), or in the policies of individual nations, such as the United States or any Central American nation, can change the productivity and stability of milk production systems rapidly and greatly.

The second group of factors includes many components. Perhaps most important are the genetic characteristics of cattle and forage species available to Central American milk producers. However, many other biological and ecological factors also come into play, such as the characteristics of tropical soils.

Sustainability as a Function of Dependency on Inputs

A major challenge faced by Central American milk producers, particularly those using pure dairy breeds, is that milk production systems fashioned after middle-latitude systems depend greatly on inputs that are not or cannot be produced on the farm, especially the inputs needed to provide a balanced diet for dairy cows. Yellow corn has become an important component in balanced feeds in recent years. While corn is a major crop in Central America, its use in animal diets has traditionally been prohibitively expensive. Because of price liberalization and increasing government authorization for the import of this grain in many Central American nations, yellow corn is now a common component in balanced feed mixes. Soybean cake is another dietary component that has assumed greater importance in dairy cattle diets in some parts of Mesoamerica. In Costa Rica, for example, the installation of a soy processing plant has produced an abundance of soy cake available at prices very competitive with the international market. Nonetheless, the soybeans themselves are imported. Animal feeds may also contain wheat byproducts that derive from national processing plants. Again, however, the wheat itself is normally an imported product.

The Costa Rican milk industry provides a good example of the dependency of milk production on these imported products. As table 20.1 shows, about one-third of Costa Rica's milk production occurs on specialized dairy farms, where concentrates account for 40 percent of daily dry matter consumption. This dependence on imported primary materials is not limited to Costa Rica or to Central America. Many countries, with very efficient milk production systems, depend heavily on imported primary materials in concentrated feeds. Israel and The Netherlands, for example, use ratios of concentrate to forage greater than 1 to 1 and rely greatly on imported grains for the concentrate portion of the diet. The difference between these nations and the tropical nations of Central America is that the forages they grow are of much higher quality than tropical forages, which permits them to achieve greater production efficiencies. In Mesoamerica, even when the genetic quality of the dairy herd is very high, the ability to take advantage of this genetic potential is not great because of the quality of the forage in the animals' diets.

Two schools of thought have resulted from this dilemma. One argues for maximum production per cow. Maximum productivity of this type requires an excellent diet. However, the tropical pastures of Central America contribute little to milk production, less than ten kilograms of milk per cow per day. Therefore, the role of concentrates is very important in achieving maximum productivity per animal in the tropics. However, as table 20.2 depicts, many studies show that dietary supplementation for cattle that are consuming tropical pastures often does not lead to large increases in daily milk production.

The other school of thought argues that tropical dairy production should take maximum advantage of the abundance of forages in the tropics. Although tropical forages are of poor quality for milk production, they produce very high quantities of biomass, which makes it possible to use rotational grazing very efficiently. In rotational grazing systems, the cattle graze each of several small paddocks intensively for one or a few days and then move to a new area. When biomass production is high, as is true in the tropics, each individual paddock recuperates from the intensive grazing in a short time. Rotational grazing systems are economically efficient in the tropics, even though production per animal is low, because production per unit of land is high (table 20.3). However, rotational grazing systems do not capitalize fully on a high genetic potential for milk production of dairy animals.

Besides their dependence on concentrates, specialized dairy farms, especially dairies found in the high-

Table 20.1 Characteristics of Costa Rican Systems of Milk Production

	Production System		
Item	Dual Purpose	Lowland Systems	Highland Systems
Percent of farms ≤ 50 ha[a]	85	86	86
Percent of milk cows[b]	60	76	76
Stocking rate (AU/ha)[b–d]	1.2–1.5	1.6–2.6	2.3–3.2
Milk production			
Kg/cow/day[b]	4.9	8.9	11.6
Kg/lactation[b–d]	478–1,086	1,900–2,395	2,720–2,937
Kg/ha/year[c,d]	401	2931	4817
Percent of country's milk	N.D.	N.D.	30
External inputs used	Minerals, pesticides, medication, equipment	Minerals, pesticides, medication, equipment, concentrate (≤ 40%), agricultural chemicals	Minerals, pesticides, medication, equipment, concentrate (≥ 40%), agricultural chemicals

[a]Villegas, 1991.
[b]Gallardo and Vargas, 1992.
[c]Murillo and Navarro, 1986.
[d]Wadsworth, 1983.

Table 20.2 Response to Concentrate Supplementation by Dairy Cattle Grazing on Tropical Pastures in Short-term Experiments

Reference	Level of Supplementation (kg/d)		Lactation Period (Weeks)	Milk Production Lowest Level	Response kg of milk/ kg of concentrate
	High	Low			
Butterworth, 1961	2.4	0.5	17–27	6.8	0.28
Aronovich, 1965	3.8	0	11–23	10.0	0.42
	1.9	0	11–23	10.0	0.37
Royal-Jeffrey, 1972	3.1	0	17–30	8.8	0.45
	1.1	0	17–30	8.8	0.64
Aronovich, 1973	4.1	0	10–22	10.8	0.35
	2.2	0	10–22	10.2	0.24
Phllips, 1973	9.0	4.5	1–10	13.3	0.27
Rodríguez, 1972	6.0	0	8–23	7.9	0.27
Guzmán, 1970	3.7	0	9–24	14.1	(-)0.16
Jeffrey, 1976	3.0	0	—	13.6	0.37
	3.0	0	—	11.4	0.50
Martínez, 1976	2.0	0	8–22	13.9	0.35
	4.0	0	8–22	13.9	0.40
Martínez, 1978	2.5	0	1–11	9.3	0.86
	4.0	0	1–11	14.9	0.19

Adapted from Martínez and García, 1983

land regions of Central America, are also highly dependent on another imported input; chemical fertilizers. These systems usually maintain relatively high animal stocking rates that require a correspondingly high availability of forage biomass. In the highlands of Costa Rica, for example, dairy farmers commonly apply 150 to 250 kilograms per hectare of nitrogen, 40 to 60 kilograms per hectare of phosphorus, and 60 to 100 kilograms per hectare of potassium annually.

In summary, specialized dairy farms that aim to achieve high productivity are greatly dependent on imported inputs, most important, concentrate and chemical fertilizers. This dependency, often as high as 60 percent of the total production costs, makes them vulnerable to political and economic changes at the national, regional, and international level. Their economic sustainability is therefore questionable.

The search for solutions to this dilemma has been oriented toward the use of internal or indigenous resources and the development of production systems that are more self-sufficient. Tropical environments are typified by vegetation with the highest rates of biomass production in the world (Stewart, 1970). They also have the highest biodiversity of forage plants, including both grasses and legumes. These species exhibit many adaptive, productive characteristics. However, no single species, planted in monoculture, has all of the desired qualities. Current research, therefore, has started to focus on the potential for planting forages in association.

The data summarized in table 20.3 permit a comparison between forages planted in association and other strategies for maintaining dairy cattle on pasture. Neither production per animal nor production per unit area of land is as high as those achieved when balanced formulas are included in the diet and nitrogen fertilizer is applied to pasture. Nevertheless, these mixed grass-legume systems are more self-sufficient than the other systems, and their potential sustainability is higher.

This strategy, however, has its problems. Maintaining a mixed grass-pasture system over the intermediate to long term is difficult. Only a few examples of successful, intensive systems of this type that have persisted for long periods exist in the world. Well-known researchers assert that the natural dynamics of mixed plant ecosystems is such that it is always possible that one species will achieve dominance sometime during a year. Research, therefore, must seek multiple potential combinations of grasses and legumes. Many alternatives will be needed to achieve success given the diverse climatic, edaphic, and technological conditions that exist in Mesoamerica.

One legume species, *Arachis pintoii*, illustrates the potential importance of legumes in tropical forage crop production systems (Argel, 1991). This promising forage legume is a perennial that reproduces both asexually, by rhizomes and stolons, and sexually. Table 20.4 shows the quantity of viable seed that *A. pintoii* produces and the number of seedlings that emerges when this legume is grown alone and when it is grown in

Table 20.3 Potential Production for Different Milk Production Systems Based on Grazing

Type of Pasture	Stocking Rate	Milk Production per		
		Cow/Day	Lactation	Ha/Year
Natural or unfertilized pasture	0.8–1.5	6–7	1,400–1,700	1,300–2,700
Fertilized pasture, not irrigated	2.5–3.3	6–8	1,600–2,000	5,300–6,800
Fertilized and irrigated pastures	2.7–4.5	7.0–8.5	1,700–2,400	6,000–9,000
Moderate potential cattle	2.0–4.0	10–14	3,000–4,500	8,500–15,000
High potential cattle	5.5–8.0	9–12	2,400–3,600	16,000–20,000
High stocking rates				
Legume-grass mixes				
Moderate potential cattle	1.0–2.0	8–9	2,100–2,400	2,700–4,700
High potential cattle	1.0–2.0	11–13	3,300–4,200	5,000–8,000

Adapted from García, 1983

association with grasses. This legume grows by runners. As a result, it has exhibited great compatibility with grass species, such as star grass, that normally quickly dominate the legumes in mixed plantings. However, *A. pintoii* is not a prolific biomass producer. Trials in Guápiles, Costa Rica, yielded only 4.1 tons of dry matter during the long rainy season typical of this area. Nevertheless, the quality of the forage is high. In the same experiments in Costa Rica, crude protein varied from 10.1 to 17.9 percent, and in vitro digestible dry matter ranged from 63.1 to 63.3 percent. As a result, the potential for increasing milk production in associations involving *A. pintoii* is high (table 20.5).

Milk consumption and the prices received for milk are also important factors in the sustainability of milk production systems. Costa Rica has the highest level of both milk production and per capita milk consumption in Central America. In 1990, milk production in this country was 436,785 million liters, an increase of 6.7 percent over 1989. Per capita consumption for 1987 and 1988 in Costa Rica, expressed in equivalents of liquid milk, was 139.3 and 141 kilograms per year, respectively. It is estimated that milk products account for about 20.4 percent of recommended protein intake (Villegas, 1991). Milk consumption is lower in other Central American nations, according to the average per capita for the region, which was estimated as 70.6 liters for 1988. Costa Rica ranked highest in that year, with 147.1 liters per person per year, and Nicaragua and Guatemala ranked lowest, with 44 and 34.8 liters, respectively. In estimating this per capita consumption, both imported milk and milk from international donations arc included. This component is significant in many countries if one considers total fluid milk production, which was 285.2 million liters for El Salvador, 275.5 for Honduras, 245.1 for Guatemala, and 163.0 for Nicaragua in 1988 (CATIE, 1990). Only Costa Rica is self-sufficient in milk production. A high percentage of its farms specialize in milk production, in contrast to the rest of the region, where the highest percentage of the milk comes from dual-purpose systems (milking beef cows or crossbreds and raising the calves to weaning).

Other Data on Milk Consumption

Costa Rica's policies for establishing milk prices illustrate the problems inherent in price determination in many Mesoamerican countries. The Ministry of the Economy fixes milk prices in Costa Rica, based on a production model prepared by both government and

Table 20.4 Seed Reserves and Number of Emerged Seedlings of *A. Pintoii* CIAT 17434 in Monoculture and Associated with Two Types of Pasture, Carimagua, Colombia

Pasture	Seeds (1983)		Seedlings (1984)
	Number/m^2	Weight (g)	Number/m^2
A. Pintoii (not grazed)	6535 ± 286	611 ± 29	—
A. Pintoii + B. Humidicola	618 ± 209	57 ± 20	128 ± 17
A. Pintoii + B. Dictyoneura	670 ± 176	48 ± 16	145 ± 16

Adapted from Argel, 1991.

Table 20.5 Persistence, Percentage of Cover, and Milk Production, *A. Pintoii* CIAT 17434 Associated with Stargrass (*C. Nlemfuensis*), Four Years of Grazing, Double-purpose Cattle, Turrialba, Costa Rica

Pasture	% Grass	% Legume	% Weeds	Milk Production (kg/cow/day)
Stargrass	51 a	0	10 a	7.7*a*
Stargrass + arachis	44 a	44	4 b	8.8*b*

Adapted from Argel, 1991.

Letters *a* and *b* indicate statistical differences ($P \leq .05$).

private sector collaborators. The price of milk is increased when increases in production costs exceed a determined level. The model used is based on a model farm, employing technologies that are highly dependent on imported inputs. As a result, market liberalization can be expected to have very significant impacts on dairy farmers. In the rest of the area, price is controlled by the government but does not take production costs into account, leading to subsidized prices or poor stimulus for producers, who have to sell at prices that are not attractive to their entrepreneurial efforts. Accepting international donations for children's nutritional improvement programs has caused a negative effect on prices and has acted as a drawback to production (CATIE, 1990).

Biological and Ecological Determinants of Sustainability

Table 20.6 summarizes the biological and ecological factors that influence the sustainability of milk production in Central America according to the author's perception. As the table shows, it is impossible to consider biological and ecological factors in isolation. Rather, they are components in systems characterized by both internal interactions and interactions between systems.

The capacity of plants with fibrous roots, such as the grasses, to provide protection for the soil and aid in soil retention is well recognized. A classic study conducted by Johnston-Wallace et al. in 1942 illustrates this positive characteristic of grass-based production systems (table 20.7). The study also shows that increasing grass cover through fertilization and including clovers in the system increases soil retention even more. Humphreys (1991) quotes similar results for tropical forages. Soil losses decreased from 18.3 tons per hectare per year on exposed soils to 3.6 tons per hectare on soils with a legume cover (*Pueraria* and *Centrosema*) and 0.01 tons per hectare on soils under natural grass cover (*Paspalum conjugatum*).

However, the presence of animals in dairy systems reduces the environmental benefits of forage coverage for the soil. An adult bovine may exert a pressure of 1.3 to 2.8 kilograms per square centimeter of hoof area (Humphreys, 1991). Their potential for compacting soils is therefore very high. This potential is increased significantly by vertical compression and rotational force when the animal moves suddenly. Thus animals compact soils, and compaction has negative effects on physical characteristics like aeration and root penetration. Animals also consume the biomass that protects the soil, slow plant growth, and destroy plants by trampling. Animal stocking rates are directly correlated to

Table 20.6 Edaphic and Environmental Factors that Influence the Sustainability of the Soil-Plant-Animal System

Factor	Soil-Plant System	Plant-Animal System	Animal-Soil System
Edaphic	+ Soil cover – Topography	– Trampling + Nutrient recycling	– Compaction + Nutrient recycling
Environmental	+ Nutrient recycling ± Competition with agriculture or forestry	± Competition with agriculture or forestry	– Water pollution

Source: Carlos A. Jiménez-Crespo, 1997 (original data from author).

The plus sign indicates a positive effect, and the minus sign indicates a negative effect.

Table 20.7 Soil Loss and Runoff from Intense Rainfall (3.5 cm/10 min.) under Different Types of Soil Cover

Slope (%)	Type of Pasture	Runoff (%)	Soil Loss (kg/ha)
23	Bentgrass + white clover (fertilized)	0.5	3.4
23	Bentgrass, unfertilized	8.1	299.0
19	bare soil	48.7	19,198.0

Source: D. J. Johnston-Wallace, J. S. Andrews, and J. R. Lamb, 1942.

both compaction and trampling. However, the effects of trampling vary greatly, depending on the forage species. Grasses suffer less than other forages (Humphreys, 1991).

Eliminating the negative effects that animals have on pasture stability and long-term sustainability of the plant-soil system is impossible. However, negative impacts can be attenuated by maintaining acceptable stocking rates, avoiding situations that cause animals to move suddenly, and eliminating pasturage in regions where climatic conditions are adverse. It seems more plausible to depend on these management practices, along with the use of more resistant species, than to try to recover soils by periodic plowing. There is evidence that only one year after plowing the soil compaction rate can increase by 50 percent.

Animal production systems can also play a positive role in long-term agricultural sustainability because of their ability to recycle nutrients (figure 20.1). It is estimated that adult milk cows can return to the soil up to 80 percent of the nitrogen, phosphorus, and potassium that they consume. This nutrient recycling helps maintain soil fertility and forage production (table 20.8). Furthermore, cattle leave almost half the biomass in areas under pasturage, material that decomposes and contributes to the reservoir of soil nutrients. The role of legumes in nitrogen fixation is well known. It now appears that some grasses also enter into symbiotic relationships that permit fixation of atmospheric nitrogen.

Despite this capacity for recycling nutrients, animal-based systems do suffer net nutrient losses. Milk and other animal products represent a form of exporting nutrients from the soil-plant-animal system. A dairy cow that produces 8,000 liters of milk per year will "export" 42 kilograms per hectare of nitrogen, 8 kilograms per hectare of phosphorus, and 11 kilograms per hectare of potassium (Humphreys, 1991). Nutrients are lost to the atmosphere and to lower soil layers, where they cannot be recaptured by forages. Some nutrients are transported outside the pasture system when cattle urinate and defecate elsewhere. Milk production systems based on pasturage are therefore not self-sufficient. Over time, they will lose nutrients which must be replaced by elemental minerals. Greater attention must be devoted to establishing strategic, rational fertilization regimes for these systems.

Dairy systems in Central America today occupy many different kinds of terrain with different potential uses for agriculture and forestry. Many scientists believe, and studies of potential land use tend to confirm, that livestock systems for both beef and dairy production in the region will have to cede land in the future. Some lands now occupied by dairies can be used for other, more profitable agricultural activities. In the future, dairies will be displaced from these land to zones that are more marginal for crop production, because of topography, drainage, rockiness, or other factors. At the same time, some of the steepest lands now used for livestock production should be returned to forest cover.

On the one hand, I believe that slopes of more than 60 percent should not be used for pasturage, although they can support livestock production when appropriate soil conservation measures are employed. These

Table 20.8 Total Dry Matter Yield (DM) and Total N, P, K, Ca, and Mg Content on Pasture with and without Manure Recycling for a Five-year Period

	DM (kg/ha)	N (kg/ha)	P (kg/ha)	K (kg/ha)	Ca (kg/ha)	Mg (kg/ha)
Recycling	33,302	864	141	977	231	65
No recycling	26,882	749	115	691	244	60
Difference	5,420	115	26	286	(-)13	5

Adapted from Mott, 1974.

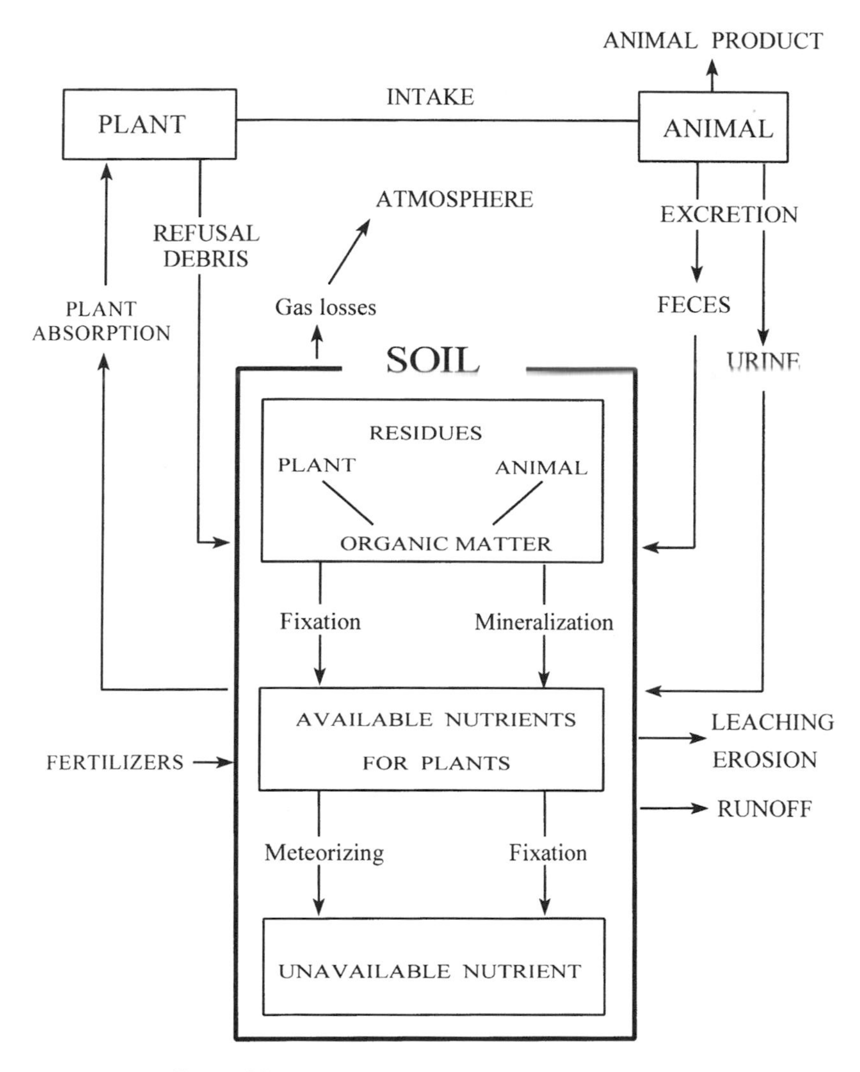

Figure 20.1 Nutrient Cycle in a Grazing Ecosystem

lands, for example, can support systems in which permanent pastures are harvested manually or mechanically and fed to animals that are stabled partially or completely. On the other hand, lands with slopes of less than 15 or 20 percent will be used for livestock production in Central America only when they are unsuitable for more intensive agriculture for some reason. In most of these cases, primary biomass productivity will be low. Nonetheless, livestock-based systems can be sustainable under these conditions when proper management techniques are used. In fact, many lands of these types are used today for milk production.

Milk production systems based on grazing generally have little potential for environmental contamination. In some nations, where the number of animals in each dairy is great and the manure produced is correspondingly high, water pollution from manure and urine is a major problem. In Central America, the water used to wash the stables and milking parlors is generally used to fertilize pastured areas and has very little potential for water pollution, with the exception of subterraneous reservoirs.

Conclusions

In Central America dependency on external resources is one of the greatest limitations to the long-term sustainability of specialized milk production systems. The greatest potential for increasing sustainability lies in developing systems based on moderate production per cow but greater productivity per unit area of land. This approach makes the best possible use of indigenous forage resources. Furthermore, milk production can contribute to the overall sustainability of agriculture because this kind of agricultural system is well suited to some marginal lands, and in fact significant milk production already occurs on these lands. Nonetheless, milk production can never be totally self-sufficient. Although there are efficient systems in which significant nutrient recycling occurs, there are inevitable losses, and the dairy products that leave the farm are in effect a nutrient export. Therefore, these systems in the future will require some outside inputs to be sustainable over the long term, more so as they become more intensively managed.

21

Globalization, Population Growth, and Agroforestry in Mesoamerica

Perspectives for the 21st Century

Lawrence T. Szott

Mesoamerican farmers face three especially important challenges. (1) The globalization of markets may negatively affect demand and prices for their products. (2) They must increase production to provide for a growing population. (3) At the same time, farmers need to maintain environmental resources that provide the basis for agricultural production and the well-being of both urban and rural dwellers. These forces are likely to change production systems, resource use, and patterns of human settlement and activities irrevocably in the coming years.

What do these changes bode for traditional or newer agroforestry practices? What is the potential of traditional systems or modifications of them to produce high-quality products at low prices to compete in global markets? Can these systems increase food production without negatively affecting environmental quality? Might they even help to mitigate local, regional, or transnational environmental problems and increase the quality of life for rural, urban, and international consumers? What impedes farmers' adoption of new or improved practices and how can these limitations be overcome?

In this chapter, I discuss the possible responses of agroforestry systems managed by small, resource-poor farmers to changes in labor availability, lower prices for agricultural products, and emerging markets. I analyze the potential of these systems to increase production and reduce or mitigate environmental degradation in the face of these changes.

Traditional Systems

Farms are small and are located on hillsides or on less fertile soils in most of Mesoamerica, mainly because of high demographic pressure and the occupation of more fertile sites by export-oriented agroindustries. Although the form and function of farming systems vary throughout the region, many have common characteristics. They are usually labor-intensive with diverse enterprises. The primary objective is to ensure food and a secure livelihood, and only secondarily to produce for local or regional markets. Small farms are also often cash-limited and depend on family labor, and land availability is sometimes a limitation as well (S. E. Carter, 1991; Lindarte and Benito, 1993). Trees invariably assume key roles in small farmers' strategies to ensure a secure livelihood, an income, conserve natural resources, and reduce risk.

Semiarid and subhumid zones of Mesoamerica include large areas of Mexico and the Pacific coast of Central America. In these zones, tree components of agroforestry systems are commonly used to restore productivity of abandoned agricultural land through vegetated fallows (see Mayne, chapter 4). They help overcome shortages in forage availability during the dry season and reduce the cost of fencing in silvopastoral systems (see Hagnauer, chapter 19). Trees also improve the climate around houses and animal pens when used in home gardens and serve as sources of fruit, timber, and firewood in many agroforestry systems (Szott, in press).

In the warm, humid areas of Mesoamerica, chiefly along the Atlantic coast of Central America and in the Peten of Mexico and Guatemala, agroforestry systems may be more structurally and functionally diverse than in dry zones. Nevertheless, trees play similar roles in farmholder strategies (Gillespie, Knudson, and Geilfus, 1993; L. A. Navarro, 1987; Somarriba, 1993). Common systems include complex home gardens; simpler, multistrata mixtures of perennial and semiperennial crops; shaded, commercial perennial crops; vegetated fallows; and small-scale woodlots or commercial plantations.

Trees also grow along farm or plot boundaries and around buildings, are scattered within pastures, and serve as living fences. Sometimes they also make important contributions to regional and international economies. Witness the historical importance in the tropical lowlands of perennial crops such as cacao (*Theobroma cacao*) and timber species such as mahogany (*Swietenia macrophylla*), tropical cedar (*Cedrela odorata*), and laurel (*Cordia alliodora*) and the widespread importance of coffee associated with fruit or timber trees in cooler, montane areas.

Complex, capital-intensive agroforestry systems are less common in Mesoamerica, although some notable examples exist. Under these conditions, agroforestry is relatively unimportant because, given a choice, land managers with access to more resources often substitute capital inputs like mechanization or pesticides for labor. They may simplify complex systems to allow management on a large scale with uniform technology, for example, replacing multistrata coffee shade with monospecific shade or even no shade.

Globalization and Population Growth

A simplified schematic diagram of the effects of globalization and increasing population growth on farming systems, environmental quality, and rural to urban migrations is shown in figure 21.1. Globalization is expected to have several results. First, lower prices will reduce income from some agricultural products, such as basic grains. Second, it may increase or eliminate off-farm employment opportunities. Third, it may create new market niches and farming systems to exploit them. Possible outcomes of increasing population growth include less land per farmer, resulting in more intensive production; increased farming on marginal lands; and rural to urban migration. More intensive use of arable and marginal lands and agricultural inputs may increase environmental degradation and lower standards of living in the short term. In the long term, they will lead to a decline in income and quality of life, forcing farmers to abandon current agricultural practices or search for nonagricultural employment in urban areas. The influx of a large unskilled population to urban areas would decrease the standard of living of urban dwellers and increase the demand for social services and jobs, which current political and economic systems will be hard-pressed to satisfy.

I believe that the effects of increasing population growth and globalization of markets on current agroforestry systems will not be uniform. These trends will affect the importance or viability of some systems negatively. They will favor other systems or components, depending on their goals, resources available to farm managers, and increases or decreases in market and employment opportunities. While considering the potential responses of agroforestry systems to these changes is worthwhile, in reality many of these factors are likely to interact, making prediction of their outcomes exceedingly difficult.

Labor Availability

Globalization is likely to affect off-farm employment opportunities. If these increase, on-farm labor availability will decline and shortages may occur—a result that may affect farms in several ways. (1) Farmers may reduce or eliminate economically noncompetitive activities. (2) Farm households may shift or increase the tasks assigned to nonemployed members. (3) Management practices may shift from those that are labor-demanding to those that are less labor-intensive and/or that are based on more intensive use of purchased inputs.

Agroforestry systems with labor-intensive annual crop components, such as alley cropping, or those based on labor-intensive perennial crops, such as coffee or some fruit tree systems, are likely to become less economically viable because of shortages and higher prices for labor (J. Carter, 1995a, b, c; Ong, 1994; Szott and Kass, 1993). Tree crops, extensive grazing, or vegetated fallows may replace annual grains as a result. Nonshaded coffee may substitute for shaded coffee. Farmers may reduce the use of trees that require frequent management, such as some fruit trees or fodder banks. Although agroforestry techniques with low labor requirements will be favored generally, changes in these systems will depend on the farmer's preference, availability of land, product prices, and availability of capital (see Box 1). Sometimes, purchased inputs may be substituted for labor, for example, herbicides or fertilizers for manual weeding or organic residue additions, which may increase environmental problems if these inputs are badly managed. An emphasis on the management of trees expressly for environmental benefits is unlikely unless these benefits have clear and rapid effects on productivity. Such benefits, however, may accrue indirectly.

Conversely, fewer off-farm employment opportunities and less land per capita, because of increasing populations, will favor the use of more diverse and more labor- and land-intensive agroforestry technologies (Raintree and Warner, 1985; Somarriba, 1993). Adaptations to these conditions by small farmers will probably include the addition of trees to unused interstitial spaces on farms. Systems may also evolve toward home gardens, where purchased input levels are low, land is intensively used in both space and time, production is diverse and oriented to sale and consumption, and trees are used as a form of savings and to reduce risk (Jose and Shanmugaratnam, 1993; Raintree and Warner, 1985; Somarriba, 1993). Some labor-demanding, special niche crops such as organic coffee

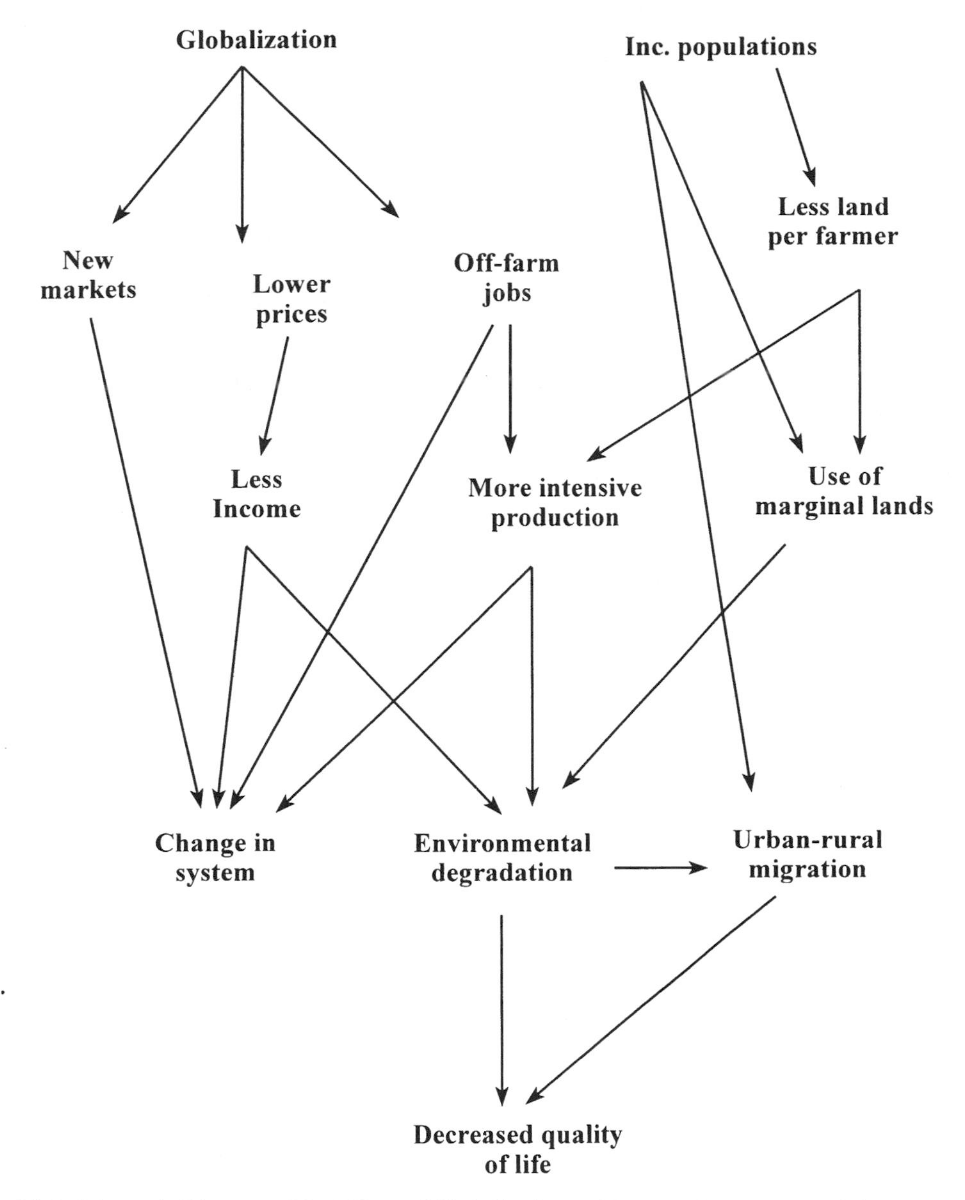

Figure 21.1 Schematic Diagram of the Effects of Globalization and Increased Human Populations on Land Use in Mesoamerica

or fruit may be at an advantage under these conditions of high labor availability (see Soto, chapter 18).

Lower Agricultural Prices

Lower prices for basic grains because of international competition will probably reduce the importance of these components in market-oriented agroforestry systems like having a component of basic grains. Agroforestry systems that include grains for domestic consumption may be less affected since undervalued family labor is frequently used in production and resource-limited farmers often place a high premium on food security. Nevertheless, even these systems may be vulnerable to competition for labor or land caused by globalization. An example is provided by the expansion of banana plantations in areas along the Atlantic coast of Costa Rica and Panama, which were previously isolated and farmed by subsistence-level farmers. The capital and knowledge available to farmholders that enable them to take advantage of economies of scale or tech-

Box 1 Changes in Coffee Management in Central America

John Beer, Coordinator
Area of Agroforestry, CATIE

Coffee is the most important crop for small- to medium-sized farms in humid regions of Central America, and most of the coffee production from the region comes from these farms. There are about 186,400 coffee farms in Central America and Panama (excluding Belize and Mexico). They occupy 7,778 square kilometers, or about 55 percent of the land area under permanent crops. Hence, the national coffee boards have an important influence on land use policy.

During the 1970s and 1980s, large increases in unit area coffee production were achieved in Central America by utilizing Green Revolution techniques: that is, improved germ plasm; high-density planting; and the concurrent and necessarily heavy applications of agrochemicals, including inorganic fertilizers (e.g., 300 kilograms of nitrogen per hectare per year). The level of diversity of traditional coffee plantations, where as many as 100 species were associated with coffee to provide shade, fruit, wood products, and so on, was greatly reduced to the extreme of coffee monocultures.

During periods of high coffee prices, professionals and farmers were persuaded that the income from these highly technified plantations compensated for their disadvantages, which included serious side effects due to contamination from high levels of pesticide and fertilizer applications, erosion, and soil degradation. However, the crash in coffee prices between 1989 and 1993 and the general conviction that prices during the coming decades will not equal the boom years of the 1980s, have provoked a reevaluation of policies by the national coffee boards.

To reduce economic, as well as ecological, risks, Central American coffee organizations are now promoting the judicious use of shade trees, along with a reduction in levels of application of agrochemicals. This does not mean that coffee farmers are encouraged to return to the traditional systems, which were highly diversified but of low productivity. Rather, the idea is to combine elements of traditional shade management with modern technology in order to achieve sustainable and competitive systems. Coffee yields from these hybrid systems may be less than those obtained in monocultures, but their profitability, taking into account diversified production (e.g., fruit and timber) and reduced input costs, can be competitive with the monocultures, especially when coffee price fluctuations are considered.

nological substitution to maintain acceptable profit margins will greatly influence the effect of lower prices on the production of these commodities.

New Market Niches

The emergence of new market niches for agroforestry products is perhaps the single most important factor that may affect income generation, sustainable resource use, and the well-being of both resource-limited farmers and urban populations. The characteristics, size, and accessibility of these markets will be decisive factors in determining the nature, extent, and economic importance of agroforestry systems. The trend of decreasing cost of production per unit of agricultural product will probably accelerate under globalization. As a result, small farmers may be able to compete in these markets with more resource-rich farmers only when these products require few purchased inputs; need labor or special skills; are oriented toward small, speciality markets; and are adaptable to production by women or children. Some examples include organic coffee, medicinal plants, and "green seal" tropical timber. Opening markets to a greater range of tropical timber species or an increase in international trade in carbon credits, to sequester carbon and reduce global warming, especially have the potential to affect positively the economic and ecological viability of small farmers. If carried out correctly, these international agreements and policies could provide incentives for planting more fruit or timber trees and would ultimately increase income and environmental benefits of agroforestry systems that are managed by small farmers.

Prerequisites for the exploitation of existing or potential markets by small farms are an adequate regional transportation infrastructure; local and regional storage facilities; and commercialization channels to regional, national, and international markets (Adriance, 1995; Murphy, 1995). The degree of organization of producers at local and regional levels will also be critical in identifying and exploiting those markets (see Rocheleau, chapter 5). Investment in the formation and/or support of such organizations by national and international donors can result in large paybacks in terms of more sustainable agricultural production and natural resource use.

Can Agroforestry Increase Production?

The supposed ability of trees to provide multiple products and a range of environmental services and benefits seems to make agroforestry technologies ideal for maintaining or increasing agricultural production and income while minimizing or even mitigating environmental degradation (Box 2). Research during the last 15 years has shown that agroforestry systems can provide high rates of timber, firewood, forage or fruit production, and economic benefits, given germ plasm properly tailored to site conditions and the effective management of plants (planting time, spacing, pruning, or thinning) and soils (Current and Lutz, 1992; Muschler, 1993; Sánchez, 1995; Sanginga et al., 1994). Plant associations in agroforestry can increase overall biological and/or economic productivity compared with monocultures, although these benefits may occur at the cost of the productivity of individual components. They can also provide more constant product and income flows and reduce farmers' risks (Arévalo, Szott, and Pérez, 1993; Beer, 1994; Current and Lutz, 1992; Diaz et al., 1993; Jensen, 1993; Somarriba, 1994, 1995). More work is necessary to overcome the methodological difficulties in quantifying the economic benefits (income or savings) of such systems since these are the chief interests of farmers (Current and Lutz, 1992; Sullivan, Huke, and Fox, 1992).

Whereas some agroforestry systems may be highly productive, others, perhaps the majority, clearly perform well below their biological and economic potential. Typically, this is due to the lack of fit between site characteristics and tree germ plasm and to inadequate management, especially competition among agroforestry components (Current and Lutz, 1992; Fernandez et al., 1994; Haggar and Beer, 1993; Sánchez, 1995). The lack of fit between site characteristics and tree germ plasm stems largely from the tendency to make blanket recommendations of species and management techniques for situations that differ from those in which the recommendations were derived. The negative effects of competition in agroforestry systems are due to the increasing competitiveness, over time, of woody perennials for light, water, and nutrients (Fernández et al., 1994; Hauser, 1993; Muschler, 1993). Management of competition entails more work for farmers. This, as well as the perceived value of the benefits that trees provide, underlies farmers' common preference to partly or wholly separate trees from annual crops or grasses in either space or time and militates against greater use of trees.

For agroforestry systems to increase productivity in the face of globalization, these weaknesses must be overcome. At the institutional level, national research institutions need to work with farmers to develop or adapt germ plasm, knowledge, and technology to local conditions. This implies continued financial support for these activities at both the international and national levels. At the research level, greater knowledge is needed of how intimate tree crop mixtures can be managed without incurring high labor or input costs. Some successful mixtures, such as shaded coffee, cocoa, or ginger, are well known (Beer, 1987; CATIE, 1995), but more work is needed to expand the range of productive plant associations. More attention should be devoted to the indigenous management of species mixtures, which may provide insights into effective species combinations and management techniques.

The point has also been raised that attributes of diverse, tree-based systems that contribute to their apparent sustainability, such as nutrient recycling and investment in supportive, perennial structures, "seem to have biological costs that are incompatible with high yields" because of the high proportion of photosynthesis respired to maintain a nonproductive supporting structure (Ewel et al., 1982). This begs the question of whether highly productive agroforestry systems are possible without breeding programs aimed at increasing the allocation of plant resources to the production of economically valuable products.

Environmental Effects

The nonsustainability of present land use in Central America is notable. Population pressure and mismanagement of these lands, in combination with natural factors, have caused widespread loss of the natural vegetation cover, soil nutrient depletion or chemical toxicity, soil compaction, low moisture retention, desertification, and weed invasion. Approximately 20 percent of Mesoamerica's soils are classified as eroded or degraded (J. Flores, 1994) and many negative impacts have resulted. Fallow periods have been reduced. Yields of the basic grain crops have stagnated or decreased during the past decade. The time and distance involved in collecting food, fodder, fuel, and water have increased. And desertification and contamination of soil and water by the inadequate use of fertilizers and pesticides are becoming increasingly common (S. E. Carter, 1991; Lindarte and Benito, 1993).

Box 2 Timber and Leguminous Trees as Shade for Cocoa on the Atlantic Coast of Costa Rica and Panama

Eduardo Somarriba
GTZ Agroforestry Project, CATIE

The substitution of timber species for traditional leguminous trees that provide only shade increased and diversified income from small holdings of cocoa in Talamanca, Costa Rica, and Bocas del Toro, Panama. This is one of the conclusions from research carried out on the management of shade in twelve young or old cocoa plantations in the region. Between 1988 and 1995, we evaluated (1) the transformation of the shade canopy in existing cocoa plantations caused by the introduction of timber species or tree legumes; (2) the biological and economic productivity of young cacao plantations shaded by either timber or leguminous species; and (3) the productivity and stability of production systems that combined cocoa, plantain, and timber.

The results indicate that the introduction of timber species in old cocoa plantations, together with the elimination of the original shade species, is simple and cheap in terms of labor and other inputs. For example, at 1.5 years of age, 54 percent of the 204 trees ha^{-1} of *Terminalia ivorensis* planted had grown above the cocoa canopy, indicating the rapidity of shade establishment. At 4 years of age, *terminalia* mortality was 24 percent, their crowns covered 70 percent of the soil, and they had accumulated a total volume of 35 m^3 ha^{-1}. Savanna oak (*Tabebuia rosea*) and laurel (*Cordia alliodora*), two other timber species, followed *terminalia* in production of shade and volume, and production was least with guaba (*Inga edulis*), a leguminous shade species introduced as a traditional control treatment.

In another study, new cocoa plantations with more intensively managed shade trees planted at a 6 m x 6 m spacing (278 trees ha^{-1}) produced 780 kg $ha^{-1}y^{-1}$ of dried cocoa with no significant differences among shade species. This suggests that timber species such as laurel, savanna oak, or *terminalia* can substitute for legumes such as guaba as shade for cocoa. Under this more intensive management, the timber species grew well. Total volumes of 5-year-old laurel, *terminalia,* and savanna oak were 90, 80, and 45 m^3 ha^{-1}, respectively. In this study, *terminalia* mortality was 68 percent.

Similar results were obtained in a replicate without savanna oak in Talamanca. Average cocoa yield was 360 kg $ha^{-1}y^{-1}$ with no significant differences among shade tree treatments. The low level of cocoa production was attributed to the high incidence of fungal diseases caused by the presence of abandoned neighboring cocoa plantations, which served as sources of inoculant. At 4 years of age, *terminalia* had a total volume of 73 m^3 ha^{-1} and laurel a volume of 70 m^3 ha^{-1}. Mortality of *terminalia* was 34 percent and that of laurel was 9 percent.

A cocoa system that includes timber species planted at low density (69 trees ha^{-1}) and plantain can produce income in the short term. This technology is very attractive to small farmers, since it avoids the need to wait 12 to 15 years in order to obtain income from fast-growing timber species. In an experiment in Charagre, Bocas del Toro, 1500 kg ha^{-1} of maize and 25 t ha^{-1} of cassava were obtained during the first year of tree establishment. Production of plantain and cocoa began in the second year, and during the first three years of production, they yielded some 400 kg ha^{-1} of dried cocoa and more than 2,000 racimes of plantain, thus providing an important source of income for the farmers. At 5 years of age, the shade trees (69 trees ha^{-1}) had a commercial volume of 26 m^3 ha^{-1}. Excellent growth of the timber trees was due to the management (especially weed control and fertilization) applied to the crops.

In spite of the excellent timber growth observed in these agroforestry systems, timber production as an element of diversification suffers from the long time required in order to obtain income. Other crops with shorter payback periods need to be included in addition to timber species in order to satisfy the immediate needs of farmers. In these systems, timber species can be conceived of as an intermediate term investment. Their strategic value derives from the increasing demand and prices for timber in international as well as local markets as a result of rapid population growth. The production of excellent quality timber during short rotation lengths and in combination with other crops of commercial interest is undoubtedly a "niche" in which small farmholders of the Mesoamerican humid tropics have a comparative advantage.

The combination of lower prices for agricultural products and greater land pressure caused by globalization and population growth may increase the mining of natural resources as farmers attempt to cut costs and maintain income levels and to expand the use of marginal lands. This kind of mining accentuates the possibility of environmental degradation. Under these circumstances, agroforestry's supposed ability to provide environmental services or benefits and reduce environmental degradation may be of equal or greater importance than its ability to increase production.

Agroforestry's environmental benefits are clearest with regard to improving the physical properties of soil and reducing soil and nutrient losses. This is due in part to the large quantities of organic residues that some systems produce, especially if trees are pruned (Szott, Fernándes and Sánchez, 1991). Living trees and their organic residues also protect the soil from rainfall impact; moderated microclimate; increased organic matter; reduced compaction; and increased pore space, aggregation, and moisture retention (Lal, 1984; Sánchez, 1987). These processes can reduce soil and water loss from runoff, wind and water erosion, and nutrient leaching (Alegre, 1991; Babbar and Zak, 1994; Lal, 1987; Lebeuf, 1993; Nair, Kang, and Kass, 1995). They can also help improve soil conditions and help reduce weeds in degraded and compacted pastures or previously cropped fields (Szott and Palm, 1986).

Although trees may effectively conserve soils and nutrients already present in the system, their ability to add significant amounts of nutrients by "pumping" from deep soil horizons, by harvesting nutrients in rain or dust, or by nitrogen fixation is likely to vary with soil type. On weathered infertile soils, the small size of these inputs or by soil chemical constraints are apt to limit nutrient additions (Giller and Wilson, 1991; Szott et al., 1991). In contrast, on less weathered, more fertile, near-neutral soils, where these constraints are reduced or eliminated, trees have greater potential to increase nutrient stocks significantly. The addition of residues from nitrogen-fixing leguminous trees may be able to substitute at least partially for nitrogen fertilizers on such soils.

Trees may not significantly increase the total quantity of nutrients in some agroecosystems. However, they may be able to improve nutrient availability by recycling large amounts of nutrients between vegetation and soil compartments via organic residues and maintaining some nutrients, such as phosphorus, in plant-available forms (H. Fassbender, 1993; Haggar et al., 1993; Meléndez, 1996; Palm, 1995; Szott et al., 1991). Increases in nutrient availability may be related to increased soil microbial biomass and soil macrofauna, which are the result of greater organic residue additions and changes in soil microclimate (Lavelle and Pashanasi, 1989; Mazzarino, Szott, and Jiménez, 1993). Organic additions may also aid in complexing aluminum in acidic soils, which may result in less toxic conditions and greater phosphorus availability (Fox and Comerford, 1992).

Despite the ability of many agroforestry systems to conserve and recycle nutrients, nutrient export in harvested products is usually larger than additions from natural sources (Fernández et al., 1994; Jensen, 1993; Palm et al., 1991). This implies that systems will be truly sustainable only under certain conditions. Harvests must be infrequent to allow the slow replenishment of nutrients from natural sources, and the product harvested must be small in quantity or low in nutrients. When these condition are not met, external inputs will be needed to replace exported nutrients (Szott and Kass, 1993). Nitrogen may provide one exception to these necessary conditions if large amounts are fixed from atmospheric sources. Systems on soils with high native fertility apparently may be sustainable (table 21.1), but on infertile soils similar levels of nutrient export are not sustainable. Many small farms are found on less fertile or marginal lands, suggesting that the ability of agroforestry systems to contribute to increased productivity will be severely constrained unless nutrients are added. In short, agroforestry systems do not provide a free lunch.

Given deforestation and land use conversion in Mesoamerica, there is justified concern about stemming the loss of biodiversity. Agroforestry systems have the potential to contribute directly to the maintenance of

Table 21.1 Estimated Annual Nutrient Budget for a Home Garden, West Java

	N	P	K	Ca	Mg
Total input (rain, N_2 fixation, stream inflow)	33	6–14	115–255	78–174	55–122
Total output (sold, leached, erosion, stream outflow)	43	11–19	151–283	275–577	83–175
Balance (negative)	10	5	27–37	196–403	28–52
% of available soil reserves lost	0.1	0.1	0.4–0.7	0.4–0.8	0.4–0.7

Source: Jensen, M., 1993. Soil conditions, vegetation structure and biomass of a Javanese home garden. *Agroforestry Systems* 24: 171–186.

Ranges of values are those calculated by two different methods.

both on-farm and regional biodiversity. Systems such as home gardens and traditional multistrata tree crops often contain over 100 species, many of them remnants from forests (Gliessman, 1990; House, 1994). The recent emphasis on incorporating native timber species into farming systems may also aid in conserving biodiversity (Butterfield and Fisher, 1994). However, indirect effects of agroforestry on biodiversity, through the establishment of more productive systems that result in less deforestation, may be its main contribution to maintaining diversity at the regional scale.

Although agroforestry systems can produce several environmental benefits, they can be as destructive as other systems when they are badly managed. Erosion may increase because of badly aligned contour rows that funnel runoff down a slope or to drip impact, as often occurs under teak (*Tectona grandis*). Drip damage from tall broadleaf trees in coffee plantations can also displace fruit from the branches. Gaesous losses of nitrogen may occur from tree-derived mulch layers (Glasener and Palm, 1991) or by burning when fire is used as a management tool (Waring and Schlesinger, 1985). Significant amounts of phosphorus and other nutrients may also be sequestered in tree biomass, thus reducing the levels of plant-available nutrients in soil (Beer, 1988; Haggar et al., 1991). Moist conditions under trees may favor some disease problems of associated crops. Aggressive, fast-growing, and rapidly propagating exotic trees may turn into weeds when managed inadequately (Hughes, 1994).

Constraints to Farmers' Adoption

Based on the potential and actual benefits provided by trees, one would expect that farmers would be readily disposed to include and manage trees in their farming systems. This is often not true because of a variety of biological, socioeconomic, and cultural factors (Kerkhof, 1990).

Perhaps the largest drawback to including and maintaining trees in agricultural systems is the economics of tree planting and maintenance, especially when trees must be guarded against damage caused by grazing animals or fire (Current and Lutz, 1992; Kerkhof, 1990). The costs of planting and managing trees are often prohibitively high for farmers who have few savings and little access to credit (Jose and Shanmugaratnam, 1993). Moreover, although planted trees may return a profit in the future, farmers are often unable or unwilling to wait until harvest to recover their investment. The discount rates that small farmers use are often very high, roughly 20 to 30 percent, which implies that investment in tree planting will be unlikely unless positive economic returns are realized within a few years. Unfortunately, rotations for marketable timber products are often 10 to 30 years long.

Opportunity costs associated with tree planting may also discourage small farmers. On small farms in dry zones, trees often compete for water and space, two resources that are critical to small farmers for production of their subsistence crops. Other problems related to competition for nutrients and light were mentioned previously. Furthermore, the social benefits or environmental services provided by trees, such as erosion control, are of little immediate value to farmers compared with the costs incurred to ensure these benefits. As a result, small farmers are unlikely to use these measures unless they receive direct incentives commensurate with the costs involved (Lindarte and Benito, 1993).

Legal problems associated with land tenure or tree use may also act as disincentives to tree planting (Fortmann, 1987). Farmers are unlikely to plant and tend trees if they have little say in management decisions or the right to harvest and sell the trees when, where, and at prices they want. These conditions are not always satisfied in Mesoamerica. Even if these requirements were satisfied, however, markets for many tree products, such as fruit, are often small, easily saturated, and ephemeral, and access to them often involves high transportation costs.

Rates of tree survival and growth are often low, especially in dry areas, and knowledge related to tree planting and maintenance may be minimal or absent. Farmers may be reluctant to undertake activities about which they have little knowledge or experience and that they consider to have a low probability of success.

Strategy

For agroforestry to contribute to increasing agricultural production and more sustainable land use in Mesoamerica, the proper political, economic, and social framework to encourage tree use on farms must be present. First, credits or incentives for planting trees, especially when they are used to reduce environmental degradation, can encourage use. Second, legal changes can simplify bureaucratic procedures and allow socially equitable ownership and use of trees. Third, prices for tree-based products must be favorable. Fourth, an adequate marketing and commercialization infrastructure is critical. Fifth, local or regional producers' organizations can help farmers use trees better. And finally, farmers need training in tree management and value-added production techniques.

We also need to reconsider the role of trees on farms and how they are managed. Managing trees to provide income and savings should receive greater emphasis (see Box 3). This implies changing the emphasis of tree use from the production of environmental benefits to that of economic benefits, with the former accruing as side-products. The use of valuable species on "wasted"

Box 3 Agroforestry in Central America and the Dominican Republic

Alberto Camero Rey
CATIE

The Agroforestry Area of the Tropical Center for Agronomic Research and Education (CATIE) is in the process of documenting agroforestry activities by institutions and projects in Costa Rica, Honduras, Nicaragua, Guatemala, El Salvador, Panama, and the Dominican Republic. A brief summary of their activities is shown below. More information can be obtained from the Agroforestry Area, CATIE 7170, Turrialba, Costa Rica.

Type of Agroforestry System (% of projects)[a]

Agrosylvopastoral	1%
Sylvopastoral	62%
Agrosylvocultural	68%

Ecological Zones (% of projects)

Arid	9%
Semi-arid	37%
Sub-humid	51%
Humid	31%

Activities (% of projects)

Extension	75%
Research	61%
Education	58%
Rural development	57%
Technical assistance	39%
Economic production	33%

Types of Institutions (% of projects)

Governmental	37%
Local non-government	31%
International	19%
Private	13%

Beneficiaries (% of projects)

Small farmholders	88%
Medium-sized farmholders	45%
Large farmholders	11%
Rural communities	46%
Development associations	42%
Private businesses	17%
Public enterprises	15%

[a]Percentages may not sum to 100% since projects or institutions may be working in more than one ecological zone or activity.

space on farms—for example, border plantings, where negative interactions with other farming components will be minimized—can increase the economic benefits of trees (Beer, 1993). Further research and promotion of the taungya or shelterbelt systems may help reduce establishment costs (Beer, 1994; Ichire, 1993). Greater recognition is needed of the site and species specificity of agroforestry components. Farmers and researchers must also understand that management, including the addition of organic or inorganic nutrients to replace those exported, is critical to their success. General "recipes" cannot be given. National research institutions have an important role to play in developing these local recommendations.

In practice, monitoring the sustainability of agroforestry systems will be difficult because of the lack of readily apparent biophysical or social indicators at appropriate scales and the need for measurement over long periods. The changes in the social, economic, and political forces that affect agriculture and land use caused by globalization and population growth are occurring rapidly. Therefore, greater attention should be given to indicators of sustainability that are more qualitative, which include farmers' perceptions of changes, benefits, problems, risks, and flexibility, rather than to those that measure changes against an unchanging baseline (Campbell, Bradley, and Carter, 1994).

Conclusions

In the face of globalization of markets and increasing population pressure, agroforestry has the potential to provide food security and increase income for rural dwellers while reducing environmental degradation. However, the realization of this potential will depend greatly on several factors: (1) new or expanded markets for tree-based products; (2) the effects of globalization on local off-farm employment; (3) the relationships among the prices of inputs, labor, and goods produced; and (4) policies that affect land use. Agroforestry practices that make use of nonutilized areas of farms, that demand few inputs, and that provide economically important products for speciality markets are likely to be most successful.

Despite the potential benefits provided by trees, biological and socioeconomic problems associated with their management often discourage their use by small, resource-limited farmers. Potential economic benefits of agroforestry may not accrue because of inappropriate germ plasm for site conditions; competition among system components; and the lack of fertilization to compensate for nutrients exported in harvested products, leading to long-term nutrient depletion. Social amenities or environmental benefits provided by trees at the farm level are rarely large or apparent enough to individual farmers to act as incentives for planting and maintenance. Low rates of survival and slow growth; long payback periods; high actual and opportunity costs; limited access to credit or technical assistance; land tenure systems that limit farmers' participation in management decisions and their legal right to tree-based products; and markets that are limited in size, are ephemeral, and are not readily accessible also limit adoption by farmers. As a result, changes in national and international markets and in national political economies will often be needed to provide incentives for tree planting and maintenance by farmers and the realization of agroforestry's potential to contribute to more sustainable land use.

22

Biologically Sustainable Agroecosystems

Using Principles of Ecology

John C. Mayne

Modern agriculture has been based on adapting the plant, soil, and climatic environment to a single desired crop, usually the most profitable crop, for many years. Farmers must grow crops for which there is a demand. Nonetheless, we now see that the agricultural system may ultimately fail if the crops grown are so unsuited to the natural environment that they need large inputs of nutrients, pesticides, herbicides, and petroleum-driven mechanical labor.

Agricultural systems that require increasing inputs over time pose particular risks for farmers with limited financial resources. The system may eventually fail for many reasons (Vandermeer, 1989). Too many nutrients brought in from outside the system may be required. Soil erosion may occur faster than topsoil is built. Soil organic matter may be depleted because of a disruption of the carbon-to-nitrogen ratio from the addition of large, frequent applications of nitrogen. The population of insect pests may explode because their predators are gone. As greater amounts of expensive inputs are needed to keep the system functioning, resource-limited farmers frequently find themselves without the cash to purchase these inputs. Many eventually cannot continue farming.

We are now trying to develop systems suited to existing environmental conditions. Agroforestry systems that use plants, trees, and combinations of plants and trees are one example. They stress farming methods suited to the soil and climate conditions of a region (Nair, 1989). Because they require less fertilizer and fewer pesticides, less soil and water degradation occur.

However, agroforestry systems must be both biologically and economically sustainable to succeed over the long run. Biological sustainability is much more difficult to assess than economic sustainability because it may take many years, even decades, to see some effects of the biological and edaphic constraints. Because of this, it is very difficult to tell farmers (particularly if they do not own the land) or analysts that a system that is functioning and earning income will, over time, require greater inputs from outside.

One way in which agroforestry contributes to the biological sustainability of farms is by using deep-rooted trees. Deep-rooted trees are an integral component of tropical agroforestry ecosystems because they extract nutrients from the soil below the rooting depth of most food crops. This is an important aspect of nutrient cycling. In many agroforestry systems, farmers cut the leaves and stems of the deep-rooted plants and apply them to the topsoil as a nutrient amendment. The topsoil is enriched with nutrients that originated from deeper in the soil profile. The result is that previously unavailable soil nutrients are made available to the shallower-rooted crop plants.

Understanding the degree and extent to which trees and crop plants compete for soil nutrients is crucial to the design of agroforestry systems. The ecological concept of competition is therefore one factor in designing better agroforestry systems.

Competition

Researchers have proposed competition as a cause of plant behavior ever since Darwin (1859). Differences in the partitioning of uptake of soil resources with depth are believed to be a result of competition. It is thought that these differences developed over time because of the coexistence of competing species (Clements, Weaver, and Hanson, 1929; Grime, 1979; Tilman, 1985, 1988; Wilson and Tilman, 1991).

Attempts to understand and quantify competition have occupied ecologists for over a century. Yet, as Caldwell et al. (1985, p. 384) state, "The nature of competition is known more by its manifestations than by its mechanisms." Definitive demonstrations of competition for soil resources among plants are rarely possible (Caldwell et al., 1985, 1987; Connell, 1983; Harper, 1977; Strong, 1983). Nonetheless, although difficult to quantify, competition is accepted as an important factor in determining species mixtures and distribution (Harper, 1968; Picket, 1980; Salisbury, 1936; Schoener, 1983; Silvertown and Law, 1987; Williams, Richards, and Caldwell, 1991; M. F. Wilson, 1973).

Much of our understanding of competition among plants has developed from studies of natural plant ecosystems in which competition was implicated as a cause of species diversity and distribution. Nutrient capture is one causative factor. Plant growth in nonagricultural systems is normally more limited by soil resources than by solar radiation (Fitter, 1986). As a result, strategies of nutrient capture among species and under varying nutrient statuses have provoked much interest (Cole and Holch, 1941; Grime et al., 1987; Gueravitch and Unnasch, 1989; G. A. Harris, 1967; Parrish and Bazzaz, 1976; Weaver, 1919; Williams et al., 1991). As attention to below-ground capture of resources grew, researchers began to investigate mechanisms by which plants of different species, using the same pool of resources, acquire these resources. The role of competition as one of the mechanisms is still being developed and debated (Connell, 1980; Grace, 1991; Grime, 1973a; Grime et al., 1987; Newman, 1973; Thompson and Grime, 1988; Tilman, 1988; Turkington, 1989).

Deep-rooting Habits: An Example of Competition?

Deep-rooted plants grow deep roots because they have evolved that habit. They are not shallow-rooted plants that grow deep roots in response to either nutrients or moisture in the subsoil. Furthermore, a deep- or shallow-rooted habit is maintained within a wide range of soil nutrient statuses. When plants with different rooting habits are planted into habitats and soils different from where they evolved, they tend to maintain their rooting habits. Ellern, Harper, and Sagar (1970) found that the root systems of both *Avena fatua* and *A. strigosa* occupied the upper soil horizon but that *A. fatua* more thoroughly occupied the deeper soil. These differences in rooting habit were found both when the two species were grown alone and when they were grown in mixed stands on soils other than those in which they had evolved. In studies of nine tropical ecosystems, Ewel et al. (1982) found the soil solum to be partitioned by both shallow-rooted and deep-rooted plants in every case.

The fine roots of deep-rooted plants predominate in the subsoil. Nevertheless, the bulk of the fine-root mass of most deep-rooted species is still found in the topsoil (Ewel et al., 1982; Gillespie, 1989; Mayne, 1993; Parrish and Bazzaz, 1976). In all these reported observations, the upper soil was occupied by fine roots of all species, but the deeper horizons were occupied exclusively by one or a few deep-rooted species.

Harper (1977) states: "That different organisms living in different places are different is easy to explain. How so many different organisms can live together in the same 'place' is more difficult to answer and is more interesting; it demands a biotic interpretation and a Darwinian solution." His observation describes root partitioning in the soil with depth. Both the biotic interpretation and Darwinian solution to the phenomenon of root partitioning of the soil are addressed by nutrient competition. It is evident that a shallow-rooted plant that is experiencing nutrient competition in the topsoil will not become a deep-rooted plant. However, competition for nutrients in the topsoil may have had a role in the evolution of plant species with different rooting habits.

As efforts to understand below-ground competition continued, experimental results suggested that increasing a resource did not always reduce competition (Abramski and Rosenzweig, 1984; Al-Mufti et al, 1977; Connell, 1978; Grime, 1973b; Snaydon, 1982; Wilson and Keddy, 1986). Using nutrient flux studies, Vaidyanathan et al. (1968) concluded that there may be more competition at higher levels of soil nutrients, because of greater nutrient mobility.

This counters the conventional view expressed in early studies that competition would decrease when resources were added (Bandeen and Buchholtz, 1967; Brenchley, 1922; Nedrow, 1937; Welbank, 1961). Donald (1958) also suggested that competition for soil resources would decrease when resources were added. While Donald's study came under almost immediate criticism for not adequately separating root from shoot competition (Aspinal, 1960), he set the stage for the debate on the intensity of competition at different resource levels.

Grime (1973b) and Newman (1973) reignited the debate on this issue, which continues to this day (Grace, 1991; Taylor, Aarssen, and Lochle, 1990; Thompson, 1987; Tilman, 1987, 1988; Wilson and Shay, 1990). Grime (1973b) attempted to explain high species densities on nutrient-poor sites, suggesting that competition is more intense on high-resource (low-stress) sites than on low-resource (high-stress) sites. Newman (1973) disputed this and proposed that high competition for light was responsible for the low species diversity found on high-resource sites. He further maintained that species diversity will increase only when competition for light is reduced by disturbance or low soil nutrient status. That is, overall competition was greater on low-resource sites. Grime (1973a) countered by stating that the high species densities found on low-nutrient soils are caused by low competition, not by greater root competition at the expense of shoot competition.

In his review of competition studies, J. B. Wilson (1988) concluded that the evidence to support the commonly held assumption that competition is reduced in the presence of increased resources is equivocal. He found, from these studies, that competition was often greater at higher-resource levels.

In spite of the lack of consensus on what factors contribute to nutrient competition and its role in the evolution of rooting habits, the partitioning of fine roots in the solum with depth is well documented in agricultural and natural ecosystems. Fine-root partitioning of the soil solum by agricultural plant species has been suggested as the reason for the phenomenon of overyielding. When plants are grown in mixtures, the total yield per unit of land area is greater than the individual yield of any of the same species grown alone (de Wit and van den Bergh, 1965; Martin and Snaydon, 1982; Vandermeer, 1989; Wilson and Newman, 1987).

The ecological research on competition has significant applications for multicrop agricultural systems, such as agroforestry systems, that use deep-rooted plants. Designing these systems requires consideration of the effects of crop density on the ability of deep-rooted plants to bring up nutrients from the subsoil. There is a tradeoff in the utilization of soil nutrients. A greater crop yield, through increased plant densities, may cause deep-rooted plants to extract fewer nutrients from topsoil and subsoil, resulting in decreased biomass production by deep-rooted plants. Fewer subsoil nutrients will be put into the nutrient cycle as crop density is maximized. As a result, fewer nutrients will be available to use as green manure or animal forage because the quantity of leaf biomass of the deep-rooted species will decrease.

The uptake of nutrients from deeper soil by deep-rooting plants is a vital component of the nutrient cycles of terrestrial ecosystems. An understanding of this process is crucial when working with systems where competition for nutrients is severe. These include both natural ecosystems on nutrient-poor sites and agricultural systems that are under ever-increasing demand to feed more people by producing greater crop yields on smaller areas of increasingly marginal land.

23

Forest Plantations in the Tropics

Richard F. Fisher

Past and continuing destruction of natural forests in the tropics have made forest plantations desirable and often necessary. Tropical forests have long been cleared for other uses or severely damaged by human use. This destruction shows few signs of abating. A Food and Agriculture Organization (FAO) survey of 62 tropical countries in 1990 showed that deforestation amounted to nearly 17 million hectares annually, or 1.2 percent of the total forest cover (Dembner, 1991). This is double the rate of the 1970s, and by all indications this rate has declined very little over the past five years.

The main causes of deforestation are well known (Head and Heinzman, 1990; Myers, 1984; Palo, 1987). Clearance for agriculture probably accounts for about half of all deforestation. Urban and industrial expansion, including mining and hydroelectric development, are another principal cause. Shifting cultivation and exploitation of charcoal and fuelwood affect a very large area of tropical forest annually. Accidental or deliberate burning of the forest adds several million hectares to the deforestation total every year. Although intensive logging accounts for only a small amount, the infrastructure of roads and navigable streams created by logging greatly eases invasion and the deforestation that results from clearing for agriculture, shifting cultivation, charcoal and fuelwood production, and other uses.

Pressures to exploit and to clear natural forest in the tropics continue despite the implementation of the Tropical Forestry Action Program and the signing of the International Tropical Timber Agreement. World demand for fine woods, construction timber, and wood fiber continues to climb. Developing countries demand an increasing quantity of wood products, especially paper products, as they develop. The emergence of China as a major paper market will undoubtedly drive the demand for wood up sharply in the next several decades. Land clearing for agriculture is also increasing in many tropical countries as the human population continues to soar.

Natural forest remains the best source of many fine woods such as mahogany, rosewood, and teak. The quality of wood found in old, mature trees from the natural forest is superior to that found in young, plantation-grown trees. Teak may be an exception as good-quality wood is beginning to be produced from mature plantations in India and parts of Southeast Asia. More natural forest is coming under sound forest management, and we may be able to produce a sufficient volume of fine woods from natural forests while avoiding deforestation in the future. However, many questions remain. Land tenure continues to be a major problem in natural forest management, and the threat of invasion by local people who wish to use the forest as agricultural or grazing land retards investment in natural forest management. In many countries land laws continue to favor invasion over management of natural forest (McGaughey and Gregerson, 1982).

The production of fine wood from the natural forest is costly. Getting preferred species to regenerate is difficult. Whether natural regeneration or enrichment planting is used to acquire seedlings of desirable species, intensive care over a period of several years is required to ensure that these seedlings become established and thrive in the forest. The cost of such care must be carried at interest for several decades before the trees can be harvested. Consequently the cost of producing the wood may exceed its value (MacKerron and Cogan, 1993). The value of fine wood has increased recently, and this has given a boost to natural forest management. Many management techniques previously thought to be uneconomical, such as underplanting, cleaning, and pruning, are now seen to be potentially useful. The tropical forest is likely to continue to produce fine woods for the foreseeable future, although their cost may become very great.

Why Plantations?

There are three principal reasons for the creation of tropical forest plantations. As the area of natural tropical forest diminishes, both local demand in tropical countries and worldwide demand for wood is increasing. This rising demand alone makes plantation forestry in the tropics an attractive venture. However, forest plantations are also a potential mechanism for rehabilitating degraded forest sites and for reestablishing natural forest on deforested sites.

The price of wood and wood products increases as wood becomes scarce, and so growing wood as a crop becomes attractive (Sedjo, 1983). The demand for and the price of fuelwood, charcoal, construction lumber, and wooden furniture has reached unprecedented heights in many tropical countries during the past decade; thus, growing wood for the local market is profitable for the first time. Producing wood and wood products for a local market is very different from producing them for an international market. This means that a quite different group of growers and producers enter the market, and a different type of forest industry emerges.

The first tropical plantations were developed to supply wood and wood products to the world market and, often, to supply a source of hard currency to the country where the plantations were found. This latter objective is often not achieved since the hard currency may flow to only a few investors in the industrial forestry scheme. Some say that "the rich get rich and the poor get seedlings" under these schemes. Large industrial forestry operations in the tropics often import goods and services from outside the country, import foreign nationals for the highly paid technical jobs, and only use local people for low-paying manual labor. This has been true even of most fuelwood and charcoal schemes.

Nonetheless, industrial tropical forest plantations are likely to be a major source of wood and wood products in the future (table 23.1) because a large volume of uniform material can be produced in a short time. Only a limited number of species have been commonly used in tropical forest plantations. Eucalypts and subtropical pines dominate this list, although the increasing number of fuelwood plantations has led to the planting of many different species in the past decade (table 23.2). There are several reasons for the limited number of species used in plantations. The most important reason is uniformity of raw material.

Table 23.1 Approximate Area of Forest Plantations in the Tropics (in 1000s of hectares)

Country	1965	1980	1990
Brazil	500.0	3,855.0	7,150.0
Cameroun	1.3	15.0	17.5
Colombia	16.3	95.0	250.0
Congo	1.3	15.0	17.5
Costa Rica	2.4	4.1	40.0
Ecuador	4.0	8.5	60.0
Guatemala	1.0	3.0	25.6
Indonesia	706.0	1,918.0	3,700.0
Jamaica	4.8	11.3	16.5
Malaysia	1.0	26.0	90.0
Myanmar	56.6	87.3	340.0
Papua New Guinea	5.2	24.5	44.3
Peru	20.0	127.0	272.0
Venezuela	1.5	112.0	350.0

Source: UN-FAO, Rome Forets Statistics, 1994.

Why the Heavy Reliance on a Limited Number of Mostly Exotic Species?

Most wood produced in industrial plantations is converted to pulp. Pulping can be carried out most efficiently if all of the wood is of one kind, which often means not only the same species but also a few select varieties that have very similar specific gravity (density) and secondary plant chemical (e.g., lignin) content. In other words, the requirements of the manufacturing process drive the selection of the species and the variety to be grown on plantations.

A second reason for the importance of a few species is that these have been thoroughly domesticated. Foresters know how to reproduce and grow them. Domestication is a lengthy process (Libby, 1973). It begins with the discovery of the basic physical, chemical, and biological properties of the species. This kind of knowledge exists for nearly all temperate zone tree species but for only a few tropical tree species. Species for industrial plantations are first selected because they have outstanding wood properties for some particular use.

However, biological properties must be taken into consideration. Some species that produce excellent wood cannot be grown in plantations because of insect or disease problems. For example, American mahogany (*Swietenia macrophylla*) and tropical cedar (*Cedrela odorata*) cannot be grown successfully in plantations in the neotropics because of attacks by the shoot borer (*Hypsipyla grandella* and *H. ferrealis*).

Species Domestication and Success in Plantations

Reproducing large quantities of uniform offspring by sexual reproduction through seed or by asexual reproduction through cuttings or tissue culture—must be simple for a species if it is to become successfully domesticated. The production of high-quality seedlings requires a thorough knowledge of pollination and seed and seedling biology. Asexual reproduction, especially by tissue culture, requires an extensive knowledge of the physiology of the tree. Our lack of knowledge in these areas for tropical trees continues to delay their domestication.

Table 23.2 Species Used in Tropical Forestry Plantations

Genus or Group	Species in Order of Importance	%
Eucalyptus	*Grandis, camaldulensis, saligna, urophylla, tereticornis, robusta, citriodora, deglupta*	38
Pinus	*Patula, caribaea, elliottii, merkusii, kesiya, oocarpa, taeda, techunumanii*	31
Tectona	*Grandis*	14
Other hardwoods	*Acacia, Gmelina, Leucaena, Grevillea, Terminalia Albizzia, Prosopis, Casuarina, Cordia, Azadiracta*	14
Other conifers	*Araucaria, Cupressus*	3

Source: Richard Fisher.

Foresters must be able to collect or produce a large volume of seed for a species to become a successful plantation species. These seeds must be capable of producing trees that are fairly uniform and that possess desirable characteristics for the end use of either the trees themselves or the forest in its entirety. It is very helpful if these seeds can be stored in a viable condition for some time, the longer the better. First, this allows seeds from trees that produce fruit at somewhat different times to be sown in the nursery at the same time. They can then be planted in the field at the same time, yielding a plantation with considerable genetic diversity. Second, this allows seeds to be stored from year to year so that reforestation is not subject to the vagaries of any one year's seed crop. Finally, this allows seeds to be transported from country to country and even continent to continent.

Successfully growing a plantation species also requires us to understand a species's genetics. Foresters are very interested in highly heritable characteristics. By this they mean characteristics of the parents, often the female parent in particular, that are very likely to show up in a high proportion of the offspring (Zobel, Van Wyk, and Stahl 1987). Size and length of branches, branch angle, foliage color, wood specific gravity, oleoresin content, and height growth rate are some characteristics that are of great interest to foresters and that are often highly heritable. Resistance to insect and disease attack and drought tolerance are other characteristics that are significantly heritable in some species. Obviously a species with many characteristics of interest that are highly heritable will become a favorite plantation species.

Teak (*Tectona grandis*), *Pinus*, and *Eucalyptus* have many features that favor them for plantation forestry. All have such characteristics as wood specific gravity, branch size and angle, and height growth rate, which are highly heritable. Thus superior trees, individuals with superior phenotypic characteristics, can be selected and their seed used to reproduce a generation of offspring with desirable and uniform characteristics. Breeding can also be carried out by using selected individuals of superior phenotype or genotype to produce offspring superior to their parents in one or more traits. Most tree seedlings planted in the temperate zone today and an increasing proportion of those planted in the tropics are either selected or bred from improved seedlings or clones.

Of course, this raises the possibility of the creation of a large area of monocultural plantations highly susceptible to disease outbreak or other calamities. To date, this has not been a major problem in forest plantations because the amount of genetic diversity for most traits in superior seedlings is still quite high. However, plantations established with clonal material do pose potential insect and disease problems. The devastation of *Leucaena leucocephala* plantations in Asia during the 1980s by the Leucaena psyllid (*Heteropsylla cubana*) is a case in point.

Clonal Forestry

Clonal forestry has become popular with species such as *Eucalyptus urophylla* and *Gemelina arborea* and the hybrid *E. urophylla* and *grandis*, which can easily be reproduced vegetatively. In this system, individuals of a superior phenotype are chosen, and thousands of genetically identical individuals are produced as rooted cuttings or by using tissue culture. These plantlets grow more uniformly in the field and respond more uniformly to cultural practices like weed control and fertilization than do seedlings. The wood from clonal plantations has uniform physical and chemical properties, and it is easier and less expensive to process than wood from the natural forest or from seedling plantations.

The clones chosen for mass reproduction in clonal forestry have good stem form, desirable wood properties, and a fast growth rate. They also display high resistance to insect and disease attack and respond well to cultural practices. However, the uniformity of all genetic traits in a large population makes it subject to attack when genetic changes in pathogen or predator populations allow them to overcome the trees' resistance to attack. To combat this problem, foresters in the tropics rely on very short rotations (the time from planting to harvest) and on the continual development of new clones of superior quality (Leakey, 1987). The technical skill and high investment that such a strategy requires make clonal forestry impractical in many areas of the tropics.

Industrial Forestry

Industrial forest plantations will continue to grow in importance in the tropics. World demand for wood and wood products is increasing, and the tropics will probably always be the low-cost producer of wood fiber, if not of solid wood products. Currently most wood produced in industrial plantations in the tropics finds its way to markets in the temperate zone, but the consumption of paper and paper products is growing rapidly in the tropics. As petroleum becomes more expensive, paper will replace plastic in many uses and the industrial forests of the tropics will produce much of their wood for local markets.

The widespread use of exotic species in industrial and even fuelwood plantations in the tropics has often been called cultural imperialism. The history of modern forestry shows that foresters, just like other immigrants, carry their seeds with them. The earliest tree plantations in North America were made with Norway spruce (*Picea abies*) and Scots pine (*Pinus sylvestris*), species that the European immigrants understood. Likewise foresters carried radiata pine (*Pinus radiata*) from California to New Zealand and Australia and *Eucalyptus* from Australia to every continent on the globe. Humans are comfortable growing what they know how to grow and using familiar raw materials for manufacture.

Lending and donor agencies also contribute to the widespread use of exotic species in forest plantations. Such agencies are very conservative, and they require a high degree of assurance that a venture will succeed before they will support it. The use of a well-tested species for forest plantations is something that they have routinely required. Only the successful domestication of more tropical tree species will increase the number of native species used in large-scale forest plantations (Evans, 1992).

It also seems that each decade of this century has had its "wonder tree," a species that grows faster; has better wood properties; or is more resistant to fire, insects, or disease than any previously known species. Subtropical pines (*Anthocephalus chinensis*), various eucalypts, (*Gmelina arborea, Lucaena leucocephela, Acacia mangium,* and *A. nelotica*), and neem (*Azadiracta indica*) have all held this honor. It seems human nature to try these species everywhere to see if they might work their magic on your site. Though the odds are low, gambling on wonder trees appears certain to continue.

Why Native Species?

Currently the interest in using native species for forest plantations in the tropics is great. This interest has arisen for two major reasons. First, native species may be better suited for the reclamation of degraded sites than exotic species. Second, they may serve as catalysts to speed the redevelopment of secondary natural forest. Most fast-growing exotic species that have been used in tropical tree plantations have high nutritional requirements. These species grow well on recently cleared natural forest sites, but they do not perform well on poor-quality sites (Ojo and Jackson, 1973). However, most tropical lands that need reforestation have undergone some site degradation and are commonly too poor for the establishment and growth of fast-growing exotic species.

In addition, the wood of native species is generally preferred in the local market, the most important market for wood in much of the tropics. Local builders and artisans know the properties of local woods and have strong preferences for them. On the one hand, farmers and other local landowners know how to assess the value of native species. They know that the local mills will or will not use logs of a particular tree. On the other hand, they are never quite sure about the acceptability of exotic species. The testing of native species in plantations and the domestication of new species is underway throughout the tropics. Recent work in Costa Rica is a good example.

Research on the establishment of native species on abandoned agricultural lands began in Costa Rica in 1985. The Organization for Tropical Studies (OTS), in collaboration with the Dirección General Forestal (DGF), established a replicated trial of 14 native species on the La Selva Biological Station (Butterfield, 1994). The early success of these experiments led to a major trial of native species known as TRIALS, or ENSAYOS. This project was divided into two parts. The first tested native and exotic species in single-tree elimination trials. Sixty-seven native species and 17 widely planted tropical exotics were tested (Butterfield, 1995). The second established replicated 0.25-hectare plots of 7 native species, 4 exotic species, and a control on abandoned pastureland to evaluate the ameliorative effects of tree plantations on degraded soils (Fisher, 1995). These plots have also been used to evaluate the growth of selected species in stands (González and Fisher, 1994).

The elimination trials identified many native species with good potential for timber plantations, agroforestry, and conservation planting on poor, degraded sites in northeastern Costa Rica (table 23.3). Also, TRIALS provided information to dispel many myths surrounding tropical hardwoods—especially regarding seed germination, growth, light requirements, soils, and wood properties (Butterfield and Fisher, 1994).

Seeds

Only 2 of the 26 species tested proved difficult to germinate when properly treated. The others had germination rates of 40 to 97 percent for fresh seed (E. González, 1991). Seed viability, however, is another issue. Most seeds prove difficult to store, and additional research into methods for storing these recalcitrant seeds is needed. New germinants can be "stored," however, in very low light. The seedlings remain alive in the cotyledon or first true leaf stages for many weeks under these conditions. This allows for seeds from many different individuals to be brought to the same height and stage of development in the nursery before outplanting to the field.

Growth

The growth potential of most tropical species is unknown. The age of large trees cannot be determined directly because they lack true annual rings. Growth rates have been measured in natural forests, but these data cannot be extrapolated to plantation conditions. Most seedlings grow in subdued light in natural forest, and many have said that tropical species do not grow well in full sunlight.

Also, TRIALS provided some of the first growth information for native species planted in full-sun conditions. Many species showed moderate height growth (1.5 to 2 meters per year), while several grew rapidly in diameter (3 to 4 centimeters per year). At the age of 3 years, *Vochysia guatemalensis* was 7.5 meters tall and 12.7 centimeters DBH (diameter at breast height), and *V. ferruginea* was 6.2 meters tall and 10.7 centimeters in DBH in the 0.25-hectare plantations of the soil improvements trial. This compared favorably with two commonly planted exotics—*Gmelina arborea*, with 8.3 meter height and 10.6 centimeters DBH, and *Pinus tecunumanii*, with 4.7 meter height and 8.8 centimeters DBH (González and Fisher, 1994). By the age of 6 years, *V. guatemalensis* had a mean annual increment (MAI) of 30.5 cubic meters per hectare per year and *V. ferruginea* a MAI of 19.5 cubic meters per hectare per year, compared with 17.4 cubic meters per hectare per year for *G. arborea* and 13.1 cubic meters per hectare per year for *P. tecunumanii.* In these plantations, only *Acacia mangium* grew more rapidly during the first six years than the two *Vochysia* species, and it died during the fifth and sixth years. Many native species can grow well on poor, degraded soils that are inhospitable for most exotics.

Light

The idea that tropical hardwoods require shade may be linked to the myth that they are slow growing. A more logical explanation might be that they are slow growing in the forest precisely because they are shaded. Many canopy emergents, large trees valued for their fine wood, are light-loving species that require a gap or opening to shoot up and compete successfully for light (Denslow, 1987). Thus, TRIALS showed that seedlings of *Dipteryx panamensis* and *V. guatemalensis* that were uprooted from the shaded forest floor and outplanted directly in pastures all died. However, seedlings of these same species reared in the nursery as containerized stock and then outplanted in pastures survived and grew very well.

Soil

Tropical hardwoods are found on all types of soil. Many, such as *Tectona grandis*, *Cedrela odorata*, and *Cordia alliodora*, prefer fertile alluvial soils. Species such as *V. ferruginea* and *V. guatemalensis* are adapted to infertile acid soils with high aluminum saturation, often toxic to other plants. Several widely planted exotics, such as *Eucalyptus* and *Gmelina*, are very sensitive to nutrient deficiencies. Eucalypts have been dubbed "prima donnas" because of their demanding soil requirements and site preparation needs (Zobel, 1988).

Wood Properties

The misconception that fast-growing trees produce low-density wood is common. By extrapolation, foresters argue that tropical hardwoods grown on a plantation will have inferior wood to trees grown in the natural forest. However, fast growth rates do not necessarily produce low-density wood (Zobel et al., 1987).

All trees produce juvenile wood for a given number of years, after which higher-density mature wood is formed. If plantation trees grow twice as fast as forest trees, a comparison of 16-centimeter DBH trees from the plantation and the forest is unfair since the plantation tree is half the age of the forest tree. The older (forest) tree might have 10 years of mature wood (i.e., 33 percent juvenile wood), while the younger (plantation) tree might have 100 percent juvenile wood and thus a lower wood density. A quality study conducted on *Hyeronima alchorneoides* from the TRIALS project showed no difference between plantation and natural-grown trees. A similar study of *V. guatemalensis*, which grows exceptionally fast on plantations, found incon-

Table 23.3 Native Species Showing Excellent Potential

Timber Species	Agroforestry Species	Conservation Species
Light to Medium Wood	Short Rotation	*Brosimum lactecens*
Cordia bicolor	*Jacaranda copaia*	*Castilla elastica*
Goethalsia meiantha	*Ochroma pyramidale*	*Cordia bicolor*
Jacaranda copaia		*Dipteryx panamensis*
Stryphnodendron microstachyum	Medium Rotation	*Genipa americana*
Vochysia ferruginea	*Calophyllum brasiliense*	*Goethalsia meiantha*
Vochysia guatemalensis	*Dipteryx panamensis*	*Hampea appendiculata*
	Laetia procera	*Hyeronima*
Medium to Heavy Wood	*Terminalia amazonia*	*alchorneoides*
Hyeronima alchorneoides	*Virola koschyni*	*Inga thibaudiana*
Laetia procera	*Alley Cropping*	*Laetia procera*
Nectandra membranacea		*Miconia multispicata*
Sclerolobium guianensis	Alley Cropping	*Minquartia guianensis*
Terminalia amazonia	*Goethalsia meiantha*	*Simarouba amora*
	Inga edulis	*Stryphnodendron microstachyum*
Unknown Wood Density		*Virola koschyni*
Albizia guachapele		*Vitex cooperi*
Pithecellobium macradenium		
Rollinia microsepala		

Source: Butterfield, R. P. and R. F. Fisher. 1994. Untapped potential: Native species for reforestation. *Journal of Foresty* 96 (6): 37–40.

clusive evidence of slightly lower wood density in very young trees (Butterfield et al., 1993).

Trees as Soil Improvers

Do trees improve the soil on which they grow? This has become a central question in agroforestry and ecosystem restoration efforts. Folk wisdom has long held that trees improve poor soil conditions, but science has recently questioned this possibility (Lundgren, 1978; Sánchez, 1987; Sánchez et al., 1985).

In temperate Europe and North America, old-field succession, resulting in reforestation and a protracted period of forest fallow, has restored good soil conditions and high productive potential to lands once badly degraded by agriculture (Billings, 1938; Griffith, Hartwell, and Shaw, 1930; Kittredge, 1948; Odum, 1960; Oosting, 1942; Ovington, 1958). Many processes by which this improvement takes place are well understood. However, the soils of these temperate sites are usually young soils with weatherable minerals and fixed-charge clays. Most of what we know about old-field successional amelioration does not transfer well to the tropics, where soils commonly contain no weatherable minerals and often have variable-charge clays.

The long-known practice of swidden agriculture, or shifting cultivation, takes advantage of the ameliorating effects of a period of forest fallow (Ewel, 1985; D. Janzen, 1973; Nye and Greenland, 1960). Gross changes in productivity and plant nutrition have been documented in shifting cultivation (Jordan, 1985; Sánchez, 1976). Nearly every researcher who has studied the effects of trees on soils has cataloged increases in soil organic matter and accompanying increases in hydraulic conductivity and cation exchange capacity during forest fallow. The importance of organic matter additions to degraded soils has been shown both in young temperate zone soils and in the old, highly weathered soils of the tropics (Nye and Greenland, 1964; Sánchez et al., 1985).

Nutrients also accumulate during a period of forest fallow. Both symbiotic and free-living nitrogen fixation may increase in the forest. However, much of the nutrient increase occurs because ions are gathered from a very large volume of soil. They are then concentrated in the surface soil through litterfall and fine root turnover. This ameliorative process works when the site has a large rootable soil volume and a measurable amount of nutrients (Golley et al., 1975; Jordan, 1985). Nutrient cycling is also very important. Successful forest fallows have very tight nutrient cycles. In addition, the decomposing litter layer and surface soil layers are intensely perfused with fine roots and mycorrhizal hyphae, which capture nutrients as they are mineralized and reduce leaching losses to near zero (Jordan, 1985; Vitousek and Reiners, 1975).

Trees can improve the site on which they grow in five general ways (Fisher, 1990). First, many species can increase soil nitrogen content by fixing atmospheric nitrogen through a symbiotic association with bacteria of the genus *Rhizobium* or actinomycetes of the genus *Frankia*. Second, trees' large root systems allow them to accumulate nutrients from a very large volume of soil. Litter fall returns these nutrients to a small volume of soil, increasing its fertility. Third, stands of trees may improve soil conditions by increasing organic matter content. Fourth, changes in above- and below-ground microclimate may ameliorate the site. Fifth, trees may alter their own site conditions through the "rhizosphere effect" (Barber, 1984; Fisher and Stone, 1968; Marschner, Romheld, and Cakmak, 1987). This is a complex series of phenomena that are not yet well understood.

The TRIALS soil improvement trial addressed four hypotheses about these mechanisms of amelioration:

1. Some trees can increase the nitrogen in the soil on which they grow (particularly N_2 fixers, e.g., *Acacia mangium, Inga edulis, Pentaclethra macroloba, Pithecellobium macradenium, and Stryphnodendron microstachyum*).
2. Trees can alter the quality and quantity of soil organic matter and change important associated properties such as bulk density, cation exchange capacity, aeration, and hydraulic conductivity (particularly heavy litter producers, e.g., *A. mangium, Gmelina arborea, Hyeronima alchorneodies, I. edulis, Pinus tecunumanii, Vochysia ferruginea,* and *V. guatemalensis*).
3. Trees can secure nutrients from a large volume of soil and concentrate them in a smaller volume of soil, leading to nutrient enrichment of the surface soil (particularly deeply rooted species, e.g., *P. tecunumanii, P. macradenium, Virola koschnyi,* and *V. guatemalensis*).
4. Trees alter soil microclimate by reducing the extremes of temperature and moisture and increasing aeration and redox potential.

Many soil properties were significantly altered after approximately three years of site occupancy by tree plantations. The effectiveness of different species in causing these changes varied widely. Only two N_2-fixing species, *I. edulis* and *A. mangium*, significantly increased the total N content beneath them. However, this does not necessarily mean that the other N_2-fixing species are ineffective. Both *I. edulis* and *A. mangium* grew very rapidly and had begun to break up by age three. Consequently, it is likely that more N mineralization was taking place beneath these species than beneath others. *Virola koschnyi* is not a N_2-fixing species, nor had it begun to break up by age three, but total soil N did significantly increase beneath it.

Soil organic carbon (C) decreased significantly over the three-year period beneath the pasture control even though the pasture was not grazed. Soil organic C increased significantly beneath *A. mangium, V. guatemalensis,* and *V. ferruginea*. The *A. mangium* plots had begun to experience considerable mortality as early as year two, and a large amount of organic matter, particularly fine and medium-sized roots, had been returned to the soil. Both *Vochysia* species are heavy litter producers, and this may account for the increase in soil organic C beneath them. Although *I. edulis* plots had also begun to deteriorate by age three, there was no significant increase in soil organic C beneath them. This may be due to the fact that a very vigorous native shrub community had begun to develop in these plots.

Although soil organic C changed under only three species and the control, soil bulk density changed significantly under all but the control and *H. alchorneodies* treatments. Bulk density increased significantly beneath *P. tecunumanii* and *G. arborea* and decreased significantly beneath all other species in the trial. The decrease in bulk density without any increase in soil organic C suggests that biological activity may have as much or more to do with changes in bulk density than does the incorporation of low-density organic material. We observed intense earthworm, arthropod, and small mammal burrowing in most forested plots. This activity was especially evident under heavy litter producers. Little such activity occurred in the control plots or under *P. tecunumanii*, which had not closed its canopy and still had a "pasture" understory after three years.

Soil acidity increased significantly beneath *P. tecunumanii* and decreased significantly beneath *G. arborea*. In all other cases it did not change. The decrease beneath *G. arborea* was possibly due to an increase in extractable calcium (Ca). However, the significant increase in soil acidity beneath *P. tecunumanii* was not accompanied by a significant decrease in extractable Ca. The change in acidity beneath *P. tecunumanii* is probably due solely to organic compounds released by the trees.

Extractable potassium (K) increased significantly beneath four species: *P. macradenium, S. microstachyum, V. koschnyi,* and *V. guatemalensis*. Three of these species—*P. macradenium, V. koschnyi,* and *V. guatemalensis*—are deep-rooted trees, which we expected might extract nutrients from a large volume of soil and concentrate it in a smaller volume of surface soil. All three of these species had captured the site by the third year and had essentially no understory and a well-developed forest floor.

Four species significantly increased the extractable phosphorus (P) in the soil beneath them: *I. edulis, V. koschnyi, V. ferruginea,* and *V. guatemalensis*. Phosphorus is tightly bound by iron (Fe) and aluminum (Al) in these soils. It appears that low molecular weight organic acids in the rhizosphere of *Vochysia* spp. release insoluble Fe, Al, and P into the soil solution. This activity is currently under study. Nitrogen also increased in the soil beneath *V. koschnyi*, and the increase in extractable P may be associated with the same rhizosphere activity that appears to produce this increase in N. The increase in extractable P beneath *I. edulis* could occur because self-thinning has occurred on these plots and considerable biomass has begun to be recycled. This species has very high foliar concentrations of both N and P, and recycling of these accumulated nutrients contributes to the high N and P content beneath them.

The tree plantations, particularly those with closed canopies, have obviously altered the microclimate within the stand. During both the wet and dry seasons, soil moisture was significantly higher at the 10-centimeters depth beneath the forest plots with a closed canopy than in the control or pasture plots. Soil moisture was significantly lower beneath the forest plots at the 30-centimeters depth during the dry season and at the 20-centimeters depth during the wet season. Although the trees are probably removing more water from the site than the pasture grasses, the increased moisture in the near surface soil beneath the trees appears to increase biotic activity in that zone. Lower temperatures may also favor this activity, but we have no data to substantiate temperature reductions.

This study set out to address four hypotheses. With respect to the first, increasing nitrogen, we can say that tree species that fix atmospheric nitrogen can significantly enrich the soil on which they grow in nitrogen. In this study, it appears that a good deal of mineralization of nitrogen derived from organic material produced by these N_2-fixing species is necessary before significant quantities of nitrogen appear in the soil's total nitrogen pool. However, several species did increase the total nitrogen. The N_2-fixing species did not appear to place excessive stress on the ability of the system to supply other nutrients in this study.

With respect to the second hypothesis, increasing organic matter, our study shows that trees can significantly increase organic matter in the soil beneath them, even in a short period. More important, this study shows that increases in soil organic matter are not necessary for there to be significant decreases in bulk density. We conclude that much of the initial decrease in bulk density that occurs when trees capture abandoned field sites is due to increased organismal activity, including root growth, burrowing by soil fauna, and the respiration of soil flora.

The third hypothesis, that deep-rooted trees may "pump" nutrients, was substantiated. As expected, *G. arborea* increased the Ca content of the soil beneath itself, and several species significantly increased the K content of the soil beneath them.

Although the fourth hypothesis, the alteration of microclimate, was not intensively studied, the changes in soil moisture content observed were significant. Once the trees had established a closed canopy, they provided a very different soil moisture regime for their associated flora and fauna. The closed plantations probably use more soil water than does the open pasture. However, the higher moisture content in the plantations' surface soil seems to favor the activity of soil fauna and flora. In turn, this increased activity seems essential in the improvement of soil conditions beneath the trees.

Native tree plantations established on degraded rainforest sites that have been in pasture significantly improve site and soil conditions (Fisher, 1995; Montagnini, 1990). Changes take place rapidly once the tree canopy closes. Some of these changes are mediated by the trees themselves, but many are mediated by the activity of fauna and flora associated with the plantation ecosystem. The exact nature of the processes involved in these changes is still unclear and deserves further study.

Plantations as Catalysts for Recovery of Biodiversity

In tropical regions, extensive degradation of forests and agricultural lands often has severe social and economic implications. Reforestation programs designed to reclaim degraded ecosystems and to satisfy local needs for forest products are likely to assume increasing importance. The traditional focus of tropical reforestation programs on maximizing productivity of planted trees for fuelwood or timber needs to be broadened if restoration forestry is to yield significant environmental and socioeconomic benefits. Recent experience in several tropical areas has shown that community-based reforestation projects aimed at the reestablishment of species-rich secondary forest on degraded land can be successful when traditional forestry practices are adapted to facilitate the regeneration of locally valued species. Forest plantations established on degraded sites long without native forest can act as catalysts for the recolonization of the site by native flora through their influence on microclimate, soil fertility, suppression of dominant grasses, and habitat for seed-dispersing wildlife (Lieth and Lohmann, 1993).

This catalytic effect of tree plantations appears to be a widespread phenomenon. A number of factors affect the rate of understory colonization by native tree and shrub species. These factors relate to initial site conditions, plantation design, and management practices. They include the degree of site degradation, particularly the availability of rootstocks and seedbanks; the proximity to native forest stands; the presence of seed-dispersing wildlife populations; tree species and spacing in the plantation; understory management practices; and the thinning regime (Parrotta, 1993). At present our knowledge of how these factors influence succession in plantation ecosystems is very limited. The TRIALS plots have been studied to help us gain an understanding of these factors in the catalytic effect of plantations.

In these 0.25-hectare plots, we sampled the tree understory after six years of plantation growth. We then determined species richness, the total number of species found beneath each species, and computed Shannon and Berger-Parker Diversity Index and the Shannon Evenness Index. A total of 68 species of trees in 27 families was found in the understory of the plantations six years after establishment. Roughly one-third of these species commonly occur in primary forest at La Selva (Hartshorn, 1983). Species richness, dominance, and evenness varied widely among the various plantation species (table 23.4). The understory beneath *V. guatemalensis* had the highest species richness (26). The highest Shannon Diversity Index (2.37), Shannon Evenness Index (0.78), and Berger-Parker Diversity Index (3.26) occurred beneath *V. ferruginea*. Six plantation species had significantly higher Shannon Indexes than the pasture control.

Even the highest values of the various diversity measures are not particularly high. However, there was a significant increase in diversity over that which would occur during normal successional conversion of the abandoned pasture back to forest in six of the plantations. The two *Vochysia* species and *A. mangium* were particularly effective in stimulating increases in species diversity. The two *Vochysia* species also have the highest number of understory species typical of primary forest, seven in each case. This compares with five beneath *S. microstachyum* and two or less in the control and beneath other species.

Clearly, monospecific plantations of trees enriched understory vegetation on these abandoned pasture sites. There was no significant relationship between species diversity and canopy coverage, our measure of light availability, or vegetation density. This was surprising, but it appears that light is not a particularly important determinant in the increase in species on these sites. Although we have no data to support our supposition, we believe that canopy structure may be the most important variable in this case.

Although some plant species in the study have wind-dispersed seeds, the seeds of the vast majority of plants in this area are dispersed by animals (Denslow and Hartshorn, 1994; Levey, Moermond, and Denslow, 1994). Many seeds are bird- or bat-dispersed. The two *Vochysia* species and *A. mangium* all quickly developed dense, closed canopies. These canopies also lifted quickly as the trees grew in height. The branching habit of these species is open, with long, nearly horizontal branches. This combination of a dense foliage coverage and open, horizontal branches should provide excellent perching and resting sites for birds and bats.

Soil improvement was also great beneath *Vochysia* and *A. mangium* (Fisher, 1995). Bulk density was significantly lower and extractable phosphorus was significantly higher beneath these species than in the pasture control. Total nitrogen also increased significantly be-

Table 23.4 Measures of Species Richness, Dominance, and Evenness of Woody Species beneath the Peje Plantations, La Selva Biological Station, Sarapiquí, Costa Rica

Treatment	Species Richness	Shannon Index[a]	Shannon Evenness	Berger-Parker Index
Vochysia ferruginea	21	2.37*a*	0.78	3.26
Acacia mangium	17	2.15*b*	0.76	3.19
Stryphnodendron microstachyum	20	2.05*b*	0.68	2.22
Vochysia guatemalensis	26	1.80*c*	0.55	1.71
Inda edulis	15	1.73*c*	0.64	1.93
Gmelina arborea	14	1.62*d*	0.61	1.65
Pinus tecunumanii	20	1.49*e*	0.54	1.63
Hieronyma alchorneoides	15	1.45*e*	0.54	1.88
Pasture control	12	1.42*e*	0.57	1.54
Virola koschnyi	14	1.13*f*	0.43	1.46
Pentaclethra macroloba	14	1.06*f*	0.40	1.36
Pithecellobium macradenium	9	0.67*g*	0.30	1.18

Source: J. Haggar, K. Wightman, and R. Fisher, 1997. The potential of plantations to foster woody regeneration within a deforested landscape in lowland Costa Rica. *Forest Ecology and Management* (in press).

[a]Values followed by the same letter are not significantly different at the 0.05 level.

neath *A. mangium*. Average soil moisture was significantly higher in the upper 10 centimeters of soil beneath the trees than in the control in both the wet and dry seasons. These conditions, coupled with greater propegule availability, may contribute to the increased vegetative diversity in plantations of these species.

Conclusion

Plantation forestry in the tropics is here to stay. Sometimes it will be solely a means of reclaiming degraded land, and the plantations will eventually become natural second-growth forest. This use of forest plantations will probably increase as we learn more about native species and as stability of land tenure encourages the conversion of degraded lands into natural areas.

In other cases, tropical plantations will be industrial forest plantations designed to meet the world's fiber needs. These plantations will use the species best suited to industrial needs. As more species are domesticated, the use of native species in these plantations will undoubtedly increase. The plantations will, however, probably continue to be intensively managed, monospecific stands grown in short rotations.

The increased domestication of native species will probably stimulate the development of a third scenario. Strong local markets in most tropical countries, coupled with the ability to grow native species, should stimulate the growth of a nonindustrial woodland owner class in tropical countries. The vast majority of forest land in the temperate zone is held by such owners, who will practice agroforestry, plantation forestry, or secondary natural forest management, depending on the size of their landholdings and their particular desires. They will plant and grow a diversity of native species on a variety of rotation lengths. This activity will provide a livelihood for many, increase biodiversity in the region, and relieve the pressure on the remaining primary, natural forest.

24

Challenges to Sustainability

The Central American Shrimp Mariculture Industry

Susan C. Stonich
John R. Bort
Luis L. Ovares

The sustainable agriculture movement was originally centered mainly on environmental conservation, farm-level production, and short-term economic strategies. Later, the movement's concerns expanded to encompass broader social, economic, and policy issues (Allen et al., 1992, 1993; Allen and Sachs, 1993; Allen and Van Dusen, 1990; Altieri, 1988). The risks and benefits to different interest groups and stakeholders in the food and agricultural system, defined by divisions of race, ethnicity, gender, and class, were analyzed. Social, economic and environmental justice were promoted. For example, *The Asilomar Declaration for Sustainable Development* called for agriculture to build and support healthy rural communities (Committee for Sustainable Agriculture, 1990). More recently, the University of California Sustainable Agriculture Research and Education Program (SAREP) defined sustainable agriculture as integrating "three main goals: environmental health, economic profitability and social and economic equity" (SAREP, 1991, 1993). According to these concepts, sustainable agriculture not only protects the environment but also meets the basic needs of all segments of society: rural and urban, farm workers and consumers, the poor, women, children, and people of color. A sustainable agriculture is "environmentally sound, economically viable, socially responsible, nonexploitative, which serves as the foundation for future generations" (Allen et al., 1993). The implications of this concept of sustainability for research include the development of interdisciplinary methodologies that reach beyond the conservation ethic. These methodologies must systemically link together the interrelated parts of the food and agricultural system at the local, regional, national, and international levels (Allen et al., 1993; Brookfield, 1991; Douglass, 1984).

Political ecology, which combines the concerns of ecology with political economy, is one such methodological approach (Blaikie, 1985, 1988; Blaikie and Brookfield, 1987; Little and Horowitz, 1987; Michael Redclift, 1984, 1987). Schmink and Wood (1987) suggest that political ecological analysis includes several crucial components: (1) Ideology orients resource use and affects which groups benefit and which are disadvantaged by different resource uses. (2) International interests such as donor agencies or private investors may support particular patterns of resource use. (3) The state's role includes defining and executing policies that favor the interests of some classes of resource users over others. (4) The class structure of the society and the lines of conflict over access to productive resources are important considerations. (5) The extent and kinds of market relations in which producers are involved determine the mechanisms by which production beyond that needed to satisfy consumption is extracted as surplus. 6) The nature of production in a region, especially the degree to which it is oriented toward simple reproduction or capital accumulation, is critical.

Political ecology emerged in the 1970s because of the need to link local land use practices to broader social, historical, and political-economic concerns (Wolf, 1972). To date, this is a loosely configured approach that encompasses the work of diverse scholars. Political ecologists have demonstrated how these interconnected social, economic, and political factors affect the way in which natural resources are distributed, managed, and exploited. They have shown how these social arrangements intensify human impoverishment and environmental destruction (see Bryant, 1992; Peet and Watts, 1993; Stonich, 1989, 1993, 1995). Political ecological analysis usually includes several scales or levels, often moving from the larger to the smaller scale. For example, analysis frequently begins with the roles of the global economy and the state, then focuses on the interrelations between groups in society who influence resource management and local resource managers, and finally examines the decisions of local resource managers.

Research in Latin America with a political ecology approach has centered on the human impoverishment and environmental destruction caused by transformations in agricultural systems (Painter and Durham, 1995). These studies have illuminated several issues that may explain deteriorating human and environmental conditions. First, environmental destruction associated with the production systems of smallholders (farmers and others) is most often a result of their impoverishment, in either absolute terms or compared with other social classes. Commonly, impoverishment is accompanied by diminished access to land or other natural resources and increased repression and violence at the hands of state authorities and more powerful individual and corporate interests engaged in land speculation. Second, smallholder producers often receive a disproportionate share of the blame for environmental destruction because of their vulnerability and lack of power (Stonich, 1989). In contrast, political ecological research shows that the activities of more powerful private, public, and corporate interests have degraded a great deal of land and other natural resources (Stonich, 1989, 1993). Destructive large-scale enterprises are usually those that use lands gained through concessions from the state and exercise sovereignty over the area in which they operate. This allows them to treat land as a low-cost input. As a result, it is more economical to move elsewhere after the environment is degraded than to try to conserve natural resources. Third, the same policies and practices that permit wealthy interests to receive concessions are also responsible for impoverishing smallholders because these policies institutionalize and exacerbate unequal access to resources. In sum, the political ecology approach applied in Latin America shows that the crucial issue that underlies environmental destruction and continuing human impoverishment is socially institutionalized inequality in the access to resources.

This chapter uses a political ecology approach to examine the environmental and social equity implications of the shrimp mariculture industry in Central America, based on the definition of sustainable agriculture given earlier. It uses the results of a comparative study of the industry conducted from 1991 through 1994 in Honduras, Panama, Costa Rica, and Nicaragua.[1] Shrimp mariculture, the cultivation of various species of shrimp in coastal, brackish ponds, is an important development strategy in many tropical areas of Asia and Latin America today. International agencies and national governments have promoted it because of its potential to augment export revenues while alleviating rural poverty through enhanced income and access to animal protein.

This chapter evaluates this development strategy, focusing on global-to-local linkages and the social and environmental consequences of increasingly capitalized and intensive maricultural production systems. First, it discusses the growth of shrimp mariculture as a component in restructuring agricultural exports from Central America and in the globalization of the industry. It then considers the political ecology of shrimp maricultural development in Central America, emphasizing the organization of the industry and the significant constraints it presents to small, resource-poor producers. The chapter goes on to examine three additional aspects of the political ecology of the industry that threaten long-term sustainability: (1) population growth and common property resource management regimes, (2) competing stakeholders and environmental justice, and (3) environmental quality and biodiversity. The chapter concludes that like previous agricultural development schemes in the region, shrimp mariculture in its present form is socially, economically, and environmentally nonsustainable.

The Restructuring of Central American Agroexports

Central American countries faced several grave facts as they entered the final decade of the 20th century. Extensive poverty remained, and the region's biophysical resources were being expended at an accelerating rate, despite the millions of dollars that financed an abundance of development projects in the previous decades. Central American nations had to rebuild their economies within a framework of crisis. Foreign debt was high. Their economies depended on the export earnings of a few primary commodities, and world prices for traditional exports were stagnant or declining. A tradition of violence was apparent in the persistent conflict in the region. The development models of the past had failed, leaving behind them undiversified economies, increased poverty, a severely disturbed environment, and unstable political systems. Moreover, although declining, population growth rates in the region remained high, making efforts to combat growing poverty and environmental destruction even more complicated (Stonich, 1993, pp. 3–8).

Central American governments turned to market-based economic policies in response to these crises. They believed that augmented foreign investment and agricultural trade would promote economic growth and ease the continuing debt crisis. They responded to pressures from international donors and lending institutions by enacting structural adjustment policies that encompassed the agricultural sector. These policies are connected to international free trade initiatives; attempts at regional economic integration among the Central American nations; and specific programs that provide preferential access to U.S. markets, such as the Caribbean Basin Initiative. The expansion of so-called nontraditional agricultural exports (NTAE), promoted by international financial and development agencies, is basic to market-oriented development approaches. Advocates maintain that this expansion will also help rural economies and the poor.[2]

Central American governments are looking for ways to generate larger amounts of foreign exchange than can be generated through traditional exports. Their strategies include increasing the nontraditional export sector and diversifying their export base. Compared with traditional staple foods such as food grains, legumes, roots, and tubers, NTAEs have higher unit values and much higher-income elasticities of demand. While representative world prices for maize, sorghum, and wheat have ranged from $75 to $175 per metric ton in recent years, prices for NTAEs have averaged $500 or more per metric ton (Jaffe, 1993). Foreign exchange earnings from nontraditional agricultural and nonagricultural exports were valued at $423 million in 1983 and accounted for approximately 12 percent of total export earnings from the region. In 1990, the value of these exports had risen to $1.3 billion in 1990. They have the potential of reaching $4 billion by 1996, making up about 50 percent of total export earnings (Stonich, Murray, and Rosset, 1994).

The NTAEs are often judged a success in terms of such economic indicators as diversifying economies, augmenting export earnings, increasing profitability, and generating jobs (Jaffe, 1993; Tabora, 1992). National economies have reduced their dependence on traditional export commodities. Some NTAE agribusinesses have been highly profitable, and many NTAE crops are labor-intensive and have generated jobs, particularly for women. In addition, international consumers have increased access to a wide array of new products at low prices.

However, while the NTAEs have certain undeniable benefits, they also incur serious costs. Evidence from several countries throughout the region reveals that the production, marketing, and export of NTAEs entail considerable economic risks and have significant environmental and social costs, especially for the poor. In addition, the benefits of NTAEs are often inequitably distributed. They are frequently enjoyed primarily by large farmers, wealthy investors, and foreign distributors because poor farmers cannot effectively compete in the market. Moreover, NTAEs may reduce local food security by competing with the production of local food crops.

Several researchers remark on the irony of promoting the production of specialty foods and flowers, consumed primarily by the national middle class and wealthy foreigners, while hunger and environmental destruction persist in the areas where they are grown. Their studies reveal that several factors characteristic of NTAE production adversely affect small farmers and hurt the rural poor. These include rising rents and land values, inadequate access to credit and capital, government policies for prices and subsidies, inadequate access to technology and technical assistance, and unfavorable insertion into the market. These studies also describe striking parallels in the social processes that accompany the recent boom in nontraditionals and the earlier expansions of other export commodities. They suggest that the expansion of NTAEs may re-create or exacerbate the social and ecological crises that resulted when commodities like coffee, cotton, sugar, and livestock were promoted in the past, and serious concerns have emerged (Barham et al., 1992; Conroy, Murray, and Rosset, 1994; Murray, 1991, 1994; Murray and Hoppin, 1992; Rosset, 1991; Stonich, 1991, 1992, 1993, 1995; Stonich, Murray, and Rosset, 1994; Thrupp, 1995). Shrimp mariculture may increase inequities in the access to resources and potential benefits from development. There are questions about whether real increases in household income have resulted, the marginalization of small producers, and the position of women workers in the NTAE processing plants. Finally, concerns also have been raised about the excessive use of pesticides and increased destruction of the natural resource base.

The Blue Revolution in Central America

Aquaculture has been called the "Blue Revolution," the counterpart to the Green Revolution in agriculture (Rubino and Stoffle, 1990; Weeks and Pollnac, 1992).[3] The comparison is apt because there are many similarities between the two. Aquaculture holds the promise of boosting the production of aquatic food species at a time when wild stocks of these species are dwindling. In addition, just as the Green Revolution was extolled as the means to end world hunger, the Blue Revolution is often touted as a way to increase the available supply of affordable food, especially for the poor in developing countries. However, it also has the potential of generating social dislocation, ecological change, and environmental destruction comparable to those caused by many Green Revolution technologies (Stonich and Bort, 1995). Further, just as the Green Revolution was necessary to establish the present global agricultural and food system, the Blue Revolution is essential in integrating many important aquatic species into that global system (Stonich and Bort, 1995).

Weeks (1992) and Weeks and Pollnac (1992, pp. 1–2) provide a useful classification of intensive, semi-intensive, and extensive shrimp production systems. Overall, intensive systems involve the most technology and are the most capital-intensive. Hatchery-raised stock, pumps, elaborate ponds, raceways, hormone treatments, and feeding systems are used to increase production in a given area. Semi-intensive systems achieve a lower level of production but involve a smaller capital investment. They frequently use both hatchery-reared and wild, captured stock and incorporate some pumping, pond fertilization, and some feeding (table 24.1). Extensive systems involve little more than digging ponds or blocking seasonal lagoons and stocking them with wild larvae. They involve a minimum of inputs. More elaborate classifications are possible, but this differentiation illustrates the potential range of options.

Aquaculture has also been promoted in the developing world on the same basis as many NTAEs. It is seen as an important strategy for increasing employment and income for the rural poor by using locally available labor and resources (Stonich, Murray, and Rosset, 1994). Aquaculture may be important in improving nutrition and augmenting employment and incomes for the rural poor in the developing world. However, the explosive growth of export-oriented, extraordinarily profitable, and capital-intensive shrimp mariculture, principally in tropical areas of Asia and Latin America, has hindered this potential. During recent years, total production of cultivated shrimp grew more than any other aquacultural product worldwide (FAO, 1993a). Although 99 percent of cultivated shrimp are raised in the developing world, they are not consumed in the countries where they are raised. Nearly all production is exported to developed countries, primarily Japan, the United States, and western Europe (computed from data in FAO, 1994b, pp. 406, 408). Unless increased shrimp production can produce significant increases in employment and income for the rural poor, improvements in their standard of living are unlikely.

The success of shrimp as a high-value international product and the reports of commercial success in other environmentally similar areas in Asia and Ecuador stimulated a great deal of interest in shrimp mariculture throughout Central America in the 1980s. The area suitable for shrimp cultivation is estimated at more than 100,000 hectares, nearly equal the area presently used for ponds in Ecuador, Latin America's largest producer (Weidner et al., 1992, p. 328). Commercial operations exist in every Central American country, based almost exclusively on Pacific white shrimp, *Panaeus vannamei*, and, to a lesser degree, on *P. stylirostris*. More than 90 percent of farms are found along the Pacific Coast, but expansion to the Caribbean Coast is imminent (Weidner et al., 1992, p. 147). A variety of farming systems have been tried, ranging from simple estuarine enclosures, used mostly by small-scale producers and cooperatives in Nicaragua and Honduras, to intensive systems most common in Guatemala. Most commercial enterprises are turning increasingly to higher-yielding semi-intensive systems based on a combination of wild and nursery-bred seed stock (Stonich and Bort, 1995).

Central American countries began to advance shrimp mariculture as a way to ameliorate the region's persistent economic problems during the 1980s but paid little attention to its long-term economic, social, and environmental effects. Between 1980 and 1992, cultured shrimp harvests in Central America grew from approximately 200 metric tons to more than 20,000 metric tons (table 24.2). This was an increase of 10,000 percent, with 75 percent of the increase occurring after 1985 (Stonich and Bort, 1995). By the mid-1990s, Honduras led the region with the production of about 8,000 metric tons, or about 50 percent of total production in the region. Panama followed with approximately 4,000 metric tons (Stonich and Bort, 1995). In Honduras, where

Table 24.1 Characteristics of Shrimp Culture Production Systems

Characteristic	Extensive	Semi-intensive	Intensive
Pond size (ha)	10–>100	5–20	0.5–2.0
Seedstock source	Naturally present	Wild/hatchery	Hatchery
Nursery ponds	No	No/yes	No
Stocking density (PL/m^3)	0.3–5.0	5–20	>10
Water management	Tidal	Pump	Pump/treatment
Exchange (%)	2–5	5–15	10–50
Aeration	No	Some	Yes
Food type	Natural food	Supplemental feed	Formulated feed
Rates (% biomass)	—	3–5	3–5
Crops per year	1–2	1.5–3.0	2–3
Yield (mt/ha/year)	0.3–0.8	0.8–5.0	5–12

Source: Weidner, Revord, Wells, and Manuar, 1992, Appendix D2, p. 152.

Note: In reality, categories are not discrete and represent a continuum.

Table 24.2 Central American Cultured Shrimp Harvests, 1980–1991 (in 1,000s of metric tons)

Country	1980	1981	1982	1983	1984	1985	1986	1987	1988	1989	1990	1991
Honduras	([a])	0.1	0.2	0.3	0.5	0.6	1.3	1.9	2.4	3.2	3.2	4.5
Panama	([a])	([a])	0.8	0.8	1.5	2.6	3.0	2.8	3.5	3.5	3.5[c]	2.7[c]
Guatemala	—	—	([b])	0.1	0.3	0.5	0.6	0.8	0.8	0.8	1.8	2.4[c]
Costa Rica	—	—	—	—	—	([b])	([b])	([b])	0.1	0.1	0.4	0.6
Nicaragua	—	—	—	([b])		([a])	([a])	([a])	([b])	0.1	0.2	([a])
Belize	—	—	—	—	([a])	([b])	([b])	([b])	([b])	0.1	0.2	([a])
El Salvador	([a])	([a])	([a])	([a])	([a])	([a])	([a])	([b])	([b])	0.1	0.1	0.1
Total	0.2[c]	0.5[c]	1.1	3.6	6.3	7.1	8.2	9.5	13.0	12.6	17.2	19.6

Source: Adapted from Weidner, Revord, Wells, and Manuar, 1992, p. 335.

[a]Not available or not applicable.
[b]Negligible.
[c]Estimated.

virtually all shrimp are exported, foreign exchange earnings from shrimp, of which approximately 70 percent were produced on farms, were exceeded only by export earnings from bananas and coffee by 1987 (Stonich, 1991, 1992). In Panama, export revenues from shrimp have been second only to bananas for several years (Europa Publications, 1995, p. 508). Except for these two nations, shrimp mariculture in Central America is limited, and major growth is projected through the 1990s.

International development assistance organizations, foreign and national entrepreneurs, and government agencies all strongly promote expansion. Enhancing nontraditional exports, improving hard currency earnings, and providing local employment opportunities are cited as potential benefits. In nations with balance-of-trade difficulties and high rural unemployment rates, the short-term economic and political promise of shrimp aquacultural expansion has a seductive appeal. Most Central American nations have not yet felt the negative environmental and social consequences of maricultural development experienced in Asia and Ecuador.[4] However, awareness of these very serious problems is increasing, as is concern for finding ways of avoiding and rectifying similar situations in Central America (Stonich and Bort, 1995).

The Political Ecology of Shrimp Maricultural Development

Environmental conditions favorable to shrimp cultivation are found in all Central American countries. Although some variation occurs, the areas appropriated for shrimp farms are usually relatively dry, low-lying zones with fringing mangroves and good tidal movement. Natural tidal movement, sometimes supplemented by pumping, provides critical water flow in impoundment tanks.

The technical requirements for all of the production areas are very similar. Pond development requires the excavation of a shallow depression from 1.0 meters to 1.5 meters deep. The excavated material is used to form a low earthen dike around the pond. The pond is linked to the ocean or brackish estuary by a shallow canal that leads to sluice gates in the dike, which permit water to flow into and out of the pond on high tides. Diesel-powered pumps are frequently incorporated into the design to add water when adjustments in the salinity levels in the pond are required. Construction is usually done with bulldozers, although dragline cranes occasionally dig some channels.

The similarities in biophysical environments and technology used for shrimp production suggest that differences in the extent of shrimp farm development between countries may be due primarily to varying social, political, and economic contexts. Some differences are readily apparent. Nicaragua went through a protracted period of economic and political turmoil in the 1980s. This strife was not conducive to economic development in the border region near the Gulf of Fonseca, the area of the country most suitable for shrimp cultivation. Similarly, the best areas in Honduras are in the south of the country near the Nicaraguan and Salvadoran borders, a militarily sensitive zone until the end of the 1980s. The forces that have helped or hindered development in Panama and Costa Rica are less apparent. A complex, slow government bureaucracy in Costa Rica has delayed shrimp aquacultural development. The country also experienced a severe economic crisis in the early 1980s, causing a great deal of economic uncertainty. Costa Rica's more rigorous environmental regulations, involving the siting of ponds, may be an additional factor in slowing development.

The general socioeconomic environment in Panama was the most conducive to entrepreneurial efforts in the 1970s. Comparatively simple import and export regulations, modest import duties, a good supply and transportation infrastructure, the use of the U.S. dollar as the local currency, and a banking system that made venture capital obtainable, all contributed to comparatively early development. However, economic difficulties in the early 1980s and the severe political and economic problems at the end of the decade have negated many early advantages.

Presently, only a few intensive systems exist in Central America. The majority of these are in Guatemala and report relatively high yields (Weidner et al., 1992). Most of the shrimp ponds in Central America are semi-intensive, and the proportion is increasing. The investment even for semi-intensive systems is substantial, however. The excavation of ponds usually costs around $5,000 per hectare and in some areas has been reported to be as high as $10,000. The costs of diesel engines and pumps can exceed $25,000 for a small 25-hectare system of ponds.[5] Ongoing diesel fuel, fertilizer, and feed costs drive the total expense up to more than $1 million for a medium-sized semi-intensive operation.

This large capital needed for pond development is a crucial factor in shaping aquacultural development throughout Central America. Shrimp ponds require investments generally beyond the reach of all but the wealthiest.[6] Wealthy elites have capital and the power and sophistication to seek external investors and obtain expert technical assistance. They also have the political power to deal with the morass of property right laws, permit systems, and import and export regulations that affect an aquacultural operation.

Producers interested in developing smaller-scale, extensive systems face a host of problems. Pond development is beyond their limited economic resources, and credit is extremely difficult to obtain. In addition, they usually lack technical sophistication. As a consequence, results of their efforts to increase production through fertilization or supplemental feeding are poor. They are typically at a disadvantage in the political arena, lacking the connections to obtain legal concessions to what are often public or state-owned areas.

The consequences are apparent in Panama, where small-scale producers first entered shrimp aquaculture. Although initially successful, they are now going out of business (Weidner et al., 1992, pp. 593–594). Bort and Sabella (1992) attribute the failure of the small-scale cooperative ventures they studied to factors both internal and external to the operations. They question the ability of resource-poor cooperatives to achieve economies of scale. Other factors also undermined success: shrimp prices fell, small producers lacked experience in dealing with the complexity inherent in even small-scale enterprises, personal conflict among members was disruptive, and U.S. military action interrupted normal activities.

Except for Nicaragua, where some small cooperatives have tried to establish shrimp farms, few small-scale farms are presently being developed. Whatever the country, four factors impinge on the success of small farmers: (1) government concession, pricing, and subsidy policies; (2) inadequate access to capital and credit; (3) inadequate access to technology and technical assistance; and (4) unfavorable insertion into the market. These are the same constraints faced by small farmers throughout the developing world, whatever they seek to produce. It is paradoxical, for example, that small producers and cooperatives in Nicaragua and Honduras face similar constraints despite more than a decade of widely divergent political and socioeconomic regimes. Moreover, given the continuing political and economic instability in the region, controlling all of the factors that affect success is impossible. In this context, the failure of small-scale maricultural projects frequently results in the transfer of natural resources and enterprises from poor social groups to elites or foreign interests (Bort and Sabella, 1992). These same four factors underlie the appropriation of maricultural enterprises by elites or foreign interests in Mexico (Cruz, 1992), the Philippines (Pomeroy, 1992), and Indonesia (Bailey, 1992).

Although the major tendency in shrimp maricultural development in Central America appears to be the establishment of larger farms, using semi-intensive methods, many uncertainties remain. Among the greatest are environmental risks. Disease and climatic fluctuation significantly affect production levels. Diseases in high-density tanks are particularly pernicious and often difficult to control and have been a serious problem in parts of Asia. Outbreaks of *vibrio* have already been reported in Central America. Especially troubling is the recent spread of the *Taura* virus, first reported in Ecuador in 1992 and spreading rapidly throughout the shrimp-farming region. Between 1993 and 1994, exports of shrimp to the United States from Honduras plummeted from 7,300 metric tons to 2,300 metric tons, largely because of the impact of the *Taura* virus (Lara, 1995). How well *Taura, vibrio*, and other diseases can be controlled is uncertain.

Environmental contamination can also present nearly uncontrollable problems. *Penaeids* have a very low tolerance to many insecticides used in agriculture. Such chemicals are already widely used in Central America, and there is concern about their use in areas next to shrimp farms (Murray, 1991; Stonich, Murray, and Rosset, 1994). The increasing use of pesticides is a trend that is very likely to continue in the future. Curbing their use near shrimp farms is unlikely because of the continuing economic importance of agriculture and the political power of agricultural interests. Environmental contamination from various diseases, agricultural pesticides, and other sources raises serious long-term quality control concerns for producers and exporters, as well as for consumers.

At a macroeconomic level, shrimp aquaculture may appear to be beneficial. The operations are export-oriented and generate badly needed foreign exchange for financially hard-pressed governments. As production increases and exports grow, more money is made and the government obtains more revenue. However, the benefits are more dubious when studied at a microeconomic level and evaluated in terms of resource allocation, environmental quality, and equity. The lion's share of the financial rewards accrues to capital, and a few wealthy national and international investors benefit. In the areas near the shrimp ponds, a very modest number of low-paying, often seasonal jobs on the farms or in the processing plants is the only real benefit. In exchange for these jobs, natural resources previously available in the areas converted to ponds are forfeited. At the same time, uncalculated environmental risks are assumed.

Other Factors Affecting the Sustainability of Mariculture

Several other interrelated aspects of the political ecology of the shrimp mariculture industry are especially relevant to its long-term sustainability. Four factors are especially important: (1) Population growth in coastal zones is increasing rapidly. (2) Common property resource management regimes that help provide livelihoods to the rural poor are collapsing. (3) The actions of competing stakeholders in the industry threaten long-term stability. (4) Negative effects on the biophysical environment, especially declines in environmental quality and biodiversity, are evident.

Population Growth and Common Property Resource Management Regimes

Various population dynamics are among the factors that are significantly jeopardizing coastal environments in Central America. Annual population growth rates in the region remain among the highest in the world, and socioeconomic conditions continue to deteriorate. Unsustainable forms of agriculture and forestry reduce the capacity of inland ecosystems to provide an adequate quality of life, resulting in an increasing movement of people to the coasts. Coastal cities and towns are now among the most rapidly growing in the region. The few remaining sparsely populated and remote coastal regions are among Central America's last settlement frontiers (Foer and Olsen, 1992).

For example, in southern Honduras, the strategies of poor households dislocated by the earlier expansion of cotton, beef cattle, and other commodities included relocation to the sparsely populated coastal region of mangrove, mud flats, estuaries, and seasonal lagoons along the Gulf of Fonseca. Unsuitable for large-scale cultivation of crops, pasture, or most other commercial agricultural crops, migrants from other highland and lowland municipalities in the south increasingly settled in this area. The families who were settling the coastal communities survived by exploiting the resources of the coast and the estuaries. They cleared areas to cultivate crops, but they also came to depend on fish, shrimp, shellfish, animals, and wood gathered from the surrounding common resource areas—lagoons, mangroves, estuaries, and the Gulf of Fonseca. Until the expansion of the shrimp industry in the early 1980s, the only major competition for these coastal resources was from commercial salt-making enterprises.

There is no reason to assume that the current management practices of these recent immigrants are "sustainable" in their own right. The humans living in the coastal zone are not a small group of remote tribal people with long-term, well-regulated social and cultural institutions to help manage the common areas in sustainable ways. To the contrary, many are recent settlers whose economic strategies appear to center around the household rather than the community and whose desperate attempts to eke out a living may contribute to environmental damage. Migration to these zones has resulted in overlapping, sometimes competing patterns of resource use. Some center around control by the community (i.e., common property resource management regimes) and some by the household (Stonich, 1995).

Contending Stakeholders and Environmental Justice

The expansion of the shrimp industry involves many diverse social actors with competing stakes in the industry. These global to local stakeholders include transnational corporate investors; international development banks; multilateral and bilateral international development assistance organizations; national government agencies; national elites; international, national, and local nongovernment environmental and social organizations; local communities; the tanning industry; artisanal and commercial salt producers; woodcutters; artisanal estuarine and near-shore fishers; commercial off-shore fishers; commercial farmers; agrarian reform communities; wild larvae collectors; women workers in the processing plants; and the growing number of landless peasants. The more powerful of these actors have begun to transform the social structure, as well as the ecology, of coastal zones where shrimp farms are constructed (Stonich, 1995).

International and national agencies and investors have not considered the local repercussions of the expanding shrimp industry to any significant extent, especially repercussions from the diverse and contending human actors involved. Local environmental consequences have received more attention, primarily in reference to the industry's long-term ability to generate foreign exchange rather than its ability to provide income and food for local residents. Clearly, local people who live in coastal communities are losing access to common property resources because of government concessions and other policies. It is less clear that these people are reaping adequate benefits through enhanced employment in the industry to offset their losses. To some extent, economies of scale operating in regard to access to concessions, land, credit, technical assistance, and marketing effectively constrain small producers and cooperatives. They favor highly capitalized investors, increasing unequal access to land resources. In Honduras, increased marginalization of the rural poor has led to increasingly violent conflict among regional stakeholders in the shrimp industry (artisanal fishers, shrimp farmers, larva gatherers, etc.), flagrant harassment, death threats, imprisonments, and murders (Stonich, 1995). Social resistance includes the formation of social and environmental movements, especially the Committee for the Defense and Development of the Gulf of Fonseca (Stonich, 1991, 1993, 1995).

Environmental Quality and Biodiversity

The conversion of mud flats, mangroves, and estuarine areas into shrimp ponds thwarts the access of local groups to the natural resources on which they depend for their livelihoods. It also results in serious environmental transformation and destruction. In their dominant, semi-intensive form, the shrimp ponds are a premier example of reduction in biodiversity. Multiple habitats that contain innumerable species are converted into an altered artificial habitat designed to be optimal for one species—shrimp.[7] Biodiversity in marine areas is reduced as well because mangrove ecosystems are nursery areas for many marine species.

Shrimp farms also raise important environmental quality issues. Many farms are related directly or indirectly to decreases in biodiversity in mangroves and adjacent coastal wetlands (Stonich, 1995). Hydrological systems may be disrupted because of the obstruction of water flow and sedimentation. Overstocking of ponds and excessive cropping may not allow adequate recuperation between harvests and discharges of shrimp farm effluent may diminish water quality. Capture of wild postlarvae and the associated indiscriminate introduction of hatchery-raised seed stock may threaten natural populations. Near shore fisheries may be reduced because of the collection of wild shrimp seed stock and loss of by-catch. Destruction and transformation of habitats, especially seasonal lagoons, and the antipredator measures taken by farmers may affect populations of migratory birds, reptiles, amphibians, and aquatic mammals. Purportedly, wetlands are contaminated by pesticides used by shrimp farm owners to kill nonshrimp species in the ponds.

The current trend seems to be to establish larger farms, which require high-energy inputs to maintain farming systems that have enormous environmental impacts, or "footprints" (Larson, Folke, and Kautsky, 1994). The environmental footprint of a semi-intensive shrimp maricultural operation on the surrounding area is estimated to be 35 to 190 times larger than the actual area under cultivation. A typical farm uses 295 joules of energy for each joule of energy produced, and 80 percent of the energy required comes from outside the farm. Larson concludes that shrimp farming is one of the most resource-intensive food-producing systems and labels it an ecologically unsustainable "throughput" system in its present form (Larson, Folke, and Kautsky, 1994).

Threats to mangroves and other coastal wetlands vary significantly throughout Central America. In the last 30 years, mangrove habitats have been reduced by more than 50 percent in some areas, and the rate of destruction is accelerating (Foer and Olsen, 1992). Regionally, the precise extent to which the reduction in mangroves is attributable to the construction of shrimp farms is unknown. For the industry leader, Honduras, the land in shrimp farms increased from 1,064 hectares to 11,515 hectares between 1982 and 1992. During the same period, the area in high-quality mangroves declined 17 percent, from 28,776 hectares to 23,937 hectares (a decline of 4,837 hectares). Forty-four percent of this decline (2,132 hectares) was due directly to the construction of shrimp ponds. An additional 2,174 hectares of dwarf or stressed stands of mangroves associated with salt flats were also transformed into shrimp farms. Thus, approximately 4,306 hectares (37 percent) of the total area in shrimp farms in 1992 had been constructed in areas previously in mangroves. Nearly all the remaining area in farms (7,209 hectares, or 63 percent of the total area in farms) were constructed in salt or mud flats (Stonich, 1995; Vergne, Hardin, and DeWalt, 1993). Although less than half the decline in high-quality mangroves can be directly attributable to shrimp farm construction, an equal area of dwarf or stressed stands of mangrove and significant areas of mud flats were also destroyed. Should the remaining 20,000 hectares of shrimp farm concessions be developed, the destruction of stress, dwarf, and mature mangroves will be more serious.

Although significant, the conversion of mud flats and mangroves to shrimp farms is only one of many land and resource uses that cause environmental destruction in coastal zones. Additional threats include population growth; urban expansion; tourism development; increased construction for infrastructural improvement (roads and port facilities); cutting mangroves for fuelwood and for making charcoal; harvesting bark for the tanning industry; clearing to construct salt ponds; pesticide residues and agricultural runoff, especially from the melon and cotton areas of the Pacific coastal plain;

and growing sediment loads that result from highland erosion. Several of these additional, so-called "threats" to mangroves also make vital contributions to the household incomes of the rural poor, revealing potential conflicts between conservation efforts and livelihood.

Conclusions

During the post–World War II period, the dominant development model in Central America has been the promotion of a series of export commodities. Their production has altered the regional ecology while diminishing access to common property resources (forests and rangelands) for most people. Export commodity production has contributed to tragic declines in environmental quality and escalating social injustice (Williams, 1986). The current promotion of NTAEs, especially shrimp, has the potential to expand environmental destruction into zones that previously had little perceived economic worth. Shrimp farming may also restrict access to the last remaining common property resources (especially coastal areas, fisheries, and surface and groundwater areas).

This chapter challenges the sustainability of the shrimp mariculture industry, especially the extent to which it can relieve poverty and environmental destruction through enhanced participation of small producers or augmented employment opportunities. The economic benefits of the shrimp industry to local communities vary significantly among the four countries studied. Nonetheless, in each country small private producers and cooperatives face decisive constraints on their effective integration into the industry. In its initial stages the industry provided many, but violently contested, low-paying, seasonal jobs to poor rural people. However, these jobs will probably decrease as pond construction peaks and as the industry seeks greater competitiveness in the global market by using hatchery-raised rather than wild-caught postlarval seed stock. Furthermore, the industry does not offer an environmentally sustainable development alternative. To the contrary, the expansion of shrimp mariculture is an important factor in the decline in biological diversity of coastal wetlands, especially along the Gulf of Fonseca in Honduras (see also Foer and Olsen, 1992). Moreover, like previous attempts to promote the production of agricultural export commodities, the expansion of shrimp farms is provoking dislocation, marginalization, and violent conflict. Honduras, often extolled as the most successful country in the region in establishing a shrimp mariculture industry, is also distinguished by having the greatest resistance and social conflict associated with the industry. The future widespread expansion of the Honduran model throughout the region could have devastating environmental and social costs.

The present model of shrimp farming promoted in the region offers significant challenges to social, economic, and environmental sustainability. Broadly conceived, sustainable maricultural development would ensure equitable opportunities for poor farmers and rural workers, guarantee food security, and promote aquacultural practices that are economically viable and environmentally sound. As with other NTAEs, some aspects of shrimp mariculture are beneficial, at least to some individual and corporate stakeholders. The significant and incompletely known social and environmental costs, however, challenge the sustainability of the existing strategy. These unknown costs show the need to integrate social, environmental, and more long-term economic considerations into agricultural and aquacultural development policy.

Undoing nonsustainable maricultural development policy and replacing it with a sustainable policy is an immense task, made more difficult by considerable political opposition, the competing interests of stakeholders, uncertain knowledge, other kinds of development in coastal zones, and population growth. The recent proliferation of grassroots groups, the emergence of social and environmental movements of the poor, and the increased attention paid to native and indigenous knowledge at the local level are positive signs and must become vital components of such efforts. The broader political ecology within which agricultural development policy and resource management take place, however, is also very important. If that continues to be based on greed and to encourage exploitation of people and biophysical environments, nothing will improve. Reform at the level of society and the state is essential to a sustainable maricultural development as well.

Acknowledgments Research for this article was funded by the University of California Pacific Rim Research Program and by the University of California Academic Senate. The authors thank the many individuals who helped field research in Central America, especially Becky Myton, Denise Stanley, and Hania Vega.

Notes

1. The research utilized a combination of ethnographic, survey, and geographic methods. Further information on the methodology and results is available from Susan Stonich.

2. Nontraditional export crops include many fresh, frozen, processed, and otherwise preserved fruits and vegetables (e.g., melons, miniature papayas, mangos, snow peas, broccoli, and eggplants), root crops, edible nuts, live plants and cut flowers, and the most commercially desirable species of crustaceans and mollusc—especially shrimp and lobster (Paus, 1988). For a discussion of the growth of nontraditional agricultural exports from Central America to the United States, see Stonich, Murray, and Rosset, 1994.

3. For statistical purposes, FAO defines aquaculture as "the *farming* of aquatic organisms, including fish, molluscs,

crustaceans, and aquatic plants. Farming implies some form of intervention in the rearing process to enhance production, such as regular stocking, feeding, protection from predators, etc. Farming also implies individual or corporate ownership of the *stock* being cultivated. For statistical purposes, aquatic organisms which are harvested by an individual or corporate body which has owned them throughout their rearing period contribute to *aquaculture*, while aquatic organisms which are exploitable by the public as a common property resource, with or without appropriate licenses, are the harvest of fisheries" (FAO, 1993a, p. iii). According to the FAO, this definition was made on the basis of common usage and parallels the distinction made between hunting and gathering, on the one hand, and agriculture, on the other.

4. For a comparative discussion of the Asian, Latin American, and Central American cases, especially issues of social equity, environmental quality, and resistance movements, see Stonich and Bort, 1995.

5. Pumping requirements vary markedly depending on tidal flow characteristics, rainfall, and so on. Pumping can very easily add $500 to $1000 per hectare in initial capital requirements.

6. Interviews with small producers and cooperatives in Honduras, Nicaragua, and Panama revealed remarkably similar constraints to the successful establishment of shrimp farms, including access to land through government concession, purchase, and/or rental; access to credit, technical assistance, and markets; and infrastructural problems.

7. As in most tropical areas, declines in biodiversity of coastal zones are due to the interactions among a number of factors, including coastal economic development; population growth in coastal zones; increased natural resource consumption (for domestic consumption and for sale), inequity of resource distribution; poorly conceived government policies, especially for concessions and other policies (which while encouraging some sectors, such as cultured shrimp and other nontraditional agricultural exports, also result in diminished biodiversity); effects of global trading systems and the weak position of Central American nations in those systems; ignorance about the biological, social, and economic value of mangroves and other coastal habitats; and the failure to understand and take into account the root causes of diminished biological diversity (Saenger, Hegerl, and Davie, 1983; WRI, 1992b).

25

Shrimp Farming in Southern Honduras

A Case for Sustainable Production

David Teichert-Coddington

Commercial shrimp farming in southern Honduras began in 1972 with the establishment of Sea Farms of Honduras near Punta Raton on the Gulf of Fonseca. Sea Farms was located on *playon* (salt flats) and private land shared with cattle ranchers. Water was pumped from Estero Los Butus, an embayment of the Gulf of Fonseca. Sea Farms expanded from 45 to about 300 hectares during the decade that followed, accounting for almost all shrimp production in Honduras.

In 1984, construction of a larger farm, Granjas Marinas San Bernardo (GMSB), began in another region of the Gulf of Fonseca. This region was remote, with few roads. The pioneering investment in GMSB stimulated rapid shrimp farm development in southern Honduras, which resulted in expansion of the pond area from 300 to 11,800 hectares between 1983 and 1995.

The region appealed to farm developers because it had salt flats on which ponds could be built without having to clear forests first. Although salt flats are bisected by estuaries fringed with dense mangrove forests, they support little vegetation because tidal inundations are infrequent and soils become excessively salty or lack nutrients for plants to prosper (Day et al., 1989; Lugo and Snedaker, 1974). Zones of stressed and dwarf mangroves are found between fringe mangroves and bare salt flats.

Ponds were built by scraping up soil from the internal peripheries of ponds to form dikes. Minimal excavation was used to reduce the possibility of contact with deeper, potentially acidic soils. Small sections of dense fringe mangroves were cleared to construct pump stations. Otherwise, mangroves generally were left intact. Construction in fringe mangroves would be extremely difficult because of soft soils, the increased possibility of acidic soils, the cost of tree clearing, and the risk of tidal inundation. Semidiurnal tides range from 1 to 3.5 meters and are among the highest in the region (Ward, 1994).

Rapid expansion of shrimp farming has led to conflict among environmentalists, shrimp producers, and the artisanal fisheries over use of land, estuaries, and gulf resources (Stanley, in press). While earlier farms were constructed on salt flats, some later farms with smaller salt flat concessions invaded fringe mangrove forests. Estuarine artisanal fishers began complaining about the lack of access to former fishing grounds now occupied by shrimp farms. Gulf artisanal fishers accused the shrimp industry of reducing their catches of fish and shrimp by removing small shrimp for pond stocking; fish larvae were killed as a by-product of capturing small shrimp. Shrimp producers, meanwhile, began complaining of thievery from their farms.

Conflicts are expected when a common resource is exploited by multiple users. Conflict resolution usually takes time and depends on the availability of credible, objective information. There is little documentation on the socioeconomic costs and benefits of the shrimp industry in this case. No fisheries' data on the catch per unit of effort exist for estuaries and the gulf. The total quantity and types of mangroves actually removed by the shrimp industry have not been unequivocally presented. The impact of shrimp farming on estuarine and gulf ecology has not been identified and quantified, except for incipient work in water quality. Yet the shrimp industry has been declared environmentally dysfunctional (Carroll and Kane, in press) and socially, environmentally, and politically unsustainable (Stonich, Bort, and Ovares, in press).

The concept of sustainability is not settled and requires definition. Sustainable agriculture has been defined as agriculture that "must produce adequate amounts of high-quality food, protect its resources and be both environmentally safe and profitable" (Reganold, Papendick, and Parr, 1990). Stonich et al. (in press) adds that social and political consequences must be included for analyses of sustainability. In this chapter, sustainability is considered a dynamic process that depends on new knowledge and understanding about the interaction between production and environment and requires collaboration among producers, educators, and the community.

The purposes of this chapter are to (1) review shrimp culture in southern Honduras; (2) describe a unique, collaborative research project with shrimp producers in Honduras, including a summary of findings to date; and (3) clarify claims that the Honduras shrimp industry is environmentally destructive and unsustainable.

Southern Honduras Shrimp Culture

Shrimp farming has great impact on the region's economy. Shrimp sales, from 11,800 hectares of shrimp ponds in Honduras, generated about $72 million in 1994. The National Association of Honduran Aquaculturists (ANDAH) estimated that 5,259 persons were directly employed on these farms for an employment ratio of 0.49 persons per hectare of pond (Wainwright, 1995a). Cooperatives account for 585 hectares, employing a total of 1,613 persons. When cooperatives are not included in calculations, employment was 0.36 persons per hectare. Eleven packing plants employ an additional 2,645 persons, although many of these are temporary employees. Many other persons find employment in supporting industries, such as shrimp larvae fishing (D. Stanley, in press), feeds, equipment, supplies, communications, and technical support. An unverified estimate of total employment in all shrimp-related businesses in Honduras is 1.5 persons per hectare.

Not much verified data are available about environmental impacts from shrimp farming. Mangrove destruction is a focal point for concern, but water quality destruction is often a more immediate danger. The shrimp-farming industry is first to lose by degrading its own water supply. As a result, there is a growing interest in how shrimp farm effluents affect estuarine water quality and the growth and survival of shrimp. These concerns have been heightened by accounts of industry difficulties due to water quality problems in the eastern hemisphere (Phillips, Lin and Beveridge, 1993; Rosenberry, 1994). Rosenberry (1994) attributed a 16 percent drop in world shrimp production between 1992 and 1993 to environmental problems and a 3 percent drop in 1995 to viral diseases (Rosenberry, 1995). Diseases are not necessarily the result of poor water quality, but they are more likely when animals are stressed by environmental factors. Water quality problems are compounded when intensively managed shrimp farms are operated in estuarine and bay areas that are also polluted by municipal wastes.

Honduran shrimp farms draw water from or discharge to Gulf of Fonseca embayments or riverine estuaries of the gulf. Most farm development occurs on riverine estuaries, where farms may recirculate each other's effluents as they pump from and discharge into the same estuary. Effluents eventually end up in the gulf, and there is concern that it will become polluted over time.

Honduras shares the Gulf of Fonseca with Nicaragua and El Salvador. Impacts on the gulf by one nation will affect the others. El Salvador has limited land for shrimp farms, but Nicaragua has about 22,000 hectares of land that has potential for shrimp farm development (Currie, 1994). Most Nicaraguan wastes would be discharged into the gulf via the Estero Real, thereby contributing to the gulf's total nutrient load.

Maintaining adequate water quality would require regulation difficult to formulate without a scientific basis. The characterization and monitoring of receiving waters would be necessary to correlate water quality changes to shrimp farm development. Differentiating between river-borne pollutants that enter the estuaries and those originating from the farms would be useful. Estimates of estuarine assimilative capacities for farm wastes would be essential to estimate the upper limits of farm development.

The collection of water quality and hydrographic data necessary to estimate estuarine assimilative capacities and to evaluate effects of farm management techniques on the effluent quality was initiated in southern Honduras in 1993. The project was a collaborative effort of producers, educational institutions, a trade organization, and the government of Honduras. Signatories to the agreement were the Ministry of Natural Resources, the government of Honduras, ANDAH, the Federation of Export Producers (FPX), the Panamerican Agricultural School (EAP), and Auburn University. The effort was funded through a U.S. Agency for International Development (USAID) grant of the Pond Dynamics/Aquaculture Collaborative Research Support Program (PD/ACRSP). A water chemistry laboratory at La Lujosa, Choluteca, analyzed water samples. An estuarine sampling program and on-farm trials were initiated with producers.

Current Shrimp-farming Practices

The Pacific white shrimp, *Penaeus vannamei*, has been the preferred culture shrimp in Honduras. Young shrimp for pond culture are captured from estuaries and produced in hatcheries. Captured larvae usually consist of 20 to 80 percent *P. stylirostris*, another Pacific white shrimp. *Penaeus stylirostris* have not been as popular as *P. vannamei* because of lower pond yields, but they may gain in popularity because they have less susceptibility to the Taura virus.

Current practices have been greatly affected by the Taura virus, which first attacked Honduras in March, 1994. The virus usually strikes shrimp during the first month of life and causes heavy mortality (Brock et al., 1995). Before the virus arrived, postlarvae shrimp were stocked in nurseries for 30 to 45 days and then transferred as juveniles to grow-out ponds. The transfer resulted in high mortalities of stressed juveniles after the virus became a problem. Most farms currently stock ponds directly with postlarvae to avoid handling them a second time. Harvest survival of hatchery-produced postlarvae ranges from 20 to 40 percent. Postlarvae captured from the wild have higher survivals, sometimes up to 80 percent. They are stocked at higher initial rates to account for expected mortality. Harvest densities of 4 to 8 per square meter are usually anticipated. Stocking densities are generally lower during the dry season than during the wet season.

Culture periods generally range from 12 to 16 weeks, although shorter and longer periods are not uncommon, depending on shrimp growth. Shrimp yields vary greatly during the year and between years (Teichert-Coddington, Rodríquez, and Toyofuku, 1994). Dry season yields may be one-third to one-half of wet season yields, probably because of lower dry season temperatures. Shrimp may not reach 10 grams during the dry season but may exceed 20 grams during the wet season (Teichert-Coddington and Rodríguez, 1995).

Shrimp are fed a pelleted diet, usually formulated to contain between 20 and 28 percent protein. Some farmers reduce protein as the shrimp increase in size, while others do the opposite. Most feed a fixed protein diet at standard rates that decrease with shrimp size. Organic fertilization is usually not used because of marketing concerns. Inorganic fertilization is used occasionally, but rates are not standardized. Pond bottom liming is used occasionally, although it is probably unnecessary because acidic soils are rare on Honduran farms. Daily water exchange usually ranges from 3 to 10 percent. More exchange generally is perceived to be better, but actual rates are usually limited by pumping capacity. Some farms exchange water according to water quality parameters, while others exchange at a set daily rate. Water exchange may decrease in the future as farms seek to reduce the likelihood of contamination by water-borne diseases.

Estuarine Water Quality

A database of estuarine water quality was established in the major shrimp-producing zones of southern Honduras (Teichert-Coddington, 1995b). Water quality was monitored weekly at 13 sites in six estuaries of the Gulf of Fonseca. Producers were responsible for obtaining a standardized sample from the intake pumps at their farms and transporting it properly to the laboratory. At the lab, water samples were analyzed for total alkalinity, chlorophyll *a*, nitrates, nitrites, total ammonia (Parsons, Maita, and Lalli, 1992), filterable reactive phosphate, total nitrogen, and total phosphorus (Grasshoff, Ehrhardt, and Kremling, 1983). Intake and discharge water for representative farms of the region was characterized as a prerequisite to determining chemical nutrient balances (Teichert-Coddington, Martínez, and Ramírez, 1996). Nitrogen and phosphorus inputs by feed, shrimp postlarvae, and fertilizer were compared to total output by water discharge and shrimp harvest. A goal of the research was to estimate the assimilation capacity of various estuaries for shrimp production, which could be integrated into a general management plan for the zone.

Water quality differences were noted among estuaries, along longitudinal transects of estuaries, and between rainy and dry seasons of the year (Teichert-Coddington, 1995b). Riverine estuaries were more fertile than gulf embayments and had less capacity to assimilate waste loads. Eutrophication of riverine estuaries was greater according to the distance upstream from the gulf. Eutrophication was greater during the dry season in riverine estuaries because of diminished freshwater inflow.

There was a mean net consumption of inorganic nitrogen and phosphorus from intake water and a mean net discharge of organic matter to estuaries. The use of inorganic fertilizers promoted the discharge of phosphorus and nitrogen. Pond discharge of both nitrogen and phosphorus increased linearly with the feed conversion ratio (Teichert-Coddington et al., 1996).

Total nitrogen and total phosphorus annual discharges from 11,500 hectares of shrimp ponds were compared to the annual discharge from the Choluteca River, the largest Honduran river discharging into the Gulf of Fonseca. The Choluteca River receives municipal wastes from the capital city of Tegucigalpa; Choluteca; several other smaller towns; and agricultural land. The Choluteca River discharges 1.8 times as much nitrogen and 4.8 times as much phosphorus as the whole shrimp industry (Teichert-Coddington et al., 1996). The water quality of the Choluteca River, in addition to shrimp farm discharge, would need to be addressed if eutrophication of the gulf were to become an issue.

Management for Sustainability

Honduran shrimp production graphed over time resembles the typical S-shaped curve of nascent industry growth (figure 25.1). Production increases were initially slow as the industry expanded in area. Expansion occurred rapidly between 1988 and 1994 but began to decrease in 1994 as problems with the Taura virus intensified. Taura infected the area around March 1994, but the effects on yields were not noticed until several months later. At least two farms were forced out of business by unanticipated low returns, which accounted for the reduction in pond area during 1995 (figure 25.1).

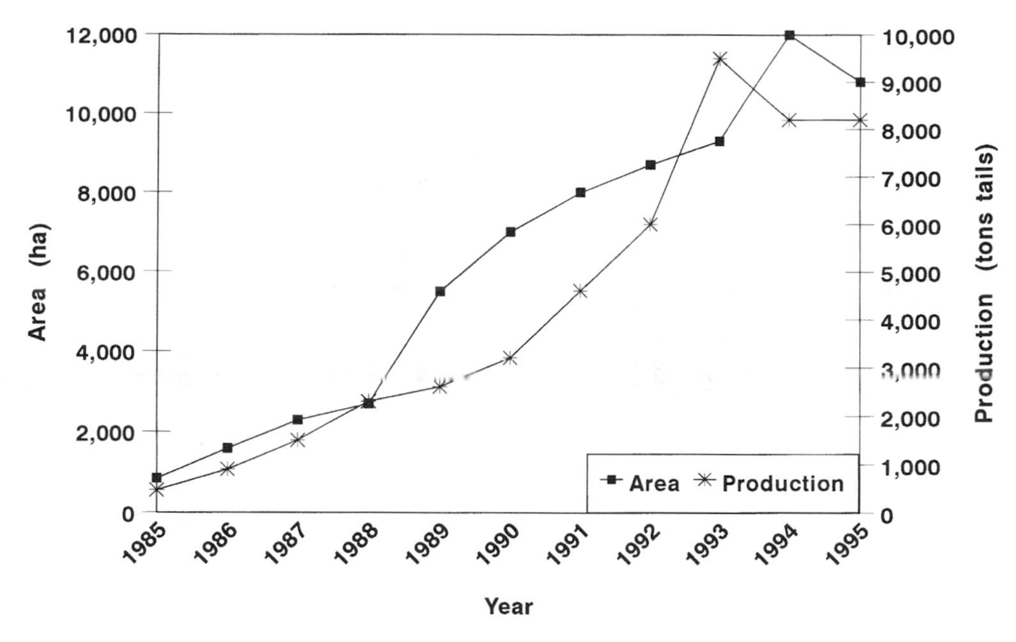

Figure 25.1 Shrimp Production and Area under Production in Honduras, 1985–1995

In addition, the Honduran government imposed a temporary moratorium on new pond development until the disease problem could be resolved. Yields in 1995 were estimated to be about the same as in 1994, despite the lower overall production area. Management was able to adjust quickly to partially compensate for low survivals by avoiding the nursery stage and doubling or tripling the stocking rates. The effects of the Taura virus emphasized the risks involved in shrimp culture. The quick reaction of the industry also demonstrated the value of information sharing among farms and of collaborative research.

Diseases like Taura may affect the industry even with perfect water quality. However, most problems are exacerbated by poor water. The focus should be switched from conserving gulf water quality to conserving estuarine water quality. The gulf has a higher assimilative capacity than estuaries because of dilution by a much greater water volume. Gulf water quality will be good if estuarine water quality is well maintained. Long-term management plans for the Gulf of Fonseca include a hydrologic study to ascertain water exchange rates between the gulf and the Pacific Ocean. This would allow for future estimates of the gulf's assimilation capacity for regional shrimp production in addition to agricultural and municipal discharge.

Management of estuarine water quality is not simple, as illustrated by a recent analysis of estuarine total nitrogen concentrations and feed used on shrimp farms. Data on monthly feed inputs and the percentage of feed protein were gathered from farms on each estuary that had been monitored for total nitrogen concentration since March 1993. The data for the Pedregal Estuary are illustrated in figure 25.2. The mean monthly nitrogen concentration in estuarine water tended to increase during the dry season and decrease during the wet season, while monthly nitrogen input from feed was higher during the wet season. There was no apparent trend in total nitrogen concentration during the sampling period. The effect of the Taura virus on pond management is seen in relatively low feed inputs during November 1994 to April 1995. However, total nitrogen concentration of the Pedregal Estuary did not demonstrate a correlated reduction. Regressions of total nitrogen concentrations on total nitrogen input were insignificant ($P > .05$), even after stratification of data by season.

Effects of nutrient discharge on estuarine water quality are usually estimated by computer models of complex, nonlinear equations that describe the capacity of an estuary to absorb given concentrations of nutrients according to the fluid dynamics of the estuary. Estuarine assimilative capacities vary widely with distance from the gulf, offshore currents, tidal fluctuation, and estuarine hydrography (Ward and Montague, 1996). Assimilative capacities are therefore localized in time and space.

The hydrography of the Gulf of Fonseca riverine estuaries is strongly affected by seasonal rains. Estuaries are fairly short, so estuarine water is totally displaced by freshwater input during heavy rainfalls and flooding. Estuarine nutrient concentrations are significantly lower during the wet season than during the dry season (Teichert-Coddington, 1995a). Maintenance of good

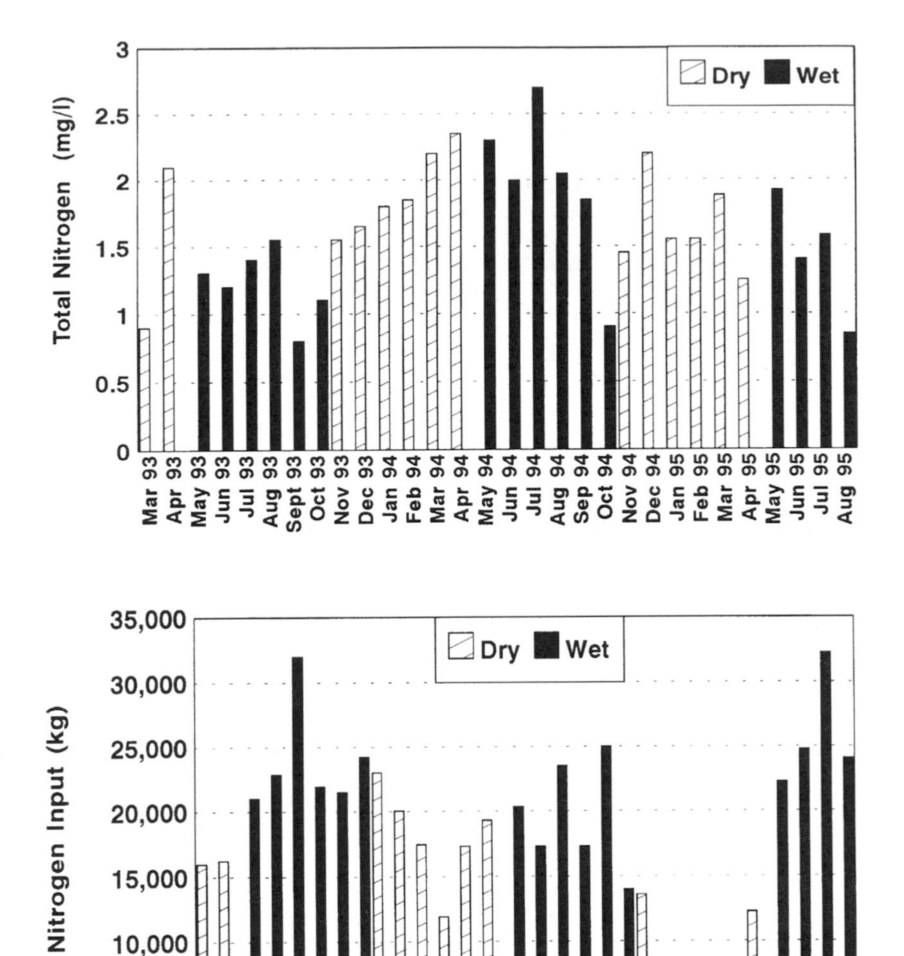

Figure 25.2 Mean Monthly Estuarine Total Nitrogen Concentrations and Total Nitrogen Input as Feed on Farms Discharging to the Pedregal Estuary, March 1933–August 1995

water quality is more critical during the dry season because assimilative capacities are lower when there is little dilution from freshwater input.

Estuarine exchange with the gulf is reduced as the distance from the gulf increases. Retention time in mangrove-fringed estuaries without freshwater input is directly proportional to the square of estuary length, according to the relationship $T_0 = L^2/B$, where L is the length of mangrove-fringed creek and B is the longitudinal diffusion coefficient, or flushing rate (Wolanski, 1992). Stagnated water in the upper reaches of estuaries becomes enriched by evaporative concentration and low exchange with relatively clean embayment water. The probability of low estuarine dissolved oxygen (DO) is higher during the dry season. Farms with intakes located closest to the gulf are most assured of good water.

Although nutrient discharge is only one factor in estuarine quality, farm management practices can be modified to minimize it, particularly during the dry season. These modifications could also increase profitability by increasing production efficiency. Inorganic fertilization can probably be eliminated in riverine estuaries and reduced in embayments because nutrient concentrations are already high enough for phytoplankton blooms (Green and Teichert-Coddington, 1990; Teichert-Coddington et al., 1996). Unnecessary fertilization merely enriches the estuaries. A self-imposed moratorium on fertilizer use in riverine estuaries has been employed by the Honduran National Association of Aquaculturists, but adherence rates are unknown.

Nitrogen appears to be a limiting factor in estuarine primary productivity. It is also a component of amino acids that compose protein, the single most expensive ingredient of shrimp feeds. Nitrogen discharge could be reduced on some farms by reducing protein levels (Teichert-Coddington and Rodríguez, 1995) and feeding rates (Rodríguez and Teichert-Coddington, 1995), particularly during the dry season.

Chances of encountering and developing poor water quality can be reduced by avoiding pond development along the upper reaches of riverine estuaries. Those areas are likely to have poor water because of their distance from the gulf, so the chances of economic success are lower. Some farms in these areas already find it necessary to close during the dry season because of poor water quality.

Long-term sustainability probably entails lower rates of economic return than those that have been realized in the past. Diseases are becoming more prevalent, and overcoming disease losses will further reduce income. There is a danger that producers will increase production to increase returns per unit area in order to offset higher operation costs. While this scheme might succeed in isolated parts of the gulf, it would probably show only short-term gains if practiced on a wide scale with existing management. Increasing production would result in greater nutrient discharge and more rapid estuarine eutrophication, particularly during the dry season.

There is an alternative to increasing stock densities while possibly decreasing nutrient discharge. Mechanical aeration, rather than water exchange, could be used to improve water quality. With this approach, oxygen is mechanically added to water, and organic material is at least partially oxidized in the pond rather than discharged as biochemical oxygen demand (BOD) in the estuary (Hopkins et al., 1993). This approach, using aeration and low water exchange to reduce nutrient discharges, has been experimentally successful at high stocking rates ($78/m^2$) (Hopkins et al., 1993, 1994) but has not been commercially proven. Higher production in Honduras would entail reductions in pond sizes to improve manageability, improvements in infrastructure like added electrical service, and a higher level of managerial expertise. In short, intensification would be risky and require high capital expenditures in aerators, shrimp seed, feed, and energy. It is doubtful that intensification could be widely applied in Honduras. On-farm research should be conducted, however, to evaluate intensification as a means for overcoming disease problems and resolving some environmental concerns.

Effects of the Shrimp Industry

Mangroves

Destruction of mangrove forests has been a focal point of opposition to the shrimp industry in Honduras and in other parts of the world. Shrimp culture has destroyed mangroves, particularly before the early 1980s, when ponds were built in fringe mangroves to take advantage of tidal water exchange. However, Csavas (1993) demonstrated that the relationship between recent shrimp development, which uses pumping to exchange water, and a country's mangrove resource base is almost an inverse one. Farm development is not related to mangrove availability. Fringe mangroves have proven to be poor sites for the construction of ponds, and recent development has been on higher ground. Some assume that mangroves once stood where ponds are now located. In Honduras, however, shrimp farming became important because ponds could be located on salt flats, where land did not have to be cleared. The situation is similar in neighboring Nicaragua, where farms are being developed on salt flats with minimal clearing of fringe mangroves.

The existence of salt flats for development does not preclude the clearing of fringe mangroves. In Ecuador, ponds were also initially built on salt flats, but later expansion included up to 20 percent of land that had been mangrove forests (Parks and Bonifaz, 1994). As indicated before, high-quality mangroves in Honduras have been cleared for pond construction on some concessions and more often for pump station construction, but the total area cleared is less than half of that cleared for other uses. A recent study for the USAID in Honduras (Vergne, Hardin, and DeWalt, 1993) evaluated mangrove destruction between 1973 to 1992 by the interpretation of aerial photographs. The study was not definitive because of poor photographic quality, but 2,132 hectares of high-quality mangroves were reported to have been cleared for occupation by 11,500 hectares of shrimp ponds. This constituted 6.9 percent of original forest cover. During the same period, 15.1 percent of original cover was removed for other uses, including firewood, construction materials, tannin production, and salt production. In another study of southern Honduras forest resources, Oyuela (1995) reported that between 1987 and 1994, mangrove forests had been reduced by 3,032 hectares (6.5%) from 46,710 hectares to 43,678 hectares. During the same period, total area of shrimp farms, including dikes, roads, and facilities, increased by 13,822 hectares from 8,291 hectares to 22,113 hectares. The expansion of shrimp farm area took place mostly on salt flats, which were reduced in area by 11,240 hectares from 14,240 hectares to 3,000 hectares. About 249 hectares of seasonal lagoon area were also converted to shrimp farm area, so a total of 2,333 hectares of mangrove forest could have been cleared for shrimp farming. Two conclusions may be drawn from these two reports. First, there has not been wide-scale destruction of mangrove forests for shrimp farm development in Honduras. Second, the conservation of mangrove forests that fringe the gulf is far more complex than simply prohibiting clearing by the shrimp industry.

The importance of fringe, high-quality mangroves to fisheries and coastal ecology cannot be contested. Mangroves recycle inorganic nutrients into organic matter, which is exported to the estuary (Lugo and Snedaker, 1974). Organic matter is a principal energy source for an aquatic food chain that often culminates in harvestable fish, shrimp, crabs, and oysters. Mangroves also provide nurseries for aquatic animal larvae, including shrimp, which are used for stocking ponds.

Shrimp farm construction cannot be done without clearing some of the fringe mangroves for pump structures. Stressed and dwarf mangroves are usually cleared on a larger scale to develop a salt flat. In this respect, shrimp culture is little different from other forms of agriculture that clear land for planting. The Honduran shrimp culture has an advantage over land agriculture in that it uses mostly salt flats, which are otherwise unproductive. Shrimp culture may arguably be the most environmentally and economically appropriate occupation of salt flats and adjacent stressed mangroves.

Replacement of some mangroves with ponds is not entirely negative. Organic productivity within ponds and organic enrichment from feeds may compensate for the clearing of mangroves that serve as primary producers of estuarine carbon. Mangrove productivity is sometimes limited by nutrient availability (Day et al., 1989; Lugo and Snedaker, 1974). Enrichment of estuaries by farm discharge might therefore stimulate the growth of nutrient-deficient mangroves.

Stressed mangroves contribute little to estuarine carbon supply. The net primary productivity of stressed mangroves is often zero (Day et al., 1989; Lugo and Snedaker, 1974). In this case, energy is used primarily for cellular maintenance, with little available for export. On the one hand, stressed mangroves might be flooded only at monthly high tides or during spring tides, thereby contributing little to larval habitat. On the other hand, stands of black and white mangroves now line the dikes of many farms. These mangroves developed naturally because of pond flooding and are maintained by producers to protect dikes from erosion. Some producers claim that they have actually reforested salt flats by building ponds. This claim is partially true, but pond-generated forests probably do not fill the same ecological niche as do naturally flooded forests.

Road construction by the shrimp industry has altered mangrove hydrology, sometimes resulting in mangrove mortality. However, mangrove reforestation of previously bare salt flats has also been observed, ostensibly caused by hydrology alteration from road construction. Documentation about the impact of road construction on mangroves is poor because much of the mangrove area was inaccessible for observation prior to construction.

There is much to be said for preserving the mangrove forest intact. There are also many socioeconomic benefits from some mangrove clearing for shrimp farming. The crucial question should be how much and where to clear, rather than whether to clear. Forest management practices should include harvesting mangroves for firewood and construction materials because these activities are economically important to local communities and mangroves are a renewable resource. Mangroves grow to firewood size in about 3 years (Wainwright, 1995b) and reach maturity in 20 to 25 years (Lugo and Snedaker, 1974).

It appears that the land most appropriate for shrimp farm development in Honduras has already been occupied. The total area was estimated at 18,000 hectares (Vergne et al., 1993), but this estimate included salt flats that are poorly situated for access to high-quality water. About 15,000 hectares have already been developed. It appears that water quality and long-term benefits to the existing shrimp industry might best be served by stopping development now and directing significant efforts toward the enforcement of existing environmental regulations and the security of protected zones.

Artisanal Fisheries

Fish capture by artisanal fishers is perceived to have decreased in recent years. Rapid expansion of the shrimp industry was associated with decreasing catches. Concern about the farms' impact on the fisheries is reasonable because high numbers of young shrimp are harvested annually for pond stocking and there is associated by-catch of larval finfish. However, available information does not support the notion of a negative impact on capture fisheries.

There are no data that demonstrate a veritable decrease in catch. There have been no time series studies of the Gulf of Fonseca fisheries, and no credible data on fish landings have been collected. Claims for decreasing catches are usually prescribed by smaller-sized fish and fewer fish caught per fisher. These are classic signs of increased fishing pressure, not lack of recruitment because of the removal of young shrimp and fish. Greater quantities of fish are removed from the gulf at a faster rate as the number of fishers increases. The result is fewer fish captured per unit of effort by the fisher and smaller fish size because fish do not have as much time to grow large. The total biomass of fish being removed from the gulf may actually be increasing despite a decrease in catch per unit of effort. Only fishery surveys can pinpoint the actual state of gulf fisheries, and these are not available. Meanwhile, methods should be implemented to decrease the mortality of larval finfish captured during the collection of young shrimp.

The harvest of young shrimp from estuaries is high, but there is reason to expect much higher levels of escape. Young shrimp are harvested with inefficient, small hand nets (*chayos*) during low tide along the edges of estuaries. The water area covered by harvesters (*larveros*) over a tidal cycle is small. This suggests that the majority of young shrimp in estuarine nurseries escape capture. Of course, the equation could change rapidly if methods for capturing young shrimp became more efficient.

The abundance of shrimp larvae appears to have occurred in cycles during the past decade, independent of fishing pressure. For example, a particularly good year was observed in 1994, after 10 years of fishing young shrimp in estuaries. More young shrimp were collected in 1994 than could be stocked in nursery ponds. One would expect abundance to suffer if shrimp stocks were being reduced by young shrimp capture. In fact, young shrimp abundance, and possibly adult capture, appears to be more related to temperature, rainfall, and river flow (freshwater discharge). For example, Wallace, Hosking, and Robinson (1991) found that river discharge, rainfall, and water temperature explained 90 percent of the historical variation in brown shrimp landings in Mobile Bay, Alabama. Increased river discharge and rainfall lowered salinities sufficiently in bay nurseries to negatively affect juvenile brown shrimp, which are sensitive to low salinity.

The shrimp-farming industry may actually improve the estuarine and gulf fisheries. Estuarine fisheries are normally more productive than ocean fisheries because estuarine water is more fertile. Organic and inorganic nutrients are exported from the land, and farm discharge further enriches estuaries with both organic matter and nitrogen and phosphorus (Teichert-Coddington, 1995a), which are essential nutrients for primary producers. Many fish feed directly on unicellular algae, which are the primary producers. Others consume zooplankton, benthos, or other fish that depend on algae and organic material for nutrition. The Peruvian anchovy fishery is a common example of an unusually productive ocean fishery that is dependent on primary production and is fertilized with nutrient-rich upwellings off the coast of Peru. However, the fishery has had dramatic failures during El Niño events when upwellings do not occur and primary productivity is low from lack of nutrients (Lowe-McConnell, 1987).

Moderate fertilization of estuaries by farm discharge can be helpful, but overfertilization could induce poor water quality. Water quality problems first become noticeable with low dissolved-oxygen concentrations that are provoked by high BOD from phytoplankton respiration and bacterial decomposition of dissolved and suspended organic matter. Farms exchange water to reduce BOD and thereby avert episodes of low oxygen concentrations in their ponds. If intake water also has high BOD, water exchange becomes ineffective. Eutrophication of estuarine water by farms is therefore self-limiting. Poor water quality for extended periods along the length of an estuary would force farms to close.

Socioeconomic Effects

The socioeconomic dimensions of shrimp farm development are perhaps the most difficult to objectively quantify because the terms of reference are varied and numerous. The livelihood and well-being of humans are not so easily determined and quantifiable as is the net productivity of mangroves or the catch per unit effort of fishers. Some of the issues are identified by Stanley (in press) and Stonich et al. (in press).

The southern province of Choluteca is the most populated and economically deprived in Honduras. Despite the positive economic aspects of the shrimp industry, there is an underlying feeling among some that the industry has been built on injustice and has exacerbated social and economic inequality. These issues are outside the purview of this chapter, but they should be analyzed and compared with the history of development in other economies and evaluated with respect to the alternative of no shrimp industry. The industry has been created with few guidelines in an isolated part of the country, with little previous infrastructure, to become the third-largest source of foreign currency for the country of Honduras. Development could not have occurred without high capital investment and commitment to long-term returns. It seems most reasonable, therefore, to work toward guaranteeing long-term social benefits for the local communities while allowing a healthy shrimp industry to operate.

Conclusion

The Honduran shrimp industry has been strong for more than a decade, thereby demonstrating its long-term commitment to the region. Sustainability requires management adjustments as new technical knowledge becomes available and the understanding of various impacts increases. The industry has collaborated on research to understand its impact on the aquatic environment. Data are collected from farms and estuaries by producers and research personnel and are analyzed and published. Estimates of estuarine assimilative capacities for shrimp farming have not been finalized, but the water quality provides evidence that supports regulation of development along the riverine estuaries.

Good water quality is particularly threatened during the dry season, when there is little dilution of estuarine water by freshwater.

The shrimp industry depends on mangrove forests for its own health and development. Mangroves are a resource utilized directly or indirectly by artisanal fishers, local firewood gatherers, and builders, in addition to shrimp producers. This use of a renewable common resource should be emphasized to develop collaborative mangrove forest management practices and regulatory policy. An objectively developed information base is needed to address the related socioeconomic issues of shrimp production, artisanal fishery, and mangrove conservation. To be credible, research should be designed, conducted, and analyzed with the participation of concerned parties.

Acknowledgments This study was made possible by the collaboration and support of the Dirección General de Pesca y Acuacultura, Secretaría de Recursos Naturales, government of Honduras; the Honduran National Association of Aquaculturists (ANDAH); the Pond Dynamics/Aquaculture Collaborative Research Support Program (CRSP), funded by the U.S. Agency for International Development; and Auburn University. AAES No. 8–965232. CRSP Accession No. 1137.

26

Understanding Conflict in Lowland Forest Zones

Mangrove Access and Deforestation Debates in Southern Honduras

Denise Stanley

Mariculture has proven to be one of the most conflictive natural-resource-based industries in the Third World. The entry of shrimp farms into wetland ecosystems has been greeted with both cynicism and enthusiasm. Stonich, Bort, and Ovares (chapter 24) describe the marginalization of the poor that accompanies many shrimp projects and the increasing conflict between fishers and mariculturalists over scarce natural resources. This case study explores economic factors and legal institutions that contribute to these problems of the "blue revolution."

Like any form of aquaculture, maricultural technology increases demands for three scarce factors of production—land, larva, and water.[1] In each case, a new maricultural enterprise creates interactions with the natural environment and with other resource users. As a result, conflicts have emerged over the control and management of natural resources. Farms may be constructed in mangrove areas or on salt flats[2] with unclear boundary lines. Vegetation must be cleared to build access roads, pumping canals, and even grow-out ponds. Some fishers and environmentalists argue that enclosure and deforestation of mangrove areas reduce access to fishing zones and fish habitats. Once the farm is constructed, larva gatherers use the same estuaries as artisanal fishers. These gatherers may capture and waste a large by-catch of other fish species. Finally, at the end of the cycle, some farms dump waste into the common estuary water that others will pump in to start another production cycle.

Each of these resource use problems is an example of an "externality." The activities of one producer entail costs for another that are not taken into account.[3] As Ruttan (chapter 2) points out, such negative spillover effects contribute to an overuse of the resource in question, a bias in the direction of technical change, and general environmental stress. I argue that the incomplete delineation and the controversial distribution of property rights over attributes of the ecosystem are at the heart of these conflicts.

The debates surrounding the socioenvironmental impacts of mariculture have created polarized divisions among the parties who are struggling over scarce natural resources—fishers and farmers. Throughout Asia and Latin America, fisher-based environmental NGOs (nongovernment organizations) have been constituted to mediate the impacts of the new mariculture industry. The Committee for the Defense of the Flora and Fauna of the Gulf of Fonseca (CODDEFFAGOLF) in southern Honduras is one example. This group's activities have included marches and road blockages near shrimp farms and education and publicity campaigns to promote mangrove preservation. Even now the Fundación Natura of Ecuador and the United States–based Mangrove Action Project have called a boycott of "environmentally damaging" shrimp products.

On the other side of the debate, exporters' associations and some international donors and local government officials have pointed to the employment and foreign exchange benefits of the industry. Shrimp exports are a leading export product of many Asian nations. Fishery products now represent Ecuador's second-largest source of foreign exchange. Eight thousand metric tons were exported from Honduras in 1994, making shrimp the third-largest export in dollar value (Espinoza, 1993). In addition, the industry has supposedly created at least 12,000 jobs, becoming the largest industrial employer in the southern region (Vergne, Hardin, and DeWalt, 1993).

However, little is known about the productive processes of mariculture that lead to ecological and social conflicts. Equally unclear is the impact of property rights and institutional arrangements that shape the liabilities and opportunities available to the different parties involved. Finally, statistics on the actual level of damage done by the whole mariculture industry (and the relative impact of different strata of farmers) are still debated.

This chapter focuses on one of the three conflicts commonly associated with mariculture, the struggle over the control and use of lowland mangrove forests and related species. The analysis draws on data from southern Honduras to disentangle the nature and causes of the externality problems that arise. It addresses Ruttan's call for "incentive-compatible institutional designs" to enhance agricultural sustainability. The emphasis on Honduran mariculture complements the results of David Teichert-Coddington's study on the water pollution debate there to provide an in-depth picture of one region and to offer lessons for other countries.

Shrimp Farming in Southern Honduras and Wetland Conflicts

A case study of Honduran mariculture offers an excellent opportunity to examine resource conflicts at the local level.[4] Southern Honduras is a 6,840-kilometer region that includes the Gulf of Fonseca and a lowland plain of mangrove swamps, small towns, and pastureland. The provinces of Valle and Choluteca are the most densely settled regions of the country, with 5 percent of its territory and 11 percent of the population (CRIES, 1984). According to government estimates, the 1982 average income was $118.50 per year, and more than 70 percent of the families in the region lived on less than $20 per month (CPSE/OEA, cited in Stonich, 1989).

The farm-raised shrimp boom began in 1985, although the first farm was built near Punta Raton in 1973 (Weidner et al., 1992). Technological and personnel support from shrimp farms in Panama and Ecuador, fiscal incentives to export industries, and a generous land concession program have contributed to the industry's high growth levels. The area under maricultural production has increased more than 1,000 percent in the last seven years, and in 1993 80 farms operated 6,400 hectares (D. Stanley, 1994).

The land concession program has been the most influential export promotion policy to affect local communities. The Honduran state remains the owner of lands within 40 kilometers of the coast. However, under Decree 968 in 1980 and Agreement 229 in 1991, the Ministry of Tourism (SECTUR) and the General Direction of Fishing (DIGPESCA) were mandated to transfer coastal land use rights to exporters at the nominal rental rate of about $5 per year. The concession areas are leased out to mariculturalists for 20-years with the possibility of renewal. To date, more than 25,000 hectares have been transferred to private users, effectively replacing the claims of traditional users and triggering an "enclosure of the commons."

The activities of different parties in the wetlands represent a particularly thorny form of interaction among competing users. The wetland zone with regular tidal flushing is a prized area for pond construction (Torres-Díaz, 1991).[5] Shrimp farms cannot be constructed in all soils and in agricultural areas, so the aquaculture frontier is limited. The mariculture industry's demand for intertidal land increased as world shrimp prices rose in the 1980s. In the south, there has been a boom to acquire wetlands along the Pacific Coast, salt and mud flats, and even mangroves. These areas were previously unoccupied but were used by coastal people for the collection of shellfish and wood products.

Table 26.1 shows the expansion in shrimp land. Although some studies claim that more than 30,000 hectares are suitable for shrimp farming (Guevara, 1991), it is now recognized that the cutting of mangroves and the availability of estuarine water pose severe limitations. Thus, a limit of 15,000 hectares in southern Honduras for shrimp production is more likely. Currently, there is a moratorium on new concessions for shrimp farming in Honduras. However, this decision may have come too late. As described below, the production dynamics of the export industry create incentives to expand into marginal areas and extend the aquacultural frontier to acquire more of the valued wetlands. The effects of this land hunger on mangrove forests parallel the trend described by Mayne (chapter 9) in which "the loss of forest is due more to the need for land than for timber or other forest products."

In Honduras, as elsewhere in Asia and Latin America, the clearing of mangroves and other vegetation for pond construction is a growing concern. Initial estimates now suggest that shrimp farms have destroyed 2,100 hectares of mangrove in southern Honduras (Vergne, et al., 1993), while other estimates place the figure at 4,300 hectares (X. Ramírez, 1994).[6] These numbers have been hotly contested by industry spokespersons, who argue that "smart" farmers would never enter a mangrove swamp zone since the overly high salinity levels and soil structure there imply lower yields (Torres-Díaz, 1991).

The relationship between the two interdependent parties, shrimp farmers and resident resource users, and the mangrove resource is a classic negative externality problem since the shrimp farmers are not incorporating the cost of off-site damage from their actions into their decisions to clear vegetation.[7] Habitat destruction by less-informed farmers represents a stock externality because it reduces the supply of many fish species and forest products to resident gatherers. Many studies have shown a direct relationship between mangrove forest cover and fishery stocks and crustacean species (R. Turner, 1989). Shrimp larva and fish stocks are linked to estuarine salinity, which depends on the man-

Table 26.1 Patterns of Land Concessions and Shrimp Production in Southern Honduras

	Land Concessioned		Land in Production		
Year	Hectares	% Change	Semi-intensive[a] Hectares	Extensive[a] Hectares	(%) Change
1985	5,800		750	0	
1986	6,800	17	1,500	0	50
1987	8,100	19	1,800	100	20
1988	13,030	61	3,000	250	67
1989	16,115	24	4,300	250	43
1990	22,200	38	5,500	250	40
1991	23,777	7	6,000	250	20
1992	24,000	1	8,000	250	11
1993	24,500	2	9,000	250	13
1994	25,780	5	10,000	250	11
1995	26,000	1	11,000	250	10
1996	26,000	0	12,000	250	9

Sources: Weidner et al., 1992; RRNN, 1993; DIGPESCA, 1991; projections 1994–1996 by author.

[a]Refers to semi-intensive technology by stocking and water-pumping rates.

grove forest cover and wetland development level (Swallow, 1994). The fencing and guard posts along farm perimeters, access roads, and some Honduran "winter lagoons" have also been harmful to fishers and gatherers by raising their operating costs and travel time.

Explanations for the Conflicts in the Mangrove Ecosystem

Several reasons have traditionally been cited for externality conflicts such as those in mariculture. These include the intrinsic characteristics of the resource, the information available about its properties and impacts from use, and the property rights governing its use. First, resources that are intrinsically collective—having the properties of rivalry and nonexcludability—are considered the most externality-prone because one person's use reduces another's and it is nearly impossible to exclude users or extract a price for the use.[8] Second, poorly defined or politically motivated property rights provide the wrong signals concerning the rate or manner in which the resource is used.[9] The type of tenure regime—state, private, common property, or nonproperty—is one important component of the property rights bundle. However, property rights also include the extent to which users can access and/or alienate the resource, acceptable management methods, the length of tenure, and the charges associated with resource use. Property rights shape the incentives that affect the discount rate, the investment behavior, and the environmental preservation attitudes of the users. There is a long-standing debate in the literature over whether acceptable user behavior is enhanced by private tenure rights or whether other components of the rights bundle have a greater influence on user actions (Bromley, 1989; Feder and Onchan, 1987). Finally, the lack of adequate information, or the inequitable distribution of knowledge, may lead users to deplete a resource too rapidly or cause harmful, unassessable offsite effects (Dixon and Sherman, 1990). Increased public awareness of the alternative uses for land and the externalities arising from the present arrangements would reduce the leeway of current users.

These factors explain much of the conflict between fishers and mariculturalists in Honduras and other tropical ecosystems. Mariculture places demands on a land area with private and common pool characteristics. The fact that deforestation should be occurring in nearly privatized areas proves the depth of the problem. Despite the leasehold nature of the tenure regime, a sense of private property and excludability has developed over the land. Nearly all the farms have fences, guard posts, and "no trespassing" signs on the perimeters. Yet, while the resource may be intrinsically private in its spatial attributes, the wetland forests also have common pool aspects. Excluding users from the off-site environmental benefits of wetlands is difficult, and the subtractability of these attributes by direct and indirect users remains unclear. In reality, the dual nature of the resource attributes of wetlands makes their management difficult, with a free-for-all and rent dissipation often the result.

Property rights problems also contribute to resource degradation and conflict. Several components of the property rights bundle need to be disentangled—in particular, the allocation of leaseholds by administrative rationing, the timing of the leases, the fee level, and the enforcement of use rules. Leaseholds were allocated on a first-come basis, with sales and transfers legally prohibited. The many steps in the leasehold administrative process, based in the capital city of Tegucigalpa, made it difficult for small farmers or fishers to participate in the rationing game. Originally, the concession acquisition process involved some 47 steps and 5 agencies, taking an estimated 14 months and costing 7,000 lempiras ($2,500 in 1990) (Guevara, 1991). Those with considerable political and economic resources could get the leases, no matter their management ability.

The appropriate length of leaseholds merits further study. Some farmers complain that a leasehold of 20 years provides insufficient security and is too short a period in which to expect farms to make investments in land improvement (Weidner et al., 1992). Debates in Ecuador have centered around whether full private tenure to shrimp farmers would enhance loan access and land management and reduce deforestation (Meltzoff and LiPuma, 1986). Numerous options are available to allow change in the leasehold tenure (such as transferability and performance-based renewals) that consider both the state and the private sector's interests (Shotton, 1987).

The pricing of the leasehold has also produced a lively debate. A simple linear programming exercise carried out in 1993 suggests that appropriate land is the biggest constraint in the industry. The production dynamics inherent in mariculture cause investors to be willing to pay a great deal to acquire this land (D. Stanley, 1994). The linear programming model used data from a representative 200-hectare farm to calculate the optimal production plan of stocking densities and harvest dates. It also provided the relevant "shadow prices" for the scarce resources of production associated with an efficient operating plan.[10] Depending on how much seed is stocked, the farm has an economic demand of $8 to $514 per hectare per four-month cycle. The highest value corresponds to a May stocking with a harvest in late October to match the "winter window" in the U.S. market.

The current leasehold rental price of $5 per hectare, a fixed price ceiling, is far below this economic value and the reported values on the black market of leasehold swapping among concession holders.[11] Excess demand exists for artificially cheap, scarce land. Fixing the land concession fee too low could be sending the wrong signal about the true value of the resource and causing investors to use the wetlands in a unsustainable manner. If the opportunity cost of the land (the foregone wetland benefits) is not considered and reflected in the concession price, the apparent benefits to society of converting the area to mariculture are overestimated and conversion occurs too rapidly.[12] Setting a fixed fee without regard to location or land quality does not provide the ideal signals to investors about which soils are optimal for mariculture and which are not.

Finally, unclear norms for appropriate user behavior have worsened the externality problem. Under Decree 85 of the Honduran Forest Code, illegal mangrove cutting is subject to a sanction of approximately $400 per hectare. By April 1989, more than 900,000 lempiras ($450,000) in fines were registered for mangrove deforestation at the local COHDEFOR (Honduran Corporation for Forest Development) office (COHDEFOR, 1989). However, there are political difficulties and high costs in regularly enforcing the rules against disturbances to mangroves and other intertidal vegetation (see Pérez and Robadue, 1989, on Ecuador). This selective enforcement creates confusing expectations about the penalties for harmful land use.

The Honduran state has been unable to police the actions of users in the mangrove areas or accept an alternative enforcement mechanism offered by private groups. As a result, community-based organizations, such as CODDEFFAGOLF, have taken the lead in denouncing mangrove destruction in the sympathetic Honduran press and international forums. In a parallel program, the National Association of Aquaculturists of Honduras (ANDAH), recently presented an area management plan to several international donors and is attempting to complete a Mariculture Business Ethics Code (ANDAH, 1994).

Another motivation for externalities is unbalanced and incomplete information. This factor also explains part of the rationale behind user behavior and the state's surprising actions in the assignment of property rights and the enforcement of environmental norms. A crucial lack of knowledge about the attributes and functioning of wetland ecosystems has plagued biologists and planners for decades. I observed two salient information debates during fieldwork and attendance at international meetings.

First, it is often claimed that shrimp farmers face atypically high risks with respect to the environment and the international marketplace. Entrepreneurs face substantial investment costs and possible bankruptcy if two or three harvests fail.[13] There are no agricultural insurance programs to cover investors' losses, and subsidized credit for this nontraditional export has been reduced since 1990. Many investors argue that premiums for bearing risks and incentives, such as cheap land, are needed to encourage wise use and offset the high sunk costs and "unusual" perils in this export industry (Kingsley, 1986). Another concern is that high land costs could lead farmers to use higher seed-stocking rates and "overintensify" the industry (Chamorro, 1993).[14] Of course, it could also be argued that the farmers enjoy the possibility of large returns and that a land concession program is not the most efficient way to reduce production risk (D. Stanley, 1994).

Second, many planners still believe that artisanal fisheries systems are inefficient and the mangrove areas are worthless. Aquaculture has been described as an effective method of privatizing fisheries to avoid their overexploitation. The controlled raising and harvesting of fish under a private property regime supposedly leads to higher productivity (Titenburg, 1994). Simultaneously, wetlands, salt flats, and mangroves are called "marginal, unproductive areas" in much of the export promotion literature and debates. This discourse overlooks the mangrove's values to traditional gatherers and the functions of wetlands in the overall coastal ecosystem.[15] Traditional uses of wetlands generate economic value.[16] The problem of insufficient information has made it difficult for land values or compensation to affected users to be determined by the market or regulatory mechanisms.

Toward Options for More Sustainable Use of the Mangrove Ecosystem

Dissecting these debates over the use of wetlands and the dynamics of mariculture is the challenge for policymakers and donors who are trying to promote a more socially and environmentally sustainable export industry. This chapter has reviewed the factors that contribute to conflicts between the mariculture industry and nearby communities. The analysis suggests that the intrinsic private and common pool characteristics of wetland attributes and insufficient knowledge about wetland services have encouraged misguided user actions. Yet in the Honduran case, the inadequate property rights bundle that guide resource use is probably the most compelling explanation for the conflicts that have occurred. The property rights institutions that could be modified include the type of tenure, the length of tenure, the price and costs for resource use, and norms about socially acceptable actions among users. While the current moratorium makes it difficult to reform the Honduran concession program at this stage, countries that are starting natural resource export programs can analyze different alternatives to guide behavior on public lands. There are many outstanding questions:

- Which users should acquire tenure and access rights to these natural resources? Should wetlands be private property, to be sold to the highest bidder? Or should the leasehold system continue, and if so, under what form of allocation rules?
- If wetlands are to remain in the leasehold arrangement, what is the appropriate length for the rental period? Should a management plan be required and inspected for compliance? Should leaseholds be transferrable?
- What is the appropriate rental fee for public lands? Should the price include a tax for the off-site damage done by a farm? Should the price be based on the marginal returns to the land, with variations by zone and land type? Or should the price merely cover the alternative uses of the land?[17]
- What are the suitable mechanisms to police inappropriate user actions in public lands? If the state cannot undertake a regulatory function, which local user groups have an advantage in this role? Is community-based management logical in coastal areas or is corporate control preferable?

The present debates call for the creation of more incentive-compatible institutions and the recognition of traditional values and uses of the wetland ecosystem. Controversies among interdependent resource users are resolvable. The Honduran case highlights the conclusion made by Ruttan: conflicts are not static or insurmountable. Site-specific details and the intricacies of fish stocks, plants, and swamps become important in determining the well-being and compatibility of natural resource users and the ecosystem.

Notes

1. Like a broiler grow-out industry, the cultivation of farm-raised shrimp follows five stages: (a) obtaining seed (postlarva); (b) acclimation and transfer of seed to grow-out ponds; (c) feeding and fattening of the shrimp in the ponds; (d) harvest from the ponds; and (e) cleaning, deheading, and packing the shrimp for export.

2. A salf flat is dry land that builds up after natural or man-made disturbances to the mangrove ecosystem.

3. Externalities represent a persistent interdependence among producers in which one user's actions hold uncompensated cost implications for another (Bromley, 1989).

4. To better understand the resource use patterns and interactions of the mariculture industry, I collected both qualitative and quantitative data in southern Honduras over a 15-month period of fieldwork in 1993. Records of repeated visits and formal interviews of 220 households were collected in three villages affected by shrimp farming to document the trends of land use changes, employment generation, and incomes. Interviews with shrimp farm managers, operating eight different-sized holdings, were undertaken to discuss input demands and costs. I also conducted numerous meetings with maricultural representatives, labor contractors, environmental activists, and other informants and attended three international conferences on maricultural issues. Secondary data sources (i.e., consultant reports and government documents) also were reviewed during the course of 1992–1993.

5. Or as Vergne et al. (1993) write, the preferred sites for farms are barren mud flats above the elevation of the leeward fringe of mangroves. Unfortunately, these characteristics of ideal pond siting are not known to all investors.

6. Parks and Bonifaz (1994) cite the destruction of 41,000 hectares of mangroves in Ecuador by 1993. Yet throughout the shrimp farm debate these numbers have been disputed by the mariculturalists, and the inability of researchers to decipher the aerial photography statistics by subzone, wetland characteristic, farm type, and tenure pattern have prevented a constructive debate on mangrove deforestation.

7. For a mathematical formulation of this externality problem, see Parks and Bonifaz (1994).

8. Rivalry (or subtractability) is a characteristic of resources when one person's use diminishes the resource's value to others. Excludable resources are those in which the owners or providers of the resource have low-cost methods to charge for or to keep others from using it (Salazar and Leonard, 1994).

9. Broadly, property rights are sets of ordered relationships among people that define their opportunities, their exposure to the acts of others, their privileges, and their responsibilities (Bromley, 1989). Within this spectrum it is common to refer to a natural resource tenure regime as a triad relationship among different parties and a resource, including the rules for access to a resource and the rules about its use.

10. A shadow price is the value of the marginal product of an input, or the price of the final product multiplied by the marginal contribution of the input. This is the maximum price a farm would be willing to pay for one more unit of a scarce input; different businesses have different shadow prices for the same resource, depending on the rent generated from its use.

11. Sales prices of $8,000 to $10,000 per hectare were quoted to me for the sale of concessioned shrimp land in the Chismuyo zone.

12. As Barbier (1994, p. 155) writes: "An analysis of trade-offs between conserving or converting tropical wetlands demonstrates that taking into account the opportunity cost of wetland loss leads to a lower level of conversion than would otherwise be the case." Or as López (1992, p. 1142) concludes: "As the shadow value of environmental resources increases, firms may want to conserve these resources more effectively. The problem is that in the reality of the developing world, few environmental resources are costed at their shadow value."

13. Shang (1990, p. 42) cites maricultural sunk investment costs of $15,000 per hectare.

14. A similar debate over lease fees and production techniques has arisen over grazing permits on public lands in the western United States. To date, the evidence shows that higher fees *reduce* overgrazing and land use intensity (Johnson and Watts, 1989; Lambert, 1995).

15. Debates between fishers and mariculturalists highlight their divergent perceptions of resource value. For instance, gatherers argue that a nearly extinct species of crab extracted for subsistence (*ponche*) is threatened by shrimp farming as mangroves are cut. Yet businesspeople reply that this animal lives only in the already-cleared salt flats, not mangroves, and its economic value is negligible.

16. The value of direct products extracted from mangroves (firewood, construction materials, and fish products) has been estimated in the range of $60 to 2400 per hectare per year (Dixon, 1989); recent calculations of the value of Nicaraguan mangroves, including their tourism benefits and direct products, are $68 to $120 per hectare per year (Windoxhel, 1994). The ecological functions of the whole mangrove ecosystem (both the trees and the "useless" salt flats) also include groundwater recharge, flood control, shoreline stabilization, and sediment retention; no study has successfully placed a monetary figure on these values.

17. Parks and Bonifaz (1994) have suggested that a land tax on the "discounted stock effects of additional mangrove habitat" (i.e., the marginal damage done to interdependent habitat through mangrove disturbance) could provide farmers with incentives to internalize the off-site costs of their actions and reduce deforestation; Schatz (1991) has suggested that the rental price of public wetlands should at least cover the alternative use value of wood products as the "opportunity cost of the land."

27

Conclusion

Upton Hatch
Marilyn Swisher

To balance the goals of preserving biodiversity and maintaining large segments of natural ecosystems with the needs and aspirations of a growing human population is a global challenge that is vividly illustrated by the experiences of Mesoamerica. The introduction to this volume discussed seven hypotheses, each of which brought into focus an unresolved issue for Mesoamerica's citizens and decision makers as they attempt to achieve the dual goals of development and resource preservation. The contributors to this volume have provided different, sometimes conflicting interpretations of how best to address these issues. Whether discussing the social aspects of development, the status of the region's natural resource base, or alternative approaches to agricultural production in the region, each author highlights one or more of the questions raised. This chapter focuses on some of the points that the authors make concerning these hypotheses and point to future directions for research and action in the region.

A Difference of Opinion

The first hypothesis raised the issue of scale, pointing to the limitations of microscale research in resolving many of the problems of Mesoamerica. Hatch and Swisher (chapter 1) point out that "most people, including farmers, tend to be concerned with their immediate environment. . . . Yet, over the long term, the larger-scale phenomena may determine whether the individual's immediate environment is sustainable"; they call for research directed at regional agroecosystems. Other authors, particularly those concerned with the status of Central America's natural resource base, echo this call for larger-scale analyses, and a general consensus seems to emerge that microscale research, while useful and necessary, will not answer many of the questions about resource use and agricultural development in the region.

Carroll and Kane (chapter 8), for example, point to the impact of larger-scale phenomena on small-scale features: "As the rural landscape is converted to agricultural uses, natural patches of habitat are reduced and isolated and their internal quality degraded." Boza (chapter 6) shares a similar concern for the maintenance of biologically functional conservation units in Mesoamerica. He describes the 17 large reserves of primary importance for conserving biodiversity in the region but notes that "despite the development of a large number of protected areas in Mesoamerica, loss of biodiversity remains a critical problem. The protected areas are often small, isolated, and fragmented." Pringle and Scatena (chapter 12) also emphasize the multiscalar nature of anthropogenic effects on natural ecosystems, describing human impacts on aquatic ecosystems, ranging from "within-stream effects [that] produce organic and inorganic pollution" to "global changes such as global warming [that] affect the entire hydrologic cycle." In a later chapter (chapter 13), they present case studies from Puerto Rico and Costa Rica that demonstrate the multiscalar nature of aquatic ecosystem deterioration.

Concentrating more on agricultural ecosystems, Rocheleau (chapter 5) also argues for multiple-scale research when she states: "Repositioning the household within the global economy affects the position of the household in the community, of individuals within the household, and of both within local ecosystems." She ends by calling for "a research framework that illuminates the complex patterns embedded in multiple and overlapping domains of resource use, access, and control. . . . Development of sustainable and just economies and ecologies will require a telephoto lens of observation across scales."

The second hypothesis argued that intensification of agricultural production on lands appropriate for agriculture is the best way to achieve the sustainability of agriculture in Central America. Contributors to the volume have widely differing views about the desirability and indeed the feasibility of intensification. Among those who argue for the need to intensify, several approaches emerge, ranging from replacing traditional systems with input-intensive industrial models to making modest changes in input use in traditional systems. The impacts of intensification are also described by several authors. Again, conclusions vary, ranging from those whose experience shows that intensification is not sustainable to others who believe that the effects of intensification are positive over the mid- to long-term.

Ruttan (chapter 2) argues strongly for intensification, opening his discussion with the following comments: "Throughout most of human history increases in agricultural output have been achieved almost entirely from increases in cultivated areas. We are in the closing decades of the 20th century, rapidly approaching the time when all increases in agricultural production will have to come from increases in the intensity of cultivation on the lands already used for agricultural production." He goes on to point out the "spillover" effects from agricultural intensification, such as soil erosion, waterlogging and salinization, contamination of water supplies, and loss of natural habitats, and he points out that these negative effects are only enhanced when agriculture expands into more fragile environments.

Vásquez (chapter 11) provides more detail about how land is actually used in Central America. His analysis indicates that about 23 percent of the land in Mesoamerica is suitable for agriculture. In 1990, however, nearly 40 percent of the region's land was already in use for agricultural production. He echoes Ruttan's call for intensification of production on suitable land: "The lands with the highest potential for agriculture are, in fact, often underexploited because they have been used as grazing lands. Moreover, the lands with the lowest potential are overexploited, mainly in the production of crops for internal consumption by small farmers."

Hagnauer (chapter 19), in contrast, takes issue with the concept that land use for grazing represents underexploitation of land resources, and he argues against intensification as a strategy for achieving sustainability. Discussing early attempts to replace cattle with crops on La Pacifica and other farms in northwestern Costa Rica, he says: "Many of these crops did survive in local gardens for home consumption, but they could never contribute significantly to the commercial income stream." He argues, in fact, that it is cattle that makes crop production sustainable: "The pasture, with careful herd management, replaced the necessary fallow for fragile tropical soils and guaranteed the sustainability of a cropping system in the dry tropics." Later, he explicitly argues against intensification of crop-based agriculture in the semiarid parts of Mesoamerica: "Small farms created by land reform . . . will have problems with sustainability and economic survival. In addition, large-scale irrigation projects . . . will encounter sustainability problems. To live well on a few hectares, the farmer has to intensify his or her operation. Intensification in the tropics brings a farm to the edge of mining the soil."

Moreno's (chapter 17) discussion of changes in the maize and bean intercrop system that was so highly characteristic of pre-Columbian Mesoamerican agriculture is especially interesting. He discusses both the benefits and the drawbacks of intensification, describing two major effects on the system, both of which increase labor productivity. Discussing the impact of these changes on maize production in one of the major maize-producing regions of Mesoamerica, he comments: "The increased labor productivity is not due primarily to an increase in yield. Monocultural maize yields are higher only by 90 to 130 kilograms per hectare than yields obtained when maize is planted in association with beans." He concludes: "Overall yields are highly reflective of the level of technology applied. Profitability, however, does not necessarily reflect the technology used." Moreno also notes that intensification has had some negative social impacts, largely because of the reduced demand for labor: "Although it is difficult to generalize the effects of the recent policies of structural adjustment and stabilization . . . it appears that agricultural investment for food crops has declined. This decline is largely due to the apparent reductions in real income among the labor class. These laborers and their families are an important component of the demand for food crops."

Rosemeyer, Schlather, and Kettler (chapter 16) also explicitly address the issues of intensification, focusing on increasing productivity of a traditional mulch-based bean production system, *frijol tapado*. They conclude that intensification is necessary and possible but that the type of intensification is crucial, and they call for the use of on-farm rather than off-farm resources. They describe the traditional system as "land-intensive" because of the need for a fairly long fallow period between bean crops, and they report on experiments conducted to reduce the fallow by applying fertilizer and planting nitrogen-fixing trees. They compare the productivity of the traditional system, their intensified *tapado* system, and a "modern" system of clean-till bean monoculture. They found that the improved *tapado* system does show higher yields than the traditional *tapado* system and that both systems are more profitable and yield as well or better than the modern clean-till system. They conclude: "If productivity of land is to be increased in a sustainable manner, inputs are critical. The type of inputs—renewable or internally or externally derived—and whether they are efficiently used by the crop plant are important criteria."

Our third hypothesis argued that the introduction of exotic species in Mesoamerica has resulted in inherently less sustainable agroecosystems than those based on indigenous biological resources. Several authors comment on the massive transformation of the

Mesoamerican landscape that occurred as a result of European settlement. Popenoe and Swisher (chapter 15) state: "The agricultural landscape that we see today is, in many ways, two landscapes, one superimposed on the other. The pre-Columbian components consist of indigenous crops. . . . Over this landscape, we find European crops such as wheat and, most important, livestock such as cattle and sheep." Vargas (chapter 10) discusses the landscape transformations that resulted from the introduction of plantation crops. He describes, for example, the changes that occurred in the 19th century as banana production became the key to many Mesoamerican agricultural economies. "1882 saw one of the most massive land use transformations. The government . . . gave a contract to Soto-Keith, permitting [him] to construct a railway from San Jose to the Atlantic. Under this contract, Keith received property rights to 6.3 percent of the national territory. Large-scale banana production began on these lowland Caribbean lands, destroying many hectares of very wet tropical rainforest." This transformation continues today. The emphasis on production for export by both the private and public sectors means that more nonindigenous crops, such as strawberries, snow peas, and carrots, are now becoming common on both small and large farms (see Rocheleau, chapter 5; Stonich, Bart and Ovares, chapter 24; and others).

Although many authors discuss the role of nonindigenous components in both food crop and forestry systems, differing and conflicting views about the role of nonindigenous crops, animals, and cropping systems in the sustainability of agriculture in Mesoamerica emerge. Swisher and Popenoe, for example, point out that pre-Columbian cropping systems were well adapted to the many ecological settings in Mesoamerica: "A large variety of crops was available to pre-Columbian farmers in Mesoamerica. . . . Relying largely on increasing knowledge, labor, and local resources, farmers evolved systems over the centuries that provided stable food supplies . . . [and] allowed major population centers to evolve." They call for a new paradigm for research and agricultural development, focused on popularizing little-known traditional crops for nontraditional markets (such as the United States); providing data on nutritional value, crop production, and pest management strategies for native species and cultivars; and developing training materials for rural extension activities. Rosemeyer, Schlather, and Kettler's approach is similar. They, too, call for a focus on traditional production, although they stress ways to enhance the productivity of mulched beans.

Fisher's (chapter 23) discussion of the role of indigenous versus introduced species in forestry systems offers another interesting perspective. While arguing that monocultural forestry plantations have a critical role to play in supplying fuel and pulpwood in the region, he criticizes the current reliance on a limited number of exotic species and calls for more use of native species. He offers two basic arguments for increased reliance on indigenous germplasm: "First, native species may be better suited for the reclamation of degraded sites than exotic species. Second, they may serve as catalysts to speed the redevelopment of secondary natural forest."

Despite this apparent agreement about the value of indigenous crops and cropping systems, differences of emphasis and interpretation are important. For example, Popenoe and Swisher's call for a focus on the export of traditional crops to nontraditional markets falls within the definition of NTAE by Rocheleau: "The phrase *nontraditional agricultural exports* applies to a wide range of crops. . . . The term simply implies any crop not traditionally grown as a cash crop for export in a given place. In fact, it refers more to a new, flexible way of organizing export production and marketing fruits and vegetables than to any specific set of crops." She is highly critical of the impact of NTAE. Although Rocheleau does not comment specifically on the paradigm that Popenoe and Swisher propose, her criticisms of the NTAE strategy suggest that she would disagree with their approach.

The greatest disagreement about the role of exotic germ plasm centers on the role of livestock in Mesoamerica. Here, opinions vary widely. Livestock production was originally concentrated in the semiarid parts of the region. Vargas (chapter 10) notes, "Forest clearing also occurred under the auspices of large landowners. Very large cattle ranches came into being . . . developed by clearing tropical dry forests." Mayne (chapter 9) states: "The major areas in Central America now undergoing deforestation are the eastern and Caribbean lowland portions of Honduras and Nicaragua. . . . It is the forest in the wettest and most inaccessible areas of Central America that have been left undisturbed the longest." He points to the "triumvirate" of road construction, cattle ranching, and subsistence farming as the major cause for deforestation in the last 40 years in Central America. Boza and Popenoe and Swisher also comment on the role that livestock, especially cattle, have played in deforestation.

Other authors, however, argue that cattle production is not destructive, especially Hagnauer and Jiménez. It is significant that both authors point to the importance of milk production in sustainable livestock production systems, and both argue for production systems based on the forage production capacity of the tropical environment.

The livestock production system in Hagnauer's La Pacifica was a low-input "dual use" in which both meat and milk were produced: "This system produces 1 kilogram of milk and 85 grams of edible meat per day of lactation. . . . This balanced production will not cause surpluses of any one of the two products on the market." He goes on, however, to argue for using livestock breeds that are not highly "improved" genetically: "No gains are obtained by upgrading the breeds to include more than 50 percent improved genetic material. This tactic [produces] resistant animals that give a sustainable and balanced yield of milk and meat produced on locally available forage resources."

Jiménez (chapter 20) argues that milk production systems and extensive ranching operations differ greatly and that cattle systems based on milk production do not contribute significantly to deforestation: "Few farmers clear land to raise cattle for milk production. . . . Milk production systems are largely more intensively managed than beef production systems. Often, small farmers, rather than large-scale ranchers, play an important role in providing milk to a nation's population." Like Hagnauer, he argues that achieving the highest productivity per unit animal—or maximizing the genetic potential of the milk cow—is not an appropriate goal for livestock systems in the tropics. Rather, he calls for an approach based on achieving maximum yield per unit area of land—or maximizing the use of indigenous forage resources: "Tropical environments are typified by vegetation with the highest rates of biomass in the world. . . . They also have the highest biodiversity of forage plants, including both grasses and legumes. These many species exhibit many adaptive, productive characteristics."

Both of these authors, then, reject midlatitude models for livestock production, but they do argue that livestock have an important, positive contribution to make to the sustainability of agriculture in Mesoamerica. Their call for livestock systems that maximize the use of "internal or indigenous resources" echoes the call for greater reliance on native species made by other authors. Thus, while there is significant disagreement about the impact of livestock production on sustainability, there is apparently a general agreement about the need for greater reliance on native germ plasm both for animal and for plant production.

The fourth hypothesis, clearly related to the previous discussion, argues that the Central American landscape is also a product of dependence on and domination by external cultures and economies, another factor that reduces the sustainability of agriculture in the region. Many authors comment directly and indirectly on this issue. They tend to agree that Mesoamerica is highly interdependent on other economies, but they disagree about the potential negative impact of interdependence. Beginning with the introduction of European crops and livestock after the conquest, the interdependence of Central America's economies increased as plantation agriculture and coffee production grew in the 19th century. Today, globalization, market liberalization, and monetary restructuring and stabilization all provide an increased impetus to integrated local and global economies.

Rocheleau describes the process well, focusing on the linkages that are forged between global and local systems:

> The latest wave of economic restructuring and the rise of nontraditional agricultural exports (NTAE) have pulled many more smallholders into more specialized, monocrop production. . . . This has, in turn, tighly linked many farm households and entire communities into global markets as consumers of both imported staple foods and manufactured goods. As a consequence, farm households have . . . dramatically increased the exchanges of energy and materials between their plots and larger national and international systems.

Rocheleau clearly argues that increased interdependence and linkage have primarily a negative impact on smallholders in the region: "Households vary greatly in their ability to participate in these restructured economies and ecologies on favorable, or at least viable, terms of control and exchange. Even those who cannot or choose not to participate in NTAE initiatives are affected." She points out that the impact of globalization on households and communities are affected by many characteristics, including gender, age, ethnicity, and social class. In a telling discussion of the negative effects, she states:

> Finally, rural people find themselves increasingly enmeshed in international commerce to satisfy basic needs for food and clothing. . . . As a result, they are increasingly vulnerable to decreases in the prices offered for traditional agricultural commodities and to increases in the cost of food. Often they must enter into NTAE ventures and/or wage labor to make ends meet. Price increases for food and other imports erode the ability of rural farmers to resist selling land to agribusiness or converting to NTAE monocrops.

Moreno shares many of Rocheleau's concerns about the impact on small, subsistence farmers: "Structural adjustment will affect small and mid-sized farmers that produce basic grains such as maize and beans to the degree that they participate in the market. Without doubt, those mid-sized farmers that, after much effort, have become creditors will be most affected." He goes on to comment that new opportunities for international markets will also affect different groups of farmers differently: "Subsistence farmers are not likely to benefit from the increased orientation to overseas markets. . . . Market opportunities are more likely to be available to larger farmers. . . . Small and medium-sized commercial farmers will often not be able to afford the increasing interest rates that have generally been associated with stabilization programs." In summary, Moreno and

Rocheleau appear to argue, although from different perspectives, that the globalization of markets is likely to have an adverse impact on both small and medium-sized landholders, and on both subsistence and commercial farmers.

Several contributors to this volume discuss the growing mariculture industry in Central America, an interesting example of differences in opinion about both the degree to which export-oriented agriculture contributes to and relies on indigenous resources and the potential benefits of export production for farmers in the region. As Stanley (chapter 26) points out: "The entry of shrimp farm enterprises into wetland ecosystems has been greeted with both cynicism and enthusiasm."

Stonich, Bort, and Ovares (chapter 24) clearly argue that the mariculture option, or "Blue Revolution," has not, by and large, had a favorable impact:

> In each country small, private producers and cooperatives face decisive constraints on their effective integration into the industry. In its initial stages the industry provided many, but violently contested, low-paying, seasonal jobs to poor rural people. However, these jobs will probably decrease. . . . Furthermore, the industry does not offer an environmentally sustainable alternative. To the contrary, the expansion of shrimp mariculture is an important factor in the decline in biological diversity of coastal wetlands.

Teichert-Coddington (chapter 25), in contrast, argues that local shrimp producers are achieving self-organization and addressing both environmental and economic issues. Discussing the Honduran industry, he says: "The . . . industry has been strong for more than a decade, thereby demonstrating its long-term commitment to the region. . . . [It] has collaborated on research to understand its impact on the aquatic environment. Data are collected from farms and estuaries by producers and research personnel and are analyzed and published." He also argues, in contraposition to Stonich et al., that sustainable management of the mangrove resource is feasible: "The shrimp industry depends on mangrove forests for its own health and development. Mangroves are a resource utilized directly or indirectly by artisanal fishers, local firewood gatherers, and builders, in addition to shrimp producers. This use of a renewable common resource should be emphasized to develop collaborative mangrove forest management practices and regulatory policy."

This debate over the role and future of shrimp mariculture captures, in essence, the ongoing discussion of the role of exotic versus indigenous resources and interdependence versus independence for Mesoamerica's rural population. Other authors in the volume make similarly contrasting comments about other sectors of the agricultural economy, including livestock production, forestry, and agroforestry.

Our fifth hypothesis stated that "from a local perspective, production for export does not mine natural resources any more than production for national, urban consumers. Both have negative impacts on the long-term sustainability of local resources and communities." Contributors to this volume demonstrate widely differing opinions concerning these statements. While some agree, others argue that production for national consumption is inherently more sustainable over the mid- to long term than production for international consumption.

On the one hand, Pringle and Scatena (chapter 13) argue that production activities destined for consumers outside the watershed, wherever they may be, greatly increase the pressure on the water resource base. On the other hand, Soto (chapter 18) suggests that organic farming makes it possible to produce for "external" markets while maintaining environmental quality and simultaneously improving incomes. She argues that small farmers are able to participate in this technology as well as large farmers. Rocheleau counters that organic farming often places greater pressure on family labor. The result is that women and children are often the primary sources of labor for organic systems. Rocheleau asserts that this added burden on family labor ultimately decreases educational opportunities for female children and off-farm employment for women.

Once again, the mariculture examples provide distinctive and differing viewpoints on this issue. Stonich asserts that shrimp mariculture is "socially, economically, and environmentally unsustainable" and argues that aquaculture has often involved a transfer of ownership from the local poor to high-income local elites or foreign interests. She stresses the emphasis on export promotion as a major factor in environmental decline and social injustice. Teichert-Coddington counters that shrimp farmers have an incentive to maintain the quality of their water supply and have begun to form resource management groups for collective action; he argues that the shrimp industry has enhanced the standard of living in areas where mariculture is practiced, without long-term degradation of the natural ecosystem.

Despite his concerns over the impact of structural adjustment policies associated with globalization and international agricultural marketing, M[illegible] (chapter 17) argues that these changes will generally decrease the importance of agriculture and improve environmental quality because small and medium-sized producers will decease their use of credit and purchased inputs, thereby moving to more environmentally friendly production practices that involve fewer chemicals. Rocheleau, however, argues almost the opposite—that increased globalization results in not only greater poverty but also increased damage to local ecosystems as smallholders are forced to produce for international markets to meet their own needs for cash.

In summary, there is little agreement about the role of production for international markets on the local resource base. Case studies provide examples of both benefits and disadvantages. Despite this disagreement, however, most contributors do foresee enormous changes coming for Mesoamerican agroecosystems as a result of the globalization of agricultural markets, particularly for small and medium-sized farmers.

Hatch and Swisher also argue that "economic sustainability depends on prices reflecting scarcity. Therefore, for long-term sustainability, economic policy must address ways to ensure that prices reflect resource scarcity." The role of institutions, especially public institutions, versus the effects of a free market economy is an important consideration in this discussion. Ruttan (chapter 2) argues that prices that reflect the scarcity of depletable and unique ecosystems have been imposed by institutions that have been developed to force consideration of the externalities caused by production. These institutions have been largely imposed through public regulatory agencies, although in some cases the initiative came from voluntary private arrangements to maintain the quality of the resource base on which an industry depends, for example, shrimp mariculture in Honduras (Teichert-Coddington, chapter 25).

The importance of public regulatory agencies versus nongovernment formal or nonformal organizations in valuing natural resources is discussed by several authors. Some argue that public agencies have been slow to develop and have lacked enforcement capabilities in much of the area. Stonich asserts, on the one hand, that shrimp mariculture has not been sufficiently regulated, resulting in an undervalued mangrove resource. The values placed on these resources by local, poor populations are generally ignored.

Stanley (chapter 26), on the other hand, describes institutional arrangements and property rights developed by local populations that are useful in ensuring that incentives exist to manage resources efficiently. Although many of the arrangements are informal in nature, they allow resource values to be captured by individuals or groups who wish to protect the quality of the resource base. Stanley's discussion supports the notion put forth by Ascher that local populations are capable of being good resource managers if institutional arrangements, among other requirements, are met. Ascher describes several critical issues for successful community resource management, most of them involving institutional commitment: "(1) government recognition of user rights, (2) specification of the kinds of exploitation that will be prohibited or regulated, (3) economic context in which community resource management will function, (4) whether the government will provide compensation for leaving some resources intact, and (5) how the user group will organize itself."

Szott (chapter 21) suggests that many forest products are undervalued relative to their scarcity. However, he is optimistic that small landowners can identify and compete in global market niches by using such agroforestry practices as organic coffee, medicinal plants, and organic tropical timber. Nonetheless, he argues that credits or incentives for tree planting by small farmers would be an effective strategy for increasing incomes and promoting environmental quality. Mayne (chapter 22) illustrates the complexities of local institutions in establishing long-term, viable agroforestry systems.

Finally, Hatch and Swisher focus on how to prioritize conservation efforts. They argue that because we cannot know future demands and preferences and because public and private resources available for conservation efforts are finite, setting priorities or conditions to guide effective conservation efforts are needed. Many authors share this concern. Boza (chapter 6) documents the need to preserve large, contiguous areas to sustain many of the important animal species in Mesoamerica. Ascher (chapter 3) asserts the viability of local community resource managers if appropriate organizational structures and requirements are to be met. Irreversible damage has been done to dryland ecosystems, largely by urban and agricultural land uses (Vargas, chapter 10); institutions must be developed to maintain the quality of the ecosystems that remain.

Several authors focus on the importance of environmental education and citizens' involvement in developing an effective balance between economic growth and environmental conservation. De la Rosa (chapter 14) calls for research institutions to facilitate this involvement and encourage scientific interchange among scientists worldwide. Rosemeyer, Schlather, and Kettler (chapter 16) and Vásquez (chapter 11) suggest the idea of delineating certain ecosystems or land classes as unsuitable for agriculture or other degrading land uses. Vásquez argues that the database needed to conduct detailed land use capability studies in the region do not exist, a major barrier to effective planning. Rosemeyer et al. are concerned that research that makes traditional systems more attractive could lead to the establishment of a system on fragile lands, especially hillsides, a concern shared by Moreno (chapter 17). Pringle and Scatena (chapters 12 and 13) describe the development of innovative programs to conserve aquatic resources on a watershed basis. They are encouraged by this strategy, but they are also discouraged by the slow progress of many parts of Mesoamerica in instituting this policy.

The issue of irreversibility as it relates to the degradation of mangrove and other coastal ecosystems by shrimp mariculture is contested by Stonich et al. (chapter 24) and Teichert-Coddington (chapter 25). The former strongly supports the notion that these ecosystems have been irreversibly degraded and that local populations have not benefited from the rapid development of an internationally competitive shrimp mariculture industry. Teichert-Coddington counters that in fact much of the land used for shrimp ponds were salt flats with little value to the local population and that destruction of mangroves has been minimal. The issue is not whether these coastal ecosystems will be developed, but how much and where. In opposition to Stonich's assertion that local groups have not benefited, he describes in detail the involvement of local farmers not only in the commercial production but also in attempts to organize themselves to protect environmental quality.

Toward the Future

Environmental degradation in Mesoamerica continues unabated, with no apparent improvement in the standard of living for most people in the region. Returns to raw resources have tended to decline over the past several decades, implying that the resource had to be extracted and exploited at an ever increasing rate to make up for the decline in unit price. Mesoamerica's exploitation of its resource base provides a classic example of the environmental, social, and economic deterioration that result from this approach to resource management and development.

Further, the megatrends discussed in the introduction to this volume are likely to exacerbate the social and economic disparities that drive unmanaged resource exploitation. In the new global economy, there will be winners and losers in every country of the world. The notion of the "industrialized nations versus the developing countries" is, to a large degree, an artifact of an earlier era. The new reality, which has been with us for a long time, is that the upper classes and well educated benefit most from the changes that are occurring. They have the best chance of competing in international markets for their consumer products. Skilled professionals will be the winners, regardless of their nationality. Low-skilled workers will generally be the losers, again regardless of nationality. However, the larger economies do have a better link to the rule making that occurs, and they will be better placed to cushion the impact of change on their less skilled, economically disadvantaged citizens. Thus, low-skilled workers in small countries—that is, much of Mesoamerica's population—will be the big losers unless appropriate institutions and policies can be developed.

The key issue for Mesoamerica is how to both protect its resources and enhance the standard of living of the population as a whole in the face of a changing global economy and structure. A safety net for the losers must be an important element in future government policy. Furthermore, the resource base of Mesoamerica must be conserved. Ecologically, the region contains a significant portion of the world's biodiversity, as Boza and others point out. Locally, as Pringle and Scatena, Rocheleau et al., and others indicate, environmental deterioration places an even greater burden on the poorest segments of the population. Agricultural production, now and in the future, depends on this resource base. Furthermore, these resources are often undervalued today in world market prices. The opportunity cost of the region's environmental resources is high, as recreation and genetic diversity increase their value and supply falls in the industrialized nations. The unique environmental assets should not be undersold. Thus, both governments and citizens in the region need to be good stewards of their natural resource base.

The contributors to this volume do not have answers to the many issues that this dilemma poses. Nonetheless, we can point to certain key components in developing a strategy for sustainable development in Central America.

First, what Rocheleau calls a "telefocus lens across scales" is critical. Too often, macroscale studies of policy and economics and microscale studies of ecology and agricultural innovation are divorced. Both suffer—especially in small, poorer nations—from the lack of an adequate database, whether that be the soils inventory that Vásquez argues for or the regional database for dryland species espoused by Vargas or a national, economic database called for by Ruttan. Furthermore, a continual interplay between macro-level and micro-level study is needed. Landscape ecology (Carroll and Kane, chapter 8) offers one approach. Experimentation with alternative local and regional forms of self-governance and resource management, such as those described by Ascher and Teichert-Coddington, offers another. Examination of macroscale changes on household structure and function, described by Rocheleau, is yet another component. The common thread among these approaches is a move to examine the relationships between local phenomena and larger-scale national and international policies, economies, and structure.

Second, greater attention must be paid to the tension between individual and societal good. The megatrends discussed in the introduction are apt to exacerbate an already highly skewed distribution of income and wealth. Wealthy members of these societies will continue to invest their savings, not in their local economies, but in investment opportunities in the industrialized world. Furthermore, resources are undervalued, and the environmental cost of [illegible] has not been incorporated into either individual or national decision making.

Most important are the problems based on differences between individual and social well-being. For example, when natural resources are undervalued, it is profitable to the individual to exploit them at current prices. However, the returns to the natural resource base can be increased by increased prices for exploitation. On the one hand, the individual needs to survive and prosper today, and these are legitimate needs. On the other hand, it is equally clear that reducing exploitation of natural resources has the best chance of improving the long-term standard of living for the nation and region. Government policies, in both the industrialized nations and the developing nations, must reconcile these conflicts and develop scenarios that permit rational exploitation today while preserving resources for the future. Otherwise, as Carroll points out, the result is either "ecologically dysfunctional but economically attractive landscapes" or "ecologically functional but economically problematic landscapes."

Finally, prioritization is key. Priorities must include both environmental and economic components. Land use planning can play a key role here, not in the traditional sense described by Rosemeyer et al. (e.g., as a way to direct activities), but rather as a way to direct public investment. Infrastructure investment should be targeted to areas with the best chance of producing returns. Similarly, areas with the greatest environmental importance should be targeted for public investment in resource protection. Boza provides many examples.

The role of indigenous resource management, described by Ascher and others, can play a key role. Indigenous populations in Mesoamerica had sophisticated property rights before colonization. It was not some romantic notion of an environmental ethic that caused them to conserve and manage their resources effectively. Rather, they depended on the resource base for survival. As a result, they developed very specific rules for resource exploitation that were essentially private in nature. Command and control methods that we often use today are generally inferior to the processes developed by many indigenous populations. However, a simple reinstatement of indigenous groups is nearly impossible, as Ascher points out. The world has changed. What is useful is to study these traditional systems, learn from them, and develop ways to incorporate indigenous management into the changed world of the peoples of Mesoamerica today.

Perhaps La Pacific serves best to illustrate many of the differing opinions and strategies for the future discussed in this volume. The 40 years of experience with this privately owned cattle ranch are one of the few cases that we have, worldwide, in which the goals of environmental and resource preservation and economic sustainability based on production for both the national and international market were studied and recorded. Certainly, Hagnauer's thesis that cattle ranching is an economically and environmentally sustainable form of agricultural development for the dry Mesoamerican tropics is controversial. Many will no doubt find his conclusions disturbing, especially his criticisms of land reform and agricultural intensification. Nonetheless, La Pacifica is an example of successful, private sector resource conservation and profitability, and the long interaction of national and international scientists with the farm and its managers certainly represents the kind of public and private collaboration that de la Rosa and others propose.

This volume has focused on the interactions between human and natural systems that require an enhanced appreciation of potentially conflicting influences of biophysical and sociocultural-economic factors. Analyses are needed that transcend scales and disciplines. Neither micro- nor macroscale insights will be useful in addressing many of the issues presented in this volume if no attention is given to how these scales interact. Similarly, disciplinary analyses must be distilled into coherent strategies that incorporate the message of each individual discipline. In addition, conflicts between the perspectives and motives of individuals and those of society are likely to increase as some members of society are able to reap benefits from globalization that are not generally available to other, large segments of society. The challenge to land use planning and resource management will be the incorporation of indigenous and individual rights that provide the appropriate incentives to pursue the often conflicting social goals of economic development and environmental quality.

References

Abramski, Z., and M. L. Rosenzweig. 1984. Tilman's predicted productivity-diversity relationship shown by desert rodents. *Nature* 309:150–151.

Abt Associates. 1990a, May. Honduras natural resource policy inventory. Executive Summary, vol. I. USAID/ROCAP RENARM Project. APAP II Technical Report no. 111. Bethesda, Md.

Abt Associates. 1990b. Honduras natural resource policy inventory. Policy Inventory, vol. II. USAID/ROCAP RENARM Project. APAP II Technical Report no. 111. Bethesda, Md.

Adler, R. W., J. C. Landman, and D. M. Cameron. 1993. *The Clean Water Act twenty years later*. Island Press, Washington D.C.

Adriance, J. 1995. Planting the seeds of a new agriculture: Living with the land in Central America. *Grassroots Development* (Inter-American Foundation) 19(1):2–17.

Alegre, J. C. 1991. Runoff and erosion losses under forest, low-input, and alley cropping on slopes in the humid tropics of Peru. *Agronomy Abstracts*, p. 58. American Society of Agronomy 1991 Annual Meeting, Denver, Col. American Society of Agronomy, Madison, Wisc.

Alfaro, R. 1994. Mejorando el Frijol Tapado. In Tapado. Los sistemas de siembra con cobertura, pp. 221–226. D. Thurston, M. Abawi Smith, and S. Kearl (eds.). CATIE-CIIFAD, Cornell University Press, Ithaca, N.Y.

Alho, C. J. R., T. E. Lacher, Jr., and H. C. Goncalves. 1988. Environmental degradation in the Pantanal ecosystem. *BioScience* 38:164–171.

Allen, Patricia, and Carolyn Sachs. 1993. *Sustainability in the balance: What do we want to sustain?* Agroecology Program and the Focused Research Activity in Agroecology and Sustainable Agriculture, University of California, Santa Cruz.

Allen, Patricia, and Debra Van Dusen. 1990. Sustainability in the balance: Raising fundamental issues. Summary paper of the conference *Sustainable Agriculture: Balancing Social, Economic, and Environmental Concerns*, sponsored June 1990 by the Agroecology Program, University of California, Santa Cruz.

Allen, Patricia, Debra Van Dusen, Jackelyn Lundy, and Stephen Gliessman. 1992. Integrating social, environmental, and economic issues in sustainable agriculture. *American Journal of Alternative Agriculture* 6:34–39.

Allen, Patricia, Debra Van Dusen, Jackelyn Lundy, and Stephen Gliessman. 1993. *Sustainability in the balance: Expanding the definition of sustainable agriculture.* Agroecology Program, University of California, Santa Cruz.

Almendares, J., M. Sierras, P. K. Anderson, and P. R. Epstein. 1993. Critical regions, a profile of Honduras. *Lancet* 342:1400–1403.

Al-Mufti, M. M., C. L. Sydes, S. B. Furness, J. P. Grime, and S. R. Band. 1977. A quantitative analysis of shoot phenology and dominance in herbaceous vegetation. *Journal of Ecology* 65:759–791.

Altieri, M. 1987. *Agroecology: The scientific basis of alternative agriculture.* Westview, Boulder, Col.

Altieri, Miguel A. 1988. Beyond agroecology: Making sustainable agriculture part of a political agenda. *American Journal of Alternative Agriculture* 3:142–143.

Alva, C. A., J. G. van Alphen, A. de la Torre, and L. Manrique. 1976. Problemas de drenaje y salinidad en la costa Peruana. *International Institute for Land Reclamation and Improvement Bulletin* 16:28.

ANDAH (National Association of Aquaculturists of Honduras). 1994. *La industria camaricultura buscando asegurar la viabilidad a largo plazo*. Forum Proceedings, Choluteca, Honduras.

Anderson, Anthony E. (ed.). 1990. *Alternatives to deforestation: A sustainable use of the Amazon rainforest*. Columbia University Press, New York.

Anderson, D., and R. Grove (eds.). 1987. *Conservation in Africa: People, policies and practice*. Cambridge University Press, London and New York.

Andrén, H. 1995. Effects of landscape composition on predation rates at habitat edges. In *Mosaic landscapes and ecological processes*, pp. 225–255. L. Hansson, L. Fahrig, and G. Merriam (eds.). Chapman and Hall, London.

Annis, S. 1990. Debt and wrong-way resource flows in Costa Rica. *Ethics in International Affairs* 4:107–121.

Aragón, A. 1981. Evaluación bioeconómica de un hato de doble propósito en el trópico monzónico de Costa Rica. Tesis M.Sc. CATIE, Turrialba, Costa Rica.

Aragón, A., and W. Hagnauer. 1980. Análisis de la ganaderia en la cuenca del arenal. In estudio ecológico integral de las zonas de afectación del proyecto arenal. Centro Cientifico Tropical, Director. San José.

Aragón, A., W. Hagnauer, and W. Kropf. 1976. Ganado y Leguminosas de doble propósito. Su manejo y sus ventajas. Mimeo. Día del Campo para el Personal de Extensión Agrícola, Hda. La Pacífica, Cañas, Costa Rica.

Aranda, M. 1991. Wild mammal skin trade in Chiapas, Mexico. In *Neotropical wildlife use and conservation*, pp. 174–177. J. G. Robinson and K. H. Redford (eds.). University of Chicago Press, Chicago.

Arango, Ochoa Raúl. 1992. Derechos territoriales indígenas y ecología en las selvas tropicales del América Latina. Gaia Foundation Bogotá, Columbia.

Araya, Pochet C. 1973. La minería y sus relaciones con la acumulación de capital y la clase dirigente: 1821–1841. *Revista de Estudios Sociales Centroamericanos* 5:31–64.

Araya, R., and W. González. 1986. *El frijol bajo el sistema tapado en Costa Rica*. CIAT, Cali, Colombia.

Arcia, G., L. Merino, and A. Mata. 1991. *Modelo interactivo de población y medio ambiente*. Procesos litográficos. Análisis y proyecciones para el Valle Central, San José, Costa Rica.

Arévalo, L. A., L. T. Szott, and J. M. Pérez. 1993. El pijuayo como componente de un sistema agroforestal. In *IV Congreso Internacional Sobre Biología, Agronomía, e Industrialización de Pijuayo*, pp. 267–286. J. Mora Urpí, L. T. Szott, M. Murillo, and V. M. Patiño (eds.). Universidad de Costa Rica, San José.

Argel, P. 1991, December 2. Arachis pintoii: A new tropical pasture legume. Paper presented at the 1991 CIAT Annual Program Review, International Center for Tropical Agriculture, Cali, Colombia.

Arias, R. 1995. Farmer from Bajo Los Arias, Acosta-Puriscal, Costa Rica. Personal communication.

Arizpe, Lourdes, Fernánda Paz, and Margarita Velásquez. 1993. *Cultura y cambio global: Percepciones sociales sobre la deforestación en La Selva Lacondona*. Miguel Angel Porrua, Mexico.

Arizpe, L., M. P. Stone, and D. Major (eds.). 1994. *Population and environment: Rethinking the debate*. Westview, Boulder, Col.

Arizpe, Lourdes, and Margarita Velasquez. 1994. The social dimensions of population. In *Population and environment: Rethinking the debate*, pp. 67–86. L. Arizpe, M. P. Stone, and D. Major (eds.). Westview, Boulder, Col.

Arriagada, Irma. 1992. Mujeres rurales de América Latina y el Caribe: Resultados de programas y proyectos. In *Una Nueva Lectura: Género en el desarrollo*. Virginia Guzman, Patricia Portocarrero, y Virginia Vargas (eds.). Ediciones Populares Feministas, CIPAF, Santo Domingo.

Ascher, William. 1994. *Communities and sustainable forestry in developing countries*. Institute of Contemporary Studies Press, San Francisco.

Aspinal, D. 1960. An analysis of competition between barley and white persicaria. II. Factors determining the course of competition. *Annals of Applied Biology* 48: 637–654.

Augspurger, C. K. 1983. Phenology, flowering asynchrony, and fruit set of six neotropical shrubs. *Biotropica* 15: 257–267.

Babbar, L., and D. R. Zak. 1994. Nitrogen cycling in coffee agroecosystems: Net N mineralization and nitrification in the presence and absence of shade trees. *Agriculture, Ecosystems, and Environment* 48:107–113.

Bachmann, F. 1983. Verkehrsmilchleistung der Mutterkuh und Wachstum des Kalbes in einem "doble propósito"—Betrieb in Costa Rica. Thesis, Federal Institut of Tecnology, Zürich.

Bailey, Conner. 1992. Coastal aquaculture development in Indonesia. In *Coastal aquaculture in developing countries: Problems and prospects*, pp. 102–121. R. Pollnac and P. Weeks (eds.). ICMRD, University of Rhode Island, Kingston.

Bandeen, J. D., and K. P. Buchholtz. 1967. Competitive effects of quackgrass upon corn as modified by fertilization. *Weeds* 15:220–224.

Barber, S. A. 1984. *Soil nutrient bioavailability*. Wiley-Interscience, New York.

Barbier, E. 1994. Valuing environmental functions: Tropical wetlands. *Land Economics* 70(2):155–173.

Barborak, J. R. 1992. History of protected areas and their management in Central America. In *Changing tropical forests: Historical perspectives on today's challenges in Central and South America*. H. K. Steen and R. P. Tucker. (eds.). Proceedings of a conference sponsored by the Forest History Society and IUFRO Forest History Group, San José, Costa Rica.

Barfield, C. S., and M. E. Swisher. 1994. Historical context and internationalization of IPM. *Food Reviews International* 10(2):215–267.

Barham, Bradford, M. Clark, E. Katz, and R. Schurman. 1992. Nontraditional agricultural exports in Latin America. *Latin American Research Review* 27(2): 43–82.

Barker, M. R., and W. A. Wunsche. 1977. Plantio direto in Rio Grande do Sul. *Outlook in Agriculture* 9:114–120.

Barquero, M. 1995, July 14. Agro espera fondos por c5,200 milliones. *La Nación*, San José.

Barrera, A. 1976. *Nomenclatura etnobotánica Maya: Una interpretación taxonómica*. Colección Científica no. 36. Instituto Nacional de Antropología e Historia, México.

Barrera, A. 1977. El manejo de las selvas por los Mayas: Sus implicaciones silvícolas y agrícolas. *Biotica* 2(2):47–61.

Bartone, C. R., and H. J. Salas. 1984. Developing alternative approaches to urban wastewater disposal in Latin America and the Caribbean. *Pan American Health Organization Bulletin* 18:323–327.

Batie, Sandra. 1989, December. Sustainable development: Challenges to the profession of agricultural economics. *American Journal of Agricultural Economics*, pp. 1085–1101.

Beer, J. 1987. Advantages, disadvantages, and desirable characteristics of shade trees for coffee, cacao, and tea. *Agroforestry Systems* 5:3–13.

Beer, J. 1988. Litter production and nutrient cycling in coffee (*Coffea arabica*) or cacao (*Theobroma cacao*) plantations with shade trees. *Agroforestry Systems* 7:103–114.

Beer, J. 1993. Consideraciones básicas para el establecimiento de especies maderables en linderos. Proyecto Agroforestal/GTZ, CATIE, Turrialba, Costa Rica.

Beer, J. 1994. Alternativas de reforestación: Taungya y sistemas agrosilviculturales permanentes vs. plantaciones puras. Serie Técnica, Informe Técnico no. 230, Proyecto Agroforestal/GTZ, CATIE, Turrialba, Costa Rica.

Bel, Ingeniería S. A., San José, Costa Rica, and Bookman and Edmunton, California. 1976. Estudio de Factibilidad de la Cuenca Baja del Rio Tempisque. vol 2.

Bellows, B. C. 1992. Sustainability of steep land bean (*Phaseolus vulgaris L*) farming in Costa Rica: An agronomic and socioeconomic assessment. Ph.D. thesis, University of Florida, Gainesville.

Belsky, A. J. 1995. Spatial and temporal landscape patterns in arid and semi-arid African savannas. In *Mosaic landscapes and ecological processes*, pp. 31–56. L. Hansson, L. Fahrig, and G. Merriam (eds.). Chapman and Hall, London.

Benavides, S. T. [1963] 1976. Fractionation of soils from the Llanos Orientales de Colombia. In *Properties and management of soils in the tropics*. P. Sanchez. Wiley, New York.

Benstead, J. P., J. G. March, C. M. Pringle, and F. N. Scatena. (In press). Effects of water abstraction and damming on mitratory tropical stream biota: Simulation modeling and mitigation strategies. *Ecological Applications*.

Berry, S. 1989. Social institutions and access to resources. *Africa* 59(1):41–55.

Bertoni, J., and F. Lombardi. 1985. *Conservaçao do solo*. Livruceres, San José. In Lal, R. 1995b. Erosion control on sloping land with conservation tillage.

BID-PNUMA (Banco Interamericano de Desarrollo y Programa de las Naciones Unidas para el Desarrollo). 1991. *Nuestra propia agenda*. BID-PNUMA, Washington, D.C.

Billings, W. D. 1938. The structure and development of old-field pine stands and certain associated physical properties of the soil. *Ecology Monographs* 8:437–499.

Birdsey, R. A., and P. L. Weaver. 1987. Forest trends in Puerto Rico. U.S. Department of Agriculture Forestry Service, Southern Forest Experiment Station Research Note SO-331. New Orleans.

Blaikie, Piers. 1985. *The political economy of soil erosion in developing countries*. Longman, London.

Blaikie, Piers. 1988. The explanation of land degradation in Nepal. In *Deforestation: Social dynamics in watersheds and mountain ecosystems*, pp. 132–158. J. Ives and D. C. Pitt (eds.). Routledge, London.

Blaikie, Piers, and Harold Brookfield. 1987. *Land degradation and society*. Methuen, London and New York.

Board on Agriculture, Board on Science and Technology for Development. 1992. *Sustainable agriculture development in the humid tropics*. National Academy Press, Washington, D.C.

Board on Agriculture, National Research Council. 1991. *Sustainable agriculture research and education in the field*. National Academy Press, Washington, D.C.

Bonner, R. 1993. *At the hand of man: Peril and hope for Africa's wildlife*. Alfred A. Knopf, New York.

Boon, P. J. 1992. Essential elements in the case for river conservation. In *River conservation and management*, pp. 11–34. P. J. Boon, P. Calow, and G. E. Petts (eds.). John Wiley, New York.

Boon, P. J., P. Calow, and G. E. Petts (eds.). 1992. *River conservation and management*. Wiley, New York.

Borchert, R. 1997. Responses of tropical trees to rainfall seasonality and its long-term changes. *Climate change* (in press).

Bort, John, and James C. Sabella. 1992. Small-scale shrimp aquaculture in the Bay of Parita, Panama. In *Coastal Aquaculture in Developing Countries: Problems and Prospects*, pp. 87–101. R. Pollnac and P. Weeks (eds.). ICMRD, University of Rhode Island, Kingston.

Boserup, E. 1965. *The conditions of agricultural growth: The economics of agrarian change under population pressure*. Aldine, New York.

Boyce, J. K., A. Fernández, E. Furst, and O. Segura Bonilla. 1994. *Café y desarrollo sostenible: Del cultivo agroquímico a la producción orgánica en Costa Rica*. Editorial Fundación UNA., Heredia, Costa Rica.

Boza, M. A. 1993. Conservation in action: Past, present and future of the national park system in Costa Rica. *Conservation Biology* 7:239–247.

Boza, M. A., D. Jukofsky, and C. Wille. 1995. Costa Rica is a laboratory, not ecotopia. *Conservation Biology* 9:684–685.

Bray, David. 1991a. The forests of Mexico: Moving from concessions to communities. *Grassroots Development* 15(3):16–17.

Bray, David. 1991b. The struggle for the forest: Conservation and development in the Sierra Juarez. *Grassroots Development* 15(3):13–24.

Bray, David, Marcelo Carreon, Leticia Merino, and Victoria Santos. 1993. On the road to sustainable forestry. *Cultural Survival Quarterly* 17(1):38–42.

Brenchley, W. E. 1922. Some factors in plant competition. *Annals of Applied Biology* 6:142–170.

Brenes, A. M. 1953. Con la comisión sustriece en Guanacaste. *Boletín Informativo del Museo Nacional* 2(3):8–45.

Brennan, P. 1996, June 14. Brewery moves to Nicaragua. *Tico Times*. San José.

Brenner, M. 1994. Lakes Salpeten and Quexil, Peten, Guatemala, Central America. In *Global geological record of lake basins*, vol. 1. E. Gierlowski-Kordesch and K. Kelts (eds.). Cambridge University Press, Cambridge.

Brock, J. A., R. Gose, D. V. Lightner, and K. Kasson. 1995. An overview on Taura Syndrome, an important disease of farmed *Penaeus vannamei*. In *Aquaculture '95*. S. Hopkins and C. Browdy (eds.). Proceedings of the Special Session on Shrimp Farming, San Diego, Calif. World Aquaculture Society, Baton Rouge, La.

Bromley, D. 1989. *Economic interests and institutions*. Basil Blackwell, New York.

Bronstein, J. L. 1995. The plant-pollinator landscape. In *Mosaic landscapes and ecological processes*, pp. 256–288. L. Hansson, L. Fahrig, and G. Merriam (eds.). Chapman and Hall, London.

Bronstein, J. L., P. H. Gouyon, and C. Gliddon. 1990. Ecological consequences of flowering asynchrony in monoecious figs: A simulation study. *Ecology* 71:2145–2156.

Brookfield, Harold. 1991. Environmental sustainability with development: What prospects for a research agenda? *The European Journal of Development Research* 3(1): 42–66.

Brul, P. B., and B. J. van Elzaker. 1995. La comercialización de productos orgánicos en Europa. In *Simposio Centroamericano sobre agricultura orgánica*. J. García and J. Najera (eds.). Marzo. Universidad Estatal a Distancia, San José.

Bryant, Raymond L. 1992. Political ecology: An emerging research agenda in Third World studies. *Political Geography* 11(1):12–36.

Buckles, D. Ponce, G. I. Sain, and G. Medina. 1994. Tierra cobarde se vuelve valiente: El uso y difusión del frijol de abono *(Mucuna deeringianum)* en laderas del litoral Atlántico de Honduras. In *Tapado. Los sistemas de siembra con cobertura*, pp. 277–301. D. Thurston, M. Abawi Smith, and S. Kearl (eds.). CATIE-CIIFAD (Centro Internacional de Mejoramiento de Maiz y Trigo) CIMMYT, El Batan, Mexico.

Bunyard, Peter. 1989a. *The Colombian Amazon: Policies for the protection of its indigenous peoples and their environment*. Ecological Press, Cornwall, England.

Bunyard, Peter. 1989b. Guardians of the Amazon. *New Scientist* 1695:38–41.

Burkey, T. V. 1993. Edge effects in seed and egg predation at two neotropical rainforest sites. *Biological Conservation* 66:139–143.

Bussing, W. A. 1987. *Peces de las aguas continentales de Costa Rica*. Editorial de la Universidad de Costa Rica, San José.

Buttel, Frederick. 1991, May 10–12. Knowledge production, ideology, and sustainability in the social and natural sciences. Paper presented at Conference on Varieties of Sustainability, Alsimolar, Calif.

Butterfield, R. P. 1994. Forestry in Costa Rica: Status, research priorities, and the role of La Selva Biological Station. In *La Selva: Ecology and natural history of a neotropical rain forest*, pp. 315–328. L. A. McDade, K. S. Bawa, and G. S. Hartshorn (eds.). University of Chicago Press, Chicago and London.

Butterfield, R. P. 1995. Promoting biodiversity: Advances in evaluating native species for reforestation. *Forest Ecology and Management* 75:111–121.

Butterfield, R. P., R. P. Crook, R. Adams, and R. Morris. 1993. Radial variation in wood specific gravity, fiber length and vessel area for two Central American hardwoods: *Hyeronima alchorneoides* and *Vochysia guatemalensis*. *IAWA Journal* 14:153–161.

Butterfield, R. P., and R. F. Fisher. 1994. Untapped potential: Native species for reforestation. *Journal of Forestry* 96:37–40.

Calderón, F., H. Sosa, V. Mendoza, G. Sain, and H. Barreto. 1990. Aspectos institucionales de la adopción y difusión de la labranza de conservación en Matalio-Guaymango. In *Análisis de ensayos regionales de agronomía*. PRM-CIMMYT, Guatemala, Guatemala.

Calderón, S., and P. C. Stanley. 1941. *Flora Salvadoreña*. Imprenta Nacional, San Salvador, El Salvador.

Caldwell, M. M., D. M. Eisenstat, J. H. Richards, and M. F. Allen. 1985. Competition for phosphorous: Differential uptake from dual-isotope-labeled soil interspaces between shrub and grass. *Science* 229:384–386.

Caldwell, M. M., J. H. Richards, J. H. Manwaring, and D. M. Eisenstat. 1987. Rapid shifts in phosphate acquisition show direct competition between neighboring plants. *Nature* 327:615–616.

Calvo, G., and C. L. de la Rosa. 1993. Estudio de impacto ambiental, Canal de Alivio Río Zapote, Upala, Provincia de Alajuela, Costa Rica. Ministerio de Obras Públicas y Transportes, Costa Rica.

Calvo, G., C. L. de la Rosa, and D. Norman. 1994. Estudio de impacto ambiental, Río Bijagua, Upala, Provincia de Alajuela, Costa Rica. Ministerio de Obras Públicas y Transportes, Costa Rica.

Calvo, J. C. 1990. The Costa Rican national conservation strategy for sustainable development: Exploring the possibility. *Environmental Conservation* 17:355–358.

Camino, R. De. 1993. El papel del bosque húmedo tropical en el desarrollo sostenible de América Central: Desafíos y posibles soluciones. *Revista Forestal Centroamericana* 2(6):7–16.

Campbell, B. M., P. N. Bradley, and S. E. Carter. 1994. Sustainability and peasant farming systems. Some observations from Zimbabwe. Stockholm Environment Institute, Stockholm.

Campbell, C. 1996. Out on the front lines but still struggling for voice: Women in the rubber tappers' defense of the forest in Xapuri, Acre, Brazil. In *Feminist Political Ecology*, pp. 27–61. D. Rocheleau, B. Thomas-Slayter, and E. Wangari (eds.). Routledge, London.

Cardoso, C. F. S. 1973. La formación de la hacienda cafetalera en Costa Rica. *Estudios Sociales Centroamericanos* 6(2):9–55.

Carney, Judith. 1988. Struggles over land and crops in an irrigated rice scheme: The Gambia. In *Women and Land Tenure in Africa*, pp. 59–78. Jean Davison (ed.). Westview, Boulder, Col.

Carrillo, M., E. Estrada, and T. C. Hazen. 1985. Survival and enumeration of the fecal indicators *Bifidobacterium adolescentis* and *Escherichia coli* in a tropical rainforest watershed. *Applied Environmental Microbiology* 50:468.

Carroll, C. R., and D. Kane. In press. Landscape ecology of transformed neotropical environments. In *Tropical managed ecosystems*. U. Hatch and M. Swisher (eds.). Oxford University Press, New York.

Carroll, C. R., and G. K. Meffe. 1994. Management to meet conservation goals: General principle. In *Principles of conservation biology*, pp. 307–335. G. K. Meffe and C. R. Carroll (eds.). Sinauer Associates, Sunderland, Mass.

Carson, Rachel. 1962. *Silent spring*. Houghton Mifflin, Boston.

Carter, J. 1995a, June. Alley farming: Have resource-poor farmers benefitted? Natural Resource Perspectives (Overseas Development Institute) vol. 3.

Carter, J. 1995b, June. Socio-economic factors influencing alley farming. Natural Resource Perspectives (Overseas Development Institute) vol. 3, suppl. A.

Carter, J. 1995c, June. The technical performance of alley farming: Farmer experiences. Natural Resource Perspectives (Overseas Development Institute) vol. 3, suppl. B.

Carter, J., G. Marrow, and V. Pryor. 1991. Aspects of the ecology and reproduction of Nassau grouper, *Epinephelus striatus*, off the coast of Belize, Central America. Gulf and Caribbean Fisheries Institute, Miami, Fla.

Carter, S. E. 1991. Análisis geográfico del uso de la tierra en Centroamerica. In *Agosto*, pp. 19–75. Agricultura Sostenible en las Laderas Centroamericanas: Oportunidades de Colaboración Interinstitucional. Memorias del Taller. IICA Coronado, Costa Rica.

Castañeda, O. 1995. La agricultura orgánica en Guatemala. In *Proceedings of the Simposio Centroamericano sobre agricultura orgánica, Marzo*. J. García, and J. Najera (eds.). Universidad Estatal a Distancia, San José.

CATIE (Centro Agronómico Tropical de Investigación y Enseñanza). 1995. Serie Técnica. Informe Técnico 243. Reforestación con maderables. CATIE, Turrialba, Costa Rica.

CATIE. 1990. *Situación actual de la producción, industrialización y comercialización de la leche en Centro América*. Centro Agronómico Tropical de Investigación y Enseñanza. Area de Ganadería Tropical. Turrialba, Costa Rica. (Elaborado para el Banco Centroamericano de Integración Económica). Serie Técnica. Boletin Técnico/CATIE no. 21.

CEDECO (Corporación Educativa para el Desarrollo Costarricense). 1992. Memoria del Ier Encuentro Nacional Campesino de Frijol Tapado, July 12–14, 1991, San Ignacio de Acosta, Costa Rica.

Cedeño, H. 1982. Grandes razgos de la evolución geomorfológica cuaternaria del area de Larigua, Cocle. M.S. thesis, Centro Regional de Cocle, Panama.

Chaboussou, F. 1995. *A teoríada trofobiose. Novos caminhos para uma Agricultura Sadia*. Fundaçao Gaia, Porto Alegre, Brazil.

Chambers, R., A. Pacey, and L. A. Thrupp. 1989. *Farmer first: Farmer innovation and agricultural research*. Intermediate Technology Publications, London.

Chamorro, R. 1993. Coordinator of Mariculture Non-Traditional Export Promotion Program, FPX. Personal communication.

Chapman, Duane, and Randolph Barker. 1991. Environmental protection, resources depletion, and the sustainability of developing country agriculture. *Economic Development and Cultural Change* 39:723–737.

Chetley, A. 1987. Bitter harvest in Costa Rica. *South* 80:107–108.

CIAT (Centro Internacional de Agricultura Tropical). 1986. Bean research pay-off in Costa Rica. *CIAT Internacional* 5:3–5.

CIAT (Centro Internacional de Agricultura Tropical). 1993. Informe anual del programa de Laderas, Cali, Colombia.

Clarke, M. R. 1982. Socialization, infant mortality and infant nonmother interaction in howling monkeys (*Alouatta palliata* in Costa Rica). Thesis, University of California, Davis.

Clarke, M. R., E. L. Zucker, and N. J. Scott. 1986. Population trends of the mantled howler groups at La Pacífica. Guanacaste, Costa Rica. *American Journal of Primatology* 11:79–88.

Clements, F. E., J. E. Weaver, and H. C. Hanson. 1929. *Plant competition: An analysis of community functions*. Carnegie Institute, Washington, D.C.

COHDEFOR (Honduran Corporation for Forest Development). 1989. Cortes Ilegales al Bosque de Manglar en el Golfo de Fonseca para la Instalación de Proyectos Acuícolas and Denuncia Forestal no. R.F.F.M. #34/89. Mimeos. Choluteca, Honduras.

Colchester, M. 1990. The international timber organization: Kill or cure for the rainforests? *Ecologist* 20:166–173.

Colchester, M. 1993. Colonizing the rainforests: The agents and causes of deforestation. In *The struggle for land and the fate of the forests*. M. Colchester and L. Lohmann (eds.). Zed Books, London.

Cole, H. E., and A. E. Holch. 1941. The root habits of certain weeds in southeastern Kansas. *Ecology* 22:141–147.

Collier, B. 1991. Poison blooms in America's foreign garden. *Austin America–Statesman* 24.

Collins, Jane. 1991. Women and the environment: Social reproduction and sustainable development. In *The women and development annual*, vol. 2, pp. 33–58. R. S. Gallin and A. Ferguson (eds.). Westview, Boulder, Col.

Committee for Sustainable Agriculture. 1990, January. *Asilomar declaration for sustainable agriculture*. Agroecology Program, University of California, Santa Cruz.

Committee on Global Change of the Commission on Geoscience, Environment and Resources. 1990. *Research strategies for the U.S. Global Change Research Program*. National Academy Press, Washington, D.C.

Committee on Science, Engineering and Public Policy. 1991. *Policy implications of greenhouse warming*. National Academy Press, Washington, D.C.

Committee on the Role of Alternative Farming Methods in Modern Production Agriculture. 1989. *Alternative Agriculture*. Board on Agriculture, National Research Council. Washington, D.C.

CONAP (Consejo Nacional de Areas Protegidas). 1990. *Reserva de la biosfera Maya*. CONAP, Guatemala, Guatemala.

Conklin, N. 1987. The potential nutritional value to cattle of some tropical browse species from Guanacaste, Costa Rica. Thesis, Cornell University, Ithaca, N.Y.

Connell, Des. W., and D. W. Hawker. 1992. *Pollution in tropical aquatic systems*. CRC Press, Ann Arbor, Mich.

Connell, J. H. 1978. Diversity in tropical rainforests and coral reefs. *Science* 199:1302–1310.

Connell, J. H. 1980. Diversity and the coevolution of competitors, or the ghost of competition past. *Oikos* 35:131–138.

Connell, J. H. 1983. On the prevalence and relative importance of interspecific competition: Rvidence from field experiments. *American Naturalist* 122:661–696.

Conroy, Michael E., Douglas L. Murray, and Peter R. Rosset. 1994. The fruits of crisis in Central America: Gambling on non-traditional agriculture. Unpublished manuscript.

Consejo Agropecuario Centroamericano. 1994. Informe sobre los efectos de la sequía en la producción de granos básicos en Centroamérica. Mimeo Consejo Agropecuario Centroamericano, San José, Costa Rica.

Constanza, Robert. 1989. What is ecological economics? *Ecological Economics* 1:1–7.

Conway, Gordon R., and Jules N. Pretty. 1991. *Unwelcome harvest: Agriculture and pollution*. Earthscan Publication, London.

Conway, K., D. Harding, I. Ruderfer, and K. Maxwell. 1991. Home gardens in Coto Brus county: Forces influencing small scale diversification. Student-directed research project for sustainable development course, School for Field Studies, Beverly, Mass.

CORBANA (Corporacion Bananera Nacional). 1994. *Annual statistics report*. CORBANA, San José.

Córdoba, H. S., H. Barreto, and J. Crossa. 1993. Impacto del desarrollo de híbridos de maíz en Centro América: Confiabilidad de las ganancias en rendimiento sobre el genotipo H5 y consideraciones para selección de testigos regionales. In *Síntesis de resultados experimentales 1992*. J. Bolaños, G. Sain, R. Urbina, and H. Barreto (eds.). PRM-CIMMYT, Guatemala.

Courtemanch, D. L. 1994. Bridging the old and new science of biological monitoring. *Journal of the North American Benthological Society* 13:117–121.

Covich, A. P., T. A. Crowl, S. L. Johnson, D. Varza, and D. L. Gertain. 1991. Post-hurricane Hugo increases in aytid shrimp abundances in a Puerto Rican montane stream. *Biotropica* 23:448–454.

Cox, G. W., and M. D. Atkins. 1979. *Agricultural ecology: An analysis of world food production systems*. W. H. Freeman, San Francisco.

CRIES (Comprehensive Research Inventory Studies—Michigan State University). 1984. *Resource assessment of the Department of Choluteca*. RRNN, RRNW (Ministry of Natural Resources), Tegucigalpa, Honduras.

Cruz, María L. 1992. Evaluation of the impact of shrimp mariculture development upon rural communities in Mexico. In *Coastal aquaculture in developing countries: Problems and prospects*, pp. 54–73. R. Pollnac and P. Weeks (eds.). ICMRD, University of Rhode Island, Kingston.

Csavas, I. 1993. Aquaculture development and environment issues in the developing countries of Asia. In *Environment and aquaculture in developing countries*. R. S. V. Pullin, H. Rosenthal, and J. L. Maclean (eds.). ICLARM Conference Proceedings 31, International Center for Living Aquatic Resources, Manila.

Cummings, R. G. 1989. *Improving water management in Mexico's irrigated agricultural sector*. World Resources Institute, Washington, D.C.

Current, D., and E. Lutz. 1992. A preliminary economic and institutional evaluation of selected agroforestry projects in Central America. World Bank, Environment Department, Policy and Research Division Working Paper no. 1992–38, Washington, D.C.

Currie, D. J. 1994. Ordenamiento de camaronicultura, Estero Real, Nicaragua. Programa Regional de Apoyo al Desarrollo de la Pesca en el Istmo Centroamericano (PRADEPESCA), Apartado Postal 645, Balboa, Ancón, Panama.

Dahlberg, Kenneth A. 1991. Sustainable agriculture: Fad or harbinger. *BioScience* 41:337–340.

Daly, Herman E. 1991, July. From empty world economics to full world economics: Recognizing an historical turning point in economics development. In Environmentally sustainable economic development: Building on Brundtland. Robert Goodland, Herman Daly, and Salah El Serafy (eds.). World Bank Environment Working Paper no. 46:18–26, Washington, D.C.

Darlin, P. 1982. *Distribución de la estación seca en los países Centroamericanos*. Centro Agronómico Tropical de Enseñanza y Investigación, Turrialba, Costa Rica.

Darwin, C. 1859. *On the origin of species by means of natural selection, or the preservation of favored races in the struggle for life*. D. Appleton, New York.

Daubenmire, R. 1972. Some ecological consequences of converting forest to Savanna in northwestern Costa Rica. *Tropical Ecology* 13.

Davison, J. (ed.). 1988. *Agriculture, women and land in Africa*. Westview, Boulder, Col.

Day, J. W. J., C. A. S. Hall, W. M. Kemp, and A. Yánez-Arancibia. 1989. *Estuarine ecology*. Wiley, New York.

Deere, Carmen. 1990a. *Households and class relations: Peasants and landlords in northern Peru*. University of California Press, Berkeley.

Deere, Carmen. 1990b. *In the shadow of the sun: Caribbean development alternatives and U.S. policy*. Westview, Boulder, Col.

Deere, Carmen Diana, and Magdalena Leon (eds.). 1987. *Rural women and state policy: Feminist perspectives on Latin American agricultural development*. Westview, Boulder, Col.

De Janvry, Alain, Elisabeth Sadoulet, and Erik Thorbecke. 1993. Introduction to state, market, and civil organizations: New theories, new practices, and their implications for rural development. *World Development* 21(4): 565–575.

de la Rosa, C. L. 1992. Introducción a la biología de ríos tropicales. Manuscript used in several workshops, Monteverde Conservation League, San José.

de la Rosa, C. L. 1993a. Julio Hernández Sequeira: Uniendo educación y conservación. Oficina de Manejo Ambiental, USAID Costa Rica. *Revista de Historia Natural, Ecología y Conservación* 1:32–33.

de la Rosa, C. L. 1993b. Lo que el río se llevó. Oficina de Manejo Ambiental, USAID Costa Rica. *Revista de Historia Natural, Ecología y Conservación* 1:32–33.

de la Rosa, C. L. 1995a. Diagnóstico sobre el estado de la biodiversidad de la cuenca del Río San Juan, Organization of American States (OAS), Proyecto de Manejo Ambiental y Desarrollo Sostenible de la Cuenca del Río San Juan.

de la Rosa, C. L. 1995b. Diagnóstico sobre el estado de la información disponible sobre biodiversidad para la Cuenca del Río San Juan, Organization of American States (OAS), Proyecto de Manejo Ambiental y Desarrollo Sostenible de la Cuenca del Río San Juan.

de la Rosa, C. L. 1995c. Middle American streams and rivers, In *River and stream ecosystems: Ecosystems of the world*, vol. 22, chap. 7. C. E. Cushing, K. W. Cummins, and G. W. Minshall (eds.). Elsevier, Amsterdam.

de la Rosa, C. L. In press. *Guía para la identificación de las sub-familias y tribus de Chironomidae (Diptera) de Costa Rica*. Editorial FIREMA, San José.

de la Rosa, C. L., and N. C. Barbee. 1993. Guía de los organismos comunes de las aguas dulces de Costa Rica. Publicaciones especiales de la Oficina de Manejo Ambiental, USAID, Upala, Costa Rica.

de la Rosa, C. L., and N. C. Barbee. 1994. Rapid bioassessment protocols for tropical streams: A study in northern Costa Rica. Internal report, USAID, San José.

de la Rosa, C. L., and A. Ramírez. 1995. A note of phototactic behavior and on phoretic associations in larvae of *Mecistogaster ornata* Rambur from northern Costa Rica (Zygoptera: Pseudostigmatidae). *Odonatologica* 24:219–224.

Dembner, S. 1991. Provisional data from the forest resources assessment 1990 project. *Unasylva* 164:40–44.

de Mendoza, Hurtado L. 1994. Managua, Nicaragua, The Nature Conservancy. personal communication.

Dengo, G. 1968. Estructura geológica, historia tectónica y morfológica de América Central. Centro Regional de Ayuda Técnica, USAID, Mexico.

Denslow, J. S., 1987. Tropical rainforest gaps and tree species diversity. *Annual Review of Ecology and Systematics* 18:431–451.

Denslow, J. S., and G. S. Hartshorn. 1994. Tree-fall gap environments and forest dynamic processes. In *La Selva: Ecology and natural history of a neotropical rain forest*, pp. 120–127. L. A. McDade, K. S. Bawa, H. A. Hespenhide, and G. S. Hartshorn (eds.). University of Chicago Press, Chicago and London.

Derrau, M. 1980. *Precis de geomorphologie*. Masson, Paris.

Desai, G. M., and V. Gandhi. 1990. Phosphorous for sustainable agricultural growth in Asia: An assessment of alternative sources and management. In *Phosphorous requirements for sustainable agriculture in Asia and Oceana*. International Rice Research Institute, Laguna, Philippines.

Dewberry, T. C., and C. Pringle. 1994. Lotic conservation and science: Moving towards common ground to protect our stream resources. *Journal of the North American Benthological Society* 13:399–404.

de Wit, C. T., and J. P. van den Bergh. 1965. Competition between herbage plants. *Netherlands Journal of Agricultural Science* 13:212–224.

Díaz, Panduro W., L. Szott, M. Arcos Sandoval, A. Arévalo López, and J. Pérez Vela. 1993. Análisis y evaluación económica del cultivo de pijuayo en sistemas agroforestales. In *IV Congreso Internacional Sobre Biología, Agronomía, e industrialización de Pijuayo*, pp. 323–346. J. Mora Urpí, L. T. Szott, M. Murillo, and V. M. Patiño (eds.). Universidad de Costa Rica, San José.

Díaz, R. 1982. Caracterización y relaciones ambiente-manejo en sistemas de frijol y sorgo asociados con maíz en Honduras. UCR/CATIE. Tesis Mg. Sc., Turrialba, Costa Rica.

DIGPESCA (Dirección General de Pesca). 1991. *Análisis técnico-legal de concesiones de Camáron*. RRNN (Ministry of Natural Resources), Tegucigalpa, Honduras.

Dixon, J. 1989. The valuation of mangroves. *Tropical Coastal Area Management Newsletter* 4(3), Metro Manila.

Dixon, J., and P. Sherman. 1990. The economics of protected areas: A new look at benefits and costs. Island Press, Washington, D.C.

Donald, C. M. 1958. The interaction of competition for light and for nutrients. *Australian Journal of Agricultural Research* 9:421–435.

Donkin, R. A. 1979. *Agricultural terracing in the aboriginal new world*. University of Arizona Press, Tucson.

Doppelt, B., M. Scurlock, C. Frissell, and J. Karr. 1993. *Entering the watershed: A new approach to save America's river ecosystems*. Island Press, Washington, D.C.

Douglass, Gordon K. 1984. The meanings of agricultural sustainability. In *Agricultural sustainability in a changing world order*, pp. 3–29. G. K. Douglass (ed.). Westview, Boulder, Col.

Douglass, Gordon K. (ed.). 1984. *Agricultural sustainability in a changing world order*. Westview, Boulder, Col.

Doxon, Lynn E. 1988. Diversity, distribution and use of ornamental and edible plants in home gardens of Honduras. Ph.D. thesis, Department of Horticulture, Kansas State University, Manhattan.

Dudgeon, D. 1994. The need for multi-scale approaches to the conservation and management of tropical inland waters. *Mitteilungen Internationale Vereinigung Limnologie* 23:11–16.

Dudgeon, D., A. H. Arthington, W. Y. B. Chang, J. Davies, C. L. Humphrey, R. G. Pearson, and P. K. Lam. 1994. Conservation and management of tropical Asian and Australian inland waters: Problems, solutions and prospects. *Mitteilungen Internationale Vereinigung Limnologie* 24:369–386.

Durham, W. H. 1979. *Scarcity and survival in Central America: Ecological origins of the Soccer War*. Stanford University Press, Menlo Park, Cal.

Edmunds, G. F. 1982. Ephemeroptera. In *Aquatic biota of Mexico, Central America and the West Indies*, pp. 242–248. S. H. Hulbert and A. Villalobos-Figueroa (eds.). San Diego State University, San Diego, Calif.

Edwards, Clark. 1988. Real prices received by farmers keep falling. *Choices* 3(47):22–23.

Egan, Jack. 1990, November 26. The fish story of the decade. *U.S. News and World Report*.

Eicher, Carl K. 1993. Inventing and sustaining productive national and international agricultural research systems. In *Agriculture, environment and health: Toward sustainable development into the 21st century*. Vernon W. Ruttan (ed.). University of Minnesota Press, Minneapolis.

EIS (Environmental Impact Statement). 1995. North Coast Superaqueduct Project, Draft Environmental Impact Statement, Puerto Rico Aqueduct and Sewer Authority.

Elizondo, L. H. 1994. Instituto Nacional de Biodiversidad, San José, Costa Rica. Personal communication.

Ellern, S. J., J. L. Harper, and G. R. Sagar. 1970. A comparative study of the distribution of the roots of *Avena fatua* and *A. strigosa* in mixed stands using a ^{14}C-labelling technique. *Journal of Ecology* 58:865–868.

EPA (Environmental Protection Agency). 1995, April. Volunteer stream monitoring: A methods manual. Field Test Draft, EPA 844 D 95 001.

Escobar, Arturo. 1995. *Encountering development: The making and unmaking of the Third World*. Princeton University Press, Princeton, N.J.

Escobar, Arturo, and Sonia Alvarez (eds.). 1992. *The making of social movements in Latin America: Identity, strategy and democracy*. Westview, Boulder, Col.

Escobar, Arturo, and Alvaro Pedrosa. In press. *Pacífico: Desarrollo o diversidad? Estado, capital y movimientos sociales en el Pacífico Colombiano*. Universidad del Valle, Cali, Colombia.

Espinoza, J. 1993, June 29. Biologist, ANDAH (National Association of Aquaculturalists of Honduras), Tegucigalpa, Honduras. Interview.

Esteva, G. 1992. Development. In *The development dictionary: A guide to knowledge as power*. Wolfgang Sachs (ed.). Zed Press, London.

Europa Publications. 1995. *South America, Central America, and the Caribbean*, 5th ed. Europa Publications, London.

Evans, J. 1992. Plantation forestry in the tropics: Tree planting for industrial, social, environmental, and agroforestry purposes, 2nd ed. Clarendon Press, Oxford.

Ewel, J. J. 1985. Designing agricultural ecosystems for the humid tropics. *Annual Review of Ecology and Systematics* 17:245–271.

Ewel, J., F. Benedict, C. Berish, B. Brown, S. Gliessman, M. Amador, R. Bermúdez, A. Martínez, R. Miranda, and N. Price. 1982. Leaf area, light transmission, roots, and leaf damage in nine tropical plant communities. *Agro-Ecosystems* 7:305–326.

Facio, R. 1972. *Estudio sobre economía costarricense*. Editorial Costa Rica, San José.

FAO (Food and Agricultural Organization of the United Nations). 1955. *World forest resources: Results of the inventory taken in 1953 by the forestry division of FAO*. FAO, Rome.

FAO. 1981. Proyecto de evaluación del los recursos forestales tropicales (en el marco SINUVIMA): Los recursos forestales de la América tropical. Informe Técnico 1. FAO. Rome.

FAO. 1990a. Centroamérica y los problemas de desarrollo en el campo. Oficina Regional para América Latina y el Caribe. FAO, Santiago, Chile.

FAO. 1990b. Interim report on forest resources assessment 1990 project. Committee on Forestry, Tenth Session. FAO, Rome.

FAO. 1992. *FAO Yearbook, Forest Products*. Forestry Series 27, Statistics Series 116. FAO, Rome.

FAO. 1993a. Aquaculture Production 1985–1991. FAO Fisheries Circular no. 815, Revision 5. FAO, Rome.

FAO. 1993b. *The challenge of sustainable forest management: What future for the world's forests?* FAO, Rome.

FAO. 1993c. Forest resources assessment 1990: FAO forestry paper 112. FAO, Rome.

FAO. 1993d. Prevention of water pollution by agriculture and related activities. *Proceedings of the FAO Expert Consultation*, Santiago, Chile, October 20–23, 1992. FAO, Rome.

FAO. 1994a. *Yearbook: Forest products, 1981–1992*. Forestry Series no. 27. FAO, Rome.

FAO. 1994b. *Yearbook of Fishery Statistics: Commodities 1992*, vol. 75. FAO, Rome.

FAO. 1996. Technical background documents for the World Food Summit. Vol. 1. Executive summary. FAO, Rome.

FAO-UNESCO. 1975. *Soil map of the world. Vol. III: Mexico and Central America*. UNESCO, Paris.

FAO-UNESCO. 1976. *Mapa Mundial de Suelos 1:5.000.000. Vol. III: México y América Central*. Unesco, Paris.

Fassbender, H. W. 1969. Estudio del fósforo en suelos de América Central. IV. Capacidad de fijación de fósforo y su relación con características edáficas. *Turrialba* 19:497–505.

Fassbender, H. 1993. *Modelos edafológicos de sistemas agroforestales*, 2nd ed. CATIE, Turrialba, Costa Rica.

Feder, G., and T. Onchan. 1987. Land ownership security and farm investment in Thailand. *American Journal of Agricultural Economics* 69 (2):311–320.

Feldstein, H., and S. Poats. 1989. *Working together: Gender analysis in agriculture*. Kumarian Press, West Hartford, Conn.

Fernández, E. C. M., D. Garrity, L. T. Szott, and C. A. Palm. 1994. Use and potential of domesticated trees for soil improvement. In *Tropical trees: Potential for domestication and the rebuilding forest resources*, pp. 137–147. R. R. B. Leakey and A. C. Newton (eds.). Institute of Terrestrial Ecology and the Edinburgh Center for Tropical Forests, Edinburgh.

Fickle, J. E. 1980. *The new South and the new competition: Trade association development in the southern pine industry*. University of Illinois Press, Urbana.

Firehock, K., and J. West. 1995. A brief history of volunteer biological water monitoring using macroinvertebrates. *Journal of the North American Benthological Society* 14:197–202.

Fisher, R. F. 1990. Amelioration of soils by trees. In *Sustained productivity of forest soils*, pp. 290–300. S. Gessel, D. S. Lacate, G. F. Weetman, and R. F. Powers (eds.). Faculty of Forestry, University of British Columbia, Vancouver, Canada.

Fisher, R. F. 1995. Amelioration of degraded rain forest soils by plantations of native trees. *Soil Science Society of America Journal* 59:544–549.

Fisher, R. F., and E. L. Stone. 1968. Increased availability of nitrogen and phosphorus in the root zone of conifers. *Soil Science Society of America Journal* 33:955–961.

Fitter, A. H. 1986. Acquisition and utilization of resources. In *Plant ecology*, pp. 375–405. M. J. Crawley (ed.). Blackwell, Oxford.

Fleming, T. H. 1988. *The short tailed fruit bat*. University of Chicago Press, Chicago.

Flor, C. A. 1985. Revisión de algunos criterios sobre la recomendación de fertilizantes en frijol. In *Frijol: Investigación y producción*. M. López, F. Fernández, and A. Van Schoonhoven (eds.). CIAT, Cali, Colombia.

Flora, Cornelia Butler, and Blas Santos. 1986. Women in farming systems in Latin America. In J. Nash, H. Safa, and contributors, *Women and change in Latin America*, pp. 208–228. Bergin and Garvey, New York.

Flores, J. 1994. El estado actual y las tendencias en la utilización de los recursos naturales en América Central. In *Ganadería y recursos naturales en América Central: Estrategias para la sostenibilidad*, pp. 21–38. E. J. Homan (ed.). CATIE, Turrialba, Costa Rica.

Flores, M. 1995. La utilización de leguminosas de cobertura en sistemas agrícolas tradicionales de Centroamérica. Informe Técnico no. 5. Centro Internacional de Información Sobre Cultivos de Cobertura (CIDICCO), Tegucigalpa, Honduras.

Flores, Villela O., and P. Gerez. 1989. Patrimonio vivo de México: Un diagnóstico de la diversidad biológica. Conservation International, Washington, D.C.

Flowers, R. W. 1992. Review of the genera of mayflies of Panama, with a checklist of Panamanian and Costa Rican species. In *Insects of Panama and Mesoamerica: Selected studies*, chap. 4, pp. 37–51. D. Quintero y A. Aiello (eds.). Oxford University Press, Oxford.

Flowers, R. W., and C. L. de la Rosa. In press. *Guía para la identificación de las efímeras (Ephemeroptera) de Costa Rica*. Editorial FIREMA, San José.

Foer, Gordon, and Stephen Olsen (eds.). 1992. *Central America's coasts*. U.S. Agency for International Development, Research and Development/Environmental and Natural Resources, Washington, D.C.

Ford, R. I. 1984. Prehistoric phytogeography of economic plants in Latin America. In *Pre-Columbian plant migration*, chap. 9. D. Stone (ed.). Papers of the Peabody Museum of Archeology and Ethnology. vol. 76. Harvard University, Cambridge, Mass.

Fortmann, L. 1987. Tree tenure: An analytical framework for agroforestry projects. In *Land, trees, and tenure*, pp. 17–34. J. B. Raintree (ed.). ICRAF, Nairobi, Kenya; Land Tenure Center, Madison, Wisc.

Fortmann, L., and D. Rocheleau. 1985. Why agroforestry needs women: The case of plan sierra. Unasylva. *Women and Forestry* 36:145.

Fox, T. R., and N. B. Comerford. 1992. Influence of oxalate loading on phosphorus and aluminum solubility in Spodosols. *Soil Science Society America Journal* 56: 290–294.

Fragoso, J. M. V. 1991. The effect of hunting on tapirs in Belize. In *Neotropical wildlife use and conservation*, pp. 154–162. J. G. Robinson and K. H. Redford (eds.). University of Chicago Press, Chicago.

Frakes, L. A. 1979. *Climates throughout geologic times*. Elsevier Scientific Publishing, Oxford.

Fuentes, C. M., and R. Quirós. 1988. Variación de la composición de la captura de peces en el río Pananá, durante el periodo 1941–1984. Serie Informes Técnicos del Departamento de Aguas Continentales 6. Instituto Nacional de Investigación y Desarrollo Pesquero, Mar del Plata, Argentina.

Gallardo, J., and J. Vargas. 1992. Diagnóstico de situación de finca. Report no. 1. Ministerio de Agricultura y Ganadería, San José.

Gámez, R. 1991. Biodiversity conservation through facilitation of its sustainable use: Costa Rica's Instituto Nacional de Biodiversidad, INBio (National Institute of Biodiversity). Trends in *Ecology and Evolution* 6: 377–378.

García, J., and J. Najera. 1995. *Simposio Centroamericano sobre agricultura orgánica. Marzo*. Universidad Estatal a Distancia, San José.

García, M., Jr. 1992. Policies and issues on Philippine fisheries and aquatic resources. In *Proceedings of the Roundtable Discussion on Philippine Fisheries Policies and the Workshop on Territorial Use Rights in Fisheries*. Philippine Council for Aquatic and Marine Resources and Development, Los Banos.

García, T. 1983. Potencial y utilización de los pastos tropicales para la producción de leche. In *Los pastos en Cuba. Tomo 2. Utilización*. Instituto de Ciencia Animal, La Habana, Cuba.

García-Barrios, R., and L. García-Barrios. 1990. Environmental and technological degradation in peasant agriculture: A consequence of development in Mexico. *World Development* 18:1569–1585.

García-Barrios, R., and P. Taylor. 1995. The dynamics of socio-environmental change and the limits of neo-Malthusian environmentalism. In *The limits to markets: Equity and the global environment*. T. Mount, H. Sue, and M. Dore (eds.). Blackwell, Oxford.

Garita, C. D. 1989. Mapa de cobertura boscosa. Report, Unidad de Cartografía y Topografía DGF, MIRENEM, San José.

General Accounting Office. 1990, February. Alternative agriculture: Federal incentives and farmers' options. US/GAO/PEMD-90–12. General Accounting Office, Washington, D.C.

Georgia Department of Natural Resources. 1994. *Georgia*

adopt-a-stream manual. Environmental Protection Division, Georgia.

Gessel, S. P., D. W. Cole, D. Johnson, and J. Turner. 1977. The nutrient cycles of two Costa Rican forests. In *Actas del IV Simposio Internacional de Ecología Tropical*, vol. II, pp. 623–643, March 1977, University of Panama, Panama City. Cited in C. F. Jordan. 1985. *Nutrient cycling in tropical forest ecosystems*. Wiley, New York.

Gibbs, J. P. 1991. Avian nest predation in tropical wet forest: An experimental study. *Oikos* 60:155–161.

Gilbert, L. E. 1975. Ecological consequences of a coevolved mutualism between butterflies and plants. In *Coevolution of animals and plants*, pp. 210–240. L. E. Gilbert and P. H. Raven (eds.). University of Texas Press, Austin.

Gilbert, R. O. 1987. *Statistical methods for environmental pollution monitoring*. Van Nostrand Reinhold, New York.

Giller, K. E., and K. J. Wilson. 1991. *Nitrogen fixation in tropical cropping systems*. CAB International, Wallingford, England.

Gillespie, A. R. 1989. Modelling nutrient flux and interspecies root competition in agroforestry interplantings. *Agroforestry Systems* 8:257–265.

Gillespie, A. R., D. M. Knudson, and F. Geilfus. 1993. The structure of four fome gardens in the Petén, Guatemala. *Agroforestry Systems* 24:157–170.

Gladwell, J. S., and M. Bonell. 1990. An international programme for environmentally sound hydrological and water management strategies in the humid tropics. In *Proceedings of the International Symposium on Tropical Hydrology and Fourth Caribbean Islands Water Resources Congress*, pp. 1–10. Technical Publication Series TPS-90-2. American Water Works Association, Herndon, Va.

Glander, K. 1980. Reproduction and population growth in free ranging mantled howling monkeys. *American Journal of Physical Anthropology* 53:25–34.

Glander, K. 1992. Dispersal pattern in Costa Rican mantled howling monkeys. *International Journal of Primatology* 13:415–436.

Glander, K. 1996. *The howling monkey of La Pacífica*. Duke University Primate Center, Durham, N.C.

Glanz, W. E. 1991. Mammalian densities at protected versus hunted sites in Central Panama. In *Neotropical wildlife use and conservation*, pp. 163–173. J. G. Robinson and K. H. Redford (eds.). Chicago University Press, Chicago.

Glasener, K., and C. A. Palm. 1991. Ammonia volatilization from tropical legume mulches. *Agronomy abstracts*, p. 61. American Society of Agronomy Annual Meeting, Denver, Col. American Society of Agronomy, Madison, Wisc.

Gliessman, S. R. 1988. Agroecology Program, University of California, Santa Cruz. Personal communication.

Gliessman, S. R. 1990. Understanding the basis of sustainability for agriculture in the tropics: Experiences in Latin America. In *Sustainable agricultural systems*, pp. 378–390. C. A. Edwards, R. Lal, P. Madden, R. H. Miller, and G. House (eds.). Soil Conservation Society, Ankeny, Iowa.

Glynn, P. W., L. S. Howard, E. Corcoran, and A. D. Freay. 1984. The occurrence and toxicity of herbicides in reef building corals. *Marine Pollution Bulletin* 15: 370–374.

Goeller, H. E., and Alvin M. Weinberg. 1976, February 20. The age of substitutability. *Science* 191:683–689.

Goenaga, C. 1991. The state of coral reefs in the wider Caribbean. *Interciencia* 16:12–20.

Goldman, C. R. 1976. Ecological aspects of water impoundment in the tropics. *Revista de Biologia Tropical* 24: 87–112.

Golley, F. B., J. T. McGinnis, R. G. Clemenets, G. I. Child and M. J. Duever. 1975. *Mineral cycling in a tropical moist forest ecosystem*. University of Georgia Press, Athens.

Gollin, J. D. 1994, November 15. Trees down, lights out in Honduras: Scientists link deforestation to disruption of the country's water supply. *Christian Science Monitor*, p. 12.

Golubev, G. N. 1984. Economic activity, water resources and the environment: A challenge for hydrology. *Hydrological Sciences Journal* 28:57.

Gómez, L. D. 1986. *Vegetación de Costa Rica*. Editorial EUNED, San José.

Gómez-Pompa, A., and A. Kaus. 1992. Taming the wilderness myth. *Bioscience* 42: 271– 279.

Gómez-Pompa, A., J. Salvador Fores, and V. Sosa. 1987. The "Pet Kot": A man-made forest of the Maya. *Interciencia* 12:10–15.

González, Ballar R. 1985. Política, derecho y medio ambiente. *Revista de Ciencias Jurídicas* 54:27–39.

González, E. 1991. Recolección y germinación de semillas de 26 species arbóreas del bosque húmdedo tropical. *Revista Biología Tropical* 39:47–51.

González, E., and R. F. Fisher. 1994. Growth of native forest species planted on abandoned pasture land in Costa Rica. *Forest Ecology and Management* 70:159–167.

Gonzálvez, V. 1995. La Agricultura Orgánica en Nicaragua. In *Simposio Centroamericano sobre Agricultura Orgánica, Marzo*. J. García and J. Najera (eds.). Universidad Estatal a Distancia, San José.

Goodland, Robert. 1991. The case that the world has reached limits. In Robert Goodland, Herman Daly, and Salah El Serafy. *Environmentally Sustainable Economic Development: Building on Brundtland*. Environmental Working Paper 46, pp. 5—17. World Bank, Washington, D.C.

Gordon, W. F., H. G. Baker, and P. A. Opler. 1974. Comparative phenological studies of trees in tropical wet and dry forest in the lowlands of Costa Rica. *Journal of Ecology* 68:881–919.

Grace, J. B. 1991. A clarification of the debate between Grime and Tilman. *Functional Ecology* 5:583– 587.

Grainger, A. 1990. *The threatening desert*. Earthscan Publication, London.

Granados, C. 1983. Gold mining and town development in Costa Rica. M.A. thesis. University of Oregon, Eugene.

Grasshoff, K., M. Ehrhardt, and K. Kremling. 1983. *Meth-*

ods of seawater analysis. Verlag Chemie, Weinheim, Germany.

Green, B. W., and D. R. Teichert-Coddington. 1990. Lack of response of shrimp yield to inorganic fertilization in grow-out ponds. In Seventh annual administrative report, Pond Dynamics/Aquaculture Collaborative Research Program 1990. H. S. Egna, J. Bowman, and M. McNamara (eds.). International Research & Development, Oregon State University, Corvallis.

Griffith, B. G., E. W. Hartwell, and T. E. Shaw. 1930. The evolution of soils as affected by the old field white pine-hardwood succession in central New England. *Harvard Forest Bulletin* 15:1 82.

Grime, J. P. 1973a. Competition and diversity in herbaceous vegetation. *Nature* 244:311.

Grime, J. P. 1973b. Competitive exclusion in herbaceous vegetation. *Nature* 242:344–347.

Grime, J. P. 1979. *Plant strategies and vegetation processes*. Wiley, Chichester, England.

Grime, J. P., J. M. L. Mackey, S. H. Hillier, and D. J. Read. 1987. Floristic diversity in a model system using experimental microcosms. *Nature* 328:420–422.

Groves, Theodore, Roy Radner, and Stanley Reiter (eds.). 1987. *Information, incentives, and economic mechanisms*. University of Minnesota Press, Minneapolis.

Gueravitch, J., and R. S. Unnasch. 1989. Experimental removal of a dominant species at two levels of soil fertility. *Canadian Journal of Botany* 67:3470–3477.

Guevara, R. 1991, April 24–26. Ponencia sobre la acuicultura en Honduras, con enfasis en el cultivo de Camáron. Presentation to the First Symposium of Farmed Shrimp in Central America, Hotel Honduras Maya, Tegucigalpa, Honduras.

Gunder, Frank. 1970. *Capitalismo y subdesarrollo en América Latina*. Siglo XXI, Buenos Aires.

Guzmán, H. M., and C. E. Jiménez. 1992. Contamination of coral reefs by heavy metals along the Caribbean coast of Central America (Costa Rica and Panama). *Marine Pollution Bulletin* 24: 554–561.

Guzmán, Virginia. 1992. Desde los proyectos de desarrollo a la sociedad. In *Una Nueva Lectura: Género en el desarrollo*, pp. 305–326. Virginia Guzmán, Patricia Portocarrero, and Virginia Vargas (eds.). Ediciones Populares Feministas, CIPAF, Santo Domingo.

Haggar, J. P. 1994. Trees in alleycropping systems: Competitors or soil improvers? *Outlook on Agriculture* 23: 27–32.

Haggar, J. P., and J. W. Beer. 1993. Effect on maize growth of the interactions between increased nitrogen availability and competition with trees in alley cropping. *Agroforestry Systems* 21:239–249.

Haggar, J. P., and J. J. Ewel. 1994. Experiments on the ecological basis of sustainability: Early findings on nitrogen, phosphorus and root systems. *Interciencia* 19:347–351.

Haggar, J. P., E. V. J. Tanner, J. W. Beer, and D. C. L. Kass. 1993. Nitrogen dynamics of tropical agroforestry and annual cropping systems. *Soil Biology Biochemistry* 25:1363– 1378.

Haggar, J. P., G. P. Warren, J. W. Beer, and D. Kass. 1991. Phosphorus availability under alley cropping and mulched and unmulched sole cropping systems in Costa Rica. *Plant Soil* 137:275–283.

Haggar, J., K. Wightman, and R. Fisher. 1997. The potential of plantations to foster woody regeneration within a deforested landscape in lowland Costa Rica. *Forest Ecology and Management* (in press).

Hagnauer, W. 1978. Reflexiones sobre el futuro del Pacífico Seco. Mimeo. Tercer Congreso de Ingenieros Agrónomos, San José.

Hagnauer, W. 1980, Diciembre. Análisis agro-meteorológico en la zona de Cañas y Bagaces en los años 1921 a 1979. Informe Semestral Instituto Geográfico Nacional San José.

Hagnauer, W. 1992. *El Sistema Agroecológico de Guanacaste. Oportunidades para la agricultura y el turismo*, p. 39. Fundación para el Desarrollo Sostenible, Cañas. Costa Rica.

Hagnauer, W. 1993. *El balance energético del distrito de Riego Arenal—Tempisque*, p. 39. Fundación para el Desarrollo Sostenible, Cañas, Costa Rica.

Hall, C. 1976. *El café y el desarrollo histórico-geográfico de Costa Rica*. Editoral Costa Rica, San José.

Hall, K. D., G. C. Daily, and P. E. Ehrlich. 1995. Knowledge and perceptions in Costa Rica regarding environment, population, and biodiversity issues. *Conservation Biology* 9:1548–1558.

Hambleton, A. 1994. A survey of United States government funded activities supporting biodiversity research and conservation in Costa Rica. Internal report, USAID, San José.

Hamilton, L. S. 1994, March 23–26. Does "deforestation" always result in serious soil erosion? Paper given at the International Symposium on Management of Rain Forest in Asia, Oslo.

Hansen-Kuhn, K. 1993. Sapping the economy: Structural adjustment policies in Costa Rica. *Ecologist* 23:179–184.

Hargreaves, G. H., J. K. Hancock, and R. H. Hill. 1975. Precipitation probabilities, climate and agricultural potential for Honduras, El Salvador, Nicaragua and Guatemala. Utah State University, Logan.

Harlan, J. R. 1971. Agricultural origins: Centers and noncenters. *Science* 174:468–474.

Harper, J. L. 1968. The regulation of numbers and mass in plant populations. In *Population biology and evolution*, pp. 139–158. R. C. Lewontin (ed.). Syracuse University Press, Syracuse, N.Y.

Harper, J. L. 1977. *Population biology of plants*. Academic Press, London.

Harris, G. A. 1967. Some competitive relationships between *Agropyron spicatum* and *Bromus tectorum*. *Ecological Monographs* 37:89–111.

Harris, L. D., and N. K. Fowler. 1975. Ecosystem analysis and simulation in Mkomazi Reserve, Tanzania. *East Africa Wildlife Journal* 13:325–345.

Harris, Stewart. 1968. OTS course, 1967. Personal communication.

Harrison, P., and B. Turner II. 1978. *Pre-Hispanic Maya agriculture*. University of New Mexico Press, Albuquerque.

Harrison, S., and L. Fahrig. 1995. Landscape pattern and population conservation. In *Mosaic landscapes and ecological processes*, pp. 293–308. L. Hansson, L. Fahrig, and G. Merriam (eds.). Chapman and Hall, London.

Harriss, J. (ed.). 1982. *Rural Development: Theories of peasant economy and agrarian change*. Hutchinson, London.

Hartmann, B. 1987. *Reproductive rights and wrongs*. Harper Collins, New York.

Hartshorn, G. 1983. Plants. In *Costa Rican natural history*, pp. 118–157. D. H. Jansen (ed.). University of Chicago Press, Chicago.

Hartshorn, G. 1984. Belize: A field study. USAID, Belize.

Hartshorn, G. S. et al. 1982. *Costa Rica country environmental profile: A field study*. Tropical Science Center, San José.

Harwood, Richard R. 1990. A history of sustainable agriculture. In *Sustainable agricultural systems*. Clive A. Edwards, Rattan Lal, Patrick Maden, Robert H. Miller, and Gar House (eds.). Soil and Water Conservation Authority, Ankeny, Iowa.

Hauptli, H., D. Kalz, B. Thomas, and R. Goodman. 1990. In *Sustainable agricultural systems*, chap. 10, pp. 378–390. C. Edwards, R. Lal, P. Madden, R. Miller, and G. House (eds.). St. Lucie Press, Del Ray Beach, Fla.

Hauser, S. 1993. Root distribution of *Dacyladenia (Acacia) barteri* and *Senna (Cassia) siamea* in alley cropping on Ultisol. I. Implications for field experimentation. *Agroforestry Systems* 24:111–121.

Hawkins, R. 1985. El sorgo en Latinoamérica: una visión general. In P. Compton and B. De Wall. *El sorgo en América Latina*, pp. 2–20. Memorias del Taller, 16–22 de September 1984, Intsormil-Cimmyt, Mexico.

Hayami, Yujiro, and V. W. Ruttan. 1985. *Agricultural development: An international perspective*, pp. 280–298. Johns Hopkins University Press, Baltimore, Md.

Hazen, T. C., H, Santiago-Mercado, G. A Toranzos, and M. Bermúdez. 1987. What does the presence of fecal coliforms indicate in the waters of Puerto Rico?: A review. *Boletin Associacion Medica Puerto Rico* 70:189.

Head, Suzanne, and Robert Heinzman. 1990. *Lessons of the rainforest*. Sierra Club Books, San Francisco.

Hecky, R. E., D. J. Ramsey, R. A. Bodaly, and N. E. Strange. 1991. Increased methylmercury contamination in fish in newly formed freshwater reservoirs, In *Advances in Mercury toxicology*, pp. 33–52. T. Suzuki, N. Imura, and T. W. Clarkson (eds.). Plenum Press, New York.

Heinzman, R., and C. Reining. 1990. *Sustained rural development: Extractive forest reserves in the northern Peten of Guatemala*. Tropical Resources Institute, Working Paper no. 37. Yale University, New Haven, Conn.

Herrick, J. 1993. Restauration of tropical pasture ecosystems and the role of cattle dung patches. Thesis, Ohio State University, Columbus.

Heyligers, P. C. 1963. *Vegetation and soils of a white sand savanna in Surinam*. Nord-Hollandsche Vitgevers Maatchappif no. 3., Amsterdam.

Hilty, S. 1982. *Perfil ambiental de El Salvador*. USAID and University of Arizona, Tucson.

Hindmarsh, Richard. 1991, September/October. The flawed "sustainable" promise of genetic engineering. *Ecologist* 21:196–205.

Hirshleifer, Jack. 1970. *Investment, interest and capital*. Prentice Hall, Englewood Cliffs, N.J.

Hodell, D. A., J. H. Curtis, and M. Brenner. 1995. Possible role of climate in the collapse of classic Maya civilization. *Nature* 375(6530):391–394.

Holdridge, L. R. 1967. *Life zone ecology*. Tropical Science Center, San José.

Holdridge, L. R., W. C. Grenke, W. H. Hatheway, T. Liang, and J. A. Tosi Jr. 1971. *Forest environments in tropical life zones. A pilot study*. Pergamon Press, Oxford.

Holl, K. D., G. C. Daly, and P. R. Ehrlich. 1995. Knowledge and perceptions in Costa Rica regarding environments, population and biodiversity issues. *Conservation Biology* 9:1548–1558.

Holzenthal, R. W. 1985. Studies of neotropical Leptoceridae (Trichoptera): Their diversity, evolution, and biogeography, with revisions of selected genera. Ph.D. thesis, Clemson University, Clemson, S.C.

Holzenthal, R. W. 1995. The caddisfly genus *Nectopsyche*: New gemma group species from Costa Rica and the neotropics (Trichoptera: Leptoceridae). *Journal of the Northern American Benthological Society* 14:61–83.

Hopkins, J. S., C. L. Browdy, P. A. Sandifer, and A. D. Stokes. 1994, January 14–18. Effect of two feed protein levels and two feed rate, stocking density combinations on waste quality and production in intensive shrimp ponds which do not utilize water exchange, p. 30. Abstract presented at the Annual Meeting of the World Aquaculture Society, New Orleans, La.

Hopkins, J. S., R. D. I. Hamilton, P. A. Sandifer, C. L. Browdy, and A. D. Stokes. 1993. Effect of water exchange rate on production, water quality, effluent characteristics and nitrogen budgets of intensive shrimp ponds. *Journal of the World Aquaculture Society* 24(3): 304– 320.

Horn, S. 1985. Preliminary pollen analysis of quaternary sediments from deep sea drilling project site 565, western Costa Rica. U.S. Government Printing Office, Washington, D.C.

Horton, D. 1987. *Potatoes: Production, marketing and programs for developing countries*. Westview, Boulder, Col.

Houghton, R. A., D. S. Lefkowitz, and D. L. Skole. 1991. Changes in the landscape of Latin America between 1850 and 1985. Progressive loss of forests. *Forest Ecology and Management* 38:143–172.

House, P. R. 1994. La biodiversidad en 10 huertos caseros en El Camelote, Copán. Proyecto Huertos Caseros, CATIE/CARE, Turrialba, Costa Rica.

Hughes, C. 1994. Risks of species introduction in tropical forestry. *Commonwealth Forestry Review* 73:243–252.

Humphreys, L. R. 1991. *Tropical pasture utilization. Grazing and the environment for pasture growth*, pp. 26–45. Cambridge University Press, Cambridge.

Hunter, J. M., and S. I. Arbona. 1995. Paradise lost: An introduction to the geography of water pollution in Puerto Rico. *Social Science Medicine* 40:1331–1355.

Hurwicz, Leonid. 1972. On informationally decentralized systems. In *Decision and organization*, pp. 297–336. C. B. McGuire and Roy Radner (eds.). North Holland Publishing, Amsterdam.

Iasa, K., and B. Jennings. 1981. Science and authority in international agricultural research. *Journal of Concerned Asian Scholars.*

ICAFE (Instituto del Café de Costa Rica). 1992. *Informe sobre la actividad cafetalera de Costa Rica.* ICAFE, San José, Costa Rica.

ICATA (Instituto Centroamericano de Tierras Agricolas). 1984. *Perfil ambiental de la República de Guatemala,* vol. II. USAID, Central American Regional Office, Guatemala.

Ichire, O. 1993. Using agroforestry for fruit tree establishment. *Agroforestry Today* 5(3):9–12.

IFOAM (International Federation of Organic Agriculture Movements). 1992. *Basic standards of organic agriculture and food processing.* IFOAM General Assembly, Saõ Paulo, Brazil.

Im, J. N., Y. K. Cho, D. H. Kim, and Y. H. Shin. 1977. Soil conservation in newly reclaimed slopeland in Korea, pp. 11–20. Report of Agricultural Science Institute, Suwon, Korea.

INIAP (Instituto de Investigaciones Agropecuarias de Panamá). 1988. Mapa general de la capacidad de uso de los suelos del área Gaymí. Escala 1:100.000. INIAP, Panamá.

Inter-American Development Bank. 1984. Water resources of Latin America. *Water International* 9:26–36.

Isreal, D. W. 1987. Investigation in the role of phosphorus in symbiotic dinitrogen fixation. *Plant Physiology* 84: 835–840.

ISSS (International Society of Soil Science), ISRIC (International Soil Reference and Information Center), FAO (Food and Agriculture Organization). 1994. *World reference base for soil resources.* Compiled and edited by O. C. Spaargaren. Graphics Service Centre Van Gils, Wageningen, Holland.

Jaffe, Steven. 1993. *Exporting high-value food commodities: Success stories from developing countries.* World Bank Discussion Papers, no. 198. World Bank, Washington, D.C.

James, D. E., P. Nijkamp, and J. B. Opschoor. 1989. Ecological sustainability in economic development. In *Economy and ecology: Toward sustainable development,* pp. 27–48. F. Archibugi and P. Nijkamp (eds.). Kluwer Academic Publishers, Dordrecht, Holland.

Janzen, D. 1973. Tropical agrosystems. *Science* 182:1212–1219.

Janzen, D., and P. Martin. 1982. Neotropical anachronisms: The fruits of the gamphotheres. *Science* 214:19–27.

Janzen, D. H. 1966. Fire, vegetation structure and the ant X acacia interaction in Central America. *Ecology* 48(1): 26–35.

Janzen, D. H. 1971. Euglossine bees as long-distance pollinators of tropical plants. *Science* 17:203–205.

Janzen, D. H. 1991. How to save tropical biodiversity. *American Entomologist* 37(3):159–171.

Janzen, D. H. 1994. *Guanacaste National Park: Tropical ecological and cultural restoration.* Editorial Universidad Estatal a Distancia, San José.

Jensen, M. 1993. Productivity and nutrient cycling of a Javanese homegarden. *Agroforestry Systems* 24:187–201.

Jiménez, S. E. 1978. Comentarios sobre la producción de frijol común (*Phaseolus vulgaris*) en Costa Rica. *Agronomía Costarricense* (Costa Rica):103–108.

Johannessen, C. 1963. *Savannas of interior Honduras.* University of California Press, Berkeley.

Johnson, R., and M. Watts. 1989. Contractual stipulations, resource use, and interest groups: Implications from federal grazing contracts. *Journal of Environmental Economics and Management,* pp. 87–96.

Johnston-Wallace, D. J., J. S. Andrews, and J. R. Lamb. 1942. The influence of periodic close grazing and pasture fertilization upon erosion control. *Journal of the American Society of Agronomy* 34:963–974.

Jones, C. 1986. Intra-household bargaining in West African farming systems. In *Understanding Africa's rural households.* J. L. Moock (ed.). Westview, Boulder, Col.

Jordan, C. G. 1985. *Nutrient cycling in tropical forest ecosystems.* Wiley, New York.

Jose, D., and N. Shanmugaratnam. 1993. Traditional homegardens of Kerala: A sustainable human ecosystem. *Agroforestry Systems* 24:203–213.

Kaimowitz, D. 1992. *El apoyo tecnológico necesario para promover las exportaciones agrícolas no tradicionales en América Central.* No. 30 Serie de Documentos de Programas IICA. San José.

Kaimowitz, D. 1994. El ajuste se hizo y estamos en lo mismo: Ahora qué hacemos? In G. Masís and F. Sancho. Comp. *La agricultura de exportación en Centroamérica: opciones de desarrollo en la década de los 90,* pp. 193–203. Producciones Gráficas, San José.

Katz, Elizabeth G. 1992. Intra-household resource allocation in the Guatemalan highlands: The impact of non-traditional agricultural exports. Ph.D. thesis, Department of Agricultural Economics, University of Wisconsin, Madison /UMI Dissertation Services, Ann Arbor.

Kemp, R. H., G. Namkoong, and F. H. Wadsworth. 1993. Conservation of genetic resources in tropical forest management: Principles and concepts. FAO Forestry Paper 107. FAO, Rome.

Kerkhof, P. 1990. *Agroforestry in Africa: A survey of project experience.* Panos Institute, London.

Kettler, J. 1995. A sustainable stategy for fragile lands: Fallow enrichment of a traditional slash-mulch system in Coto Brus, Costa Rica. Ph.D. thesis, University of Georgia, Athens.

Kimber, C. T. 1988. *Martinique revisited: The changing plant geographies of a West Indian island.* Texas A & M University Press, College Station.

Kingsley, R. 1986. Aquaculture—Understanding the risk factors. *Infofish Marketing* 6:17–18.

Kittredge, J. 1948. *Forest influences.* McGraw-Hill, New York.

Klinge, H. 1976. Bilanzierung von Hauptnahrstoffen im Okosystem tropisher regenwald (Manaus)-vorlaufige daten. *Biographica* 7:59–76.

Koechlin, J. 1961. *La vegetation des savannes dans le sud de la Republique du Congo*. Orstom, Montpellier, France.

Koechlin, J. 1978. *Vegetation et mise en valeur dans le sud do Mato Grosso*. Travaux et Documents de Geographie Tropicale, no. 35, Bordeaux, France.

Korten, A. 1995. A bitter pill: Structural adjustment in Costa Rica. Food First Institute for Food and Development Policy, Oakland, Calif.

Kramer, L. 1996. Temporal change and patch dynamics of Costa Rican tropical dry forest remnants in an agricultural landscape assessed using remotely sensed thermal data. Ph.D. thesis, University of Georgia, Athens.

Krishna, J. H. 1990, July 22–27. Tropical hydrology and Caribbean water resources. In *Proceedings of the International Symposium on Tropical Hydrology and the Fourth Caribbean Islands Water Resources Congress, San Juan, Puerto Rico*. J. J. Krishna, V. Quiñones-Aponte, F. Gómez-Gómez, and G. L. Morris (eds.). American Water Resources Associations, Bethesda, Md.

Kropf, W. 1978. Influence de facteurs genetiques et non-genetiques ainsi que de la selection sur le niveau de performance ette rendement d_un troupeau de bovin a viande en zone semi-aride. These, Ecole Politecnique Federal, Zürich.

Kropf, W., A. Aragón, N. Künzi, and W. Hagnauer. 1983. El sistema de doble propósito. *Zootenia* 45:23–27.

La Gaceta. 1995, November 13. *Producción ecológica*. Cap. XVI. La Gaceta no. 215. Gobierno de Costa Rica, San José. Decreto de la Comisión Nacional de Agricultura Orgánica. Periódico del Gobierno de Costa Rica, San José.

Lageaux, C. J. 1991. Economic analysis of sea turtle eggs in a coastal community on the Pacific coast of Honduras. In *Neotropical wildlife use and conservation*, pp. 136–144. J. G. Robinson and K. H. Redfords (eds.). University of Chicago Press, Chicago.

Lagemann, J. 1983. *Sistemas de producción en Acosta-Puriscal*. CATIE, Turrialba, Costa Rica.

Laidlaw, T. 1996. Adopt a Stream. Master's thesis, University of Georgia, Institute of Ecology, Athens.

Lal, R. 1984. Soil erosion from tropical arable lands and its control. *Advances in Agronomy* 37:183–248.

Lal, R. 1987. *Tropical ecology and physical edaphology*. Wiley, New York.

Lal, R. 1995a, December 4–8. Biophysical factors in the choice of tillage systems for sloping lands. In *Conference Proceedings*. Red Latinoamericano de Labranza Conservacionista (RELACO), San José. Centro de Investigaciones Agronómicas of the Universidad de Costa Rica.

Lal, R. 1995b, December 4–8. Erosion control on sloping land with conservation tillage. In *Conference Proceedings*. Red Latinomericana de Labranza Conservacionista (RELACO), San José. Centro de Investigaciones Agronómicas of Universidad de Costa Rica.

Lamb, B. L., and E. Lord. 1992. Legal mechanisms for protecting riparian resource values. *Water Resources Research* 28:965–978.

Lambert, D. 1995. Grazing on public rangelands: An evolving problem of property rights. *Contemporary Economic Policy* 13:119–128.

Lanly, J., K. D. Singe, and K. Janz. 1991. FAO's 1990 reassessment of tropical forest cover. *Nature and Resources* 27:21–26.

Lara, Carlos G. 1995, July. Estadísticas sobre la producción de camarones cultivados en Costa Rica. Paper presented at Taller de Consulta para Acuicultores Nacionales, San José.

Larson, Jonas, Carl Folke, and Nils Kautsky. 1994, September/October. Ecological limitations and appropriations of ecosystem support by shrimp farms in Colombia. *Environmental Management* 18(5):663–676.

Lassere, G. 1977. *Les Ameriques du Centre*. Presse Universitaire de France, Paris.

Laurent, Sanabria, R. 1992. Interacción sistema natural-asentamiento humano: El caso de Coto Brus. Engineering degree thesis, University of Costa Rica, San Pedro.

Lavelle, P., and B. Pashanasi. 1989. Soil macrofauna and land management in Peruvian Amazon (Yurimaguas, Loreto). *Pedobiologia* 33:283–291.

Lawrence, M. S. 1993. Jamaica's water resources: Some threats to its good quality. In *Prevention of water pollution by agriculture and related activities*, pp. 113–125. Proceedings of the Food and Agriculture Organization of the United Nations, Santiago, Chile, 1992. FAO, Rome.

Leakey, R. R. B. 1987. Clonal forestry in the tropics. *Commonwealth Forestry Review* 66:61–75.

Lebeuf, T. I. 1993. Sistemas agroforestales con *Erythrina fusca* y su efecto sobre la pérdida de suelo y la escorrentía superficial en tierras de ladera, San Juan Sur, Turrialba, Costa Rica. In Erythrina *in the New and Old Worlds*, pp. 175–184. Nitrogen Fixing Tree Research Reports Special Issue. S. B. Westley and M. H. Powell (eds.). Nitrogen Fixing Tree Association, Paia, Hawaii.

Lee, J. A., David Norse, and Martin Parry. 1994. Monitoring the effects of acid deposition and climate change on agricultural potential. In *Agriculture environment and health: Sustainable development in the 21st century*, pp. 284–307. Vernon W. Ruttan (ed.). University of Minnesota Press, Minneapolis.

Lee, T. R. 1988. The evolution of water management in Latin America. *Water Resources Development* 4:160–168.

Lee, T. R. 1991. *Water resources management in Latin America and the Caribbean*. Studies in Water Policy and Management, no. 16, Westview, Boulder, Col.

Lehman, M. P. 1992. Deforestation and changing land use patterns in Costa Rica. In *Changing tropical forests: Historical perspectives on today's challenges in Central and South America*. H. K. Steen and R. P. Tucker (eds.). Forest History Society, Durham, N.C.

Lele, Sharachchandra M. 1991. Sustainable development: A critical review. *World Development* 19(6):607–621.

Leonard, H. J. 1986. *Recursos naturales y desarrollo en América Central: Un perfil ambiental regional*. (Trad. del Inglés por G. Budowski y T. Maldonado, 1987.) CATIE, San José.

Leonard, H. J. 1987. *Natural resources and economic development in Central America: A regional environmental profile*. International Institute for Environment and Development. Transaction Books, Oxford.

Levey, D. J., T. C. Moermond, and J. S. Denslow. 1994. Fugivory: An overview. In *La Selva: Ecology and natural history of a neotropical rainforest*, pp. 282–294. L. A. McDade, K. S. Bawa, H. A. Hespenhide, and G. S. Hartshorn (eds.). University of Chicago Press, Chicago and London.

Libby, W. J. 1973. Domestication strategies for forest trees. *Canadian Journal of Forest Research* 3:265–277.

Lieth, H., and M. Lohmann. 1993. *Restoration of tropical forest ecosystems*. Kluwer Academic Publishers, Dordrecht, Boston, and London.

Lindarte, E., and C. Benito. 1991a, August 13–16. Agricultura sostenible de ladera en América Central: Instituciones, tecnología políticas. In Taller agricultura sostenible en las laderas Centroamericanas: Oportunidades de colaboración interinstitucional. International Institute of Cooperation in Agriculture, San José.

Lindarte, E., and C. Benito. 1991b. Instituciones, tecnología y políticas en la agricultura sostenible de laderas en América Central. In *Agricultura sostenible en las laderas Centroamericanas. Oportunidades de colaboración interinstitucional*, pp. 77–189. Memorias del Taller. IICA/CIAT, San José.

Lindarte, E., and C. Benito. 1993. *Sostenibilidad y agricultura de laderas en América Central*. Serie Documentos de Programas 33. IICA, San José.

Lipton, Michael. 1991. Accelerated resource degradation by Third World agriculture: Created in the commons, in the West or in bed? In *Agricultural sustainability, growth, and poverty alleviation: Issues and policies*, pp. 213–241. Stephen A. Vosti, Thomas Reardon, and Winfried von Urff (eds.). International Food Policy Research Institute, Washington, D.C.

Little, E. L. 1970. Relationships of trees of the Luquillo Experimental Forest, pp. 47–58. In *A tropical rain-forest*. H. T. Odum and R. F. Pigeon (eds.). USAEC Division of Technical Information Extension, Oak Ridge, Tenn.

Little, Peter, and Michael Horowitz (eds.). 1987. *Lands at risk in the Third World: Local level perspectives*. Westview, Boulder, Col.

Lloyd, J. L. 1963. *Historia tectónica del orogeno Suramericano*. Informe Semestral of the International Geophysical Union, San José, Costa Rica.

López, R. 1992. Environmental degradation and economic openness in LDCs: The poverty linkage. *American Journal of Agricultural Economics* 74:1130–1143.

Lowe-McConnell, R. H. 1987. *Ecological studies in tropical fish communities*. Cambridge University Press, New York.

Lucke S. O. 1993, October 18–22. *Bases de un marco conceptual y lineamientos generales para el diseño de un sistema de planificación ambiental y ordenamiento territorial en Costa Rica*. Memorias de IX Congreso Nacional Agronómico y de Recursos Naturales, San José.

Lugo, A. E. 1985. Water and ecosystems of the Luquillo Experimental Forest. U.S. Department of Agriculture, Forest Service, Research Paper SO-63, Southern Forest Experiment Station, New Orleans, La.

Lugo, A. E. 1988. The future of the forest: Ecosystem rehabilitation in the tropics. *Environment* 30:17–45.

Lugo, A. E., and F. N. Scatena. 1992. Epiphytes and climate change research in the Carribberan: A proposal. *Selbyana* 13:123–130.

Lugo, A. E., and S. C. Snedaker. 1974. The ecology of mangroves. *Annual Review of Ecology and Systematics* 5: 39–64.

Lundell, C. L. 1945. The vegetation and natural resources of British Honduras. In *Plants and plant science in Latin America*. F. Verdoorn (ed.). Chronica Botanica, New Bedford, Mass.

Lundgren, B. 1978. *Soil conditions and nutrient cycling under natural and plantation forests in Tanzanian highlands*. Swedish University of Agricultural Sciences, Uppsala.

MacFarland, C., and R. Morales. 1994, October 19–21. Conservation of biodiversity as the critical support for social development and sustainable natural resource use in Mesoamerica. Paper presented at the RENARM Technical Collaborators Workshop, Guatemala. USAID, Guatemala.

MacKerron, C. B., and D. G. Cogan. 1993. *Business in the rain forest*. Investor Responsibility Research Center, Washington, D.C.

Mahler, R. 1993. *Guatemala: A natural destination*. John Muir Publications, Santa Fe, N.M.

March, J. G., J. P. Benstead, and C. M. Pringle. 1996. Migration of freshwater shrimp larvae: elevational and diel patterns in two tropical river drainages, Puerto Rico. Abstract. *Bulletin of the North American Benthological Society* 13:161.

Marschner, H., V. Romheld, and I. Cakmak. 1987. Root-induced changes in nutrient availability in the rhizosphere. *Journal of Plant Nutrition* 10:1175–1184.

Martin, M. P. L. D., and R. W. Snaydon. 1982. Root and shoot interactions between barley and field beans when intercropped. *Journal of Applied Ecology* 19:263–272.

Martínez, R. O., and R. García. 1983. Alimentación con concentrados a vacas lecheras en pastoreo. In *Los pastos en Cuba. Tomo 2. Utilización*. Instituto de Ciencia Animal, La Habana, Cuba.

Mata, A., and O. Blanco. 1994. *La cuenca del Golfo de Nicoya*. Editorial de la Universidad de Costa Rica, San José.

Mateo, N., A. Díaz, and R. Nolasco. 1981. El sistema maíz maicillo en Honduras. Centro Agronómico de Investigación y Enseñanza. Turrialba, Costa Rica.

Mather, A. S. 1987. Global trends in forest resources. *Geography* 72(3/4):1–15.

Mathewson, K. 1984. *Irrigation horticulture in highland Guatemala—The Tablón System of Panajachel*. Westview, Boulder, Col.

Mayne, J. C. 1993. An investigation of nutrient uptake partitioning by depth as a response to competition. Ph.D. thesis, University of Florida, Gainesville.

Mazzarino, M. J., L. Szott, and M. Jiménez. 1993. Dyanmics of soil total C and N, microbial biomass, and water-soluble C in tropical agroecosystems. *Soil Biology Biochemistry* 25:205–214.

Mbogoh, S. G. 1991. Crop production. In Environmental change and dryland management in Machakos District, Kenya, 1930–1990: Production profile. M. Tiffen (ed). Working Paper no. 55. Overseas Development Institute. London.

McAffee, Katherine. 1987. *Storm signals: Structural adjustment and development alternatives in the Caribbean.* Zed Books, London.

McCarry, Nancy S. 1990. Variation among home gardens in Guatemala: Reflections of household characteristics. Master's Thesis. Department of Geography, University of Texas, Austin.

McGaughey, S. E., and H. M. Gregerson. 1982. *Forest-based development in Latin America.* Inter-American Development Bank, Washington, D.C.

McNeely, J. 1994. Lessons from the past: Forest and biodiversity. *Biodiversity and Conservation* 3:3–20.

McNeely, J., M. Gadgil, C. Leveque, C. Padoch, K. Redford. 1995. Human influences on biodiversity. In V. Heywood and R. Watson. *Global biodiversity assessment*, pp. 711–821. UNEP/Cambridge University Press, Cambridge.

Meléndez, G. 1996. Transformaciones de C, N, y P en sistemas agroforestales. M.S. thesis, Facultad de Agronomía, Universidad de Costa Rica, San José.

Meltzoff, S., and E. LiPuma. 1986. The social and political economy of coastal zone management: Shrimp mariculture in Ecuador. *Coastal Zone Management Journal* 14:349–380.

Méndez, A. 1988. *Population growth, land scarcity and environmental deterioration in rural Guatemala.* Universidad del Valle, Guatemala, Guatemala.

Meybeck, M., D. Chapman, and R. Helmer. 1989. *Global freshwater quality. A first assessment*, Basil Blackwell, Oxford.

Michon, G., and F. Mary. 1985, December 2–9. Transforming home gardens and related systems in West Java (Bogor) and West Sumatra (Maninjau): Facts and consequences. Paper presented at the International Workshop on Home Garden Systems, Bandung, Indonesia.

MIDEPLAN (Ministerio de Planificacion Nacional). 1982. *Plan nacional de desarrollo 1982–1986: Volvamos a la tierra.* Imprenta Nacional. San José.

Mikesell, Raymond F. 1991, July. Project evaluation and sustainable development. In *Environmentally sustainable economic development: Building on Brundtland*, pp. 54–60. Robert Goodland, Herman Daly, and Salah El Serafy (eds.). World Bank Environment Working Paper 46. World Bank, Washington, D.C.

Ministerio de Agricutura (MAG)–Recursos Naturales, Energía y Minas (MIRENEM). 1995. Metodología para la determinación de la capacidad de uso de las tierras de Costa Rica, MAG, San José.

Mojica, F. 1995. La agricultura orgánica en Costa Rica. In *Simposio Centroamericano sobre Agricultura Orgánica. Marzo.* J. García and J. Najera (eds.). Universidad Estatal a Distancia, San José.

Momsen, Janet. 1993. Women, work and the life course in the rural Caribbean. In *Full circles: Geographies of women over the life course.* Cindi Katz and Janice Monk (eds.). Routledge, London.

Monge, H. 1994. Funcionamiento e implicaciones de la producción agrícola no tradicional: El caso de Costa Rica. In *La agricultura de exportación en Centroamérica: Opciones de desarrollo en la década de los 90*, pp. 371–457. G. Masís and F. Sancho (eds.). Producciones Gráficas, San José.

Monge, J. 1985. Descripción de la siembra de frijol bajo el sistema tapado en la zona de Acosta. Tesis Ingeniero Agronomo Universidad de Costa Rica, San José.

Montagnini, Florencia. 1990. Impacts of native trees on tropical soils: A study in the Atlantic lowland of Cost Rica. *Ambio* 19:386–390.

Montoya, Maquín, J. M. 1966. Notas fitogeográficas sobre *Quercus oleoides. Revista Turrialba* 16(1):57–66.

Moock, J. 1986. *Understanding Africa's rural households.* Westview, Boulder, Col.

Moock, J. L., and R. E. Rhoades. 1992. *Diversity, farmer knowledge, and sustainability.* Cornell University Press. Ithaca, N.Y.

Moreno, A. 1985. *Estructura y función del sistema agroforestal gliricidia sepium-bovinos-gramíneas. Estudio de Caso finca La Pacífica.* CATIE, Turrialba, Costa Rica.

Moreno, R. (ed.). 1980. *Localización de sistemas de cultivos en Centroamérica.* CATIE, Turrialba, Costa Rica.

Moreno, R., and R. Hart. 1978. Intercropping with cassava in Central America. In *Proceedings of the Workshop Intercropping with Cassava*, Trivandrum, India, November 27–December 1, 1978. E. Weber, M. Campbell, and B. Nestel (eds.). IDRC, Ottawa.

Morris, L. M. 1994. Ten concepts on water supply and drought in Puerto Rico. *Dimensions, Segundo Trimestre*, pp. 7–14.

Moser, Carolyn. 1989. Gender planning in the Third World: Meeting practical and strategic gender needs. *World Development* 17(11):1799–1825.

Mott, G. 1974. Nutrient recycling in pastures. In *Forage fertilization.* D. A. Mays (ed.). American Society of Agronomics, Madison, Wisc.

Murillo, O., and L. Navarro. 1986. Validación de prototipos de producción de leche en la zona Atlántica de Costa Rica. Technical Report no. 90. CATIE, Turrialba, Costa Rica.

Murphy, E. C. 1995. La Selva and the magnetic pull of markets: Organic coffee-growing in Mexico. *Grassroots Development* (Inter-American Foundation) 19(1):27–34.

Murray, Douglas. 1991, Fall. Export agriculture, ecological disruption, and social inequality: Some effects of pesticides in southern Honduras. *Agriculture and Human Values* 8(4):19–29.

Murray, Douglas. 1994. *Cultivating crisis: Pesticides and development in Latin America.* University of Texas Press, Austin.

Murray, Douglas, and Polly Hoppin. 1992, April. Recurring contradictions in agrarian development: Pesticides and

Caribbean Basin nontraditional agriculture. *World Development* 20(4): 597–608.

Muschler, R. G. 1993. Biomass production, light transmission, and management of *Erythrina bertoana*, *Erythrina fusca*, and *Gliricidia sepium* used as living supports in Talamanca, Costa Rica. In Erythrina *in the New and Old Worlds*, pp. 192–199. Nitrogen Fixing Tree Research Reports Special Issue. S. B. Westley and M. H. Powell (eds.). Nitrogen Fixing Tree Association, Paia, Hawaii.

Myers, N. 1984. *The primary source: Tropical forests and our future*. Norton, New York.

Myers, N. 1995. The world's forests: Need for a policy appraisal. *Science* 268:823–824.

Naiman, R. J. (ed.). 1992. *Watershed management: Balancing sustainability and environmental change*. Springer-Verlag, New York.

Naiman, R. J., J. J. Magnuson, D. M. McKnight, and J. A. Stanford. 1995. *The freshwater imperative: A research agenda*. Island Press, Washington, D.C.

Nair, P. K. R. 1989. Ecological spread of major agroforestry systems. In *Agroforestry systems in the tropics*. P. K. R. Nair (ed.). Kluwar Academic Publishing, Dordrecht, Holland.

Nair, P. K. R., B. T. Kang, and D. C. L. Kass. 1995. Nutrient cycling and soil-erosion control in agroforestry systems. In *Agriculture and environment: Bridging food production and environmental protection in developing countries*. A. S. R. Juo and R. D. Freed (eds.). American Society of Agronomy Special Publication no. 60. American Society of Agronomy, Madison, Wisc.

NAS (National Academy of Sciences). 1972. *Genetic vulnerability of major crops*. NAS Washington, D.C.

Nash, L. 1993. Water quality and health. In *Water in crisis: A guide to the world's fresh water resources*, pp. 25–39. P. H. Gleick (ed.). Oxford University Press, New York.

National Research Council. 1989. *Alternative agriculture*. National Academy Press, Washington, D.C.

Nations, J. D., and D. J. Komer. 1983. Central America's tropical rainforests: Positive steps for survival. *Ambio* 12:232–238.

Nations, J. D., and D. J. Komer. 1987. Rainforests and the hamburger society. *The Ecologist* 17(4/5):161–167.

Nature Farming. 1990. *Proceedings of the 1st international Conference on Nature Farming*. Pacific Culture Center and Nature Farming International Research Foundation, Khon Kaen, Thailand.

Nature Farming. 1992. *Standards of nature farming systems and practices*, 2nd ed. Nature Farming International Research Foundation, Atami, Japan.

Naumann, M. 1994. A water-use budget for the Caribbean National Forest of Puerto Rico. Special Report, U.S.D.A. Forest Service.

Navarro, E. 1993 and 1995. Consejo Nacional de Producción, San Vito, Costa Rica. Personal communication.

Navarro, L. A. 1987. Characteristics of farms producing basic grains in four areas of Central America. In *Advances in agroforestry research*, pp. 309–317. J. W. Beer, H. W. Fassbender, and J. Heuveldop (eds.). CATIE, Turrialba, Costa Rica.

Nedrow, W. W. 1937. Studies on the ecology of roots. *Ecology* 18:27–52.

Newman, E. I. 1973. Competition and diversity in herbaceous vegetation. *Nature* 244:310.

Norgaard, Richard B. 1991, June. Sustainability as intergenerational equity: The challenge to economic thought and practice. Report no. DP 7. World Bank, Washington, D.C.

Norgaard, Richard B., and Richard B. Howarth. 1991. Sustainability and discounting the future. In *Ecological economics: The science and management of sustainability*, pp. 88–101. Robert Costanza (ed.). Columbia University Press, New York.

Nye, P. H., and D. J. Greenland. 1960. *The soil under shifting cultivation*. Technical Communication no. 5o. Commonwealth Agricultural Bureaux, Harpenden, England.

Nye, P. H., and D. J. Greenland. 1964. Changes in the soil after clearing tropical forest. *Plant and Soil* 21:101–112.

Nygren, Anja. 1993. *El bosque y la naturaleza en la percepción del campesino costarricense: Un estudio de caso*. CATIE, Turrialba, Costa Rica.

Obeng, L. E. 1981. Man's impact on tropical rivers. In *Perspectives in running water ecology*, pp. 265–288. M. A. Lock and D. D. Williams (eds.). Plenum Press, New York.

OCIA (Organic Crop Improvement Association). 1995. *Estándares Internacionales de Certificación*. OCIA, Belle Fontaine, Ohio.

Odum, E. P. 1960. Organic production and turnover in old field succession. *Ecology* 41:34–49.

OFIPLAN (Oficina de Planificacion Nacional). 1977. *La Costa Rica del año 2000*. Imprenta Nacional, San José.

OFIPLAN. 1979. *Plan nacional de desarrollo forestal 1979–1982*. Imprenta Nacional, San José.

Ojo, G. O. A., and J. K. Jackson. 1973. The use of fertilizer in forestry in the drier tropics. In *Proceedings of the FAO/IUFRO International Symposium on Forest Fertilization*, Paris, pp. 351–364.

Oldfield, Margery, and Janis B. Alcorn. 1991. *Biodiversity: Culture, conservation and ecodevelopment*. Westview, Boulder, Col.

OMA (Oficina de Manejo Ambiental). 1993–1995. *Natural History, Ecology and Conservation Magazine* 1(1–4). Programa de Educación Ambiental, Proyecto de Consolidación de la Zona Norte, USAID-MIDEPLAN, Upala, Costa Rica.

Ong, C. 1994. Alley cropping—Ecological pie in the sky? *Agroforestry Today* 6:8–10.

Ongley, E. D. 1993. Water quality data programmes for developing land use and resource policies: Latin America. In *Prevention of water pollution by agriculture and related activities*, pp. 263–272. Proceedings of the Food and Agriculture Organization of the United Nations, Santiago, Chile, 1992. FAO, Rome.

Oosting, H. J. 1942. An ecological analysis of the plant communities of Piedmont, North Carolina. *American Midland Naturalist* 28:1–126.

Opler, P. 1969. Unpublished list.

Ostrom, Elinor. 1990. *Governing the commons: The evolution of institutions for collective action*. Cambridge University Press, Cambridge.

Ostrom, Elinor. 1992. *Crafting institutions for self-governing irrigation systems*. Institute for Contemporary Studies Press, San Francisco.

Ovington, J. D. 1958. Studies of the development of woodland conditions under different trees. *Journal of Ecology* 46:391–405.

Oyuela, D. O. 1995. Los recursos forestales y la vida silvestre, en la zona sur de Honduras. Serie Miscelánea no. 9. Escuela Nacional de Ciencias Forestales (ESNACIFOR), Siguatepeque, Comayagua, Honduras.

Pagan, F. A., and H. M. Austin. 1970. Report on a fish kill at Laguna Joyuda, Western Puerto Rico in the summer 1967. *Caribbean Journal of Science* 10:203–208.

PAHO (Pan American Health Organization). 1990a. *The situation of drinking water supply and sanitation in the American region at the end of the decade 1981–1990, and prospects for the future*, vol. 1, PAHO, Washington, D.C.

PAHO. 1990b. Wastewater disposal in the Caribbean: Status and strategies. *Bulletin of PAHO* 24:252–255.

PAHO. 1991. Drinking water supply and sanitation in the America's: Status and prospects. *Bulletin of PAHO* 25: 87–96.

PAHO. 1992. Health and the environment. *Bulletin of PAHO* 26:370–378.

Painter, Michael, and William Durham (eds.). 1995. *The social causes of environmental destruction in Latin America*. University of Michigan Press, Ann Arbor.

Palladino, Paolo S. A. 1989. Entomology and ecology: The ecology of entomology. Upublished thesis, University of Minnesota Graduate School, Minneapolis.

Palloni, Alberto. 1994. The relation between population and deforestation: Methods for drawing causal inferences from macro and micro studies. In *Population and environment: Rethinking the debate*, pp. 125–168. L. Arizpe, M. P. Stone and D. Major (eds.). Westview, Boulder, Col.

Palm, C. A. 1995. Contributions of agroforestry trees to nutrient requirements of intercropped plants. *Agroforestry Systems* 30:105–124.

Palm, C. A., A. J. McKerrow, K. M. Glasener, and L. T. Szott. 1991. Agroforestry in the lowland tropics: Is phosphorus important? In *Phosphorus cycles in terrestrial and aquatic ecosystems*, pp. 134–141. H. Tiessen, D. Lopez-Hernández, and I. Salcedo (eds.). Regional Workshop 3. South and Central America. SCOPE and United Nations Environmental Program. Saskatchewan Institute of Pedology, Saskatoon, Canada.

Palo, M. 1987. Deforestation perspectives for the tropics: A provisional theory with pilot applications. In *The global forest sector: An analytical perspective*, pp. 57–90. M. Kallio, D. P. Dykstra, and C. S. Binkley (eds.). Wiley, New York.

Panama. 1988–1990. *Contraloría general de la República*. Dirección de Estadísticas y Censos, Sección Agropecuaria.

Parks, P., and M. Bonifaz. 1994. Nonsustainable use of renewable resources: Mangrove deforestation and mariculture in Ecuador. *Marine Resource Economics* 9(1): 1–18.

Parrish, J. A. D., and F. A. Bazzaz. 1976. Underground niche separation in successional plants. *Ecology* 57:1281–1288.

Parrotta, J. A. 1993. Secondary forest regeneration on degraded tropical lands: The role of plantations as "foster ecosystems." In *Restoration of tropical forest ecosystems*, pp. 63–73. H. Lieth and M. Lohmann (eds.). Kluwer Academic Publishers, Dordrecht, Boston, and London.

Parson, J. J. 1976. Forest to pasture: Development or destruction? *Revista de Biología Tropical* 24:121–138.

Parsons, J. J. 1955. The Miskito pine savanna of Nicaragua and Honduras. *Annals of the Association of American Geographers* 45(1):36–63.

Parsons, T. R., Y. Maita, and C. M. Lalli. 1992. *A manual of chemical and biological methods for seawater analysis*. Pergamon Press, New York.

Parsons Corporation, Marshall and Stevens Incorporated, International Aero Service Corporation. 1971. *Soil survey of the Pacific region of Nicaragua*. Tax Improvement and Natural Resources Inventory Project. Managua, Nicaragua.

Pasos, R. 1994. El ultimo despale . . . : La frontera agrícola Centroamericana. Fundación para el Desarrollo Económico y Social para Centroamérica. Managua, Nicaragua.

Pasos, R., P. Girot, M. Laforge, P. Torrealba, and D. Kaimowitz. 1994. *El ultimo despale . . . : La frontera agrícola Centroamericana*. Fundación para el Desarrollo Económico y Social para Centroamérica, Managua, Nicaragua. Imprenta Comercial, San José.

Patiño, Víctor M. 1965. *Historia de la actividad agropecuaria en América Equinoccial*. Imprenta Departamental, Cali, Valle, Colombia.

Paus, E. (ed.). 1988. *Struggle against Dependence: Nontraditional export growth in Central America and the Caribbean*. Westview, Boulder, Col.

Pearce, David, Anil Markandya, and Edward B. Barbier. 1989. *Blueprint for a green economy*. Earthscan, London.

Pedroni, L., and J. Flores Rodas. 1992. *Diagnóstico forestal regional para Centroamérica y propuestas de trabajo*. IC/UICN/ORCA/COSUDE, San José.

Peet, Richard, and Michael Watts. 1993. Introduction: Development theory and environment in an age of market triumphalism. *Economic Geography* 69(3):227–253.

Peet, R., and M. Watts (eds.). 1996. *Liberation ecologies*. Routledge, London.

Pena-Torrealba, H. 1993. Natural water quality and agricultural pollution in Chile. In *Prevention of water pollution by agriculture and related activities*, pp. 67–76. Proceedings of the Food and Agriculture Organization of the United Nations, Santiago, Chile, 1992. FAO, Rome.

Penrose, D., and S. M. Call. 1995. Volunteer monitoring of benthic macroinvertebrates: Regulatory biologists' perspectives. *Journal of the North American Benthological Society* 14:203–209.

Pérez, E., and D. Robadue. 1989. Institutional issues of shrimp mariculture in Ecuador. In S. Olsen and L. Arriaga. *Establishing a sustainable shrimp mariculture industry in Ecuador*. Coastal Resources Center, University of Rhode Island, Narragansett.

Petritz, D. J. 1996, June 14. Country's experts discuss food security: Self-sufficiency in decline for basic staples. *Tico Times*. San José.

Petts, G. E., J. G. Imhoff, B. A. Manny, J. F. B. Maher, and S. B. Weisberg. 1989. Management of fish populations in large rivers: A review of tools and approaches. In *Proceedings of the International Large River Symposium*, pp. 578–588. D. P. Dodge (ed.). Canadian Special Publication of Fisheries and Aquatic Sciences.

Phillips, M. J., C. K. Lin, and M. C. M. Beveridge. 1993. Shrimp culture and the environment: Lessons from the world's most rapidly expanding warmwater aquaculture sector. In *Environment and aquaculture in developing countries*. R. S. V. Pulling, H. Rosenthal, and J. L. Maclean (eds.). ICLARM Conference Proceedings 31. International Center for Living Aquatic Resources, Manila.

Picket, S. T. A. 1980. Non-equilibrium coexistence of plants. *Bulletin of the Torry Botanical Club* 107:238–248.

Pico, R. 1974. *The geography of Puerto Rico*. Aldine, Chicago.

Piñeda, N. 1984. *Geografía de Honduras*. Editorial E.S.P., Tegucigalpa, Honduras.

Pingali, Prabhu. 1988. *Intensification and diversification of Asian rice farming systems*. International Rice Research Institute (IRRI), AF, 88–41. IRRI, Los Banos, Philippines.

Pingali, P., Y. Bigot, and H. P. Binswanger. 1987. *Agricultural mechanization and the evolution of farming systems in sub-Saharan Africa*. Johns Hopkins University Press, Baltimore, Md.

Pla, S. I. 1995, December 4–8. *La erodibilidad de los Andisoles en Latinoamérica*. Conference proceedings, Red Latinomericana de Labranza Conservacionista (RELACO), San José. Centro de Investigaciones Agronómicas of the Universidad de Costa Rica, San José.

Plafkin, J. L., M. T. Barbour, K. D. Porter, S. K. Gross and R. M. Hughes. 1989. *Rapid bioassessment protocols for use in streams and rivers: Benthic macroinvertebrates and fish*. EPA/440/4–89/001. U.S. Environmental Protection Agency, Washington, D.C.

Plucknett, Donald H., and Nigel J. H. Smith. 1986. Sustaining agricultural yields. *BioScience* 36(1):40–45.

Poats, S., H. Feldstein, and D. Rocheleau. 1989. Gender and intra-household analysis in on-farm research and experimentation. In *The household economy: Reconsidering the domestic mode of production*, pp. 245–266. R. Wilk (ed.). Westview, Boulder, Col.

Poats, Susan, Marianne Schmink, and Anita Spring (eds.). 1988. *Gender issues in farming systems research and extension*. Westview, Boulder, Col.

Pohl, R. W. 1980. Flora costarricensis. Fieldiana Botany. *Gramineas*, series 4.

Pomeroy, Robert S. 1992. Aquaculture development: An Alternative for small-scale fisherfolk in developing countries. In *Coastal aquaculture in developing countries: Problems and prospects*, pp. 73–86. R. Pollnac and P. Weeks (eds.). ICMRD, University of Rhode Island, Kingston.

Poore, D. 1989. *No timber without trees*. Earthscan, London.

Popenoe, H. 1984, Summer. Commentary on "Social Ethics and Land Reform: The Case of the El Salvador" by L. R. Simon. *Agriculture and human values*, p. 36.

Porras, A. 1980. Análisis general de la legislación forestal. *Revista de Ciencias Jurídicas* 14:164–173.

Porras, A. 1981. Derecho ambiental de Costa Rica. *Revista Judicial* 5(20):83–86.

Porras, A., and B. Villarreal. 1986. *Deforestación en Costa Rica: Implicaciones sociales, económicas y legales*. Editorial Costa Rica, San José.

Posey, Daryll. 1985. Indigenous management of tropical forest ecosystems: The case of the Kayapo Indians of the Brazilian Amazon. *Agroforestry Systems* 3:139–158.

Posner, J. L., G. A. Antonini, G. Montañez, R. Cecil, and M. Grigsby. 1983. Land systems of hill and highland tropical America. *Revista Geográfica* no. 98.

Postel, S. 1987, September. Defusing the toxics threat: Controlling pesticides and industrial waste. *Worldwatch* 79:11.

PRASA (Puerto Rico Aqueductos and Sewage Authority). 1995a. North coast superaqueduct project draft environmental impact statement. PRASA, San Juan, Purto Rico.

PRASA. 1995b. Supplementary preliminary environmental impact statement for the Carraizo Reservoir Dredging Project. DIA-P JCA-92-0006. PRASA, San Juan.

Pray, Carl E. 1987. Private agricultural sector research in Asia. In *Policy for agricultural research*, pp. 411–432. Vernon W. Ruttan and Carl E. Pray (eds.). Westview, Boulder, Col.

Pretty, Jules N. 1990. Sustainable agriculture in the Middle Ages: The English manor. *Agricultural History Review* 3(1):1–19.

PRIAG (Programa Regional de Investigacion de Apoyo a los Granos)–Universidad Libre de Amsterdam. 1993. El desarrollo tecnológico en los sistemas de producción de granos básicos en Centroamérica. El Caso de Panamá. El Caso de El Salvador. El Caso de Honduras. PRIAG–Universidad Libre de Amsterdam, Amsterdam, Holland.

Price, Colin. 1991, January. Do high discount rates destroy tropical forests? *Journal of Agricultural Economics* 42: 77–83.

Primack, R. 1993. *Essentials of conservation biology*. Sinnauer Associates, Sunderland, Mass.

Pringle, C. M. 1988. History of conservation efforts and initial exploration of the lower extension of Parque Nacional Braulio Carrillo. In *Tropical rainforests: Diversity and conservation*, pp. 225–241. F. Almeda and C. Pringle (eds.). Allen Press, Lawrence, Kan.

Pringle, C. M. 1991. Geothermal waters surface at La Selva Biological Station, Costa Rica: Volcanic processes introduce chemical discontinuities into lowland tropical streams. *Biotropica* 23:523–529.

Pringle, C. M. 1996. Atyid shrimp (Decapoda: Atyidae) influence spatial heterogeneity of algal communities over different scales in tropical montane streams, Puerto Rico. *Freshwater Biology* 35:125–140.

Pringle, C. M. 1997. Exploring hos disturbance is transmitted upstream: Going against the flow. *Journal of the North American Benthological Society* 16:425–438.

Pringle, C. M., and A. Ramirez. In press. Use of both benthic and drift sampling techniques to assess tropical stream invertebrate communities along an altitudinal gradient, Costa Rica. *Freshwater Biology.*

Pringle, C. M., and N. G. Aumen. 1993. Current efforts in freshwater conservation. *Journal of the North American Benthological Society* 12:174–176.

Pringle, C. M., and G. A. Blake. 1994. Quantitative effects of atyid shrimp (Decapoda: Atyidae) on the depositional environment in a tropical stream: Use of electricity for experimental exclusion. *Canadian Journal of Fisheries and Aquatic Sciences* 51:1443–1450.

Pringle, C. M., G. A. Blake, A. P. Covich, K. M. Buzby and A. Finley. 1993. Effects of omnivorous shrimp in a montane tropical stream: Sediment removal, disturbance of sessile invertebrates and enhancement of understory algal biomass. *Oecologia* 93:1–11.

Pringle, C. M., G. L. Rowe, F. J. Triska, J. F. Fernández, and J. West. 1993. Landscape linkages between geothermal activity, solute composition and ecological response in streams draining Costa Rica's Atlantic slope. *Limnology and Oceanography* 38:753–774.

Pringle, C. M., and F. J. Triska. 1991. Effects of geothermal waters on nutrient dynamics of a lowland Costa Rican stream. *Ecology* 72:951–965.

Pulliam, H. R., and J. B. Dunning. 1994. Demographic processes: Population dynamics on heterogeneous landscapes. In *Principles of conservation biology*, pp. 179–205. G. K. Meffe and C. R. Carroll (eds.). Sinauer Associates, Sunderland, Mass.

Pulsipher, Lydia. 1993. He won't let she stretch she foot: Gender relations in traditional West Indian houseyards. In *Full circles: geographies of women over the life course*, pp. 122–137. Cindi Katz and Janice Monk (eds.). Routledge, New York.

Quesada, C. 1990. *Estrategia de conservación para el desarrollo sostenible de Costa Rica*. MIRENEM, San José.

Quijada, B. 1995. La agricultura orgánica en Panamá. In *Simposio Centroamericano sobre Agricultura Orgánica, Marzo*. J. García and J. Najera (eds.). Universidad Estatal a Distancia, San José.

Quiñones, Márquez F. 1989. Perspectivas futuras de los recursos de agua de Puerto Rico. *Acta Científica* 3: 26–35.

Quirós, M. 1994. Consejo Nacional de Producción, San Vito, Costa Rica. Personal communication.

Quirós, R. 1993. Inland fisheries under constraint by other uses of land and water resources in Argentina, pp. 29–94. In *Prevention of water pollution by agriculture and related activities*. Proceedings of the Food and Agriculture Organization of the United Nations, Santiago, Chile, 1992. FAO, Rome.

Raintree, J. B., and K. Warner. 1985. Agroforestry pathways for the integral development of shifting cultivation. Paper presented at the Ninth World Forestry Congress, Mexico City.

Ramírez, A. 1992. Description and natural history of Costa Rican dragonfly larvae. I: *Heteragrion erythrogastrum* (Selys, 1886) *(Zygoptera: Megapodagrionidae). Odonatologica* 21:361–365.

Ramírez, A. 1994. Descripción e historia natural de las larvas de odonatos de Costa Rica. III: *Gynacantha tibiata* (Karsch, 1891) (*Anisoptera: Aeshnidae*). *Bulletin of American Odonatology* 2:9–14.

Ramírez, A. In press. *Manual para la identificación de los géneros de libélulas (Insecta: Odonata) de Costa Rica.* Editorial FIREMA, San José.

Ramírez, X. 1994. CODDEFFAGOLF: Los defensores de los manglares del Golfo de Fonseca, Honduras. *Revista Forestal Centroamericana* 9:27–32.

Rathcke, B., and E. P. Lacey. 1985. Phenological patterns of terrestrial plants. *Annual Review of Ecology and Systematics* 16:179–214.

Raup, Hugh M. 1964, March. Some problems in ecological theory and their relation to conservation. *Journal of Ecology* 52:19–28.

Raynolds, L. 1994. The restructuring of Third World agro-exports: Changing production relations in the Dominican Republic. In *The global restructuring of agro-food systems*. P. McMichael (ed.). Cornell University Press, Ithaca, N.Y.

Redclift, M. 1988. Agriculture and the environment: The Mexican experience. In *The Mexican economy*. G. Phillip (ed.). Croom Helm, London.

Redclift, Michael. 1984. *Development and environmental crisis*. Methuen, New York.

Redclift, Michael. 1987. *Sustainable development: Exploring the contradictions*. Methuen, New York.

Redford, K. H., and J. G. Robinson. 1991. Subsistence and commercial uses of wildlife in Latin America. In *Neotropical wildlife use and conservation*, pp. 6–23. J. G. Robinson and K. H. Redford (eds.). University of Chicago Press, Chicago.

Reganold, J. P., R. I. Papendick, and J. F. Parr. 1990, June. Sustainable agriculture. *Scientific American* 262:112–115.

Reich, P., and R. Borchert. 1982. Phenology and ecophysiology of the tropical tree, Tabebuia Neochrysantha (*Bignoniaceae*). *Ecology* 63:294–299.

Reiff, F. 1993. Health impacts related to irrigated agriculture in Latin America. In *Prevention of water pollution by agriculture and related activities*, pp. 327–340. Proceedings of the Food and Agriculture Organization of the United Nations, Santiago, Chile, 1992. FAO, Rome.

Reining, C., and R. Heinzman. 1992. Nontimber forest products in the Petén, Guatemala: Why extractive reserves are critical for both conservation and development. In *Sustainable harvest and marketing of rain forest products*. M. Plotkin and L. Famolare (eds.). Island Press, Washington, D.C.

Reining, L. 1992. *Erosion in Andean hillside farming*. Hohenheim Tropical Agriculture Series. University of Hohenheim, Stuttgart, Germany.

República de Colombia. 1991. Constitución de la República de Colombia. Bogotá.

Rice, B. 1986. The fresh connection. *Audubon* 88:104–107.

Richards, E. M. 1991. Plan Piloto Forestal. *Commonwealth Forestry Review* 70(4):290–311.

Rocheleau, D. 1984. An ecological analysis of soil and water conservation in hillslope farming systems: Plan Sierra, Dominican Republic. Ph.D. thesis, University of Florida, Gainseville.

Rocheleau, D. 1987. The user perspective and the agroforestry research and action agenda. In *Agroforestry: Realities, possibilities and potentials*, pp. 59–88. H. Gholz (ed.). Martinus Nujhoff Publishers, Dordrecht, Netherlands.

Rocheleau, D. 1995a. Gender and biodiversity: A feminist political ecology perspective. *IDS Bulletin* 26(1):9–16.

Rocheleau, D. 1995b. Gendered resource mapping. In *Power, process and participation: Tools for change*, pp. 110–120. R. Slocum, L. Wichart, D. Rocheleau, and B. Thomas-Slayter (eds.). Intermediate Technology Publications, London.

Rocheleau, D. 1995c. Maps, numbers, text and context: Mixing methods in feminist political ecology. *Professional Geographer* 47(4):458–467.

Rocheleau, D., P. Benjamin, and A. Duang'a. 1995. The Ukambani region of Kenya. In *Critical environmental zones: Regional case studies*. R. Kasperson, J. Kasperson, and B. Turner II (eds.). United Nations University Press, Tokyo.

Rocheleau, D., and D. Edmunds. 1995. Men, women, trees and tenure: Gender, power and property in forest and agrarian landscapes. GENPROP International E-Mail Conference on Property and Gender, International Food Policy Research Institute, October 1995–March 1996.

Rocheleau, D., and L. Ross. 1995. Trees as tools, trees as text: Struggles over resources in Zambrana-Chacuey, Dominican Republic. *Antipode* 27(4):407–428.

Rocheleau, D., L. Ross, J. Morrobel, and R. Hernández. 1996. Forests, gardens, and tree farms: Gender, class and community at work in the landscapes of Zambrana-Chacuey. ECOGEN Project. Clark University. Worcester, MA.

Rocheleau, D., L. Ross, and J. Morrobel, with Ricardo Hernández, C. Amparo, C. Brito, and D. Zevallos. 1996. From forest gardens to tree farms: Women, men and timber in Zambrana-Chacuey, Dominican Republic. In *Feminist political ecology: Global issues and local experiences* pp 224–250. D. Rocheleau, B. Thomas-Slayter, and E. Wangari (eds.). Routledge, London.

Rocheleau, D., K. Schofield, and N. Mbuthi. 1992. Gender, power, poverty and parks: A story of men, women, water and trees at Pwani. ECOGEN Working Paper. International Development Program, Clark University, Worcester, Mass.

Rocheleau, D., and R. Slocum. 1995. Participation in context: Who, where, when, how and why? In *Power, process and participation: tools for change*, pp. 17–31. L. Slocum, L. Wichart, D. Rocheleau, and B. Thomas-Slayter (eds.). Intermediate Technology Publishers, London.

Rocheleau, D., P. Steinberg, and P. Benjamin. 1995. Environment, development, crisis and crusade: Ukambani, Kenya 1890–1990. *World Development* 23(6):1037–1051.

Rocheleau, Dianne, Barbara Thomas-Slayter, and David Edmunds. 1995. Gendered resource mapping: Focusing on women's spaces in the landscape. *Cultural Survival Quarterly* 18 (4):62–68.

Rocheleau, D., B. Thomas-Slayter, and E. Wangari. 1996. Gender and environment: A feminist political ecology perspective. In *Feminist political ecology: Global perspectives and local experiences*. D. Rocheleau, B. Thomas-Slayter, and E. Wangari (eds.). Routledge, London.

Rodhe, H., and R. Herrera. 1988. *Acidification in tropical countries*. Scientific Committee on Problems of the Environment (SCOPE) Publication no. 36. Wiley, New York.

Rodríguez, R., and D. R. Teichert-Coddington. 1995. Substitución de nutrientes inorgánicos por alimento en la producción comercial de *Penaeus vannamei* durante la época de invierno y verano de Honduras. In *Proceedings of the Third Centroamerican Shrimp Symposium*, Tegucigalpa, Honduras.

Rosemeyer, M. E. 1990. The effects of different management strategies on the tripartite symbiosis of bean (Phaseolus vulgaris L) with vesicular-arbuscular mycorrhizal fungi and Rhizobium phaseoli, in two agroecosystems in Costa Rica. Ph.D. thesis, University of California, Santa Cruz.

Rosemeyer, M. E. 1994. Comparison of yield, mycorrhizae and nodulation of beans grown under the slash/mulch and *espequeado* systems with fertilizer addition. In *Slash/mulch: how farmers use it and what researchers know about it*. D. Thurston, M. Smith, G. Abawi, and S. Kearl (eds.). CIIFAD and CATIE, Cornell University, Ithaca, N.Y.

Rosemeyer, M. E. 1995, March 6–11. El mantillo vivo en un sistema orgánico de frijol tapado. In *Las Memoria de Simposio Centroamericano Sobre Agricultura Orgánica*, pp 141–162. J. García and J. Nájera (eds). Universidad Estatal a la Distancia (UNED), San José.

Rosemeyer, M. E. 1995b. Eficiencia de aplicationes de fósforo en los sistemas frijol tapado y espequeado a treves de tresaños. In *Taller Internacional Sobre*. R. Araya and D. Beck (eds.). Bajo Fósforo en Frijol Comun. CIAT (Centro de Investigación de Agricultura Tropical), San José, Costa Rica.

Rosemeyer, M. E. In press. *Avances en investigación de frijol tapado con énfasis en fuentes alternativas de fósforo*. Memoria del Primer Encuentro Nacional Campesino de Sistemas de Coberturas, October 4–7, 1993, San José. Corporacion Educativa para el Desarrollo Costarricense (CEDECO), San José.

Rosemeyer, M. E., and S. R. Gliessman. 1992. Modifying traditional and high-input agroecosystems for optimization of microbial symbioses: A case study of dry beans in Costa Rica. *Agriculture, Ecosystems and Environment* 40:61–70.

Rosemeyer, M. E., and J. Kettler. (unpublished manuscript). Nutrient dynamics of an enriched fallow system in south Costa Rica.

Rosenberg, D. M., R. A. Bodaly, and P. J. Usher. 1995. Environmental and social impacts of large scale hydroelectric development: Who is listening? *Global Environmental Change* 5:127–148.

Rosenberg, Elliot, and Ludwig M. Eisgruber. 1992, May. Sustainable development and sustainable agriculture: A partially annotated bibliography with emphasis on economics. Graduate Faculty of Economics Working Paper 92-101, Corvallis, Ore.

Rosenberry, B. 1994, December. World shrimp farming 1994. *Shrimp News International*, San Diego, Calif.

Rosenberry, B. 1995, December. World shrimp farming 1995. *Annual Report*, *Shrimp News International*, San Diego, Cal.

Ross, Laurie. 1995. When a grassroots social movement enters a development partnership with an NGO: Overcoming the barriers that prevent local control over modern technology in Zambrana-Chacuey, Dominican Republic. Master's thesis, International Development, Clark University, Worcester, Mass.

Rosset, Peter. 1991, Fall. Sustainability, economies of scale, and social instability: Achilles heel of nontraditional export agriculture? *Agriculture and Human Values* 8(4): 30–37.

RRNN (Natural and Renewable Resource Ministry of Honduras). 1993. Importación de Larvas y post-Larvas de Camáron 1992. Mimeo. Department of Animal Health, RRNN, Tegucigalpa, Honduras.

Rubino, Michael, and Richard Stoffle. 1990. Who will control the blue revolution? Economic and social feasibility of Caribbean crab mariculture. *Human Organization* 49:386–394.

Runge, C. Ford. 1992. A policy perspective on the sustainability of production environments: Toward a land theory of value. In *Future challenges for national agricultural research: A policy dialogue*. International Service for National Agricultural Research, The Hague.

Runge, C. Ford, Robert D. Munson, Edward Lotterman, and Jared Creason. 1990. *Agricultural competitiveness, farm fertilizer, chemical use and environmental quality*. Center for International Food and Agricultural Policy, St. Paul, Minn.

Ruttan, Vernon W. 1971, December. Technology and the environment. *American Journal of Agricultural Economics* 53:707–717.

Ruttan, Vernon W. 1988, Spring/Summer. Sustainability is not enough. *American Journal of Alternative Agriculture* 3:128–130.

Ruttan, Vernon W. (ed.). 1989a, December. Biological and technical constraints on crops and animal productivity: Report on a dialogue. Staff Paper P89-45. Department of Agriculture and Applied Economics, University of Minnesota, St. Paul.

Ruttan, Vernon W. 1989b, January. Why foreign economic assistance? *Economic Development and Cultural Change* 37:411–424.

Ruttan, Vernon W. (ed.). 1992. *Sustainable development and the environment: Perspectives on growth and constraints*. Westview, Boulder, Col.

Ruttan, Vernon W. (ed.). 1994. *Agriculture, environment and health: Toward sustainable development into the twenty-first century*. University of Minnesota Press, Minneapolis.

Sachs, Wolfgang (ed.). 1992. *Development dictionary*. Zed Books, London.

Sachs, Wolfgang (ed.). 1993. *Global ecology: A new arena for international conflict*. Zed Books, London.

Saenger, P., E. J. Hegerl, and J. D. S. Davie (eds.). 1983. Global Status of Mangrove Ecosystems. Commissission on Ecology Papers no. 3, International Union for Conservation of Nature and Natural Resources. *The Environmentalist* 3.

Salazar, D., and A. Leonard. 1994. Conservation and the nature of goods. *Society and Natural Resources* 7:331–348.

Salazar, R. 1993. *El derecho a un ambiente sano: Ecología y desarrollo sostenible*. Asociación Libro Libre, Serie Jurídica, San José.

Salisbury, R. S. 1936. Natural selection and competition. *Proceedings of the Royal Society of London* 121:47–49.

Sánchez, P. A. 1976. *Properties and management of soils in the tropics*. Wiley, New York.

Sánchez, P. A. 1987. Soil productivity and sustainability in agroforestry systems. In *Agroforestry: A decade of development*. H. A. Steppler and P. K. R. Nair (eds.). ICRAF, Nairobi.

Sánchez, P. A. 1995. Science in agroforestry. *Agroforestry Systems* 30:5–55.

Sánchez, P. A., C. A. Palm, C. B. Davey, L. T. Szott, and C. E. Russell. 1985. Tree crops as soil improvers in the humid tropics? In *Attributes of trees as crops plants*, pp. 327–358. M. G. R. Cannell and J. E. Jackson (eds.). Institute of Terrestrial Ecology, Huntingdon, England.

Sanders, G. 1962. *La colonización agrícola de Costa Rica*, tomo I y II. Instituto Geográfico Nacional. San José.

Sandner, G. 1972. *La colonización agrícola de Costa Rica*. Instituto Geográfico de Costa Rica, Ministerio de Obras Públicas y Transportes, San José.

Sanginga, N., S. K. A. Danso, F. Zapata, and G. D. Bowen. 1994. Field validation of intraspecific variation in phosphorus use efficiency and nitrogen fixation by provenances of *Gliricidia sepium* grown in low P soils. *Applied Soil Ecology* 1:127–138.

San José, J. J., and M. R. Farinas. 1978. Estudio sobre los cambios de la vegetación protegida de la quema y el pastoreo en la estación biológica de los Llanos. *Bulletin of the Venezuelan Society for Natural Science*, no. 135, Caracas.

Santana, Roberto. n.d. Los espacios cerealeros Centroamericanos: Un ensayo de regionalización. Temas de Seguridad Alimentaria 2, Programa de Seguridad Alimentaria, CADESCA, Panama.

Santiago-Rivera, L. 1992. Low-flow characteristics at selected sites on streams in eastern Puerto Rico. Water Resources Investigations Report 92–4063. Prepared for USDI Geological Survey, San Juan, Puerto Rico.

SAREP (University of California Sustainable Agriculture Research and Education Program). 1991. *What is sustainable agriculture?* University of California, Davis.

SAREP. 1993. *What is sustainable agriculture, Progress report, 1990–1993*. University of California, Davis.

Scatena, F. 1989. An introduction to the physiography and history of the Bisley Experimental Watersheds in the Luquillo Mountains of Puerto Rico. USDA Forest Service General Technical Report SO-72.

Scatena, F. N. 1990. Selection of riparian buffer zones in humid tropical steeplands. In *Research needs and application to reduce erosion and sedimentation lin tropical steeplands*, pp. 328–337. Publication 192. International Association of Hydrological Sciences, Wallingford, Oxon, UK.

Scatena, F. N. 1995. Channel morphology and the minimum flow question in montane streams of the Luquillo Mountains of Puerto Rico. Abstract. International Association of Geomorphologists Symposium, Singapore.

Schatz, R. 1991. Economic rent study for the Philippines Fisheries Sector Program. Report to the Asian Development Bank. RDA International, Placerville, Calif.

Schelhas, J. In press. Land use and change: Intensification and diversification in the Lowland Tropics of Costa Rica. In *Human Organization*.

Schmidheiny, S. 1992. *Changing course—A global perspective on development and the environment*. M.I.T. Press, Cambridge, Mass.

Schmink, Marianne. 1994. The socioeconomic matrix of deforestation. In *Population and environment: Rethinking the debate*, pp. 253–276. L. Arizpe, M. P. Stone, and D. Major (eds.). Westview, Boulder, Col.

Schmink, Marianne, and Charles H. Wood. 1987. The political ecology of Amazonia. In *Lands at risk in the Third World: Local level perspectives*, pp. 38–57. P. Little and M. Horowitz (eds.). Westview, Boulder, Col.

Schmink, Marianne, and Charles Wood. 1992. *Contested frontiers in Amazonia*. Columbia University Press, New York.

Schneider, D. 1996, May. The more species, the merrier. *Scientific American*.

Schnell, R. 1976. *Introduction a la phytogeographie des pays tropicaux: La flora et la vegetation de la Afrique tropicale*, vol. 3. Gauthiers-Villars, Paris.

Schoener, T. W. 1983. Field experiments on interspecific competition. *American Naturalist* 122:240–285.

Schroeder, R. 1993. Shady practices: Gender and the political ecology of resource stabilization in Gambian garden/orchards. *Economic Geography* 69(4):349–365.

Schwartz, H. F., and G. E. Gálvez. 1980. *Bean production problems: Disease, insects, soil and climatic constraints of Phaseolus vulgaris*. CIAT, Cali, Colombia.

Scoones, Ian, and John Thompson (eds.). 1994. *Beyond farmer first: Rural peoples' knowledge, agricultural research and extension practice*. Intermediate Technology Press, London.

SECPLAN (Secretaría de Planificación (Planning Ministry)). 1989. *Perfil ambiental de Honduras*. Direccion General de Planificacion Territorial, Tegucigalpa, Honduras.

Sedjo, R. A. 1983. *The comparative economics of plantation forestry*. Johns Hopkins University Press, Baltimore, Md.

Seeliger, U., L. D. deLacerda, and S. R. Patchineelam (eds.). 1988. *Metals in coastal environments of Latin America*. Springer-Verlag, New York.

Seghers, B. H. 1992. The rivers of northern Trinidad: Conservation of fish communities for research. In *River Conservation and Management*, pp. 81–90. P. J. Boon, P. Calow and G. E. Petts (eds.). Wiley, London.

Seligson, M. 1984. *El campesino y el capitalismo agrario de Costa Rica*. Editorial Costa Rica, San José.

Sen, Gita. 1994. Women, poverty and population: Issues for the concerned environmentalist. In *Population and environment: Rethinking the debate*, pp. 67–86. M. P. Stone and D. Major (eds.). Westview, Boulder, Col.

SENARA (Servicio Nacional de Aguas Subterraneas Riego y Avenamiento). 1996. Una sintesis. Proyecto Riego Arenal Tempisque. Mimeo. Visita Mision del Banco Interamericano de Desarrollo BID. San José, Costa Rica.

Sequeira, W. 1985. *La hacienda ganadera en Guanacaste: Aspectos económicos y sociales: 1858–1900*. Editorial EUNED, San José.

Shang, Y. 1990. *Aquaculture economic analysis: An introduction*. World Aquaculture Society, Baton Rouge, La.

Shaw, P. 1989. Rapid population growth and environmental degradation: Ultimate and proximate factor. *Environmental Conservation* 16(3):199–208.

Shotton, R. 1987. Honduras—Shrimp aquaculture in the Gulf of Fonseca. UNCTAD/GATT Consultant Report Document ITC/DTC/88/900. International Trade Centre, Geneva.

Silliman, P., and P. Hazelwood. 1981. *Perfil ambiental de Honduras*. USAID and University of Arizona, Tucson.

Silva, Paola. 1991. Mujer y medio ambiente en América Latina y el Caribe: Los desafíos hacia el año 2000. In *Mujer y medio ambiente en América Latina y el Caribe*, pp. 5–20. Fundación NATURA-CEPLAES (ed.) Fundación Natura-Ceplaes, Quito, Ecuador.

Silvertown, J. W., and R. Law. 1987. Do plants need niches? Some recent developments in plant community ecology. *Trends in Ecology and Evolution* 2:24–26.

Sinclair, R., and A. Romero. 1990. Utilzación de rastrojos en alimentación de animales en Honduras. *Pasturas Tropicales* (Columbia) 13:20–22.

Slocum, L., L. Wichart, D. Rocheleau, and B. Thomas-Slayter (eds.). 1995. *Power, process and participation: Tools for change*. Intermediate Technology Publishers, London.

Snaydon, R. W. 1982. Weeds and crop yield, pp. 729–739. In *Proceedings of the 1982 British Crop Protection Conference*, Brighton, UK.

Sodikoff, G. 1996. Plunder, fire and deliverance: A study of forest conservation, rice farming and ecocapitalism in Madagascar. Master's thesis, International Development Program, Clark University, Worcester, Mass.

Solorzano, R., R. de Camino, R. Woodward, J. Tose, V. Watson, A. Vásquez, C. Villalobos, J. Jiménez, R. L. Repetto, and W. Cruz. 1991. *Accounts overdue: Natural resource depreciation in Costa Rica.* World Resources Institute, Washington, D.C.

Solow, Robert M. 1974, May. The economics of resources or the resources of economics. *American Economic Review* 64:1–14.

Solow, Robert M. 1991, June 14. Sustainability: Economists perspective. J. Seeward Johnson Lecture. Woods Hole Oceanographic Institution, Marine Policy Center, Woods Hole, Mass.

Somarriba, E. 1993. Allocation of farm area to crops in an unstable Costa Rican agricultural community. Ph.D. thesis, University of Michigan, Ann Arbor.

Somarriba, E. 1994. Maderables como alternativa para la substitución de sombra de cacaotales establecidos. Serie Técnica, Informe Técnico no. 238, Proyecto Agroforestal/GTZ. CATIE, Turrialba, Costa Rica.

Somarriba, E. 1995. Cacao bajo sombra de maderables en Puerto Viejo, Talamanca, Costa Rica. Serie Técnica, Informe Técnico no. 249, Proyecto Agroforestal/GTZ. CATIE, Turrialba, Costa Rica.

Somner, A. 1976. Attempt at an assessment of the world's tropical moist forests. *Unasylva* 28:5–24.

Sorre, M. 1928. *Mexique, Amerique Centrale*, vol. 14. Librairie Armad Colin and Geographie Universelle, Paris.

Soto, G. 1993. Introducción a la agricultura orgánica. In *Memoria primer taller seminario sobre intercambio de experiencias de uso y manejo sostenible de recursos naturales*. C. Jones (comp.). Proyecto de Conservación y Desarrollo de Arenal. Area de Conservación Arenal, Guanacaste, Costa Rica.

Stamp, Patricia. 1989. *Technology, gender and power in Africa.* International Development Research Centre, Ottawa.

Stanley, Denise. 1991. *Demystifying the tragedy of the commons. Grassroots Developent* 15/3:27–35.

Stanley, D. 1994. Políticas estatales e impactos económicos y ecológicos de las nuevas exportaciones: El Caso de Camáron NTX en Honduras. POSCAE-UNAH Working Paper no. 7. Central American Postgraduate Economics and Development Program, Tegucigalpa, Honduras.

Stanley, D. In press. Understanding conflict in lowland forest zones: mangrove access and deforestation debates in southern Honduras. In *Managed ecosystems*, U. Hatch and M. Swisher (eds.). Oxford University Press, New York.

Stanley, L. J., and B. W. Sweeney. 1995. Organochloride pesticides in stream mayflies and terrestrial vegetation of undisturbed tropical catchments exposed to long-range atmospheric transport. *Journal of the North American Benthological Society* 14(1):38–49.

Stanley, P. C. 1968. *Flora of the Panama Canal Zone.* Verlag-Van J. Cramer, New York.

Stanley, P. C., and C. Steyermark. 1945. The vegetation of Guatemala. In *Plants and plant sciences in Latin America.* F. Verdoon (ed.). Chronica Botanica, New Bedford, Mass.

Statistical Abstract of Latin America for 1957. 1959. Committee of Latin American Studies. Regents of UCLA. Los Angeles.

Statistical Abstract of Latin America, vol. 30. 1993. UCLA Latin American Center Publications. University of California, Los Angeles.

Stavrakis, O. 1978. Ancient Maya agriculture and future development. *Culture and Agriculture* 5:1–8, Published by Anthropological Study Group on Agrarian Systems, 307 Extension Hall, Oregon State University, Corvallis.

Stewart, G. A. 1970, June. High potential productivity of the tropics for cereal crops, grass forage crops and beef. *The Journal of the Australian Institute of Agricultural Science*, 85–101.

Stiles, G. F. 1980. The annual cycle in a tropical wet forest hummingbird community. *Ibis* 122:322–343.

Stonich, Susan C. 1989. Social processes and environmental destruction: A Central American case study. *Population and Development Review* 15(2):269–296.

Stonich, Susan C. 1991. The promotion of nontraditional exports in Honduras: Issues of equity, environment, and natural resource management. *Development and Change* 22(4):725–755.

Stonich, Susan C. 1992, March. Struggling with Honduran poverty: The environmental consequences of natural resource based development and rural transformation. *World Development* 20(3):385–399.

Stonich, Susan C. 1993. *I am destroying the land! The political ecology of poverty and environmental destruction in Honduras.* Westview, Boulder, Col.

Stonich, Susan C. 1995. The environmental quality and social justice implications of shrimp mariculture in Honduras. *Human Ecology* 23(2):143–168.

Stonich, Susan C., and John Bort. 1995, May 22–24. The human and environmental consequences of shrimp mariculture in Central America. Agrarian Questions: The Politics of Farming. Paper presented at the International Congress, Wageningen, Holland.

Stonich, S. C., J. R. Bort, and L. L. Ovares. In press. Challenges to sustainability: The Central American shrimp mariculture industry. In *Managed Ecosystems.* U. Hatch and M. Swisher (eds.). Oxford University Press, New York.

Stonich, Susan C., Douglas L. Murray, and Peter M. Rosset. 1994. Enduring crises: The human and environmental consequences of nontraditional export growth in Central America. *Research in Economic Anthropology* 15: 239–274.

Strasma, J. D., and R. Celis. 1992. Land taxation, the poor, and sustainable development. In *Poverty, natural resources, and public policy in Central America.* S. Annis (ed.). Transaction publishers, New Brunswick, Canada.

Strong, D. R. 1983. Natural variability and the manifold mechanisms of ecological communities. *American Naturalist* 122:636–660.

Sullivan, G. M., S. M. Huke, and J. M. Fox (eds.). 1992. *Financial and economic analysis of agroforestry systems.* Nitrogen Fixing Tree Association, Morillton, Ark.

Swallow, S. 1994. Renewable and nonrenewable resource theory applied to coastal aquaculture, forest, wetland, and fishery linkages. *Marine Resource Economics* 9: 291–310.

Swift, M. J., B. T. Kang, K. Mulongoy, and P. Woomer. 1993. Organic matter management for sustainable soil fertility in tropical cropping systems. In *Evaluation for sustainable land management in the developing world, vol. 2: Technical papers*. IBSRAM Proceedings no. 12(2), 1991. International Board for Soil Research and Management, Bangkok, Thailand.

Swisher, M. E. 1982. An investigation of the potential for the use of organic fertilizer on small, mixed farms in Costa Rica. Thesis, University of Florida, Gainesville.

Swisher, M. E., C. Vázquez, B. Bergmann, R. Linder, N. Rank, and M. Rosemeyer. 1985. Energy analylsis of the farm La Pacífica. OTS Agroecology Course 85-4, Tropical Agroecology, San José, Costa Rica.

Szott, L. T. In press. Sustainability of agroforestry systems in sub-humid and semi-arid Latin America. In *Sustainable soil management in semi-arid and sub-humid zones of Latin America*. UNEP-SCOPE-Secretaria del Medio Ambiente, Recursos Naturales, Pesca, Mexico.

Szott, L. T., and C. A. Palm. 1986. Soil and vegetation dynamics during shifting cultivation fallows. In *Proceedings of the First Symposium on the Humid Tropics*, pp. 360–378. Ministerio de Agricutura, Belem, Brazil.

Szott, L. T., E. C. M. Fernándes, and P. A. Sánchez. 1991. Soil-plant interactions in agroforestry systems. *Forest Ecology and Management* 45:127–152.

Szott, L. T., and D. C. L. Kass. 1993. Fertilizers in agroforestry systems. *Agroforestry Systems* 23:157–176.

Szott, L. T., and C. A. Palm. In press. Nutrient stocks in managed and natural humid tropical fallows. *Plant and Soil*.

Tabora, Panfilo C. 1992. Central America and South America's Pacific Rim countries: Experience with export diversification. In *Trends in agricultural diversification: Regional perspectives*, pp. 93–105. S. Barghouti, L. Garbus, and D. Umali (eds.). World Bank Technical Paper no. 180. World Bank, Washington, D.C.

Tamayo, F. 1956. Contribución al estudio de la flora llanera. *Bulletin of the Venezuelan Society for Natural Sciences*, Caracas.

Taylor, D. R., L. W. Aarssen, and C. Lochle. 1990. On the relationship between r/K selection and environmental carrying capacity: A new habitat templet for plant life history strategies. *Oikos* 58:239–250.

Teichert-Coddington, D. R. 1995a. April 26–29. Characterization of shrimp farm effluents in Honduras and chemical budget of selected nutrients. In *Proceedings of the Third Central American Shrimp Symposium*, Tegucigalpa, Honduras.

Teichert-Coddington, D. R. 1995b. Estuarine water quality and sustainable shrimp culture in Honduras. In *Aquacultre '95*. Proceedings of the Special Session on Shrimp Farming, San Diego, Calif. S. Hopkins and C. Browdy (eds.). World Aquaculture Society, Baton Rouge, La.

Teichert-Coddington, D. R., D. Martínez, and E. Ramírez. 1996. Characterization of shrimp farm effluents in Honduras and chemical budgets of selected nutrients. In *Thirteenth Annual Administrative Report*. Pond Dynamics/Aquaculture Collaborative Research Program, 1995. Office of International Research and Development, Oregon State University, Corvallis.

Teichert-Coddington, D. R., and R. Rodríguez. 1995. Semi-intensive commercial grow-out of *Penaeus vannamei* fed diets containing differing levels of crude protein during wet and dry seasons in Honduras. *Journal of the World Aquaculture Society* 26(1):72–79.

Teichert-Coddington, D. R., R. Rodríguez, and W. Toyofuku. 1994. Cause of cyclical variation in Honduran shrimp production. *World Aquaculture* 25(1):57–61.

Tejeira, R. 1975. Mapa de vegetación. In *Atlas Nacional de Panamá*. Instituto Geográfico Nacional Tony Guardia, Panama City, Panama.

Terán, F., and J. B. Incer. 1964. *Geografía de Nicaragua*. Editorial del Banco Central de Nicaragua, Managua.

Thomas, P. 1992. Untersuchungen zu anthropogen bedingten Veranderungen im Mikroklima der wechselfeuchlen Tropen. Thesis, University of Augsburg, Augsburg, Germany.

Thomas-Slayter, Barbara, and Dianne Rocheleau. 1995a. *Gender, environment and development in Kenya: Perspectives from the grassroots*. Lynn Reinner, Boulder, Col.

Thomas-Slayter, Barbara, and Dianne Rocheleau. 1995b. Research frontiers at the nexus of gender, environment, and development: Linking household, community, and ecosystem. In *The Women and Development Annual*, vol. 4, pp. 79–118. R. S. Gallin, A. Ferguson, and J. Harper (eds.). Westview, Boulder, Col.

Thomet, P., and M. Hadorn. 1996. System Kurzrasenweide im Praxistest. *UFA-Revue* (Bern) 3/96:14–16.

Thomlinson, J. R., M. I. Serrano, T. del M. López, T. M. Aide, and J. K. Zimmerman. In press. Land-use dynamics in a post-agricultural Puerto Rican landscape (1936–1988). *Biotrópica*.

Thompson, K. 1987. The resource ratio hypothesis and the meaning of competition. *Functional Ecology* 1:297–303.

Thompson, K., and J. P. Grime. 1988. Competition reconsidered—A reply to Tilman. *Functional Ecology* 2:114–116.

Thomsen, J. B., and A. Brautigam. 1991. Sustainable use of neotropical parrots. In *Neotropical wildlife use and conservation*, pp. 359–379. J. G. Robinson and K. H. Redford (eds.). University of Chicago Press, Chicago.

Thrupp, L. Ann. 1995. *Bittersweet harvests for global supermarkets: Challenges in Latin America's agricultural export boom*. World Resources Institute, Washington, D.C.

Thurston, D., M. Smith, G. Abawi, and S. Kearl. 1995. *Tapado. Los sistemas de siempra con cobertura*. CATIE/CIIFAO, Cornell University, Ithaca, N.Y.

Thurston, H. D. 1994. Slash/mulch systems: Neglected sustainable tropical agroecosystems. In *Slash/mulch: How farmers use it and what researchers know about it*. H. D. Thurston, M. Smith, G. Abawi, and S. Kearl (eds.). CIIFAD, Cornell University, Ithaca, N.Y.

Tietenberg, T. 1996. *Environmental and natural resource economics*. HarperCollins, New York.

Tiffen, M., M. Mortimore, and C. Gichuki. 1994. *More people, less erosion*. Wiley, New York and London.

Tilman, D. 1985. The resource-ratio hypothesis of plant succession. *American Naturalist* 125:827–852.

Tilman, D. 1987. On the meaning of competition and the mechanisms of competitive superiority. *Functional Ecology* 1:304–315.

Tilman, D. 1988. *Plant strategies and the dynamics and structure of plant communities*. Princeton University Press, Princeton, N.J.

Titenburg, T. 1994. *Environmental economics and policy*. HarperCollins, New York.

Torres-Díaz, A. 1991. *Manual práctico de cultivo de Camáron en Honduras*. FPX, San Pedro Sula, Honduras.

Tosi, J. A., Jr. 1974. *Los recursos forestales de Costa Rica*. Tropical Science Center, San José.

Tossi, J. 1971. Una base ecológica para investigaciones silvícolas e inventarios forestales en la República de Panamá. Technical Report no. 2. PNUD-FAO, Panama City, Panama.

Townsend, Janet. 1993. Housewifization in the Colombian rainforest. In *Different places, different voices*, pp. 270–277. Janet Momsen and Vivian Kinnaird (eds.). Routledge, London.

Townsend, Janet. 1995. *Women's voices from the rainforest*. Routledge, London.

Townsend, Janet, Ursula Arrevilla Matias, Socorro Cancino Cordova, Silvana Pacheco Bonfil, and Elia Pérez Nasser. 1994. *Voces femeninas de las selvas*. Centro de Estudios del Desarrollo Rural, Mexico and University of Durham, England.

Trejos, C. J. E. 1995. La agricultura orgánica en El Salvador. In *Simposio Centroamericano sobre Agricultura Orgánica, Marzo*. J. García and J. Najera (eds.). Universidad Estatal a Distancia, San José.

TSC (Tropical Science Center). 1982. *Costa Rica country environmental profile: A field study*. Tropical Science Center, San José.

Tucker, R. P. 1992. Foreign investors, timber extraction, and forest depletion in Central America before 1941. In *Changing tropical forests: Historical perspectives on today's challenges in Central and South America*. Proceedings of a conference sponsored by the Forest History Society and IUFRO Forest History Group. H. K. Steen and R. P. Tucker (eds.). San José, Costa Rica.

Turkington, R. 1989. The growth, distribution and neighbor relationships of *Trifolium repens* in a permanent pasture. V. The coevolution of competitors. *The Journal of Ecology* 77:717–733.

Turner, B. L. 1974. Prehistoric intensive agriculture in the Maya lowlands. *Science* 185(4146):118–124.

Turner, B. L. 1976. Population density in the classic Maya lowlands: New evidence for old approaches. *Geographical Review* 66(1):73–82.

Turner, R. 1989. Factors affecting the relative abundance of shrimp in Ecuador. In S. Olsen and L. Arriaga. *Sustainable Shrimp Mariculture Industry for Ecuador*. CRMP (Coastal Resource Management Project) Technical Report Series TR-E-6. Narragansett, R.I., University of Rhode Island Coastal Resource Center.

Ugalde, A., and J. C. Godoy. 1992. Areas protegidas de Centroamérica. Report to IV Congreso Mundial de Parques Nacionales y Areas Protegidas. International Union for the Conservation of Nature, San José.

UNECLAC (United Nations Economic Commission). 1985. Water resources of Latin America and their utilization. Estudios e informes no. 53. ECLAC, Santiago, Chile.

UNECLAC. 1990a. *Latin America and the Caribbean: Inventory of water resources and their use, Vol. I: Mexico, Central America and the Caribbean*. ECLAC, Santiago, Chile.

UNECLAC. 1990b. *Latin America and the Caribbean: Inventory of water resources and their use, Vol. II: South America*. ECLAC, Santiago, Chile.

UNECLAC. 1990c. *The water resources of Latin America and the Caribbean. Planning, hazards and pollution*. ECLAC, Santiago, Chile.

UNFPA (United Nations Population Fund). 1991. *Population, resources and the environment: The critical challenges*. UNFPA, London.

U.S. Agency for International Development. 1992. *Keys to soil taxonomy*, 5th ed. SMSS Technical Monograph no. 19. Soil Survey Staff. Virginia Polytechnic Institute and State University, Blacksburg.

U.S. Army Corps of Engineers. 1993. Water need for Puerto Rico: San Juan metropolitan region. Report Number 2040-12 to the Puerto Rican Water and Sewage Authority.

U.S. Bureau of the Census. 1996. *International data base*. U.S. Bureau of the Census, Washington, D.C.

USDA (U.S. Department of Agriculture). 1975. *Soil Taxonomy. Agricultural Handbook*, no. 436. Soil Survey Staff. U.S. Government Printing Office, Washington, D.C.

U.S. Department of Commerce Bureau of Census. 1992. *Census of the population 1990. General population characteristics of Puerto Rico*. U.S. Government Printing Office, Washington D.C.

U.S. Forest Service. 1984. Procedure for quantifying channel maintenance flows. *Water information management system handbook*, chap. 30. Forest Service Handbook FSH 2509. U.S. Forest Service, Washington, D.C.

U. S. National Research Council. 1989. *Lost crops of the Incas*. National Academy Press, Washington, D.C.

Utting, P. 1993. *Trees, peoples and power: Social dimension of deforestatoin and forest protection in Central América*. Earthscan, London.

Vaidyanathan, L. V., M. C. Drew, and P. H. Nye. 1968. The measurement and mechanism of ion diffusion in soils. IV. The concentration dependence of diffusion coefficients of potassium in soils at a range of moisture levels and a method for the estimation of the differential diffusion coefficient at any concentration. *The Journal of Soil Science* 19:94–107.

van Bath, S. H. Slicker. 1963. *The agrarian history of western Europe, A.D. 500–1850*. Edward Arnold, London.

van Bemmelen, C. 1993. *Memoria del taller cobre comercialización de productos orgánicos*. Fundación Guilombé, San José.

van Bemmelen, C. 1995. Comercialización de productos orgánicos: El caso de Costa Rica. In *Simposio Centroamericano sobre agricultura orgánica, Marzo*. J. García and J. Najera (eds.). Universidad Estatal a Distancia, San José.

Vandermeer, J. 1989. *The ecology of intercropping*. Cambridge University Press, Cambridge.

Vannote, R. L., G. W. Minshall, K. W. Cummins, J. R. Sedell, and C. E. Cushing. 1980. The river continuum concept. *Canadian Journal of Fisheries and Aquatic Sciences* 37:130–137.

Vargas, G. 1979. Les formations végétales de Costa Rica d_aprés Holdridge et leur transformations. *Memoire du DEA*. Université de Bordeaux III, Bordeaux, France.

Vargas, G. 1981. La chaine volcanique de Tilarán et le bassin inferieur du fleuve Bebedero: Conditions écologiques, végétation et mise en valeur. These de doctorate de 3 ème cycle. Université de Bordeaux III, Bordeaux, France.

Vargas, G. 1983. Origen y flora de las sabanas de la Provincia de Guanacaste, Costa Rica: Un análisis de fitogeografía histórica. *Revista Geográfica América Central* 17:57–67.

Vargas, G. 1986. La colonización agrícola en la cuenca del río San Lorenzo: Desarrollo y problemas ecológicos. *Revista Geográfica IPGH* 103:69–86.

Vargas, G. 1987a. Análisis fitogeográfico y ecológico de una sabana arbustiva en el Parque Nacional de Santa Rosa, Costa Rica. *Revista Geográfica* 106.

Vargas, G. 1987b. Caracterización fisionómica y ecológica de la vegetación y el uso del suelo de la cordillera de Tilarán, Costa Rica. *Geoistmo* 1(1):67–88.

Vargas, G. 1987c. Estudio cuantitativo y bioclimatológico de la vegetación leñosa de sabana en el Parque Nacional de Santa Rosa, Costa Rica. *Yearbook of the Association of Latin American Geographers*, vol. 13. Louisiana State University, Baton Rouge, La.

Vargas, R. J. 1995. History of municipal water resources in Puerto Viejo, Sarapipuí, Costa Rica: A socio-political perspective. Master's thesis, Institute of Ecology, University of Georgia, Athens.

Vásquez, M. A. 1989. Cartografía y clasificación de suelos de Costa Rica. Proyecto GCP/COS/009/ITA. Mimeo. Organización de las Naciones Unidas para la Agricultura y la Alimentación. Rome.

Vega, J. L. 1975. La evolución agro-económica de Costa Rica: Un intento de periodización y síntesis, 1560–1930. *Revista de Costa Rica*. Universidad de Costa Rica, San José.

Vergne, Philippe, Mark Hardin, and Billie R. DeWalt. 1993. *Environmental study of the Gulf of Fonseca*. Tropical Research and Development, Gainesville, Fla.

Villegas, Z. L. 1991. La situación actual de la ganadería de carne y leche en Costa Rica. In Roundtable Discussion of Milk and Beef Production in Tropical America. FAO Regional Office for Latin America and the Caribbean, Santo Domingo, República Dominicana.

Viola, H. J., and C. Margolis. 1991. *Seeds of change*. Smithsonian Institute Press, Washington, D.C.

Vitousek, P. M., and W. A. Reiners. 1975. Ecosystem succession and nutrient retention: A Hypothesis. *BioScience* 25:376–381.

von Braun, J., D. Hotchkiss, and M. Immink. 1989. Nontraditional export crops in Guatemala: Effects on production, income and nutrition. Research Report 73. International Food Policy Research Institute, Washington, D.C.

Vosti, Stephen, Thomas Reardon, and Winfried von Urff (eds.). 1991. *Agricultural sustainability, growth and poverty alleviation: Issues and policies*, pp. 273–294. International Food Policy Research Institute, Washington, D.C.,

Wadsworth, F. H. 1949. The development of the forest land resources of the Luquillo Mountains of Puerto Rico. Ph.D. thesis, University of Michigan, Ann Arbor.

Wadsworth, J. 1983. *Manual para el diagnóstico de finca*. Department de Relaciones Ganaderas. Escuela Centroamericana de Ganadería, Balsa, Atenas, Costa Rica.

Wagner, P. L. 1964. Natural vegetation of middle America. In *Handbook of Middle American Indians*, vol. I. R. Wanchope (ed.). University of Texas Press, Austin.

Wainwright, F. W. 1995a. Boletín Informativo Técnico no. 1. Asociación Nacional de Acuicultures de Honduras, Choluteca, Choluteca.

Wainwright, F. W. 1995b, Noviembre–Diciembre. Evalución de plantaciones de *Avicennia germinans* en finca acuícola CUMAR. S. A. Revista Informativa no. 5. Asociación Nacional de Acuicultures de Honduras, Choluteca.

Wall, J., and W. Ross. 1975. *Producción y usos de sorgo*. Hemisferio Sur, Buenos Aires.

Wallace, R. K., W. M. Hosking, and C. L. Robinson. 1991. Environmental factors explain shrimp harvest fluctuation. *Highlights of Agricultural Research* 38(1):5.

Ward, G. H. 1994, May 19–20. Estuary hydrography and assimilative capacity: Procedures for determination of limits on shrimp aquaculture in Golfo de Fonseca. Paper presented at Encuentro Regional Sobre el Desarrollo Sostenido del Golfo de Fonseca y Sus Cuencas, Choluteca, Honduras. Center for Research in Water Resources, BRC-119, University of Texas, Austin, Texas.

Ward, G. H., and C. L. Montague. 1996. *Estuaries. Handbook of water resources*. L. W. Mays (ed.). McGraw-Hill, New York.

Ward, R. C., J. C. Loftis, and G. B. McBride. 1990. *Design of water quality monitoring systems*. Van Nostrand Reinhold, New York.

Waring, R. H., and W. H. Schlesinger. 1985. *Forest ecosystems: Concepts and management*. Academic Press, Orlando, Fla.

Wattel, J., and R. Ruben. 1992. El impacto del ajuste estructural sobre los sistemas de producción de granos básicos en Centroamérica. Informe a PRIAG. Universidad Libre de Amsterdam, Amsterdam.

WCED (World Commission on Environment and Development). 1987. *Our common future*. Brundtland Commission and Oxford University Press, New York.

Weaver, J. E. 1919. *The ecological relations of roots.* Publication no. 286. Carnegie Institute of Washington, Washington, D.C.

Weeks, Priscilla. 1992. Fish and people: Aquaculture and the social sciences. *Society and Natural Resources* 5:345–357.

Weeks, Priscilla, and Richard B. Pollnac. 1992. Introduction: Coastal aquaculture in developing countries: Problems and perspectives. In *Coastal aquaculture in developing countries: Problems and prospects*, pp. 1–13. R. Pollnac and P. Weeks (eds.). ICMRD, University of Rhode Island, Kingston.

Weidner, Dennis, Tome Revord, Randy Wells, and Amir Manuar. 1992, September. *World shrimp culture. Vol. 2, Part Two: Central America.* NOAA Technical Memo. NMFS-F/SPO-6. U.S. Department of Commerce, National Marine Fisheries Service, National Oceanic and Atmospheric Administration, Silver Spring, Md.

Weinberg, W. J. 1991. *War on the land: Ecology and politics in Central America.* Zed Books, London.

Welbank, P. J. 1961. A study of the nitrogen and water factors in competition with *Agropyron repens* L. Beauv. *Annals of Botany, N.S.* 25:116–137.

West, R., and J. P. Augelli. 1976. *Middle America. Its land and people.* Prentice-Hall, Englewood Cliffs, N.J.

Western, D., and M. Pearl (eds.). 1989. *Conservation for the twenty-first century.* Oxford University Press, New York.

Weyl, L. R. 1960. Las ignimbritas Centroamericanas. Informe Semestral no. 1. International Geophysical Union, San José.

Weyl, L. R. 1980. *Geology of Central America.* 2nd ed. Gebruden Borntraegen, Bonn, Germany.

Whitacre, D. F., J. Madrigal, C. Marroquin, M. Schulze, L. Jones, J. Sutter, and A. J. Baker. 1993. Migrant songbirds, habitat change and conservation prospects in northern Peten, Guatemala: Some initial results. In *Status and management of neotropical migratory birds.* D. M. Finch and P. W. Stangel (eds.). Department of Agriculture, Fort Collins, Col.

Wilken, Gene. 1988. *Good farmers: Traditional agricultural resource management in Mexico and Central America.* University of California Press, Berkeley.

Wilkin, G. C. 1971. Food-producing systems available to the ancient Maya. *American Antiquity* 36(4):432–428.

Wilks, Richard (ed.). 1989. *The household economy: Reconsidering the domestic mode of production.* Westview, Boulder, Col.

Wille, C. 1995, November/December. The shrimp trade boils over. *International Wildlife*, pp. 18–22.

Williams, K., J. H. Richards, and M. M. Caldwell. 1991. Effect of competition on stable carbon isotope ratios of two tussock grass species. *Oecologia* 88:148–151.

Williams, Robert. 1986. *Export agriculture and the crisis in Central America.* University of North Carolina Press, Chapel Hill.

Wilson, J. B. 1988. Shoot competition and root competition. *Journal of Applied Ecology* 25:279–296.

Wilson, J. B., and E. I. Newman. 1987. Competition between upland grassess: Root and shoot competition between *Deschampsia fluxuosa* and *Fesuca ovina.* Acta Oecologia. *Oecologia Generalis* 8:501–509.

Wilson, M. F. 1973. Evolutionary ecology of plants, a review. Part IV. Niches and competition. *Biologist* 55: 74–82.

Wilson, S. D., and P. A. Keddy. 1986. Measuring diffuse competition along an environmental gradient: Results from a shoreline plant community. *American Naturalist* 127:862–869.

Wilson, S. D., and J. M. Shay. 1990. Competition, fire, and nutrients in a mixed-grass prairie. *Ecology* 71:1959–1967.

Wilson, S. D., and D. Tilman. 1991. Interactive effects of fertilization and disturbance on community structure and resource availability in an old-field plant community. *Oecologia* 88:61–71.

Windoxhel, N. 1994. Valoración económica de los manglares: Demostrando la rentabilidad de su aprovechamiento sostenible. *Revista Forestal Centroamericana* 9:18–26.

Wiseman, F. M. 1978. Agricultural and historical ecology of the Maya lowlands. In *Pre-Hispanic Maya agriculture.* P. D. Harrison and B. L. Turner II. (eds.). University of New Mexico Press, Albuquerque.

Witt, V. M. 1984. The water and sanitation decade in Latin America and the Caribbean: Strategies for its success. *Water Quality Bulletin* 9:188–194.

Witt, V. M., and F. M. Reiff. 1991. Environmental health conditions and cholera vulnerability in Latin America and the Caribbean. *Journal of Public Health Policy* 12:450–463.

Wolanski, E. 1992. Hydrodynamics of mangrove swamps and their coastal waters. *Hydrobiologia* 247:141–161.

Wolf, Eric R. 1972. Ownership and political ecology. *Anthropological Quarterly* 45(3):201– 205.

World Bank. 1993. *The environmental data book: A guide to statistics on the environment and development.* World Bank, Washington, D.C.

WRI (World Resources Institute). 1991. *Accounts overdue: Natural resource depreciation in Costa Rica.* WRI, Washington D.C.

WRI. 1992a. *A guide to the global environment.* Oxford University Press, Oxford.

WRI. 1992b. *World Resources: 1992–93.* Oxford University Press, New York and Oxford.

WRI. 1993. *Biodiversity prospecting: Using genetic resources for sustainable development.* WRI, Baltimore, Md.

Yates, S. 1988. *Adopt a stream: A northwest handbook.* Adopt-A-Stream Foundation, University of Washington Press, Seattle and London.

Yingcharoen, D., and R. A. Bodaly. 1993. Elevated mercury levels in fish resulting from reservoir flooding in Thailand. *Asian Fisheries Science* 6:73–80.

Young, A. 1989. *Agroforestry for soil conservation.* C.A.B. International, International Council for Research in Agroforestry. Wallingford, Oxon, UK.

Young, M. D. 1992. *Sustainable investment and resource use: Equity, environmental integrity and economic efficiency*. Parthenon, Park Ridge, N.J.

Zack, A., and M. C. Larsen. 1993. Puerto Rico and the U.S. Virgin Islands. Research and exploration, National Geographic Society. *Water Issue* 9:126–134.

Zeledón, R. 1994. *Código ecológico*. Editorial Porvenir, S.A., San José.

Zloty, J., G. Pritchard, and C. Esquivel. 1993. Larvae of the Costa Rican *Hetaerina* (Odonata: Calopterigidae) with comments on their distribution. *Systematic Entomology* 18:253–265.

Zobel, B. J. 1988. Eucalyptus in the forest industry. *Tappi Journal* 71:42–46.

Zobel, B. J., G. Van Wyk, and P. Stahl. 1987. *Growing exotic forests*. Wiley, New York.

Index

Note: *f* indicates "figure," *n* indicates "note," and *t* indicates "table."